SET U
LOWE
72 Ft.
77.7
(6 FT. 5.7 IN.)
114.1
(9 FT. 6.1 IN.)
3.8
USE LEVELING PLATE TO
SET SWASHPLATE 1/2° UP
AT FWD. SIDE

GEOMETRY FOR ENGINEERS

GEOMETRY FOR ENGINEERS

James H. Earle

Texas A&M University

Addison-Wesley Publishing Company
Reading, Massachusetts • Menlo Park, California
London • Amsterdam • Don Mills, Ontario • Sydney

Cover illustrations are reproduced through the courtesy of Bell Helicopter.

Library of Congress Cataloging in Publication Data

Earle, James H.
Geometry for engineers.

Rev. ed. of: Descriptive geometry. 2d. ed. c1978.
Includes index.
1. Geometry, Descriptive. 2. Engineering design.
I. Title.
QA501.E36 1983 604.2′01516 83-2793
ISBN 0-201-11315-5

EFGHIJ-HA-93210

Dedicated to my mother,
Edna Webb Earle

PREFACE

Design is a major function of the engineer and technologist, and descriptive geometry and engineering graphics are the fundamental tools of the design process. Descriptive geometry is presented in this textbook as a problem-solving tool and as a means of developing solutions to technical problems. A generous number of photographs of products and equipment are included to show some of the many applications of descriptive geometry to various projects.

Our treatment of descriptive geometry motivates the student by exposing him or her to engineering examples taken from real-life situations. Instead of boring the student with synthetic projects, the approach taken in this text is intended to stimulate interest in engineering and technology as creative professions.

Many of the illustrations and the accompanying text have been designed to enable the student to grasp key principles on his or her own; thus, less of the instructor's time is required. Throughout the book, a second color is used in the illustrations to highlight significant steps and notes. Also, the more complex problems are presented by the *step method*, whereby the steps leading to the solution of a problem are presented in sequence, with the instructional text closely related to each step. This method of presentation was tested throughout a semester's work; 2800 student samples were taken during the study.* The step method was found to be 20 percentage points superior to the conventional textbook approach. These results prompted the author to introduce the step method in this volume.

Sufficient material is included in this text for a full course in descriptive geometry for the engineering and technology student. The design process is introduced early in the book and is referred to throughout.

The problems provided at the end of each chapter permit the student to test his or her mastery of the principles covered in that chapter. However, it is highly recommended that a laboratory manual be used in conjunction with this book. Too much layout time is required when the chapter problems are used. Printed problem sheets in most laboratory manuals are much more efficient and can increase the content covered in a course by as much as 100 percent. Problem books that can be used with this textbook are *Geometry for Engineers 1, Geometry for Engineers 2,* and *Geometry for Engineers 3.* They can be obtained from Creative Publishing Company, Box 9292, College Station, Texas 77840, Phone 409-775-6047.

*James H. Earle, *An Experimental Comparison of Three Self-Instruction Formats for Descriptive Geometry.* Unpublished dissertation, Texas A&M University, College Station, Texas, 1964.

Thanks are due to the hundreds of industries who provided photographs, drawings, and examples included in this book. Appreciation is also due to Professor Michael P. Guerard, formerly of Texas A&M University, for his assistance in preparing the section on nomography. Also, we thank Professor William Zaggle of Texas A&M for his work on the chapter on computer graphics. Credits would not be complete without mentioning the encouragement, confidence, and assistance given to the author by the staff of Addison-Wesley Publishing Company.

College Station, Texas J. H. E.
January 1984

CONTENTS

1 INTRODUCTION

1.1 INTRODUCTION

Essentially all our daily activities are assisted by products, systems, and services made possible by the engineer. Our utilities, heating and cooling equipment, automobiles, machinery, and consumer products have been provided at an economical rate to our population by the engineering profession.

The engineer must function as a member of a team composed of other related, and sometimes unrelated, disciplines. Many engineers have been responsible for innovations of life-saving mechanisms used in medicine, which were designed in cooperation with members of the medical profession. Other engineers are technical representatives or salespeople who explain and demonstrate applications of technical products to a specialized segment of the market. Even though there is a wide range of activities within the broad definition of engineering, the engineer is basically a *designer*.

This book is devoted to the introduction of elementary design concepts related to the field of engineering and to the application of descriptive geometry to the design process. Examples are given that have an engineering problem at the core, and that require organization, analysis, problem-solving graphical principles, communication, and skill (Fig. 1.1).

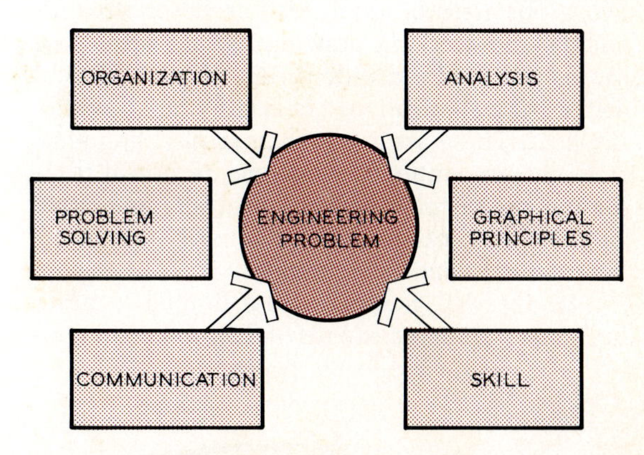

Fig. 1.1 Problems in this text require a total engineering approach with the engineering problem as the central theme.

Albert Einstein, the famous physicist, said that "Imagination is more important than knowledge, for knowledge is limited, whereas imagination embraces the entire world . . . stimulating progress, or, giving birth to evolution. . . ." (Fig. 1.2).

Fig. 1.2 Albert Einstein, the famous physicist, said "Imagination is more important than knowledge. . . ."

1.2 ENGINEERING GRAPHICS

Engineering graphics is considered to be the total field of graphical problem solving and includes two major areas of specialization, descriptive geometry and working drawings. Other areas that can be utilized for a wide variety of scientific and engineering applications are also included within the field. These areas are nomography, graphical mathematics, empirical equations, technical illustration, vector analysis, data analysis, and other graphical applications associated with each of the different engineering industries.

Engineering graphics is not limited to drafting, since it is more extensive than the communication of an idea in the form of a working drawing. Graphics is the designer's method of thinking, solving, and communicating ideas throughout the design process.

Humankind's progress can be attributed to a great extent to the area of engineering graphics. Even the simplest of structures could not have been designed or built without drawings, diagrams, and details that explained their construction (Fig. 1.3). Gradually, graphical methods were developed to show three related views of an object to simulate its three-dimensional representation. A most significant development in the engineering graphics area was descriptive geometry.

Descriptive Geometry

Gaspard Monge (1746–1818) is considered the "father of descriptive geometry" (Fig. 1.4). Young Monge used this graphical method to solve design problems related to fortifications and battlements while a military student in France. He was scolded by his headmaster for not solving a problem by the usual, long, tedious mathematical process traditionally used for problems of this type. It was only after long explanations and comparisons of the solutions of both methods that he was able to convince the faculty that his graphical methods could be used to solve the problem in considerably less time. This was such an improvement over the mathematical solution that it was kept a military secret for fifteen years before it was allowed to be taught as part of the technical curriculum. Monge became a scientific and mathemati-

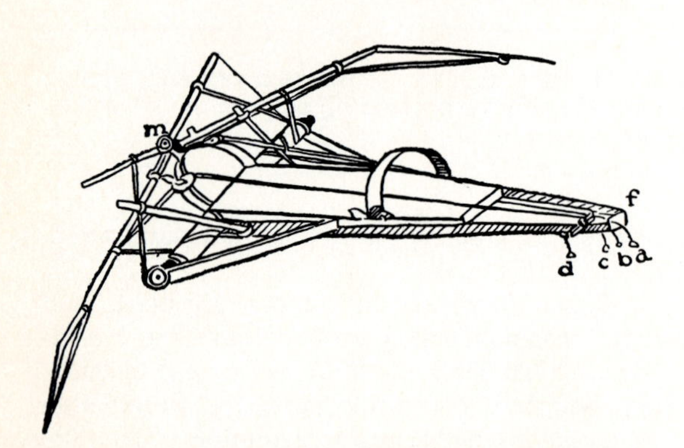

Fig. 1.3 Leonardo da Vinci developed many creative designs through the use of graphical methods.

Fig. 1.4 Gaspard Monge, the "father of descriptive geometry."

cal aide to Napoleon during his reign as general and emperor of France.

Descriptive geometry can be defined as the projection of three-dimensional figures onto a two-dimensional plane of paper in such a manner as to allow geometric manipulations to determine lengths, angles, shapes, and other descriptive information concerning the figures.

1.3 THE DESIGN PROCESS

The act of devising an original solution to a problem by a combination of principles, resources, and products is design. Design is the most distinguishing responsibility that separates the engineer from the scientist and the technician. The engineer's solutions may involve a combination of existing components in a different arrangement to provide a more efficient result, or they may involve the development of an entirely new product; but in either case the work is referred to as the act of designing.

This book emphasizes a six-step design process that is a composite of the most commonly employed steps in solving problems. The six steps are (1) problem identification, (2) preliminary ideas, (3) problem refinement, (4) analysis, (5) decision, and (6) implementation (Fig. 1.5). Although designers work sequentially from step to step, they may recycle to previous steps as they progress.

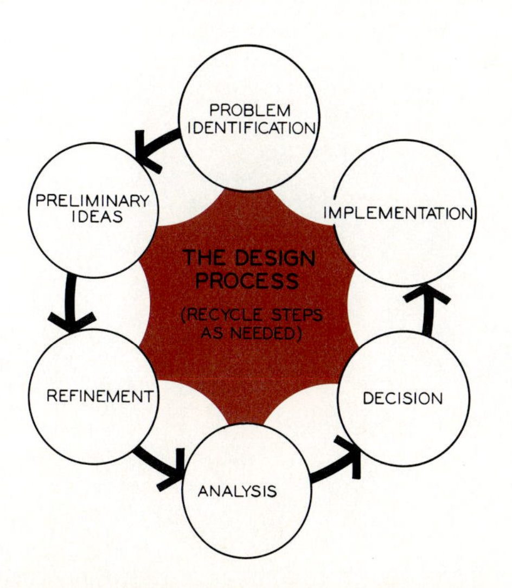

Fig. 1.5 The steps of the design process. Each step can be recycled when needed.

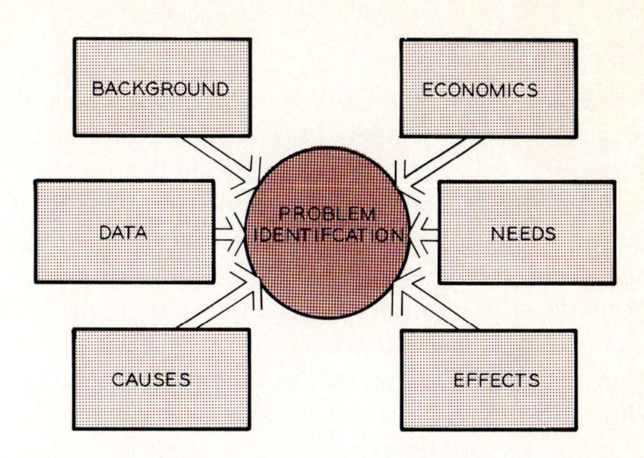

Fig. 1.6 Problem identification requires the accumulation of as much information as possible before a solution is attempted by the designer.

Engineering graphics and descriptive geometry have been integrated into these steps to stress their role in the creative process of designing. These areas are probably more essential to the design process than any other single field of study.

Problem Identification

Most engineering problems are not clearly defined at the outset; consequently they must be identified before an attempt is made to solve the problem (Fig. 1.6). For example, a prominent concern today is air pollution. Before this problem can be solved, you must identify what air pollution is and what causes it. Is pollution caused by automobiles, factories, atmospheric conditions that harbor impurities, or geographic features that contain impure atmospheres?

When you enter a bad street intersection where traffic is unusually congested, do you identify the reasons for it being congested? Are there too many cars, are the signals poorly synchronized, or are there visual obstructions resulting in congested traffic?

Problem identification requires considerable study beyond a simple problem statement like "solve air pollution." You will need to gather data of several types: field data, opinion surveys, historical records, personal observations, experimental data, and physical measurements and characteristics (Fig. 1.6).

Preliminary Ideas

Once the problem has been identified, the next step is to accumulate as many ideas for solution as possible (Fig. 1.7). Preliminary ideas should be sufficiently broad to allow for unique solutions that could revolutionize present methods. All ideas should be recorded in written form. Many rough sketches of preliminary ideas should be made and retained as a means of generating original ideas and stimulating the design process. Ideas and comments should be noted on the sketches as a basis for further preliminary designs.

ANGLES AND LENGTHS
WEIGHTS AND VOLUMES
SCALE DRAWINGS
REFINEMENT
INTERSECTIONS
PHYSICAL PROPERTIES
SHAPES AND FORMS

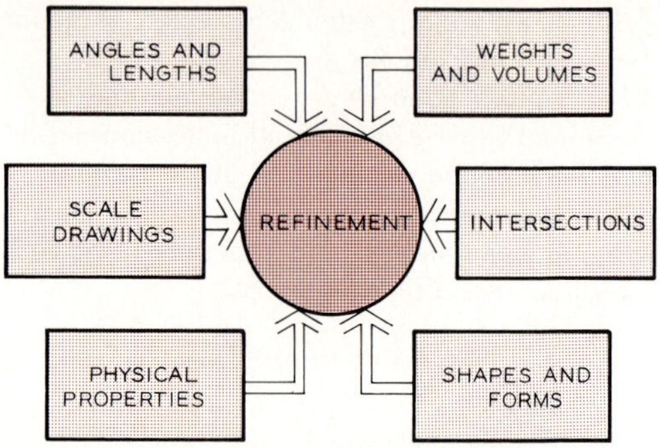

Fig. 1.8 Refinement begins with the construction of scale drawings of the better preliminary ideas. Descriptive geometry and graphical methods are used to find the necessary geometric characteristics.

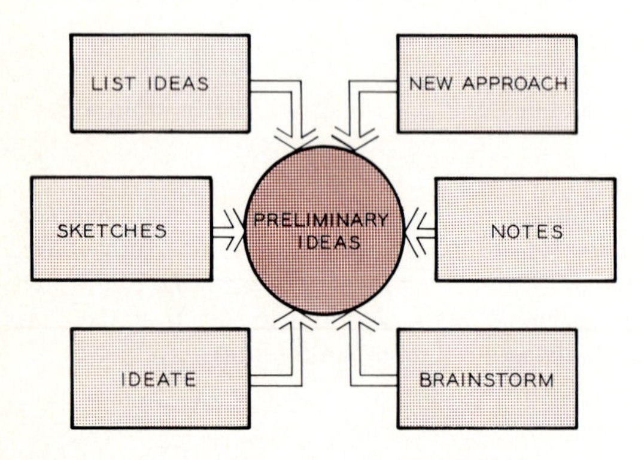

Fig. 1.7 Preliminary ideas are developed after the identification process has been completed. All possibilities should be listed and sketched to give the designer a broad selection of ideas from which to work.

Problem Refinement

Several of the better preliminary ideas are selected for further refinement to determine their true merits. Rough sketches are converted to scale drawings that will permit space analysis, critical measurements, and the calculation of areas and volumes affecting the design (Fig. 1.8). Consideration is given to spatial relationships, angles between planes, lengths of structural members, intersections of surfaces and planes.

Descriptive geometry is a very valuable tool for determining information of this type, and it precludes the necessity for tedious mathematical and analytical methods.

An example of a problem of this nature is illustrated in the landing gear of the lunar vehicle shown in Fig. 1.9. It was necessary for the designer to make many freehand sketches of the design and finally a scale drawing to establish clearances with the landing surface. The configuration of the landing gear was drawn to scale in the descriptive views of the landing craft. It was necessary, at this point, to determine certain fundamental lengths, angles, and specifications that are related to the fabrication of the gear. The length of each leg of the landing apparatus and the angles between the members at

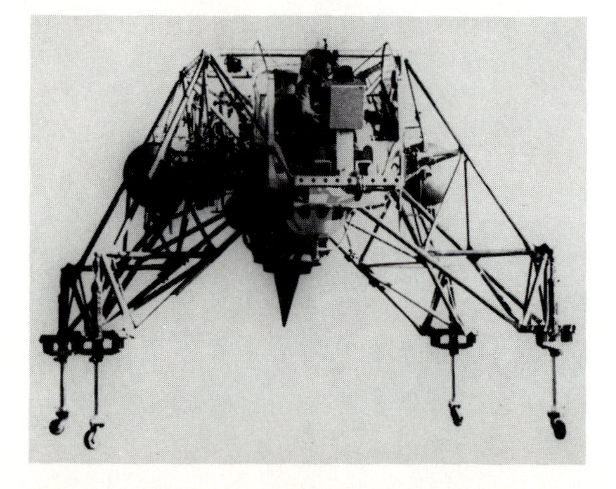

Fig. 1.9 The refinement of the lunar vehicle required the use of descriptive geometry and other graphical methods. (Courtesy of Ryan Aeronautics, Inc.)

the point of junction had to be found to design a connector, and the angles the legs made with the body of the spacecraft had to be known in order to design these joints. All of this information was easily and quickly determined with the use of descriptive geometry.

Analysis

Analysis is the step of the design process where engineering and scientific principles are used most (Fig. 1.10). Analysis involves the evaluation of the best designs to determine the comparative merits of each with respect to cost, strength, function, and market appeal. Graphical principles can also be applied to analysis to a considerable extent. The determination of forces is somewhat simpler with graphical vectors than with the analytical method.

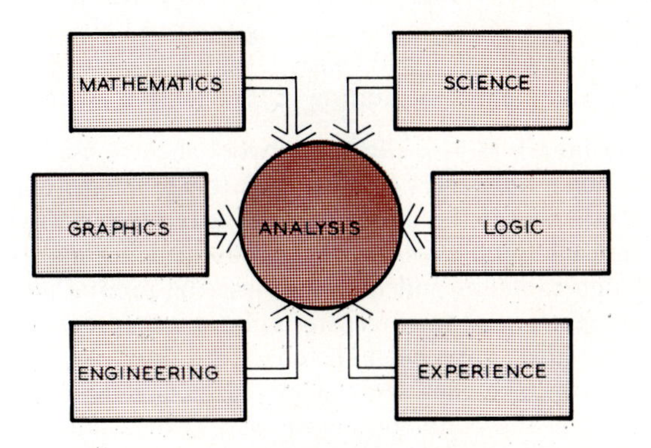

Fig. 1.10 The analysis phase of the design process is the application of all available technological methods from science to graphics in evaluating the refined designs.

Functional relationships between moving parts will also provide data that can be obtained graphically more easily than by analytical methods

Graphical solutions to analytical problems offer a readily available means of checking the solution, therefore reducing checking time. Graphical methods can also be applied to the conversion of functions of mechanisms to a graphical format that will permit the designer to convert this action into an equation form that will be easy to utilize. Data that would otherwise be difficult to interpret by mathematical means can be gathered and graphically analyzed. For instance, empirical curves that do not fit a normal equation are often integrated graphically when the mathematical process would involve unwieldy equations.

Models constructed at reduced scales are valuable to the analysis of a design to establish relationships of moving parts and outward appearances, and to evaluate other design characteristics. Full-scale prototypes are often constructed after the scale models have been studied for function.

Decision

A decision must be made at this stage to select a single design that will be accepted as the solution of the design problem (Fig. 1.11). Each of the several designs that have been refined and analyzed will offer unique features, and it will probably not be possible to include all of these in a single final solution. In many cases, the final design is a compromise that offers as many of the best features as possible.

The decision may be made by the designer on an independent, unassisted basis, or it may be made by a group of associates. Regardless of the size of the group making the decision as to which design will be accepted, graphics is a primary means of presenting the proposed designs for a decision. The outstanding aspects of each design usually lend themselves to presentation in the form of graphs that compare costs of manufacturing, weights, operational characteristics, and other data that would be considered in arriving at the final decision.

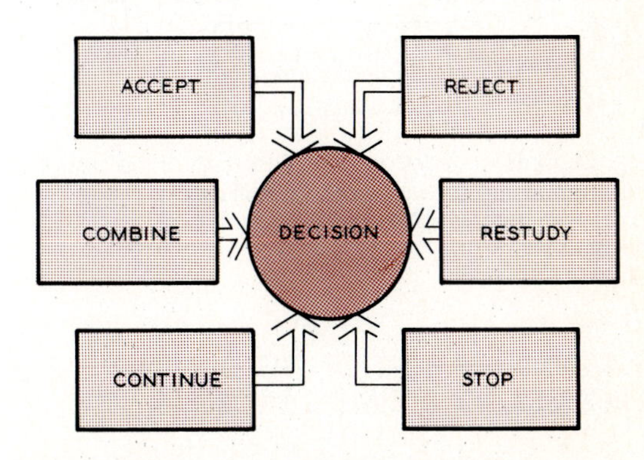

Fig. 1.11 Decision is the selection of the best design or design features to be implemented.

Implementation

The final design concept must be presented in a workable form. This type of presentation refers primarily to the working drawings and specifications that are used as the actual instruments for fabrication of a product, whether it is a small piece of hardware or a bridge (Fig. 1.12). Engineering graphics fundamentals must be used to convert all preliminary designs and data into the language of the manufacturer, who will be responsible for the conversion of the ideas into a reality. Workers must have complete detailed instructions for the manufacture of each single part, measured to a thousandth of an inch to facilitate its proper manufacture. Working drawings must be sufficiently detailed and explicit to provide a legal basis for a contract that will be the document for the contractor's bid on the job.

Designers and engineers must be sufficiently knowledgeable in graphical presentation to be able to supervise the preparation of working drawings even though they may not be involved in the mechanics of producing them. They must approve all plans and specifications prior to their release for production.

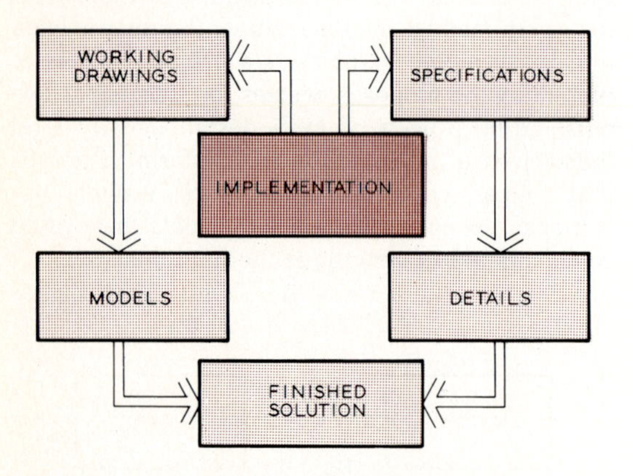

Fig. 1.12 Implementation is the final step of the design process, where drawings and specifications are prepared from which the final product can be constructed.

1.4 APPLICATION OF THE DESIGN PROCESS TO A SIMPLE PROBLEM

In order to illustrate the steps of the design process as they would be applied to a simple design problem, the following example is given.

Swing-Set Anchor Problem

A child's swing set has been found unstable during the peak of the swing. The momentum of the swing causes the A-frame to tilt with a possibility of overturning and causing injury. The swing set has swings attached to accommodate three children at a time. Design a device that will eliminate this hazard and have market appeal for owners of swing sets of this type.

Problem Identification As a first step, the designer writes down the problem statement (Fig. 1.13) and a statement of need. This overt action stimulates flow of thought and assists the designer in attacking the problem systematically. The limitations and desirable features are listed, along with necessary sketches, to enable the designer to develop a better understanding of the problem requirements. Much of the information and notes in the problem identification step may be obvious to the designer, but the act of writing statements about the problem and making freehand sketches helps to get off "dead center," which is a common weakness at the beginning of the creative process.

Preliminary Ideas A second work sheet is used to sketch preliminary solutions to the problem (Fig. 1.14). This is the most creative part of the process and has the fewest restraints. The designer makes rough sketches and notes to describe preliminary thoughts, without dwelling upon a single design. After ideas have been sketched, the designer goes over the sketches and makes additional notes to indicate the better points of each design, narrowing down the ideas to those with the most merit.

Problem Refinement The best designs—two or more unless one is highly superior to all others—are drawn to scale as orthographic drawings, as a means of refining preliminary designs. Sufficient notes are used to describe the designs without becoming too involved with details (Fig. 1.15). The refinement provides the physical properties and overall dimensions that must be considered during the earlier stages of the design process.

Orthographic projection, working drawing principles, and descriptive geometry may be used, depending upon the particular problem being refined. In this example (Fig. 1.15), simple orthographic views with auxiliary views depict the two designs. These drawings, even though refined, are still subject to change throughout the entire process.

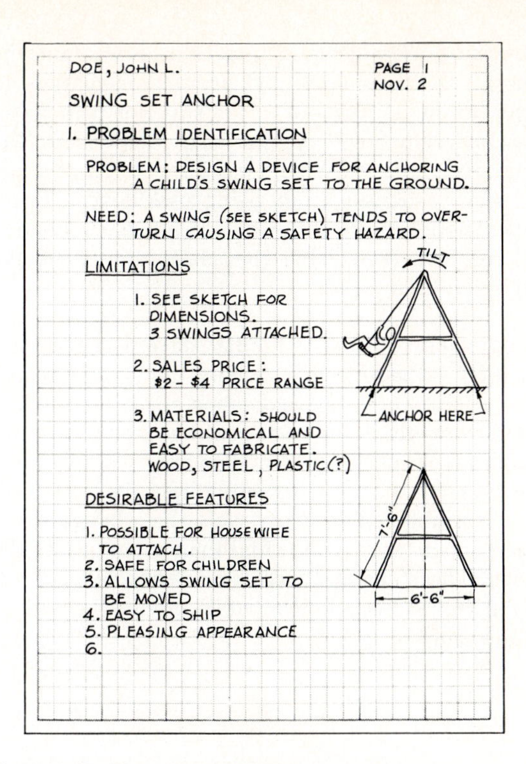

Fig. 1.13 Problem identification work sheet.

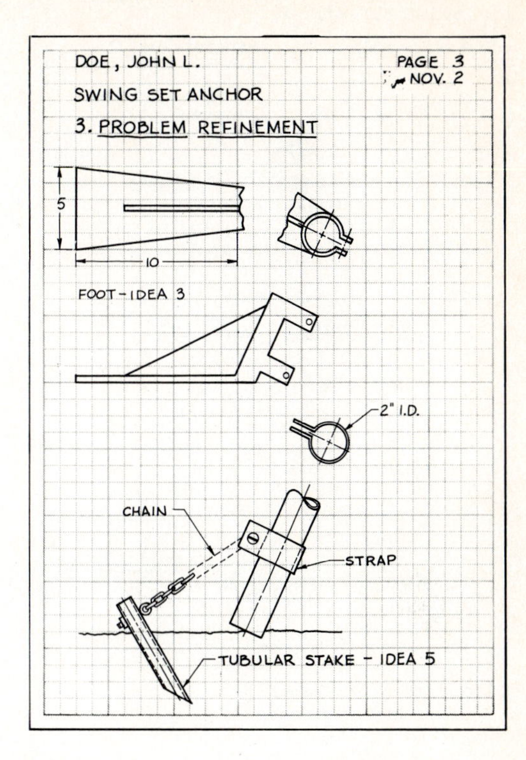

Fig. 1.15 Problem refinement work sheet.

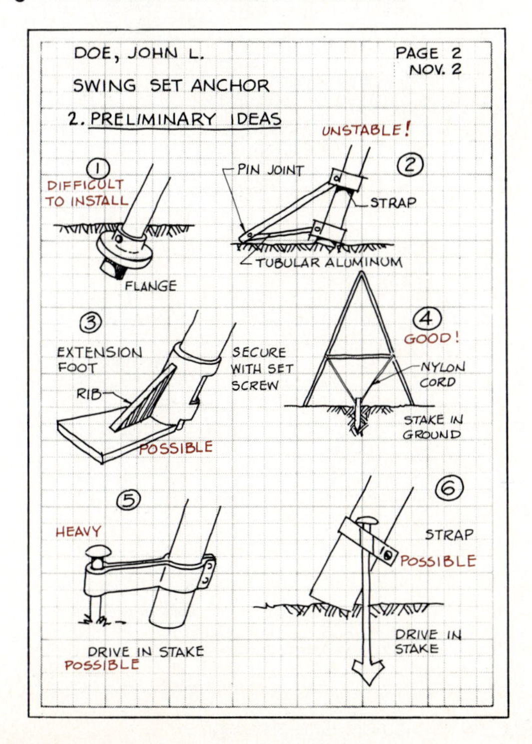

Fig. 1.14 Preliminary ideas work sheet.

Analysis Once a preliminary design has been refined to establish fundamental dimensions and relationships, the designs must be analyzed to determine their suitability and other criteria. The maximum angles of swing must be established by observations of a child swinging under average conditions. The force F at the critical angle can be calculated mathematically or estimated by observation (Fig. 1.16). Since three swings may be in use at once, the maximum condition will exist when all three swings are in phase, resulting in a triple pull, or 150 lb in this example.

The danger zones are graphically indicated in the space diagram to show the effects of the foot design and to establish the dimension that it must have in order for it to eliminate the tilting tendency at the maximum angle. The force diagram is drawn to scale to analyze the reaction forces at the extreme condition. A vector polygon is drawn with the vectors parallel to the forces in the force diagram, where the only known force is $F = 150$ lb. The magnitude of resultant R is found to be 130 lb, which is the maximum force that must be overcome at the base of the swing set.

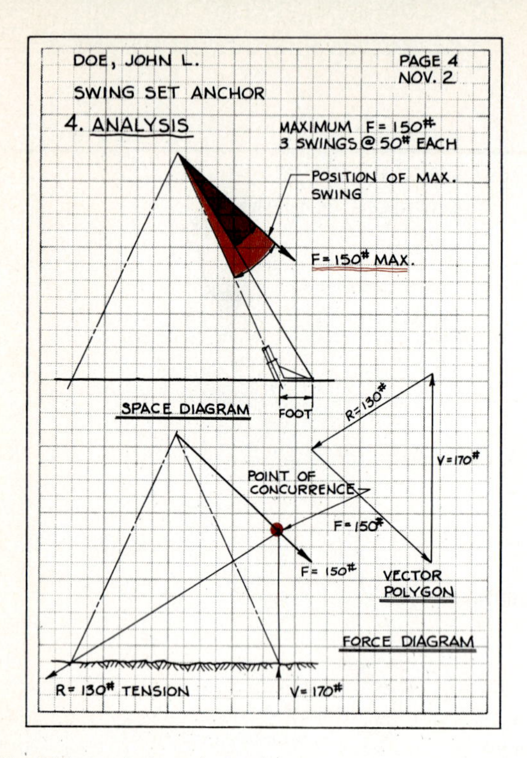

Fig. 1.16 Analysis work sheet.

Decision The designs must be evaluated and the better of the two selected for implementation. In this example (Fig. 1.17), the superior features of each design are listed for easy comparison. The disadvantages of each are listed also to prevent any design weakness from being overlooked. These tabulated lists are reviewed and a final conclusion is reached. A decision is made to implement the tubular stake design.

Implementation The tubular stake design is presented in the form of a working drawing, in which each individual part is detailed and dimensioned, and from which the parts could be made. All principles of graphical presentation are used, including a freehand sketch illustrating how the parts will be assembled (Fig. 1.18). Note that changes have been made since the initial refinement of this design. These changes were believed to be more operational and economical while serving the desired function.

DOE, JOHN L.
PAGE 5
NOV. 2
SWING SET ANCHOR

5. DECISION

SUPERIOR FEATURES

FOOT	TUBULAR STAKE
A. EASY TO ATTACH	A. POSITIVE ANCHORING
B. EASY MOVEMENT OF SET	B. EASIER TO PACK
C.	C. CHEAPER
	D. STOCK MATERIALS
	E. SIMPLE DESIGN

DISADVANTAGES

FOOT	TUBULAR STAKE
A. EXPENSIVE CONSTR.	A. SAFETY – PROJECTS OUT OF GROUND
B. BULKY	B. TRIP ON CHAIN ?
C. HEAVY	
D. LIMITED RANGE OF PROTECTION	

CONCLUSION: THE TUBULAR STAKE IS THE SUPERIOR DESIGN. IMPLEMENT.

Fig. 1.17 Decision work sheet.

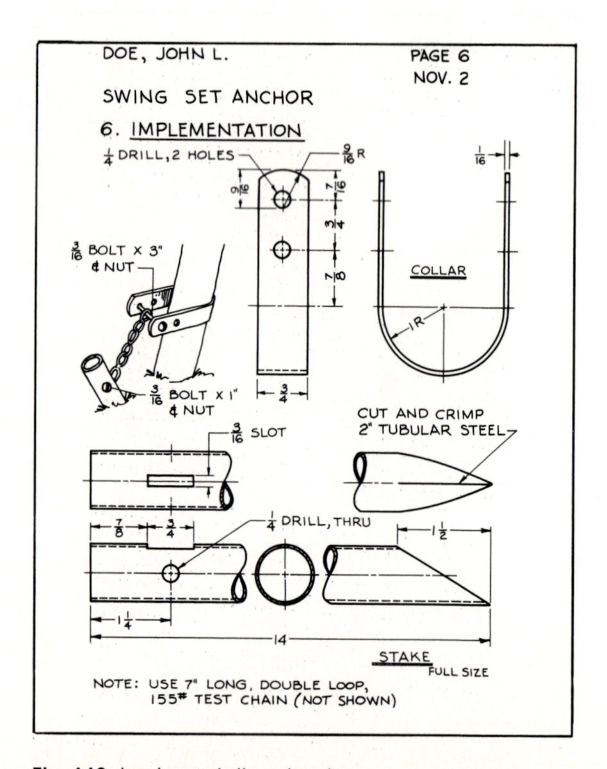

Fig. 1.18 Implementation drawing.

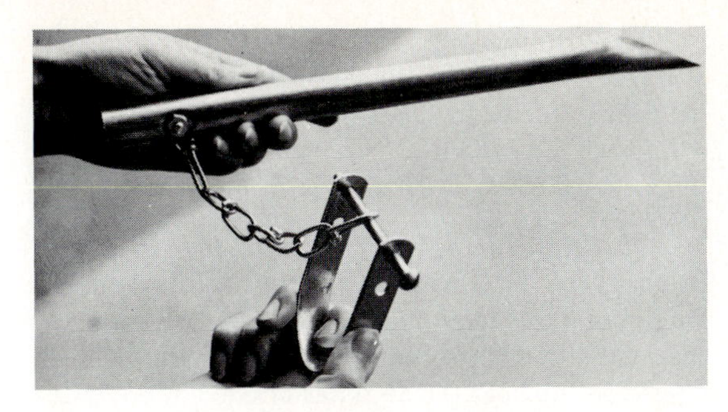

Fig. 1.19 The completed swing-set anchor.

Fig. 1.20 The anchor attached to a swing set.

Standard parts, such as nuts, bolts, and the chain, need not be drawn, but merely noted, since they are parts that will not be specially fabricated. With this drawing, the designer has implemented the design as far as can be done without actually building a prototype, model, or the actual part.

The example work sheets shown in Figs. 1.13 through 1.18 illustrate the typical approach to a simple design problem. As simple as this problem is, it would be virtually impossible for it to be designed without the utilization of graphical methods.

A photograph of the actual part is shown in Fig. 1.19 as it would be available for the market. It is shown attached to the swing set in Fig. 1.20 with the stake driven into the ground. No mention has been made of a market analysis or an evaluation of the commercial prospects of the item. This would be an ultimate requirement for the implementation of any device that is produced for consumption by the general market.

PROBLEMS

The problems at the end of each chapter are provided to afford students an opportunity to test their understanding of the principles covered in the preceding text.

Most problems are to be solved on 8½″ × 11″ paper, using instruments or drawing freehand as specified. The paper can be printed with a ¼-in. grid to assist in laying out the problems, or plain paper can be used. The grid of the given problems in later chapters represents ¼-in. or 0.20-in. intervals that can be counted and transferred to a like grid paper or scaled on plain paper.

Each problem sheet should be endorsed. The endorsement should include the student's seat number, and name, the date, and the problem number. Guidelines should be drawn with a straightedge to aid in lettering, using ⅛-in. letters. All points, lines, and planes should be lettered using ⅛-in. letters with guidelines in all cases. Reference planes should be noted appropriately when applicable.

Problems of an essay type should have their answers lettered, using single-stroke, Gothic lettering, as introduced in Chapter 3. Each page should be numbered and stapled in the upper left corner if turned in for review by the instructor.

1. List engineering achievements that have demonstrated a high degree of creativity in the following areas: (a) the household, (b) transportation, (c) recreation, (d) educational facilities, (e) construction, (f) agriculture, (g) power, (h) manufacture.

2. Make an outline of your plan of activities for the weekend. Indicate areas in your plans that you feel display a degree of creativity or imagination. Explain why.

3. Write a short report on the engineering achievement or the person who you feel has exhibited the highest degree of creativity. Justify your selection by outlining the creative aspects of your choice. Your report should not exceed three typewritten pages.

4. Test your creativity in recognizing needs for new designs. List as many improvements for the typical automobile as possible. Make suggestions for implementing these improvements. Follow this same procedure in another area of your choice.

5. List as many systems as possible that affect your daily life. Separate several of these systems into component parts or subsystems.

6. Subdivide the following systems into components: (a) a classroom, (b) a wrist watch, (c) a movie theater, (d) an electric motor, (e) a coffee percolator, (f) a golf course, (g) a service station, (h) a bridge.

7. Indicate which of the items in Problem 6 are systems and which are products. Explain your answers.

8. Make a list of new products that have been introduced within the last five years with which you are familiar.

9. Make a list of products and systems that you would anticipate for life on the moon.

10. Assume that you have been assigned the responsibility for organizing and designing a go-kart installation on your campus. This must be a self-supporting enterprise. Write a paragraph on each of the six steps of the design process to explain how the steps would be applied to the problem. For example, what action would you take to identify the problem?

11. You are responsible for designing a motorized wheelbarrow to be marketed for home use. Write a paragraph on each of the six steps of the design process to explain how the steps would be applied to the problem. For example, what action would you take to identify the problem?

12. List and explain a sequence of steps that you feel would be adequate for the design process, yet different from the six given in this chapter. Your version of the design process may contain as many of the steps discussed here as you desire.

13. Can you design a device for holding a fishing pole in a fishing position while you are fishing in a rowboat? This could be a simple device that will allow you freedom while performing other chores in the boat. Make notes and sketches to describe your design.

14. Assume that you are responsible for designing a car jack that would be more serviceable than present models. Review the six steps of the design process given in Section 1.3 and make a brief outline of what you would do to apply these steps to your attempt to design a jack. Write the sequential steps and the methods that would be used to carry out each step. List the subject areas that would be used for each step and indicate the more difficult problems that you would anticipate at each step.

15. As an introductory problem to the steps of the design process, design a door stop that could be used to prevent a door from slamming into a wall. This stop could be attached to the floor or the door and should be as simple as possible. Make sketches and notes as necessary to give tangible evidence that you have proceeded through the six steps, and label each step. Your work should be entirely freehand and rapid. Do not spend longer than 30 minutes on this problem. Indicate any information you would need in a final design approach that may not be accessible to you now.

16. List areas that you must consider during the problem identification phase of a design project for the following products: a new skillet design, a lock for a bicycle, a handle for a piece of luggage, an escape from prison, a child's toy, a stadium seat, a desk lamp, an improved umbrella, a hotdog stand.

17. Make a series of rough, freehand sketches to indicate your preliminary ideas for the solution of the following problems: a functional powdered soap dispenser for washing hands, a protector for a football player with an injured elbow, a method of positioning the cross-bar at a pole vault pit, a portable seat for waiting in long lines, a method of protecting windshields of parked cars during freezing weather, a pet-proof garbage can, a bicycle rack, a door knob, a seat to support a small child in a bathtub.

18. Evaluate the sketches made in Problem 4 above and briefly outline in narrative form the information that would be needed to refine your design into a workable form. Use freehand lettering; strive for a neat, readable paper.

19. Many automobiles are available on the market. Explain your decision for selecting the one that would be most appropriate for the following activities: a trip on a sightseeing tour in the mountains, a hunting trip in a wooded area for several days, a trip from coast to coast, the delivering of groceries, a business trip downtown. List the type of vehicle, model, its features and why you made your decision to select it.

2

APPLICATION OF DESCRIPTIVE GEOMETRY

2.1 INTRODUCTION

Descriptive geometry is the graphical discipline that has the greatest application to the refinement step of the design process. In this step, it is necessary to make scale drawings with instruments to check critical dimensions that cannot be accurately shown in sketches (Fig. 2.1).

Refinement is the first departure from unrestricted creativity and imagination. Practicality and function must now be given primary consideration. Therefore, several better ideas should be selected and refined in order to make a comparison among them during the analysis and decision steps of the design process.

Fig. 2.1 The designer's first step in the refinement stage of the design process is to draw preliminary ideas as scale drawings with instruments. (Courtesy of Chrysler Corporation.)

2.2 PHYSICAL PROPERTIES

One of the important concerns of a design's refinement is the determination of the physical properties of the proposed solutions. For example, scale drawings of six configurations of the space shuttle were made in Fig. 2.2. These scale drawings evolved from many preliminary sketches and ideas that were developed during the preliminary ideas step of the design process. As simple as these scale drawings are, they give a good comparison of the physical dimensions of the different design studies.

Many additional refinement drawings must be drawn to scale to determine such properties as sizes, volumes, and the inside configurations that will house the astronauts and the gear that is necessary to operate the space shuttle. All mechanisms and functional parts of the shuttle must be drawn and refined by scale drawings to understand better the practicality of these features.

An example of a refinement drawing of the exterior of an automobile with the appropriate overall dimensions is shown in Fig. 2.3. This series of orthographic views illustrates the overall appearance and gives the sizes of the design. Ad-

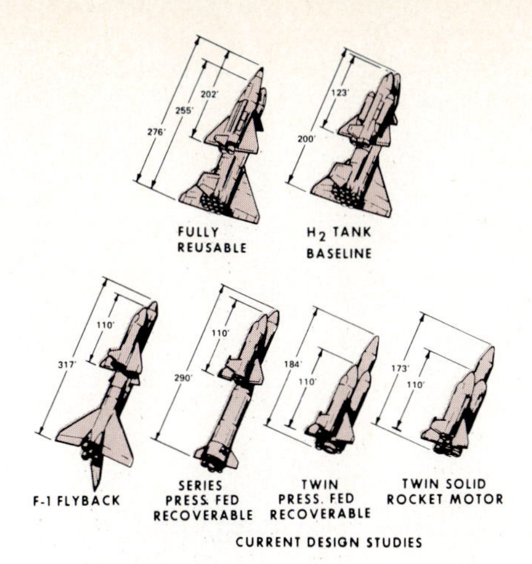

Fig. 2.2 Several refinement drawings of the space shuttle are shown here with only the major dimensions given for comparison. (Courtesy of NASA.)

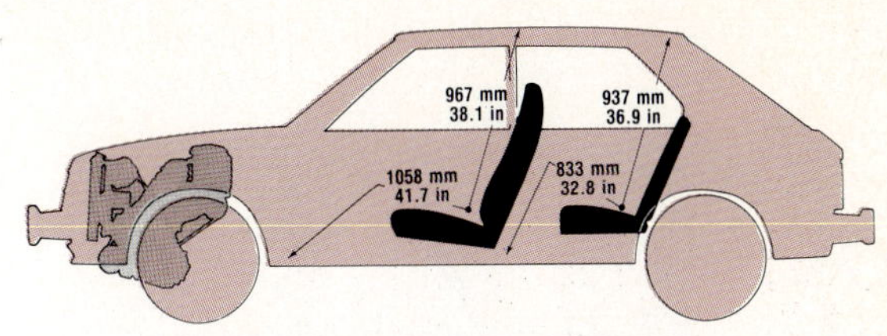

Fig. 2.4 This scale drawing gives the important dimensions and clearances necessary for a comfortable interior of an automobile. (Courtesy of Chrysler Corporation.)

ditional scale drawings can also be used to present graphically dimensions and functions that are essential to good automobile design, such as seating space as shown in Fig. 2.4.

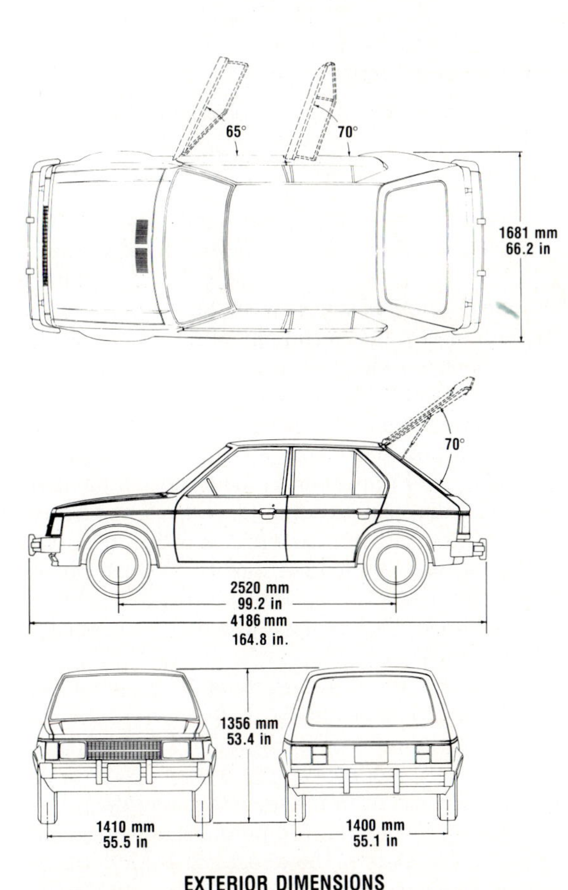

Fig. 2.3 A series of orthographic views were drawn to describe the features and dimensions of a car design. (Courtesy of Chrysler Corporation.)

2.3 APPLICATION OF DESCRIPTIVE GEOMETRY

Descriptive geometry is the study of points, lines, and surfaces in three-dimensional space. The calculation of practically any given property begins with basic geometric elements—points, lines, areas, volumes, and angles.

Before descriptive geometry can be applied, a series of orthographic views must be drawn to scale from which auxiliary views can be projected with accuracy. An example problem has been solved in Fig. 2.5 where the clearance between a hydraulic cylinder and the fender of an automobile has been determined with descriptive geometry.

Similarly, the angle between the planes of a windshield design (Fig. 2.6) can be found by descriptive geometry when the top and front views of the windshield are first drawn to scale as orthographic views. You should consider how a problem of this type would be solved without the use of descriptive geometry. Additional views could be constructed to determine the area and shape of the windshield when laid out in a flat plane.

Descriptive geometry is used in Fig. 2.7 to determine the opening size for an automobile fender to allow a clearance between the fender and the wheel. The wheel is turned to its maximum steering angles to locate lines of interference, which will determine the minimum opening of the front fender.

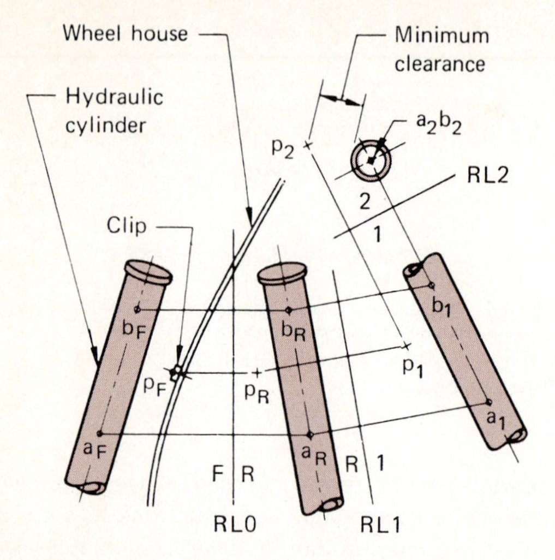

Fig. 2.5 Descriptive geometry is an effective means of determining clearances between components as shown in this example where a hydraulic cylinder's clearance with a fender is determined. (Courtesy of General Motors Corporation.)

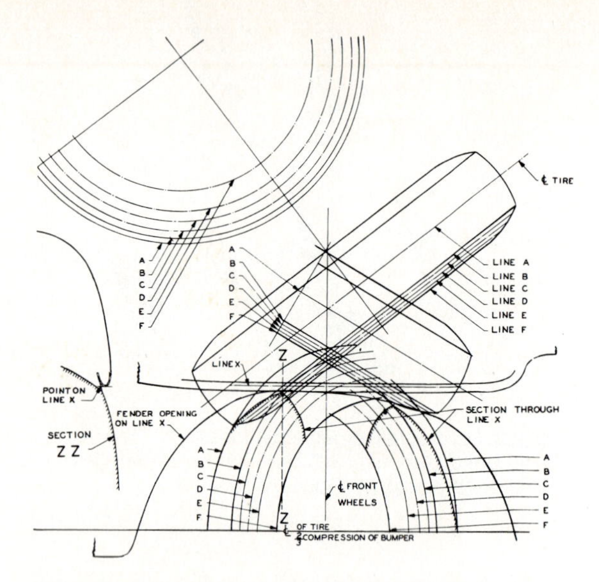

Fig. 2.7 The clearance between a fender and a tire must be found to determine the fender opening of an automobile. The clearance is found by descriptive geometry in this example. (Courtesy of Chrysler Corporation.)

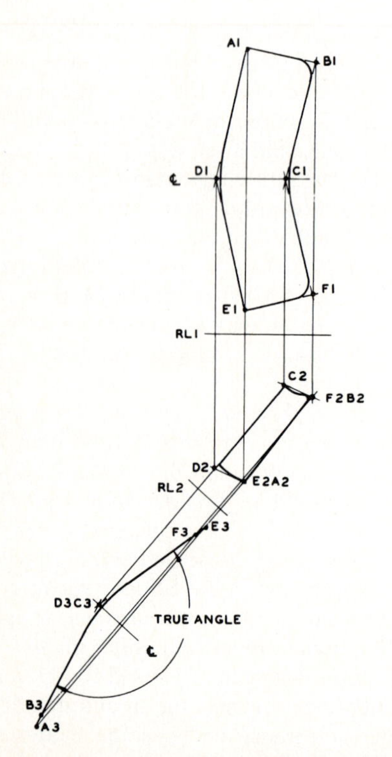

Fig. 2.6 The angle between the planes of a windshield can be determined by descriptive geometry. (Courtesy of Chrysler Corporation.)

The design of a surgical light (Fig. 2.8) is a problem that involves the application of geometry at a sophisticated level. The light fixture had to be designed in such a manner so as to provide the maximum of light on the operating area. A scaled refinement drawing is shown in Fig. 2.8(a) where you can observe the converging beams of light that are emitted from the reflectors. The beams are very narrow at their centers and are positioned at shoulder level to minimize shadows cast by interference of the surgeon's shoulders, arms, and hands.

Additional scale drawings are drawn in Fig. 2.8(b) so that the geometry of the fixture can be studied in detail. Measurements, angles, areas, and other geometry can be determined from these scale drawings (Fig. 2.9).

Figure 2.10 is an example of a refinement drawing that shows critical dimensions of the surgical lamp. Eventually this geometry will be tested by construction of a working model to confirm the information found in refinement drawings.

Another example of a problem refined by descriptive geometry is the structural frame for a 10-foot-diameter underwater sphere (Fig. 2.11). Before working drawings can be made, the physical properties and dimensions of the spherical pentagons must be determined through a series of auxiliary

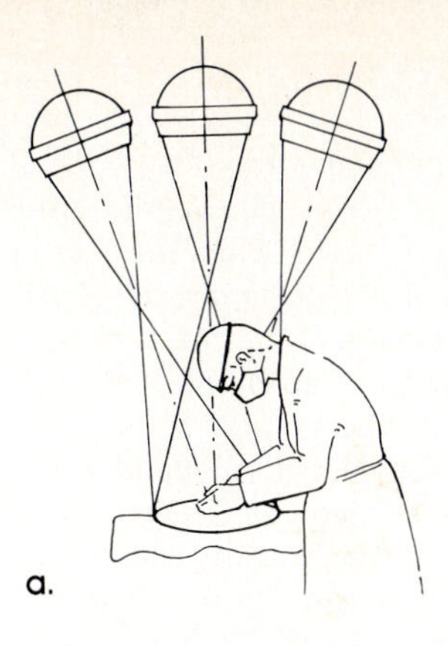

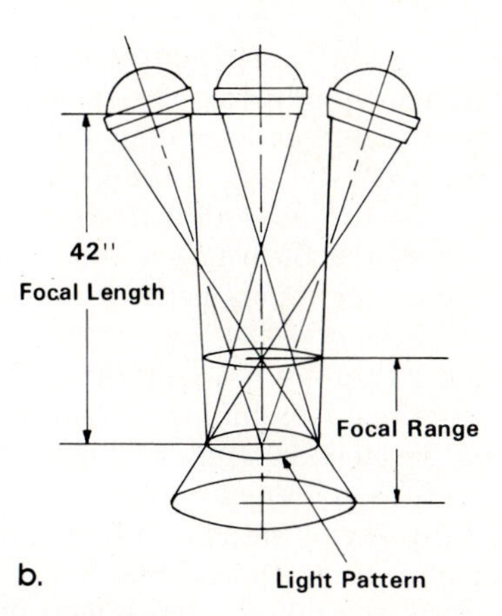

Fig. 2.8 A well-adapted surgical lamp emits light that passes around the surgeon's shoulders with the mimimum of shadow. The focal range of this surgical lamp is between 30" and 60". (Courtesy of Sybron Corporation.)

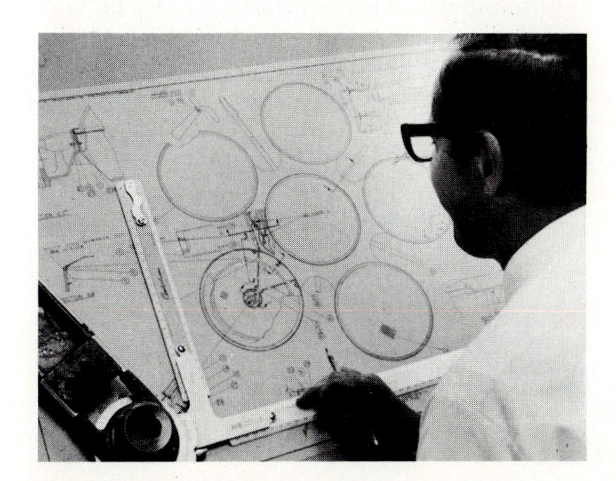

Fig. 2.9 The geometry of a surgical lamp can be studied by using scale drawings developed in the refinement step of the design process. (Courtesy of Sybron Corporation. Photograph by Brad Bliss.)

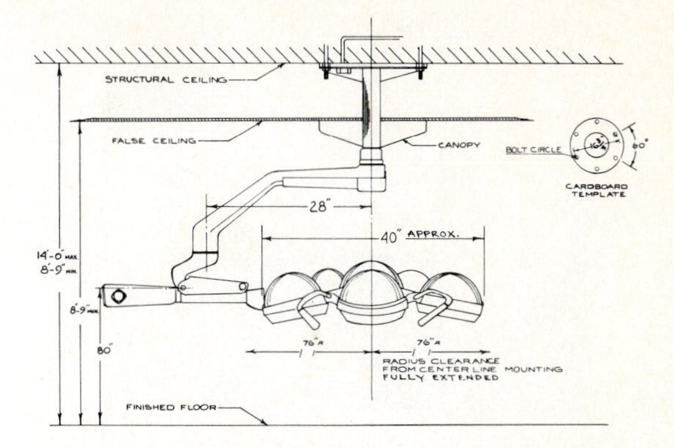

Fig. 2.10 The overall dimensions of the final design of the surgical lamp are shown in this refinement drawing. (Courtesy of Sybron Corporation.)

Fig. 2.11 This 10-foot diameter underwater sphere could not have been designed without utilizing descriptive geometry methods. (Courtesy of the U.S. Navy.)

views. The angles between the members had to be found before the joints could be detailed to fit properly (Fig. 2.12). Also, the geometry had to be determined for designing the jigs that were necessary for holding the structural parts in position during assembly.

2.4 REFINEMENT CONSIDERATIONS

Advanced designs, such as the development of a new model automobile, have numerous features that must be refined and optimized at this stage. The car's interior must be developed and presented in

Fig. 2.12 The compound joints of the structural members of the underwater sphere were designed and fabricated with the use of descriptive geometry. (Courtesy of the U.S. Navy.)

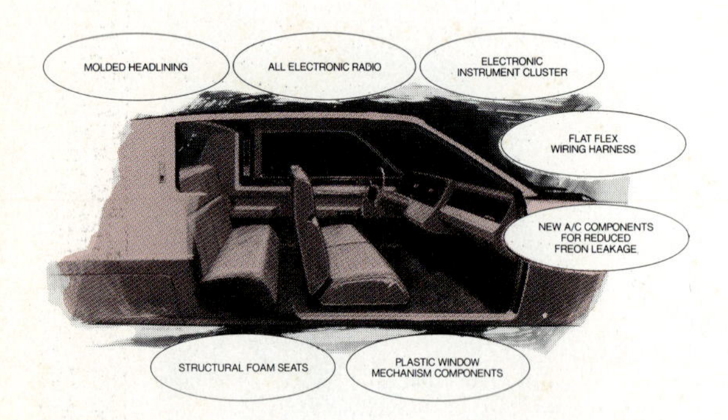

Fig. 2.13 The various features of an automobile's interior are indicated in this illustration. (Courtesy of Ford Motor Company.)

an understandable form (Fig. 2.13). The various features are shown pictorially or are listed in the surrounding balloons when they are not visible in the drawing. The same approach is used to present the various details of the chassis design (Fig. 2.14).

The overall dimensions of the automobile are shown in the orthographic view in Fig. 2.15. Also shown are the weight reductions over the 1977 models. The suspension system design of an automobile is shown graphically after it has been refined (Fig. 2.16) to explain its installation. In Fig. 2.17, the arrangement of the exhaust system is shown pictorially.

You can easily understand that many intermediate refinement drawings were required before these general pictorials were drawn, which show only the overall locations of the various components. For example, descriptive geometry had to be used to determine the bend angles and clearances in the exhaust pipe to fit a particular chassis (Fig. 2.18).

Computer graphics is a powerful tool that can be used to refine a preliminary idea. In this example, the designer is using a "light pen" on a cathode-ray tube to refine a windshield wiper system (Fig. 2.19). The image on the CRT can be shown in either two or three dimensions.

Similar to the automobile design, the Space Shuttle Orbiter was refined using a combination of orthographic drawings and notes. A profile view of the Orbiter (Fig. 2.20) gives the overall size of the craft plus a series of notes that show the locations of various payload accommodations. Each of these

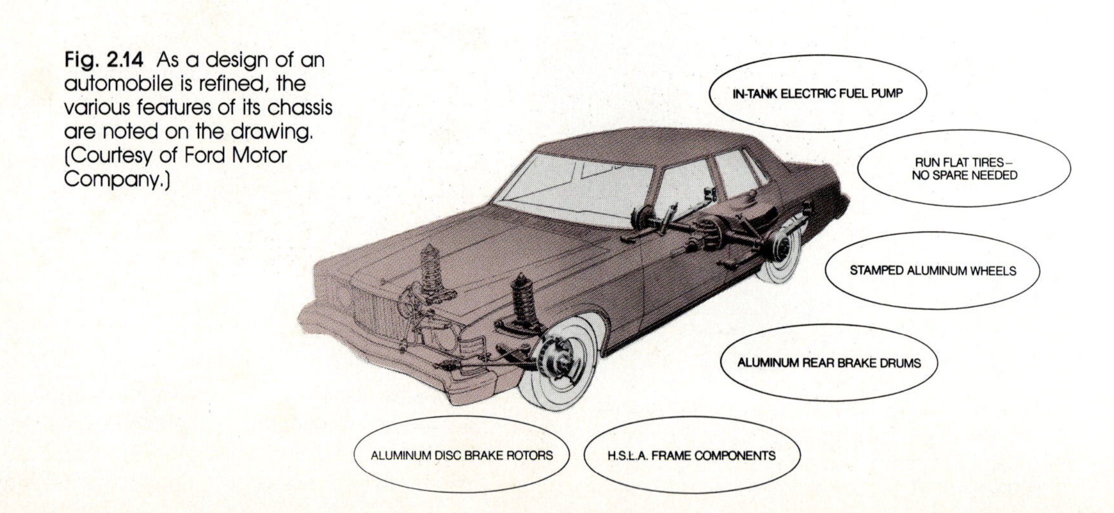

Fig. 2.14 As a design of an automobile is refined, the various features of its chassis are noted on the drawing. (Courtesy of Ford Motor Company.)

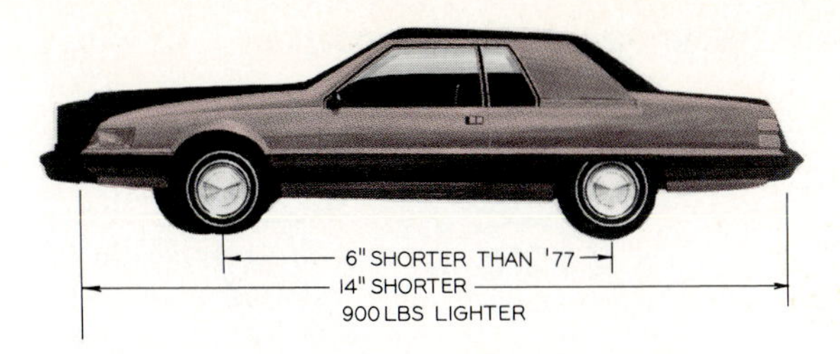

Fig. 2.15 The side view of an automobile design with the significant dimensions given. (Courtesy of Ford Motor Company.)

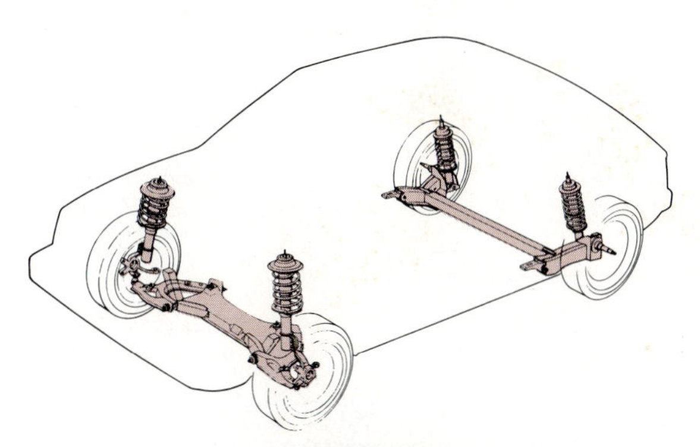

Fig. 2.16 A pictorial drawing that illustrates the suspension system of an automobile. (Courtesy of Chrysler Corporation.)

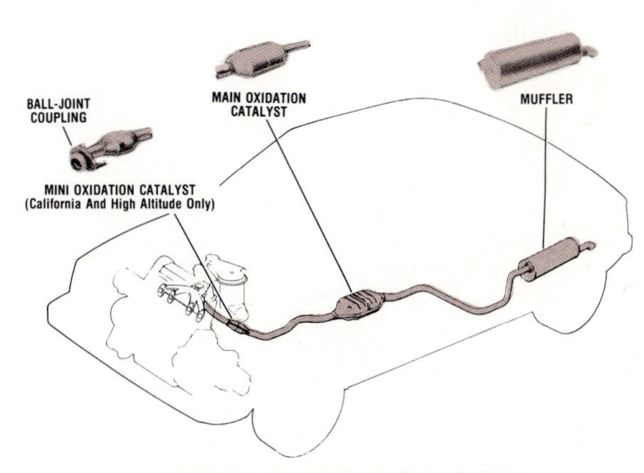

Fig. 2.17 The exhaust system of an automobile must be developed by using descriptive geometry to determine the lengths and angles necessary to clear the structural members of the chassis. (Courtesy of Chrysler Corporation.)

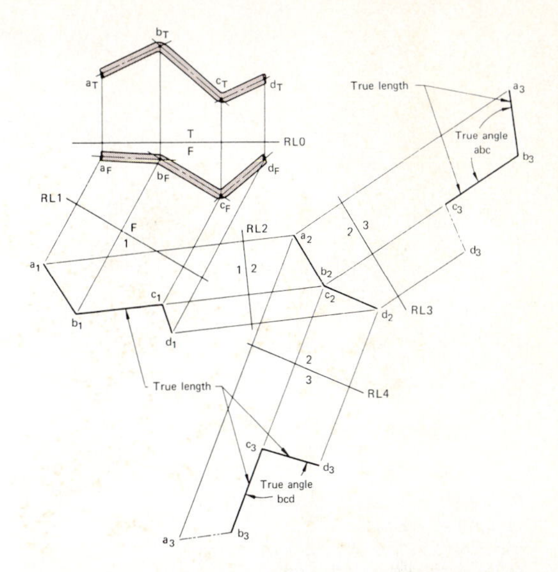

Fig. 2.18 An application of the descriptive geometry is shown in this example where the lengths and angles between pipe segments are found. (Courtesy of General Motors Corporation.)

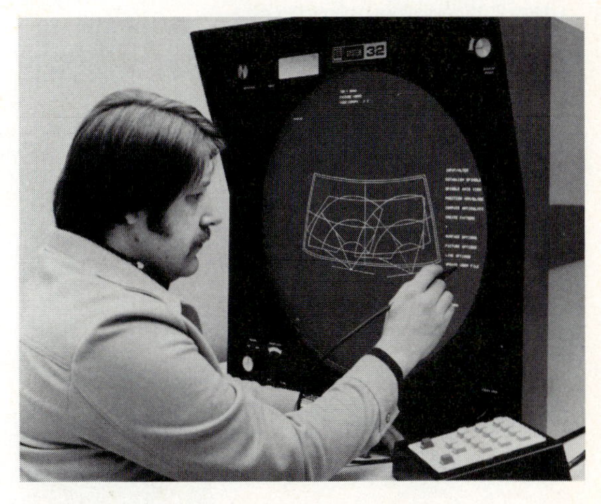

Fig. 2.19 A windshield system can be refined by plotting it on a cathode-ray tube, using computer graphics. (Courtesy of Ford Motor Company.)

parts of the design had to be developed with extensive drawings early in the design process.

More detailed physical properties of the vehicle system design are shown in the orthographic views of Fig. 2.21. Although not fully detailed, the overall dimensions and weights are given, even if they are

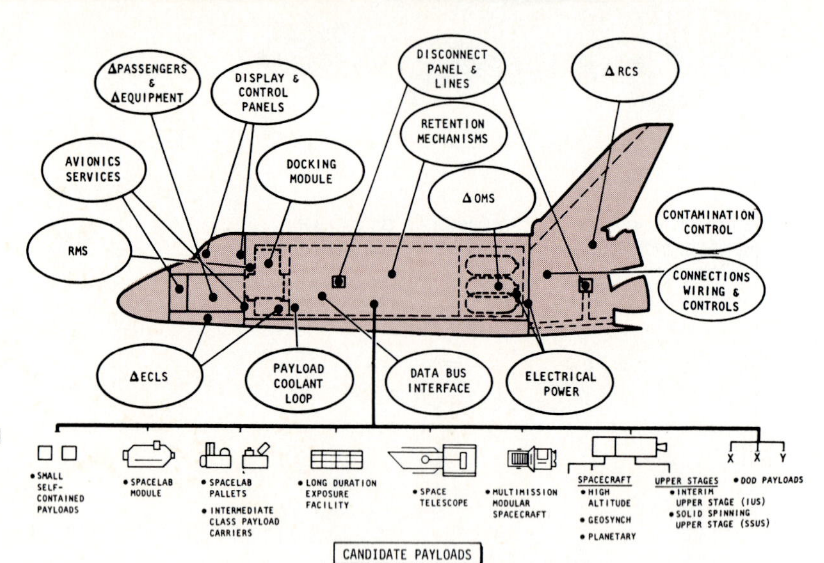

Fig. 2.20 A scale drawing and a number of features are identified in this example of a refined drawing of the space shuttle orbiter. (Courtesy of NASA.)

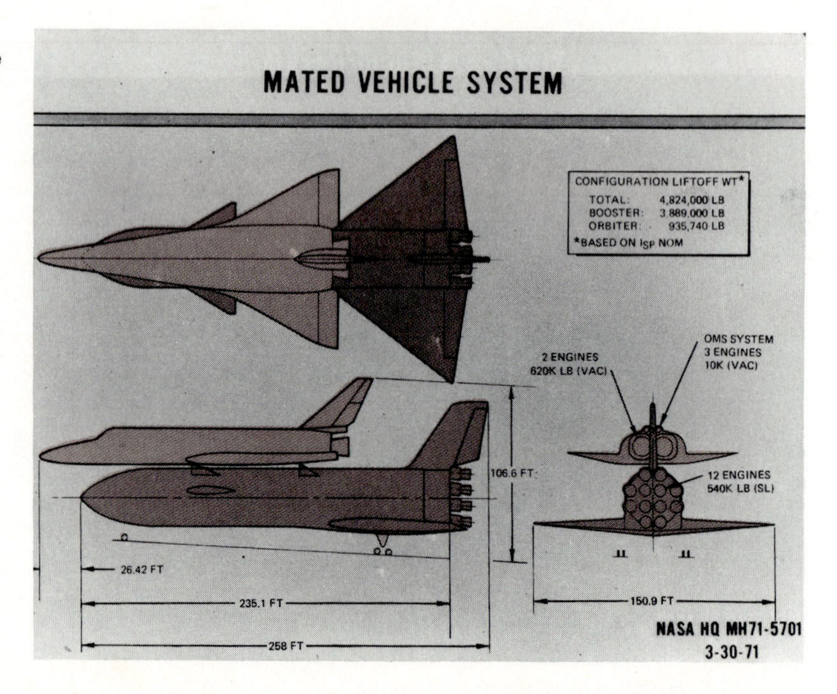

Fig. 2.21 A refined drawing of the space shuttle is shown here with its overall dimensions. (Courtesy of NASA.)

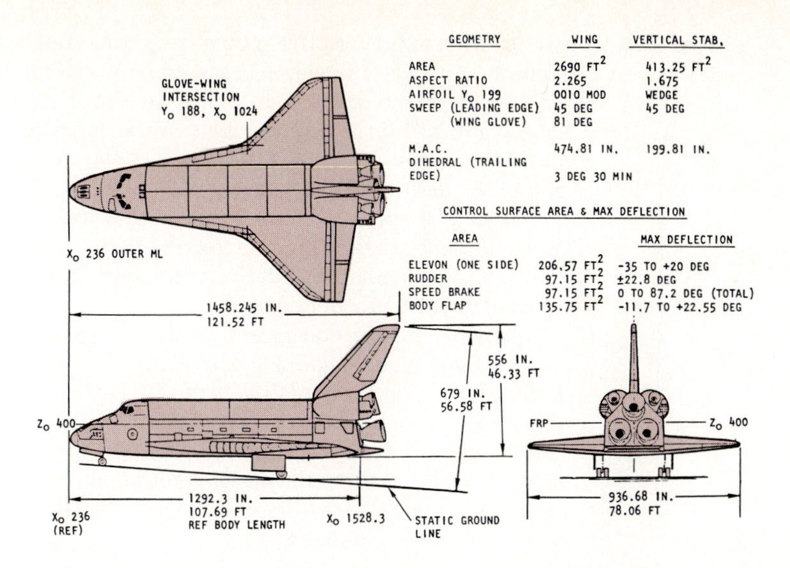

Fig. 2.22 Additional information and specifications for the orbiter vehicle are given in this refinement drawing. (Courtesy of NASA.)

general estimates at this point. The more complete specifications of the Orbiter vehicle are given in Fig. 2.22.

2.5 HUNTING SEAT—PROBLEM IDENTIFICATION

To illustrate the method of refining a preliminary product design, the hunting seat problem is continued as an example.

HUNTING SEAT: Many hunters, especially deer hunters, hunt from trees to obtain a better vantage point. Design a seat that would provide the hunter with comfort and safety while hunting from a tree and that would meet the general requirements of economy and hunting limitations.

Refinement We have selected two preliminary concepts for development, but in practice, several should be refined. A list of the design features that are to be incorporated into the design are listed on a work sheet (Fig. 2.23).

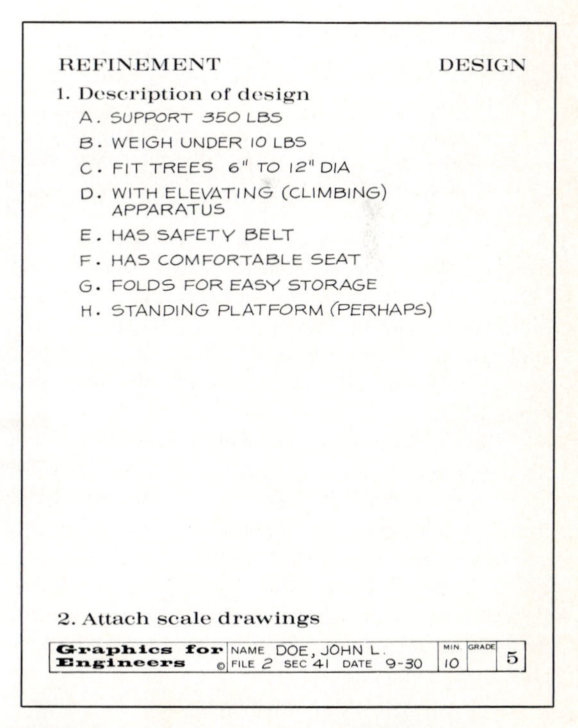
REFINEMENT DESIGN

1. Description of design
   A. SUPPORT 350 LBS
   B. WEIGH UNDER 10 LBS
   C. FIT TREES 6" TO 12" DIA
   D. WITH ELEVATING (CLIMBING) APPARATUS
   E. HAS SAFETY BELT
   F. HAS COMFORTABLE SEAT
   G. FOLDS FOR EASY STORAGE
   H. STANDING PLATFORM (PERHAPS)

2. Attach scale drawings

Graphics for Engineers © NAME DOE, JOHN L. FILE 2 SEC 41 DATE 9-30 MIN. 10 GRADE 5

Fig. 2.23 A list of a design's specifications and desirable features are listed on a work sheet of this type. In this example, the hunting seat is refined.

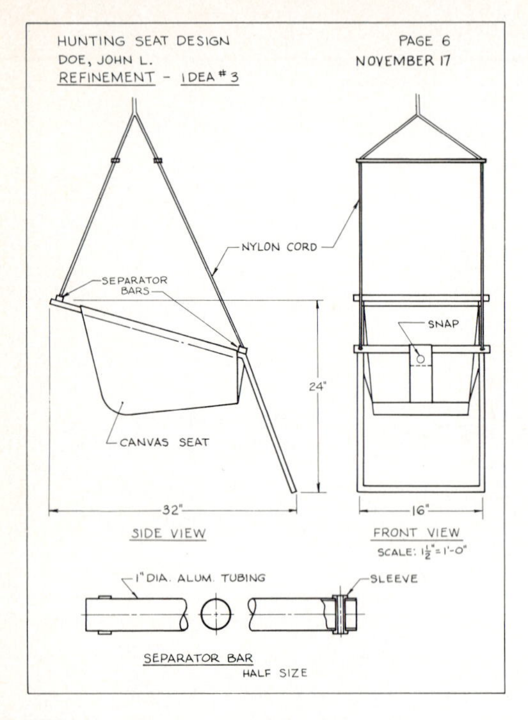

Fig. 2.24 A refinement drawing of idea #3 for a hunting seat previously developed. Only general dimensions are given on the scale drawing.

Idea #3 is refined in Fig. 2.24 where a scale drawing of the seat is given in orthographic views. Structural components are blocked in, but no attempt was made to detail each part. Tubular parts, such as the separator bars, are blocked in to expedite the drawing process. Also, some hidden lines are omitted.

The important requirement of the refinement drawings is that they be made to scale to give an accurate proportion of the design and to serve as a basis for finding angles, lengths, shapes, and other geometric specifications. Overall dimensions are given on the drawing to give general overall sizes. Specific details are shown for the separator bar to explain an idea for a sleeve to protect the nylon cord from being cut.

The canvas seat is developed as a flat pattern in Fig. 2.25. Stitch lines are shown to explain the details of assembly. The lengths of the nylon cords are found using descriptive geometry by projecting from the orthographic views. A means of adjusting and securing the footrest is refined in Detail A, where a wing nut is used to tighten the tubular members against a friction washer to lock the footrest in the desired position.

A note is given in colored pencil to indicate that additional details are required to clarify this joint. Also, a method of attaching the nylon cord near the footrest junction is needed.

Another design concept is shown developed as a refinement drawing in Fig. 2.26. Only the major dimensions are given on the scaled instrument drawing.

These two example sheets do not represent a complete refinement of the design; they are merely examples of the type of drawings required in this step of the design process.

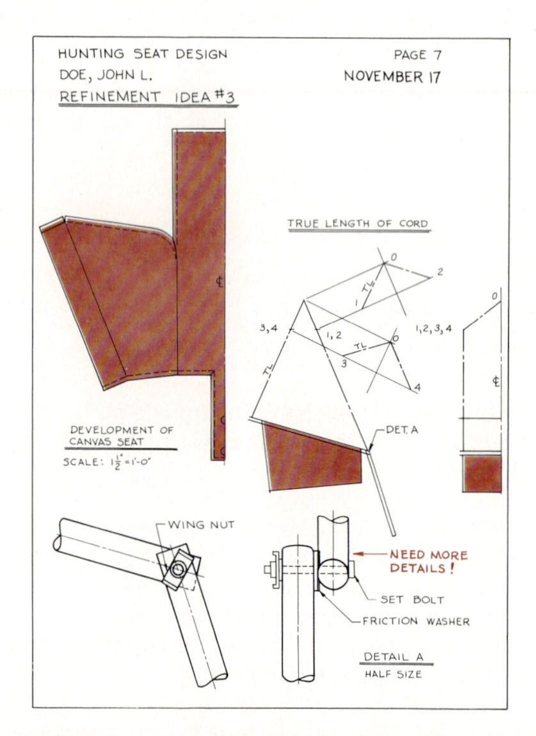

Fig. 2.25 The hunting seat design is refined using descriptive geometry and working drawings. Many additional drawings of this type are required to completely refine a preliminary concept.

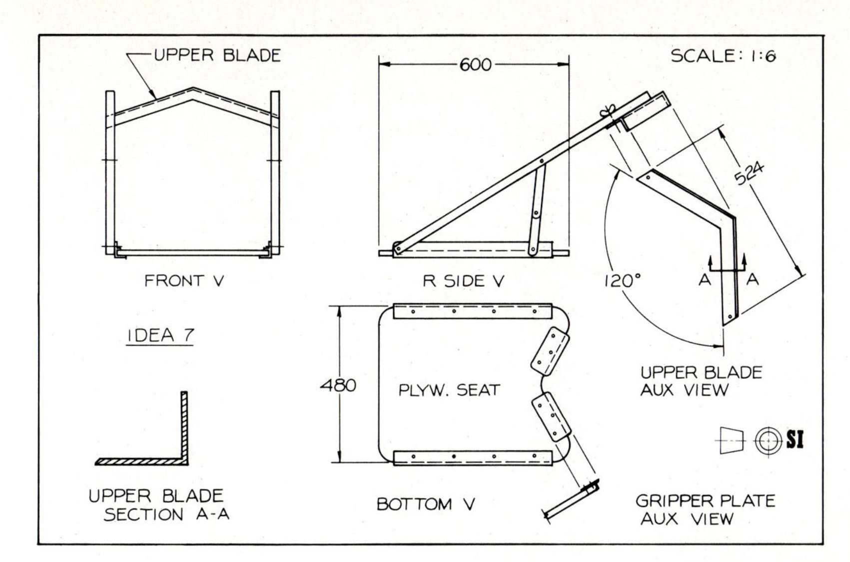

Fig. 2.26 Another design concept for a hunting seat is shown as a refinement drawing.

PROBLEMS

Problems should be presented on 8½ × 11 inch or 11 × 17 inch paper or vellum. All notes, sketches, drawings, and graphical work should be neatly prepared in keeping with good practices. Written matter should be legibly lettered using ⅛″ guidelines.

1. When refining a design for a folding lawn chair, what physical properties would a designer need to determine? What physical properties would be needed for the following items: a TV-set base, a golf cart, a child's swing set, a portable typewriter, an earthen dam, a shortwave radio, a portable camping tent, a warehouse dolly used for moving heavy boxes?

2. Why should scale drawings be used in the refinement of a design rather than freehand sketches? Explain.

3. List five examples of problems that involve spatial relationships that could be solved by the application of descriptive geometry. Explain your answers.

4. Make a freehand sketch of two oblique planes that intersect. Indicate by notes and algebraic equations how you would determine the angle between these planes mathematically.

5. What is the difference between a working drawing and a refinement drawing? Explain your answer and give examples.

6. How many preliminary designs should be refined when this step of the design process is reached? Explain.

7. Make a list of refinement drawings that would be needed to develop the installation and design of a 100-foot radio antenna. Make rough sketches indicating the type of drawings needed with notes to explain their purposes.

8. Make a list of refinement drawings that would be needed to refine a preliminary design for a rearview mirror that will attach to the outside of an automobile.

9. After a refinement drawing has been made, the design is found to be lacking in some respects so that it is eliminated as a possible solution. What should be the designer's next step? Explain.

10. Would a pictorial be helpful as a refinement drawing? Explain your answer.

11. List several design projects that an engineer or technician in your particular field of engineering would probably be responsible for. Outline the type of refinement drawings that would be necessary in projects of this type.

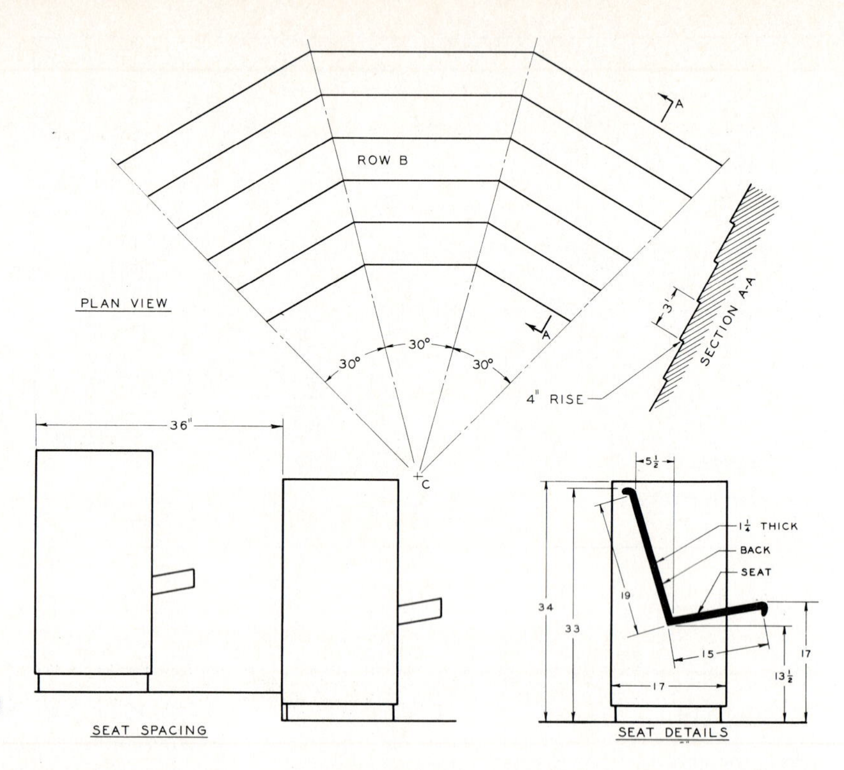

Fig. 2.27 Auditorium seating refinement (Problem 13).

12. For the hunting seat design covered in Section 2.5, what refinement drawings are necessary that were not given on the example work sheets? Make freehand sketches of the drawings needed, with notes to explain what they would reveal.

13. A seating layout is shown in Fig. 2.27 for continuous pews, which are shown in detail. Prepare the necessary refinement drawings to determine the following by descriptive geometry.

a) The angle between the seats of row B.
b) The angle between the pew backs of row B.
c) The miter angles of the seats and the backs.
d) A scale drawing of each seat and back.

Select the appropriate scale and sheet size for solving this problem.

Refinement Problems

The following sketches show products that a designer has developed as preliminary ideas. You are to prepare scaled refinement drawings with instruments of each to understand better how the products are to be made and detailed. The types of refinement drawings that you can make are:

a) Orthographic views of each individual part of the design.
b) Orthographic views of assemblies and sub-assemblies of the products.
c) Pictorials to explain relationships of parts.

You must develop and design the details as they are drawn since the preliminary sketches are just that—preliminary. Devise solutions that will work by adding your own inventiveness to the refinement drawings.

14. Sit-up device. A device that clamps to an interior door to hold one's feet while doing sit-ups for physical fitness (Fig. 2.28).

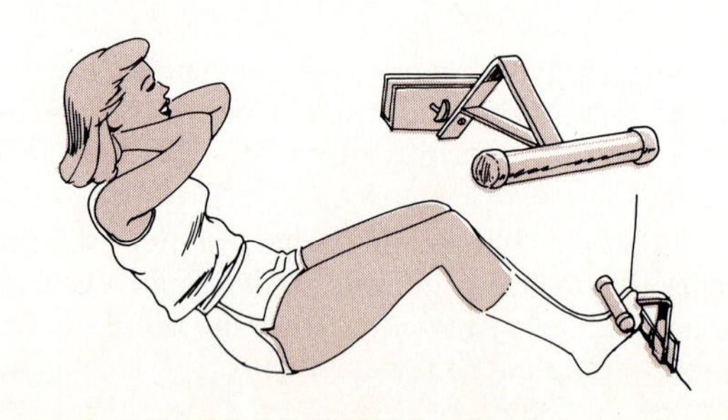

Fig. 2.28 Sit-up device refinement (Problem 14).

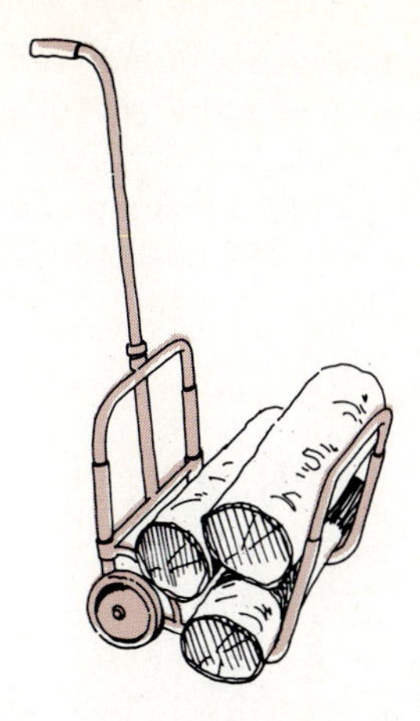

Fig. 2.29 Fireplace caddy refinement (Problem 15).

15. Fireplace caddy. A small hand cart for carrying firewood to an indoor fireplace that is decorative enough to be used as a holder of the wood once it is brought inside (Fig. 2.29).

16. Woodworking clamp. A clamp that is to be permanently attached to a workbench for holding down wood pieces up to 6″ (150 mm) in thickness to aid the woodworker (Fig. 2.30). The design involves the refinement of the mechanism and the collar that attaches to the workbench to permit height adjustment and easy removal.

17. Hold-down clamp. This hold-down clamp is designed to attach to a workbench by drilling and counterboring a hole for a mounting bolt that fits in the T-slot of the clamp. When the clamp is removed, the bolt will drop below the surface of the workbench. (Fig. 2.31).

18. Pointer mount. A mount that connects on the top of a drawing table to hold a rotational pencil pointer (Fig. 2.32).

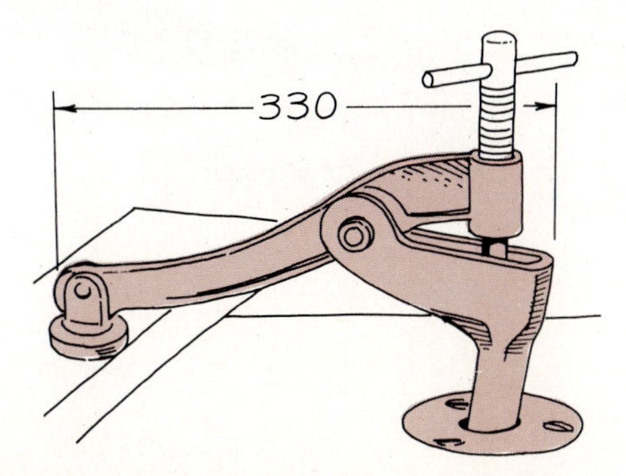

Fig. 2.30 Woodworking clamp refinement (Problem 16).

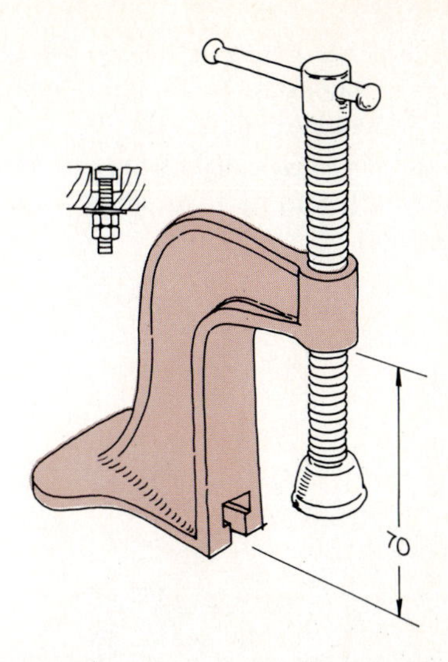

Fig. 2.31 Hold-down clamp refinement (Problem 17).

Fig. 2.32 Pointer mount refinement (Problem 18).

19. Sharpener guide. A device for holding a chisel edge at a constant angle while being sharpened on a whetstone (Fig. 2.33).

20. Luggage carrier. A portable cart for carrying luggage that will fold up to as small a size as possible (Fig. 2.34).

21. Side view mirror. A fully adjustable mirror that is to be mounted on the side of an automobile (Fig. 2.35).

22. Rotary pump. A rotary pump that operates by squeezing a liquid through a flexible tube (Fig. 2.36).

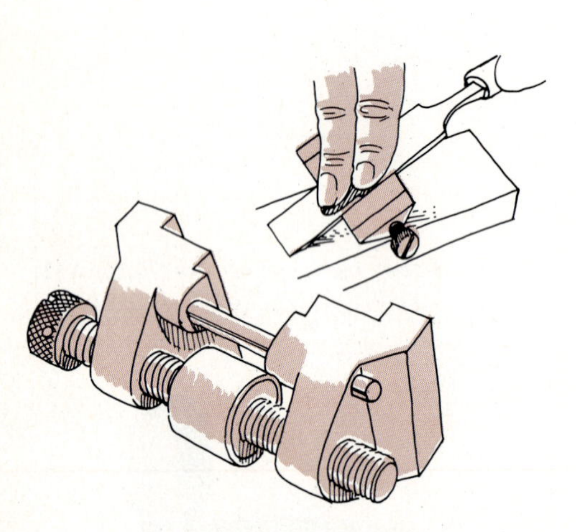

Fig. 2.33 Sharpener guide refinement (Problem 19).

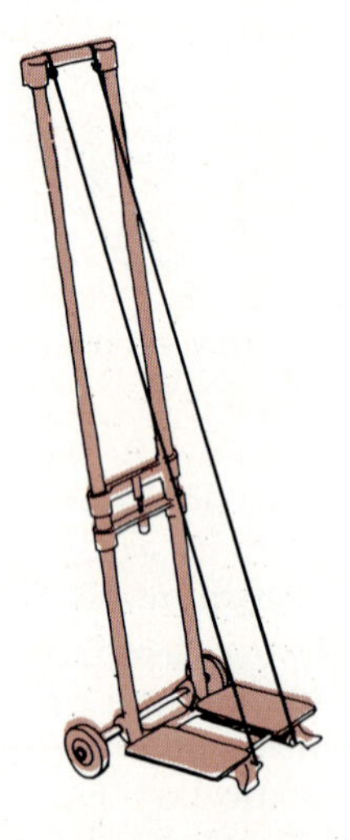

Fig. 2.34 Luggage carrier refinement (Problem 20).

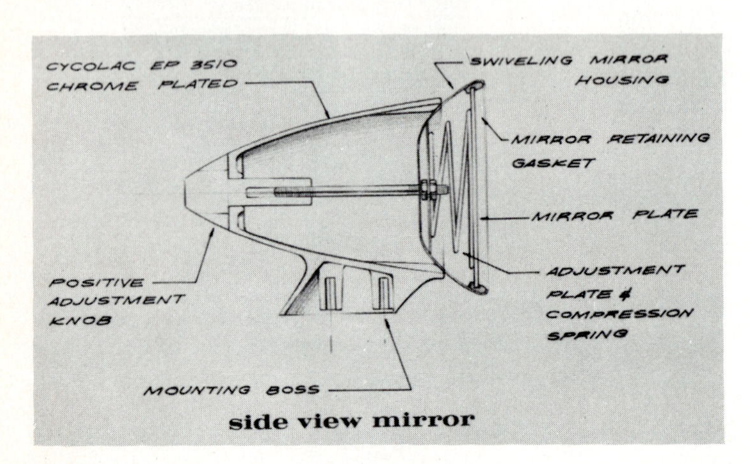

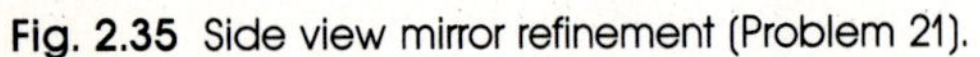

Fig. 2.35 Side view mirror refinement (Problem 21).

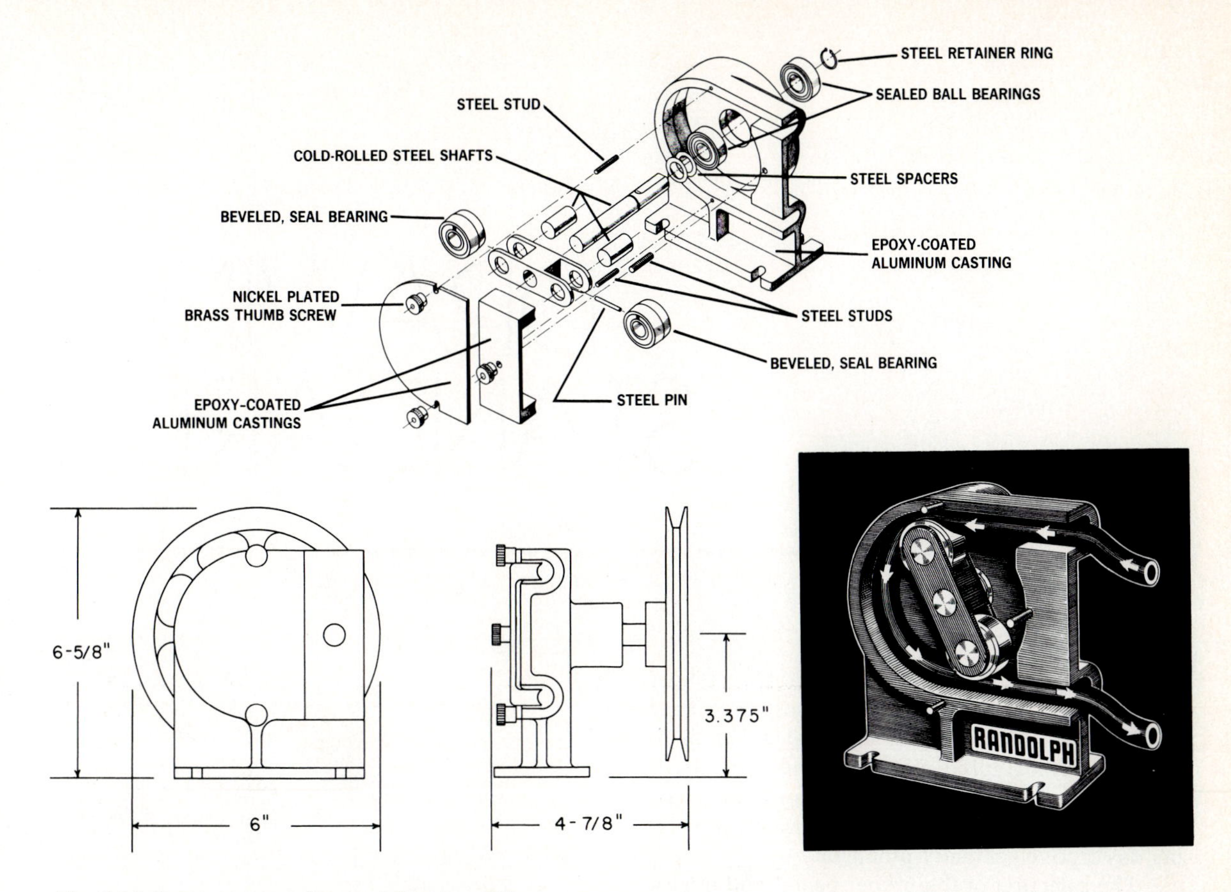

Fig. 2.36 Rotary pump refinement (Problem 22).

3

DRAWING STANDARDS

3.1 GENERAL

Descriptive geometry can be mastered only through the solution of problems that utilize principles of orthographic projection and good drafting techniques. This chapter reviews the basic drawing practices that should be followed in solving descriptive geometry problems.

The selection of the proper pencil and how it is sharpened and used is very basic to all descriptive geometry and graphics problems. Similarly, good lettering is necessary to give a drawing a professional appearance.

3.2 THE PENCIL

Since any drawing begins with the pencil, the proper pencil must be selected and sharpened correctly to yield the desired results. Pencils may be the conventional wood pencil or the lead holder, which is a mechanical pencil (Fig. 3.1). Lead

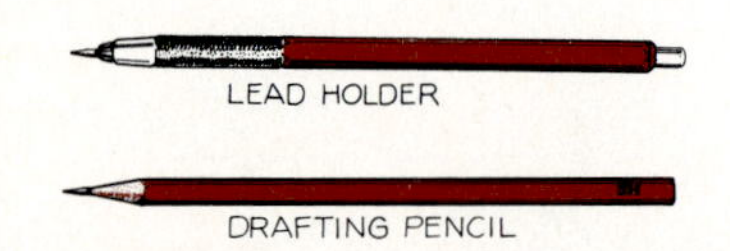

Fig. 3.1 The mechanical pencil (lead holder) or the wood pencil can be used for mechanical drawing. The ends of the lead and the wood pencil are labeled to indicate the grade of the pencil lead.

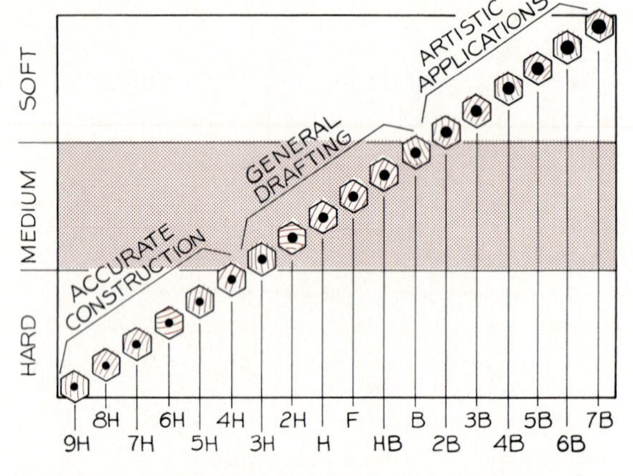

Fig. 3.2 The hardest pencil lead is 9H and the softest is 7B. Note the diameter of the hard leads is smaller than the soft leads.

grades are identified by a number and/or letter at the ends of both types. Sharpen the end opposite these markings so you will not sharpen away the identity of the grade of lead.

Pencil grades are shown graphically in Fig. 3.2, ranging from the hardest, 9H, to the softest, 7B. The pencils in the medium grade range of 3H–B are the pencils most often used for drafting work.

After you have selected your pencil, it is important that it be properly sharpened. This can be done with a small knife or a drafter's pencil sharpener, which removes the wood and leaves ap-

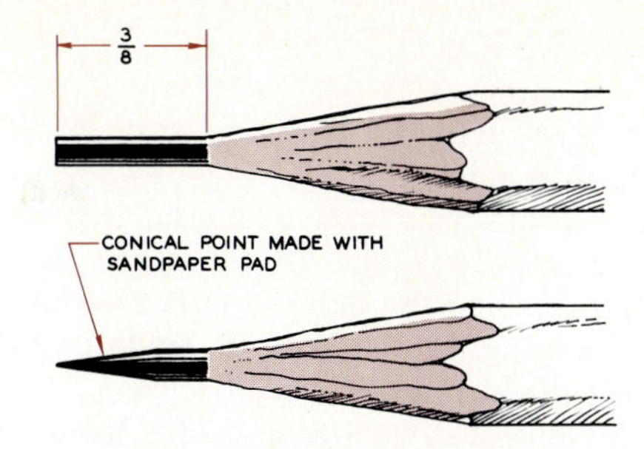

Fig. 3.3 The drafting pencil should be sharpened to a tapered conical point (not a needle point) with a sandpaper pad or other type of sharpener.

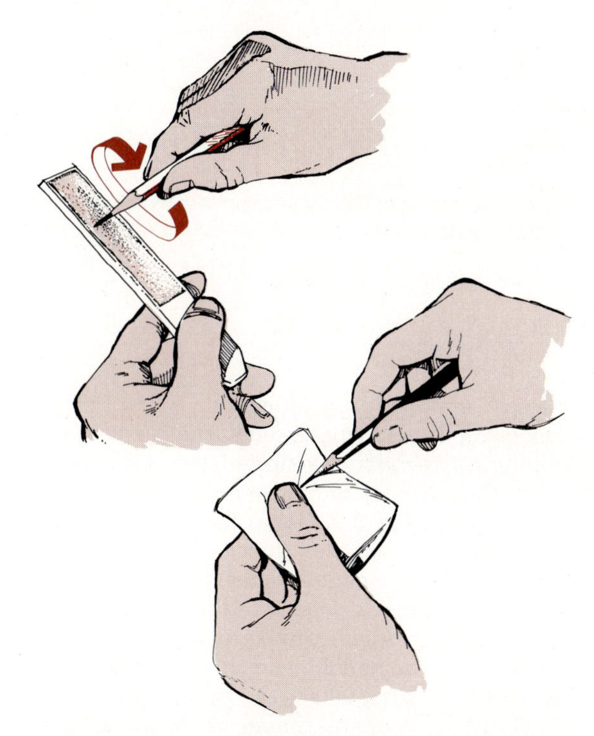

Fig. 3.4 The drafting pencil is revolved about its axis as you stroke the sandpaper pad to form a conical point. The graphite is wiped from the sharpened point with a tissue or a cloth.

proximately ⅜ inch of lead exposed (Fig. 3.3). The point can then be sharpened with a sandpaper pad to a conical point, by stroking the sandpaper with the pencil point while the pencil is being revolved between the fingers (Fig. 3.4). The excess graphite is wiped from the point with a cloth or tissue.

A pencil pointer that is used by professional drafters (Fig. 3.5) can be used to sharpen either wood or mechanical pencils. Insert the pencil in the hole and revolve it to sharpen the lead to a conical point. Other types of small hand-held point sharpeners that work on the same principle are available.

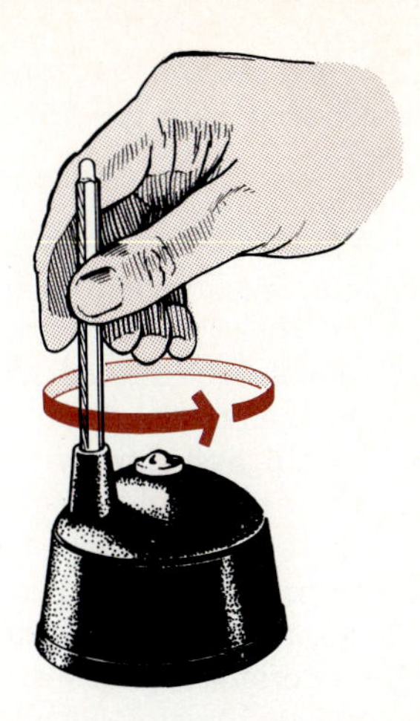

Fig. 3.5 The professional drafter will often use a pencil pointer of this type to sharpen pencils.

3.3 PAPERS AND DRAFTING MEDIA

Sizes The surface on which a drawing is made must be carefully selected to yield the best results for a given application. This usually begins by selecting the sheet size for making a set of drawings. Sheet sizes are specified by letters such as Size A, Size B, and so forth. These sizes are multiples of either the standard 8½″ × 11″ sheet or the 9″ × 12 ″ sheet, which are listed below:

Size A	8½″ × 11″	9″ × 12″
Size B	11″ × 17″	12″ × 18″
Size C	17″ × 22″	18″ × 24″
Size D	22″ × 34	24″ × 36″
Size E	34″ × 44	36 × 48

Detail Paper When drawings are not to be reproduced by the diazo process, (which is a blue-line print) an opaque paper, called *detail paper*, can be used as the drawing surface.

The higher the rag content (cotton additive) of the paper, the better will be its quality and durability because it contains cotton rather than just wood pulp.

Preliminary layouts can be drawn on detail paper and then traced onto the final tracing surface.

Tracing Paper A thin translucent paper that is used for making detail drawings is *tracing paper* or *tracing vellum*. These papers are translucent to permit the passage of light through them so drawings can be reproduced by the diazo process (blue-line process). The highest quality tracing papers are very translucent and yield the best reproductions.

Vellum is tracing paper that has been treated chemically to improve its translucency. Vellum does not retain its original quality as long as does high-quality, untreated tracing paper.

Tracing Cloth *Tracing cloth* is a permanent drafting medium that is available for both ink and pencil drawings. It is made of cotton fabric that has been covered with a compound of starch to provide a tough, erasable drafting surface that yields excellent blue-line reproductions. This material is more stable than paper, which means that it does not change its shape with variations in temperature and humidity as much as does tracing paper.

Erasures can be made on tracing cloth repeatedly without damaging the surface. This is especially important when drawing with ink.

Polyester Film An excellent drafting surface is polyester film. It is available under several trade names such as *Mylar film*. This material is highly transparent (much more so than paper and cloth), very stable, and is the toughest medium available. It is waterproof and is very difficult to tear.

Mylar film is used for both pencil and ink drawings. The drawing is made on the matte side of the film; the other side is glossy and will not take pencil or ink lines. Some films specify that a plastic-lead pencil be used and others adapt well to standard lead pencils. India ink and inks especially made for Mylar film can be used for ink drawings.

Ink lines will not wash off with water and will not erase with a dry eraser; but erasures can be made with a dampened hand-held eraser. An electric eraser is not recommended for use with this medium unless it is equipped with an eraser of the type recommended by the manufacturer of the film.

3.4 GOTHIC LETTERING

The standard type of lettering that is recommended for engineering drawings is *single-stroke Gothic lettering*. This form of lettering is given this name because the letters are made with a series of single strokes and the letter form is a variation of Gothic lettering. The strokes are made uniformly with no variation in line weight.

Two general categories of Gothic lettering are *vertical* and *inclined* lettering (Fig. 3.6). Each is equally acceptable; however, these types should not be mixed on the same drawing.

Fig. 3.6 Two types of Gothic lettering recommended by engineering standards are vertical and inclined lettering.

3.5 GUIDELINES

The most important rule of lettering is: *Use guidelines at all times*. This applies whether you are lettering a paragraph or a single letter or numeral. The method of constructing and using guidelines can be seen in Fig. 3.7. Use a sharp pencil in the 3H–5H grade range and draw these lines very lightly, just dark enough for them to be seen.

Most lettering is done with the capital letters ⅛″ high (3 mm high). The spacing between the lines of lettering should be no closer than half the height of the capital letters, ¹⁄₁₆″ in this case.

Vertical guidelines should be drawn at random to serve as a visual guide in addition to the horizontal lines. Slanted guidelines should be used for inclined lettering.

Lettering Guides

The two most-used instruments for drawing guidelines for lettering are the *Braddock-Rowe lettering triangle* and the *Ames lettering instrument*.

The Braddock-Rowe triangle is pierced with sets of holes for spacing guidelines (Fig. 3.8). The numbers under each set of holes represent thirty-

Fig. 3.7 Lettering guidelines

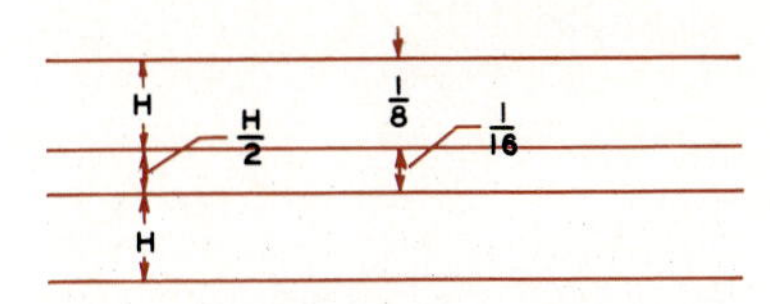

Step 1 Letter heights, *H*, are laid off and thin construction lines are drawn with a 4H pencil. The spacing between the lines should be no closer than *H*/2, or 1/16″ when 1/8″ letters are used.

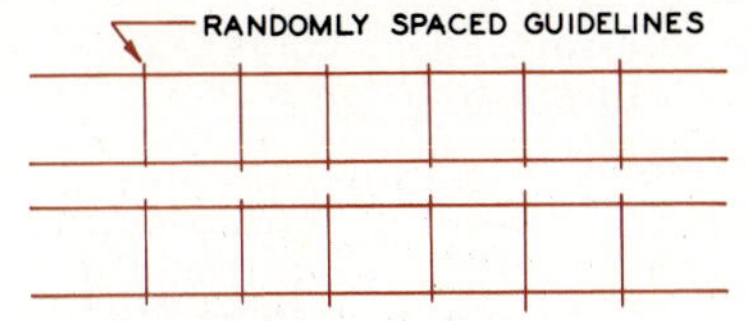

Step 2 Vertical guidelines are drawn as light, thin lines. These are randomly spaced to serve as visual guides for lettering.

Step 3 The letters are drawn with single strokes using a medium-grade pencil, H-HB. The guidelines need not be erased since they are drawn lightly.

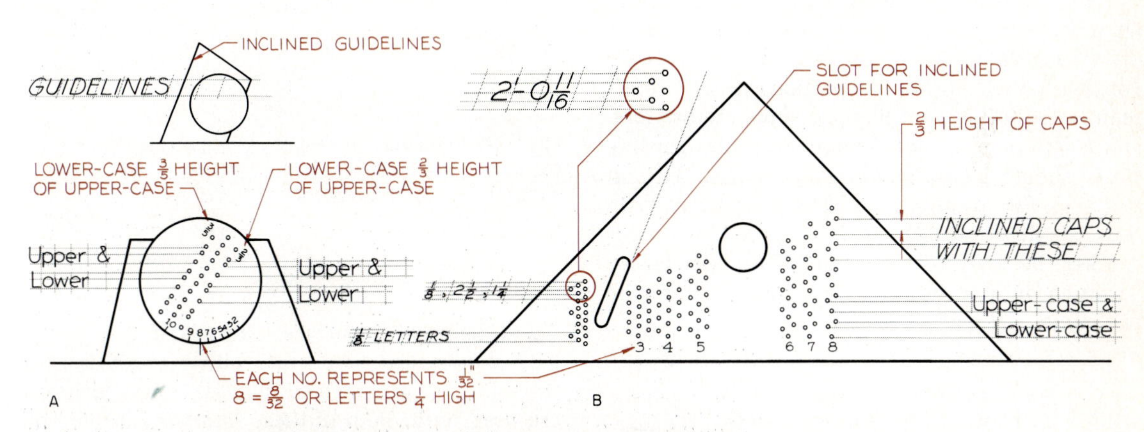

Fig. 3.8 A. The Ames lettering guide can be used for drawing guidelines for uppercase and lowercase letters, vertical or inclined. The dial is set to the desired number of thirty-seconds of an inch for the height of uppercase letters. B. The Braddock-Rowe triangle can be used as a 45° triangle as well as an instrument for constructing guidelines. The numbers designating the guidelines represent thirty-seconds of an inch. For example, the number 4 represents 4/32 or 1/8 inch for the height of uppercase letters.

seconds of an inch. For example, the numeral 4 represents 4/32″, or guidelines that are placed 1/8″ apart for making uppercase (capital) letters. Some triangles are marked for metric lettering in millimeters. Note in Fig. 3.8 that intermediate holes are provided for guidelines for lowercase letters, which are not as tall as the capital letters.

The Braddock-Rowe triangle is used in conjunction with a horizontal straightedge, such as a T-square, held firmly in position with the triangle placed against its edge. A sharp 4H pencil is placed in one hole of the desired set of holes to contact the drawing surface, and the pencil point is guided across the paper to draw the guideline while the triangle slides against the straightedge. This process is repeated as the pencil point is moved successively to each hole until the desired number of guidelines are drawn.

A slanted slot for drawing guidelines for inclined lettering is cut in the triangle. These slanting guidelines are spaced randomly by eye.

The Ames lettering guide is a very similar device with a circular dial for selecting the proper spacing of guidelines (Fig. 3.8). Again, the numbers around the dial represent thirty-seconds of an inch. The number 8 represents 8/32″, or guidelines for drawing capital letters that are 1/4″ tall. Metric guides are labeled in millimeters.

The Ames instrument is used with a pencil and straightedge, as previously explained for using the Braddock-Rowe triangle. Be sure to keep the guidelines very light so that they will not interfere with the legibility of the lettering.

3.6 VERTICAL LETTERS

Vertical Capital Letters

Capital letters (uppercase letters) are commonly used on working drawings. They are very legible and result in a word or phrase that is easy to read.

The capital letters for the *single-stroke Gothic* alphabet are shown in Fig. 3.9. Each letter is drawn inside a square box of guidelines to help you learn their correct proportions. Some letters require the full area of the box; some require less, and a few require more space. Each straight-line stroke should be drawn as a single stroke without stopping. For example, the letter A is drawn with three single strokes. Letters composed of curves can best be drawn in segments. The letter O can be drawn by joining two semicircles to form the full circle.

Fig. 3.9 The uppercase letters used in single-stroke Gothic lettering. Each is drawn inside a square to help you learn the proportions of each letter.

The shape and proportion of letters is important to good lettering. Memorize the shape of each letter given in this alphabet. Small wiggles in your strokes will not detract from your lettering if the letter forms are correct.

Examples of poor lettering are shown in Fig. 3.10. Observe the reason given for the lettering being poor in each example.

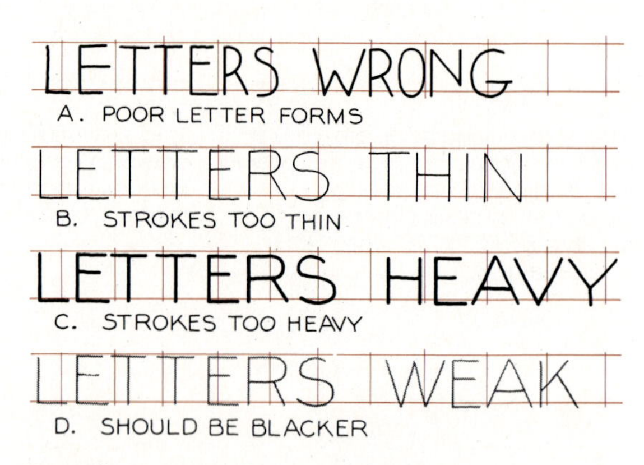

Fig. 3.10 There are many ways to letter poorly. A few of them, and the reasons why the lettering is inferior, are shown here. *Do not* make these mistakes.

Vertical Lowercase Letters

An alphabet of lowercase letters is shown in Fig. 3.11. Lowercase letters are either two-thirds or three-fifths as tall as the uppercase letters that they are used with. Both of these ratios are labeled on the Ames guide. Only the two-thirds ratio is available on the Braddock-Rowe triangle.

Fig. 3.11 The lowercase alphabet used in single-stroke Gothic lettering. The body of each letter is drawn inside a square to help you learn the proportions.

Some lowercase letters have ascenders that extend above the body of the letter such as the letter b; and some have descenders that extend below the body of the letters such as the letter y. The ascenders are the same length as the descenders.

The guidelines in Fig. 3.11 form perfect squares about the body of each letter to illustrate the proportions. A number of these letters have bodies that are perfect circles that touch all sides of the squares.

Capital and lowercase letters are used together in Fig. 3.12 as in a title. You can see the difference between the lowercase letters that are two-thirds the height of capitals and those that are three-fifths the height of capitals.

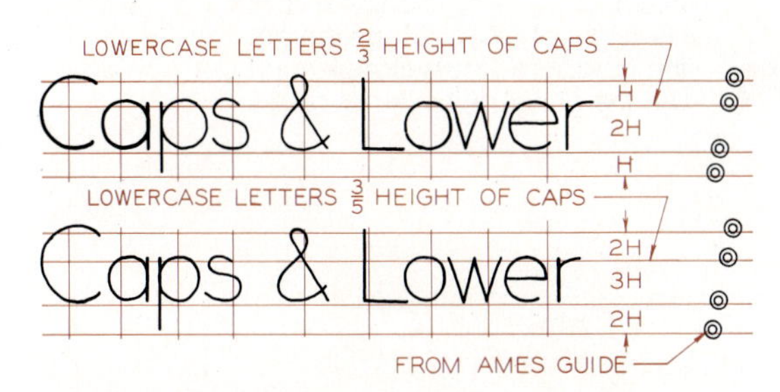

Fig. 3.12 Uppercase and lowercase letters are sometimes used together. The ratio of the lowercase letters to the uppercase letters will be either two-thirds or three-fifths. The Ames guide has both, and the Braddock-Rowe triangle has only the two-thirds ratio.

Vertical Numerals

Vertical numerals are shown in Fig. 3.13 where each number is enclosed in a square box of guidelines. As with lettering, you must learn the proportions of the numerals in order to use them properly. Each number is made the same height as the capital letters being used; usually ⅛″ high. The numeral zero is an oval and the letter O is a perfect circle in vertical lettering.

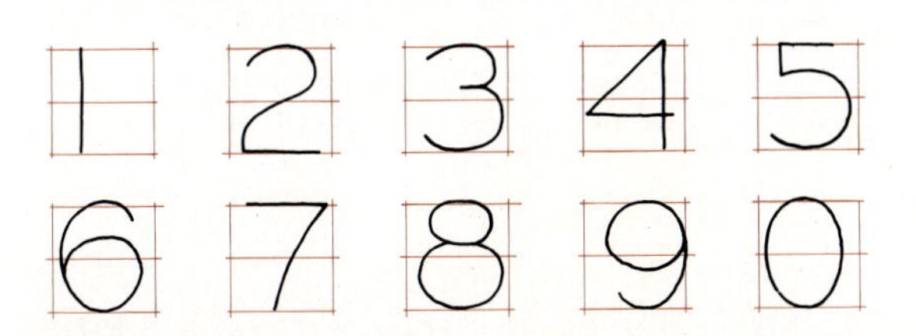

Fig. 3.13 The numerals for single-stroke Gothic lettering. Each is drawn inside a square to help you learn the proportions.

3.7 INCLINED LETTERS

Inclined Capital Letters

Inclined uppercase letters (capitals) have the same heights and proportions as vertical letters, the only difference being their inclination of 68° to the horizontal. The inclined alphabet is shown in Fig. 3.14.

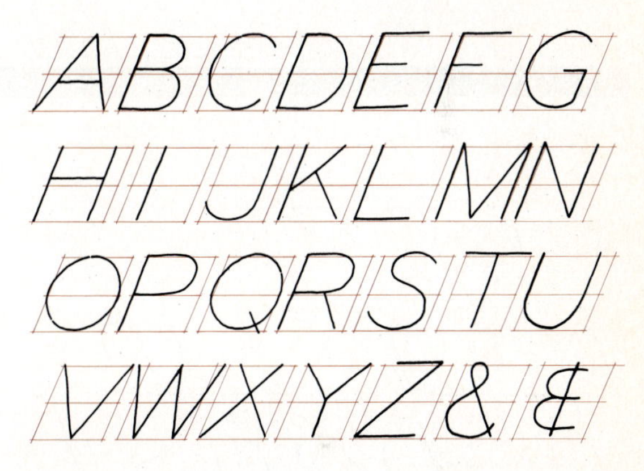

Fig. 3.14 The uppercase alphabet for single-stroke inclined Gothic lettering.

Inclined guidelines should be drawn using the Braddock-Rowe triangle or the Ames guide, as illustrated in Fig. 3.8. When these are not available, a 2 × 5 angle can be constructed and parallel guidelines drawn as shown in Fig. 3.15.

Lettering features that appear as circles in vertical lettering will appear as ellipses when inclined lettering is used.

Inclined Lowercase Letters

Lowercase inclined letters are drawn in the same manner as the vertical lowercase letters. This alphabet is shown in Fig. 3.16. Ovals (ellipses) are used instead of the circles used in vertical lettering. The angle of inclination is 68°, the same as is used for uppercase letters.

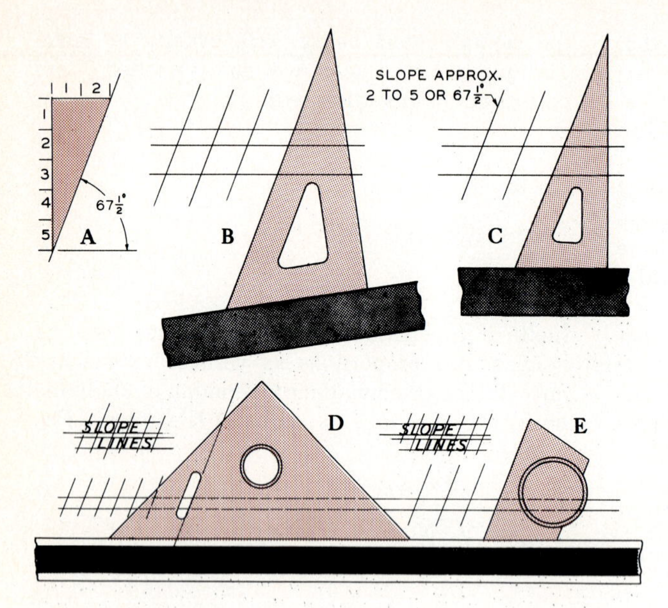

Fig. 3.15 Inclined guidelines can be constructed by any of the following: (A) draw a 2 × 5 right triangle for establishing the angle of inclination of 67.5°; (C) use a specially designed lettering triangle that has an angle of 67.5°; (D) use the slot in the Braddock-Rowe triangle; or (E) use the angle of the Ames guide.

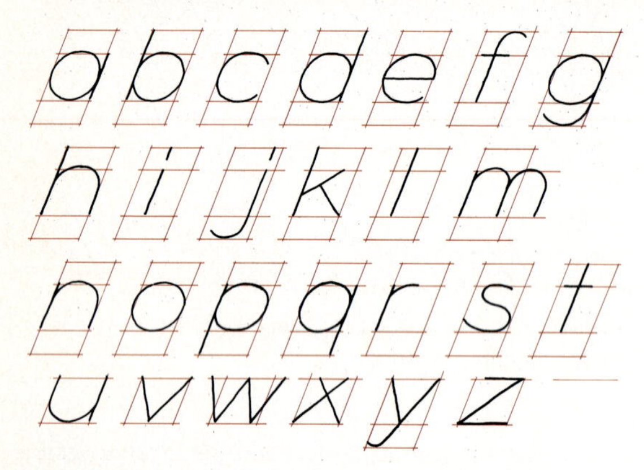

Fig. 3.16 The lowercase, single-stroke alphabet for inclined Gothic lettering. The body of each letter is drawn inside a rhombus to help you learn the proportions.

Inclined Numerals

The inclined numerals that should be used in conjunction with inclined lettering are shown in Fig. 3.17. Except for the inclination of 68° to the horizontal, they are drawn the same as vertical numbers.

The use of inclined letters and numbers in combination is seen in Fig. 3.18. The guidelines in this example were constructed using the Braddock-Rowe triangle (Fig. 3.8).

Fig. 3.17 The numerals for single-stroke Gothic inclined lettering. Each number is drawn in a rhombus to help you learn the proportions.

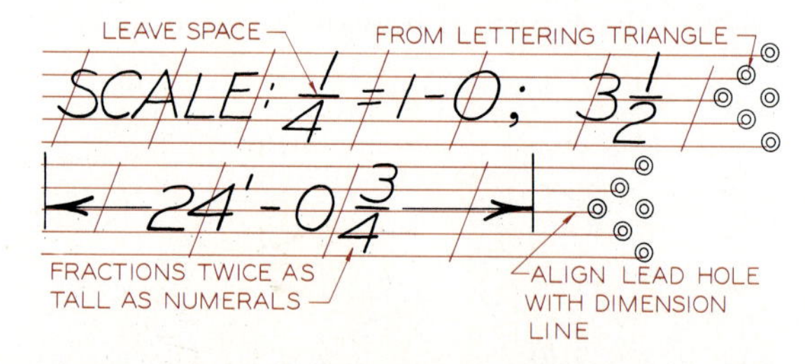

Fig. 3.18 Inclined common fractions are twice as tall as single numerals. Inch marks are omitted when numerals are used to show dimensions.

3.8 SPACING NUMERALS AND LETTERS

Common fractions are twice as tall as single numerals (Fig. 3.19). The fractions will be ¼″ tall when they are used with ⅛″ lettering. A separate set of holes for common fractions is given on the Braddock-Rowe triangle and on the Ames guide.

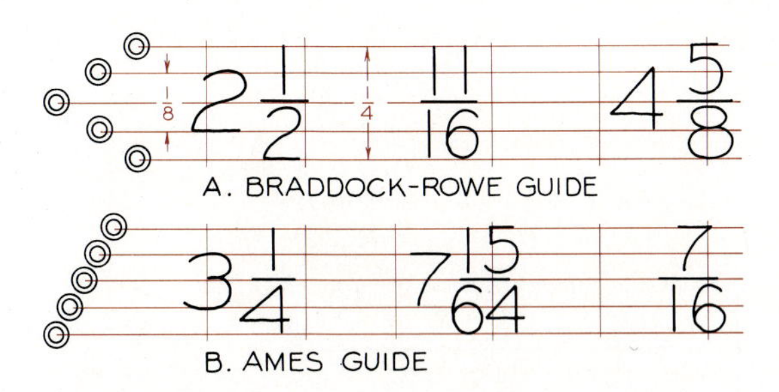

Fig. 3.19 Common fractions are twice as tall as single numerals. Guidelines for these can be drawn by using the Ames or the Braddock-Rowe triangle.

These are equally spaced 1/16″ apart with the center line being used for the fraction's crossbar. Examples of these holes can be seen in Fig. 3.19 and Fig. 3.8.

When numbers are used with decimals, space should be provided for the decimal point (Fig. 3.20). Common mistakes of spacing decimal fractions are shown at B and C. The correct method of drawing common fractions is illustrated at D, and several of the often-encountered errors are shown at E, F, and G.

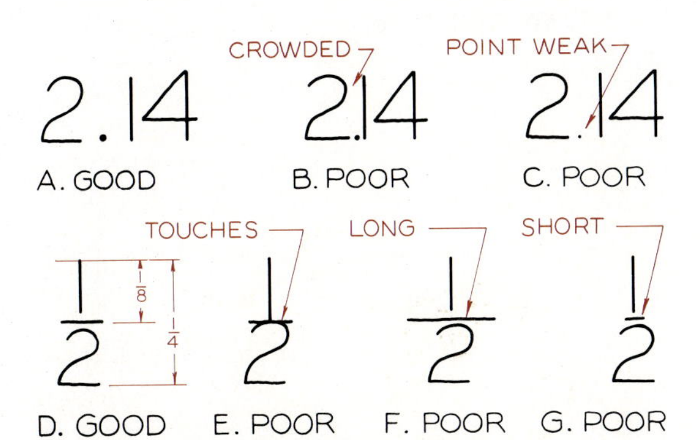

Fig. 3.20 Examples of poor spacing of numerals that result in poor lettering.

When letters are grouped together to spell words, the areas between the letters should be approximately equal for the most pleasing result (Fig. 3.21). The incorrect use of guidelines and other violations of good lettering practice are shown in Fig. 3.22. Avoid making these errors when you are lettering.

Fig. 3.21 Proper spacing of letters is necessary for good lettering and good appearance. The areas between letters should be approximately equal.

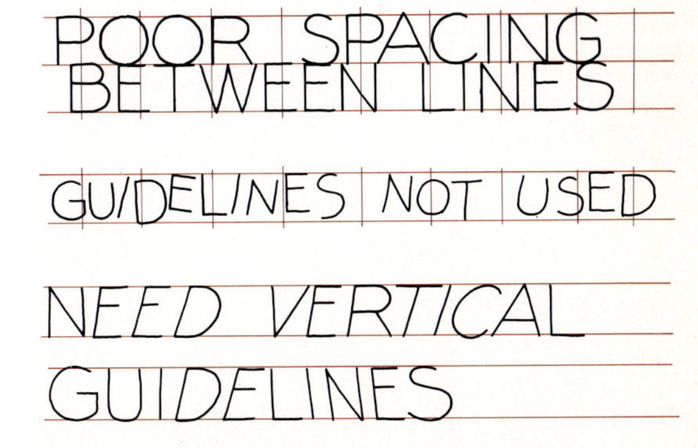

Fig. 3.22 Always leave space between lines of lettering. After constructing guidelines, *use them.* Use vertical guidelines to improve the angle of your vertical strokes.

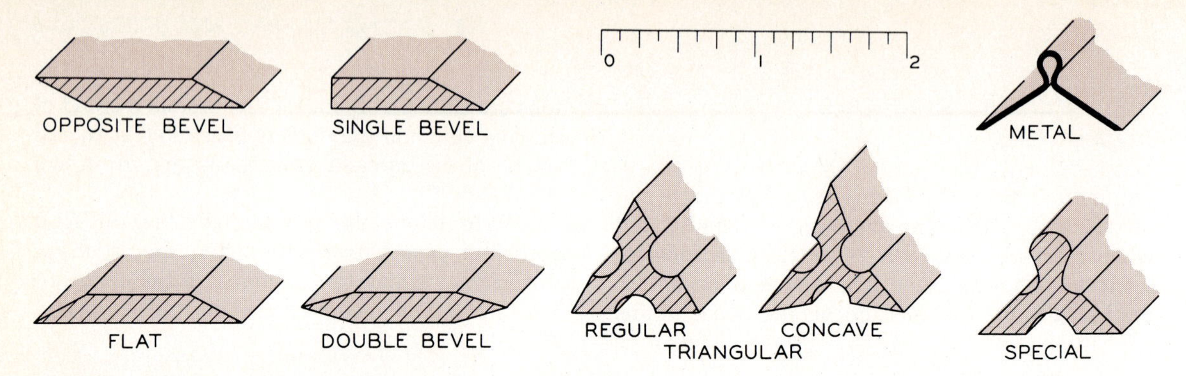

Fig. 3.23 Scales are made of wood, plastic, bamboo, and metal, and are available in any of the shapes shown here.

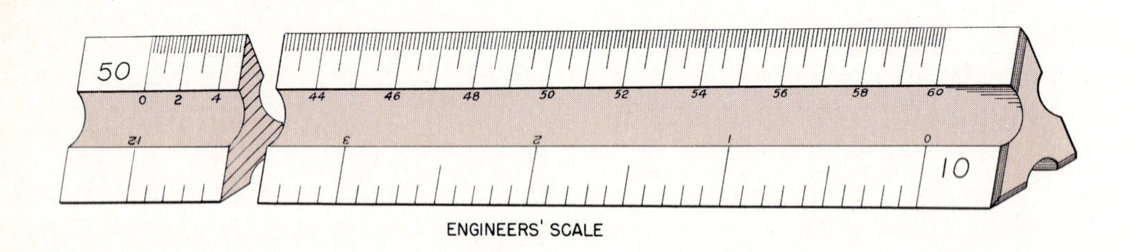

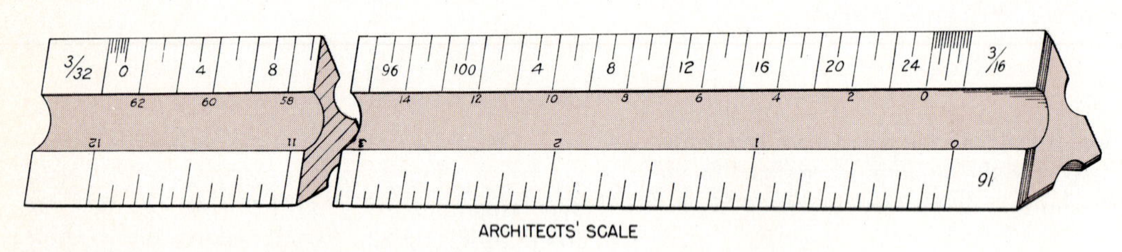

Fig. 3.24 The architects' scale is used to measure in feet and inches, whereas the engineers' scale measures in decimal units.

ARCHITECTS' SCALE

FROM END OF SCALE

BASIC FORM SCALE: $\frac{X}{X}$ = 1'-0

TYPICAL SCALES

SCALE: FULL SIZE (USE 16-SCALE)

SCALE: HALF SIZE (USE 16-SCALE)

SCALE: 3 = 1'-0

SCALE: $1\frac{1}{2}$ = 1'-0

SCALE: $\frac{1}{2}$ = 1'-0

SCALE: $\frac{3}{16}$ = 1'-0

SCALE: $\frac{3}{32}$ = 1'-0

SCALE: $1\frac{1}{2}$ = 1'-0

SCALE: $\frac{3}{4}$ = 1'-0

SCALE: $\frac{3}{8}$ = 1'-0

SCALE: $\frac{1}{8}$ = 1'-0

Fig. 3.25 The basic form of indicating the scale on the architects' scale, and the variety of scales available.

3.9 SCALES

All engineering drawings require the use of scales to measure lengths, sizes, and other measurements. The types of scales may be flat or triangular as shown in Fig. 3.23, and they are made of wood, plastic, and metal. Triangular engineers' and architects' scales are shown in Fig. 3.24.

Most scales are either 6″ or 12″ long. The scales covered in this section are the architects', engineers', mechanical engineers', and metric scales.

The Architects' Scale

The architects' scale is used to dimension and scale features encountered by the architect such as cabinets, plumbing, and electrical layouts. Most indoor measurements are made in feet and inches with an architects' scale.

The basic form of indicating on a drawing the scale that is being used is shown in Fig. 3.25. This form should be used on the drawing in the title block or in some prominent location so that the scale of the drawing can be known.

Since the dimensions made with the architects' scale are in feet and inches, it is very difficult to handle the arithmetic associated with these dimensions. It is necessary to convert all dimensions to decimal equivalents (all feet or all inches) before the simplest arithmetic can be performed.

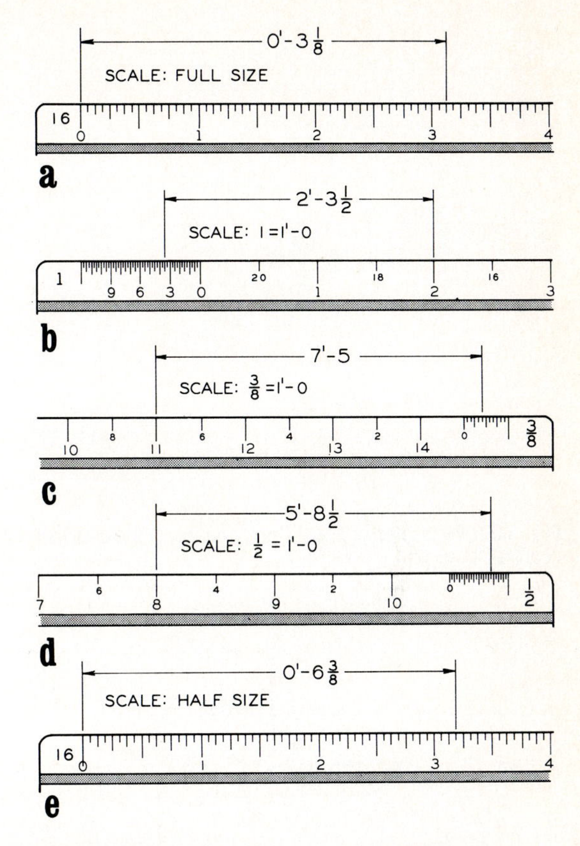

Fig. 3.26 Examples of lines measured using an architects' scale.

Scale: Full Size The 16 scale is used for measuring full-size lines (Fig. 3.26a). An inch on the 16 scale is divided into sixteenths to match the ruler used by the carpenter. This example is measured to be 3⅛″. Note that when the measurement is less than one foot, a zero may be used to precede the inch measurements, and the inch marks are omitted in all cases.

Scale: 1 = 1′–0 In Fig. 3.26b, a line is measured to its nearest whole foot (2 ft in this case) and the remainder is measured in inches at the end of the scale (3½″) for a total of 2′–3½. At the end of each architects' scale, a foot has been divided into inches for measuring dimensions that are less than a foot.

The scale 1″ = 1′–0 is the same as saying 1″ is equal to 12″, or 1/12th size.

Scale: ⅜ = 1′–0 When this scale is used ⅜″ is used to represent 12″ on a drawing. Figure 3.26c is measured to be 7′–5.

Scale: ½ = 1′–0 A line is measured to be 5′–8½ in Fig. 3.26d.

Scale: Half Size The 16 scale is used to measure or draw a line that is half size. This is sometimes specified as Scale: 6 = 12 (inch marks omitted). The line in Fig. 3.26e is measured to be 0′–6⅜.

Other lines measured on the architects' scale are shown in Fig. 3.27. When locating dimensions using any scale, hold your pencil in a vertical position for the greatest accuracy when marking measurements (Fig. 3.28).

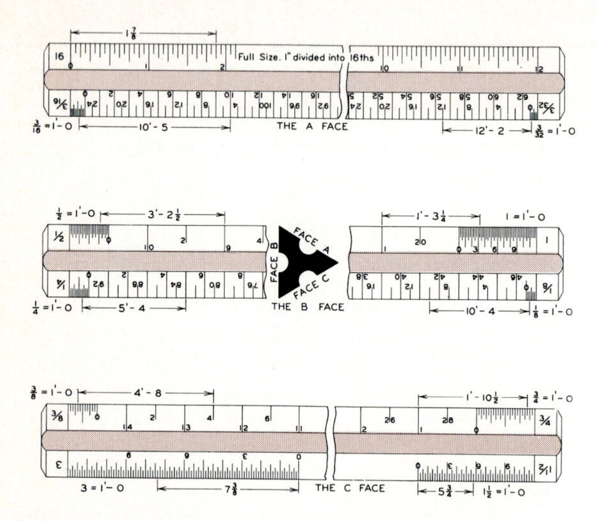

Fig. 3.27 Examples of lines measured on a three-sided architects' scale.

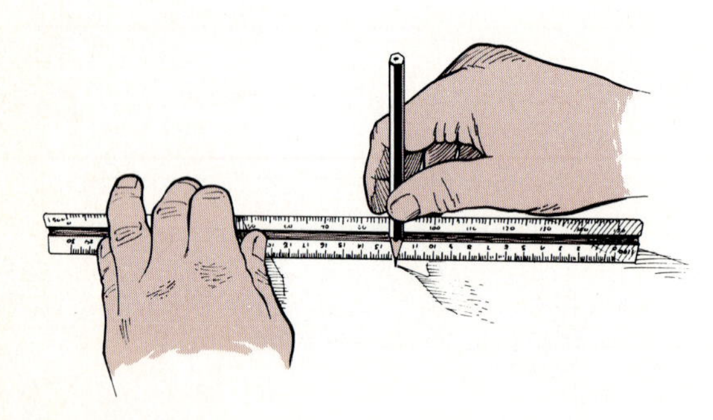

Fig. 3.28 When marking off measurements along a scale, be sure to hold your pencil vertically for the most accurate measurement.

Fig. 3.29 Inch marks are omitted according to current standards, but foot marks are shown. A leading zero is used when the inch measurements are less than a whole inch. When representing feet, a zero is optional if the measurement is less than a foot.

When indicating dimensions in feet and inches, they should be in the form shown in Fig. 3.29. Notice that the fractions are twice as tall as the whole numerals.

The Engineers' Scale

The engineers' scale is a decimal scale on which each division is a multiple of 10 units. It is used for making drawings of engineering projects that are located outdoors, such as streets, structures, land measurements, and other large dimensions associated with topography. For this reason, it is sometimes called the civil engineers' scale.

Since the measurements are in decimal form, it is easy to perform arithmetic operations without the need of converting feet and inches as when the architects' scale is used. Areas and volumes can be found easily.

The form of specifying scales on the engineers' scale is shown in Fig. 3.30, such as Scale: 1 = 10'. Each end of the scale is labeled 10, 20, 30, etc. This indicates the number of units per inch on the scale. Many combinations may be obtained by moving the decimal places of a given scale, as indicated in Fig. 3.30.

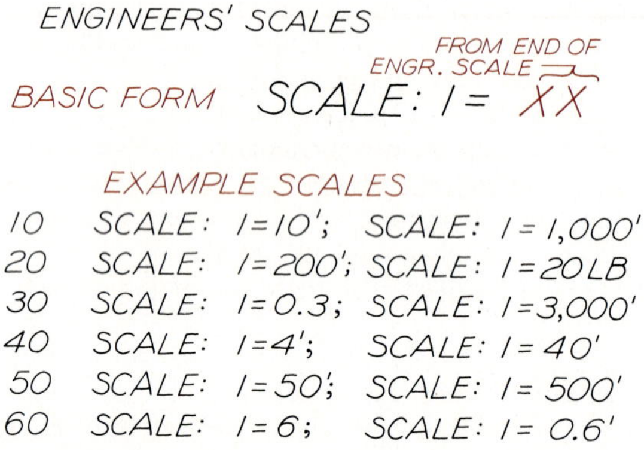

ENGINEERS' SCALES

FROM END OF ENGR. SCALE

BASIC FORM SCALE: 1 = XX

EXAMPLE SCALES

10	SCALE: 1=10';	SCALE: 1 = 1,000'
20	SCALE: 1=200';	SCALE: 1=20LB
30	SCALE: 1=0.3;	SCALE: 1=3,000'
40	SCALE: 1=4';	SCALE: 1 = 40'
50	SCALE: 1=50';	SCALE: 1 = 500'
60	SCALE: 1=6;	SCALE: 1 = 0.6'

Fig. 3.30 The basic form for indicating the scale when the engineers' scale is used, and the variety of scales that are available on this scale.

10 Scale In Fig. 3.31a, the 10 scale is used to measure a line at the scale of 1 = 10'. The line is 32.0 feet long.

20 Scale In Fig. 3.31b, the 20 scale is used to measure a line drawn at a scale of 1 = 200.0'. The line is 540.0 feet long.

30 Scale A line of 10.6 (inch marks omitted) is measured using the scale of 1 = 3.0 in Fig. 3.31c.

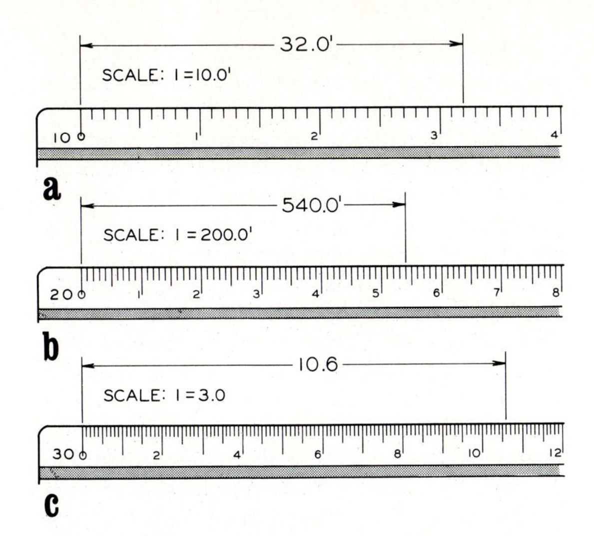

Fig. 3.31 Examples of lines measured with the engineers' scale.

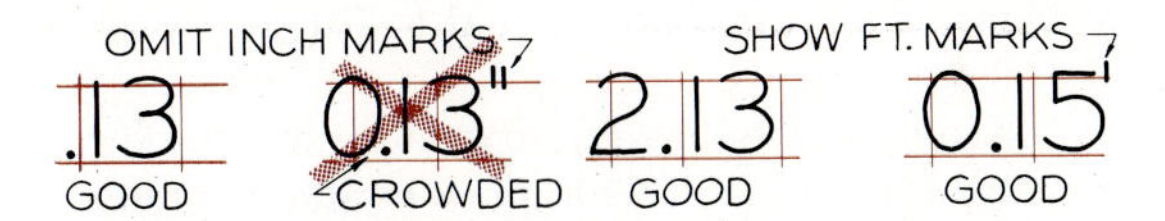

Fig. 3.32 When using English units (inches), decimal fractions do not have leading zeros and inch marks are omitted. Be sure to provide adequate space for decimal points between the numbers. Foot marks are shown.

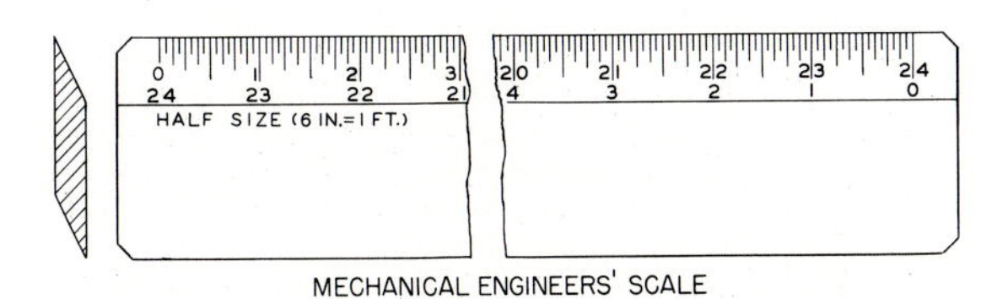

Fig. 3.33 The mechanical engineers' scales are used for measuring small parts at scales of half size, quarter size, and one-eighth size. These units are in inches with common fractions.

The format for indicating measurements in feet and inches is shown in Fig. 3.32. It is customary to omit zeros in front of decimal points when dimensioning an object using English (Imperial) units and inch marks are always omitted if the dimensions are given in inches.

Mechanical Engineers' Scale

The mechanical engineers' scale is used to draw small parts (Fig. 3.33). This scale is used to represent drawings in inches using common fractions. These scales are available in ratios of half size, one-quarter size, and one-eighth size. For example, on the half-size scale, 1 inch is used to represent 2 inches. On a quarter-size scale, 1 inch would represent 4 inches.

The Metric System—SI Units

The English system (Imperial system) of measurements has been used in the United States, Britain, and Canada since these countries were established. Presently a movement is underway to convert to the more universal metric system.

The English system was based on arbitrary units of the inch, foot, cubit, yard, and mile (Fig. 3.34). There is no common relationship between these units of measurement; consequently the system is cumbersome to use when simple arithmetic is performed. For example, finding the area of a rectangle that is 25 inches by 6¾ yards is a complex problem.

The metric system was proposed by France in the fifteenth century. In 1793, the French National Assembly agreed that the meter (m) would be one ten-millionth of the meridian quadrant of the earth (Fig. 3.35). Fractions of the meter were expressed as decimal fractions. Debate continued until an international commission officially adopted the metric system in 1875. Since a slight error in the first measurement of the meter was found, the meter

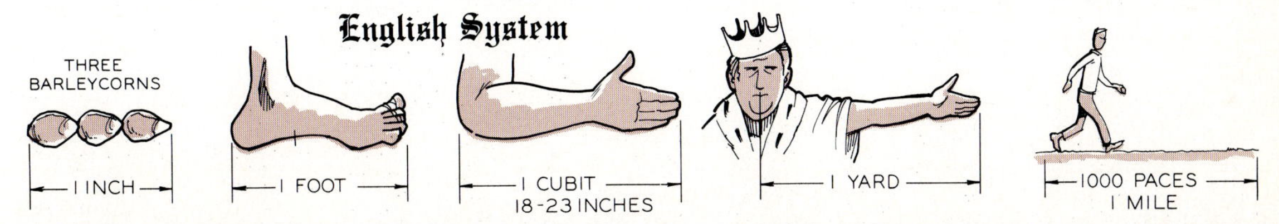

Fig. 3.34 The units of the English system were based on arbitrary dimensions.

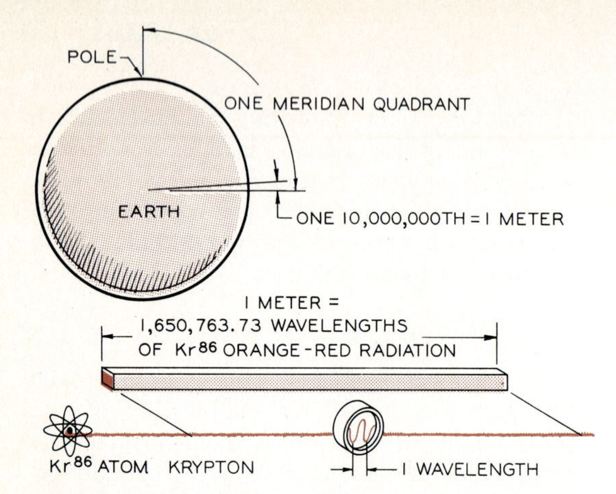

Fig. 3.35 The origin of the meter was based on the dimensions of the earth, but it has since been based on the wavelength of krypton-86. A meter is 39.37 inches.

was later established as equal to 1,650,763.73 wavelengths of the orange-red light given off by krypton-86 (Fig. 3.35).

The international organization charged with the establishment and the promotion of the metric system is called the *International Standards Organization (ISO)*. The system they have endorsed is called *Système International d'Unités* (International System of Units) and is abbreviated SI. The basic SI units are shown in Fig. 3.36 with their abbreviations. It is important that lowercase and uppercase abbreviations be used properly as shown.

Several practical units of measurement have been derived (Fig. 3.37) to make them easier to use in many applications. These unofficial SI units are widely used. Note that degrees Celsius (centigrade) is recommended over the official temperature measurement, Kelvin. When using Kelvin, the freezing and boiling temperatures are 273.15°K and 373.15°K respectively. Pressure is measured in bars, where one bar is equal to 0.1 megapascal or 100,000 pascals.

Many SI units have prefixes to indicate placement of the decimal. The more common prefixes and their abbreviations are shown in Fig. 3.38.

Several comparisons of English and SI units are given in Fig. 3.39. Other conversion factors are given in Appendix 1.

SI UNITS			DERIVED UNITS		
LENGTH	METER	m	AREA	SQ METER	m^2
MASS	KILOGRAM	kg	VOLUME	CU METER	m^3
TIME	SECOND	s	DENSITY	KILOGRAM/CU MET	kg/m^3
ELECTRICAL CURRENT	AMPERE	A	PRESSURE	NEWTON/SQ MET	N/m^2
TEMPERATURE	KELVIN	K			
LUMINOUS INTENSITY	CANDELA	cd			

Fig. 3.36 The basic SI units and their abbreviations. The derived units are units that have come into common usage.

PARAMETER	PRACTICAL UNITS		SI EQUIVALENT
TEMPERATURE	DEGREES CELSIUS	°C	0°C = 273.15 K
LIQUID VOLUME	LITER	l	$l = dm^3$
PRESSURE	BAR	BAR	BAR = 0.1 MPa
MASS WEIGHT	METRIC TON	t	$t = 10^3$ kg
LAND MEASURE	HECTARE	ha	$ha = 10^4 m^2$
PLANE ANGLE	DEGREE	°	$1° = \pi/180$ RAD

Fig. 3.37 These practical metric units are a few of those that are widely used because they are easier to deal with than the official SI units.

VALUE		PREFIX	SYMBOL
1,000,000	$= 10^6 =$	MEGA	M
1,000	$= 10^3 =$	KILO	k
100	$= 10^2 =$	HECTO	h
10	$= 10^1 =$	DEKA	da
1	$= 10^0 =$		
.1	$= 10^{-1} =$	DECI	d
.01	$= 10^{-2} =$	CENTI	c
.001	$= 10^{-3} =$	MILLI	m
.000 001	$= 10^{-6} =$	MICRO	μ

Fig. 3.38 The prefixes and abbreviations used to indicate the decimal placement for SI measurements.

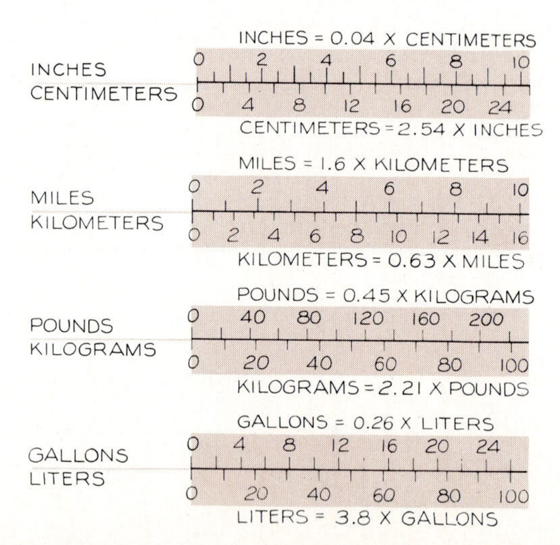

Fig. 3.39 A comparison of metric units with those used in the English system of measurement.

3.10 METRIC SCALES

The basic unit of measurement on an engineering drawing is the millimeter (mm), which is one-thousandth of a meter, or one-tenth of a centimeter. Millimeters are understood unless otherwise specified on a drawing. The width of the fingernail of your index finger can serve as a convenient gage to approximate the dimension of a centimeter, or ten millimeters (Fig. 3.40).

Metric scales are indicated on a drawing in the form shown in Fig. 3.41. A colon is placed between the numeral 1 and the ratio of the drawing size. The units are not specified since the millimeter is understood to be the unit of measurement.

Decimal units are unnecessary on most metrically dimensioned drawings; consequently the dimensions are usually rounded off to whole numbers except for these measurements that are dimensioned with specified tolerances. For metric units less than 1, a leading zero is placed in front of the decimal. In the English system, the zero is omitted from inch measurements (Fig. 3.42).

In some industries (e.g., the automotive), it is recommended that all millimeter dimensions be given with one number after the decimal point, or a zero if there is no fraction. They recommend that toleranced dimensions be carried out to three decimal places when the unit is the millimeter.

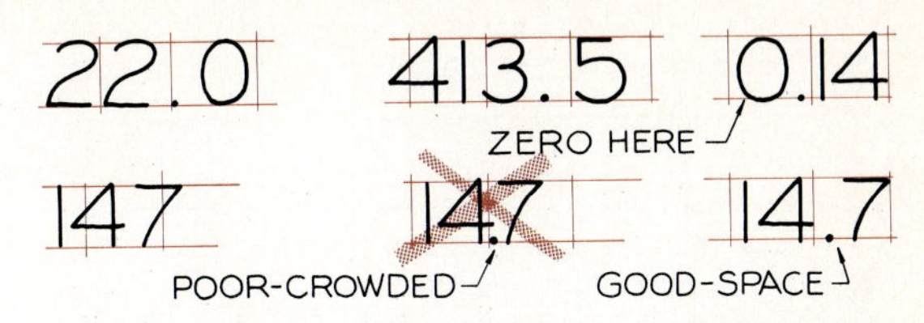

Fig. 3.42 When decimal fractions are shown in metric units, a zero is used to precede the decimal. Be sure to allow adequate space for the decimal point when numbers with decimals are lettered.

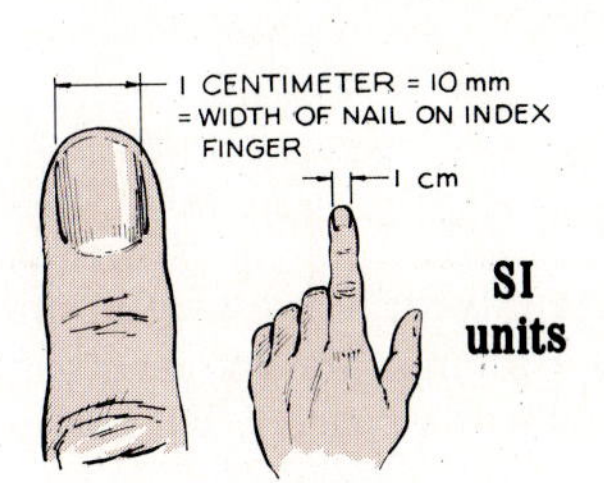

Fig. 3.40 The width of the nail on your index finger is approximately equal to a centimeter or ten millimeters.

METRIC SCALES

FROM END OF SCALE

BASIC FORM SCALE: 1:2

TYPICAL SCALES

SCALE: 1:1 (1 mm=1 mm; 1 cm=1 cm: ETC)

SCALE: 1:20 (1 mm=20 mm; 1 mm=2 cm)

SCALE: 1:300 (1 mm=300 mm; 1 mm=0.3 m)

OTHERS: 1:125; 1:250; 1:500

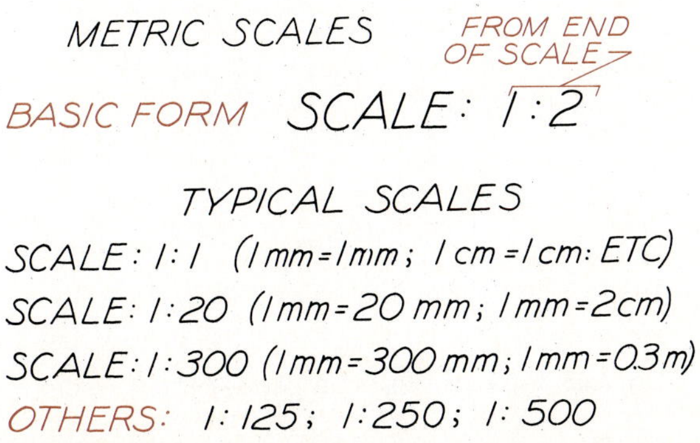

Fig. 3.41 The basic form for indicating scales when the metric scale is used, and the variety of scales that are available.

Scale 1:1 The full-size metric scale (Fig. 3.43) shows the relationship between the metric units of the dekameter, centimeter, millimeter, and the micrometer. There are 10 dekameters in a meter; 100 centimeters in a meter; 1000 millimeters in a meter; and 1,000,000 micrometers in a meter. A line of 59 mm is measured in Fig. 3.44a.

Scale 1:2 This scale is used when 1 mm is equal to 2 mm, 20 mm, 200 mm, etc. The line in Fig. 3.44b is 106 mm long.

Scale 1:3 A line of 165 mm is measured in Fig. 3.44c where 1 mm is used to represent 30 mm.

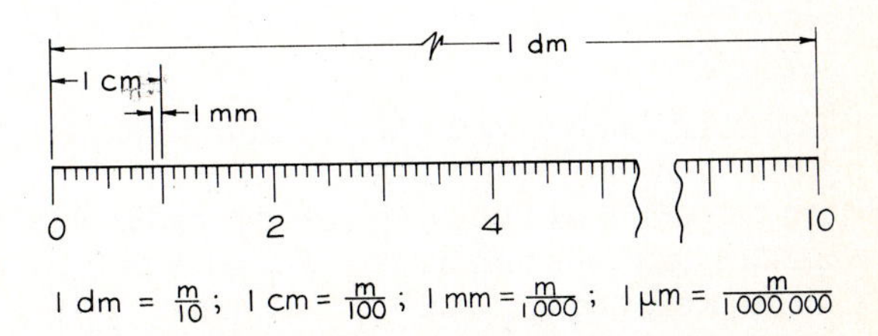

Fig. 3.43 The dekameter is one-tenth of a meter; the centimeter is one-hundredth of a meter; a millimeter is one-thousandth of a meter; and a micrometer is one-millionth of a meter.

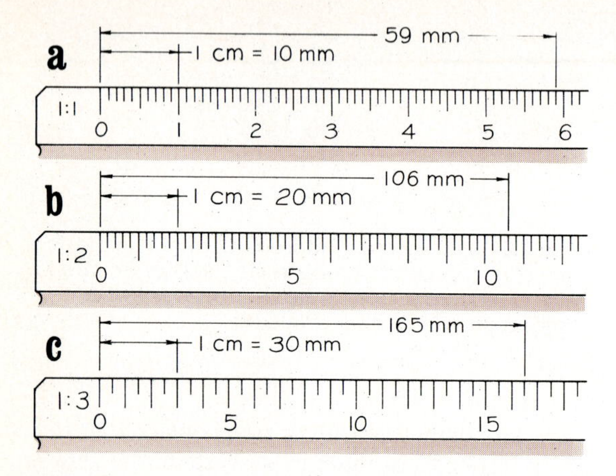

Fig. 3.44 Examples of lines measured with metric scales.

Other Scales

Many other metric (SI) scales are used: 1:250, 1:400, 1:500, and so on. The scale ratios mean that one unit represents the number of units on the right of the colon. For example, 1:20 means that one millimeter equals 20 mm, or one centimeter equals 20 cm, or one meter represents 20 m.

Metric Symbols

When drawings are made in metric units, this can be noted in the titleblock or elsewhere using the SI symbol (Fig. 3.45). The large SI indicates Système International. The two views of the partial cone are used to denote whether the orthographic views were drawn in the U.S. system (third-angle projection) or the European system (first-angle projection).

Scale Conversion

Tables for converting inches to millimeters are given in Appendix 1; however, this conversion can be performed by multiplying decimal inches by 25.4 to obtain millimeters. For example, 1.5 inches would be 1.5 × 25.4 = 38.1 mm.

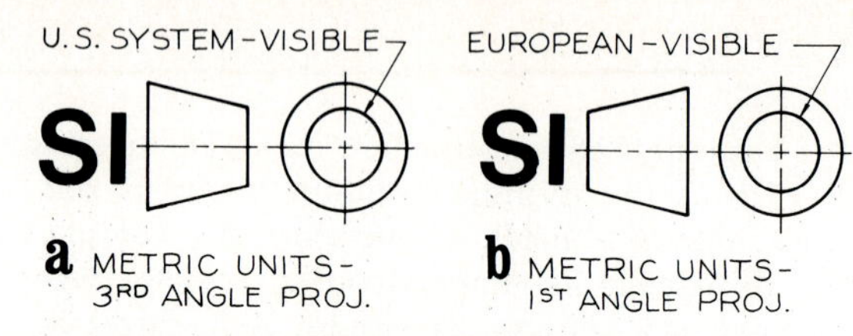

Fig. 3.45 The large letters SI indicate that the units of measurement are in metric units. The partial cones indicate that the views are arranged using the third-angle projection (the U.S. system) or the first-angle projection (the European system).

To convert an architect's scale to an approximate metric scale, the scale must be multiplied by 12. For example, Scale: ⅛ = 1′–0 is the same as ⅛ inch = 12 inches, or 1 inch = 96 inches. This scale closely approximates the metric scale of 1:100. Many of the scales used in the metric system cannot be converted to exact English scales. The scale of 1 = 5′ converts exactly to the metric scale of 1:60.

Expression of Metric Units

The general rules for expressing SI units are given in Fig. 3.46. Commas are not used between sets of zeros, but instead a space is left between them.

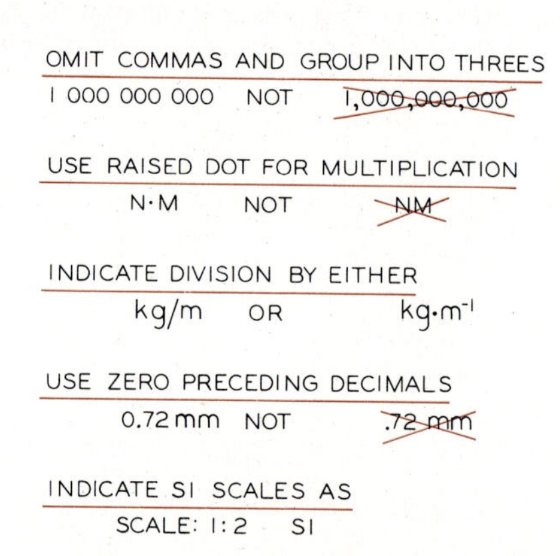

Fig. 3.46 General rules to be used with the SI system.

4

POINTS, LINES, AND PLANES

4.1 INTRODUCTION

Points, lines, and planes are the basic geometric elements that comprise the physical world in which we live. These elements are used extensively in descriptive geometry, which is the discipline that deals with graphically describing the three-dimensional geometry of our technological environment.

The method of presenting descriptive geometry is through the application of orthographic projection.

The following rules of solving and labeling descriptive problems are illustrated in Fig. 4.1.

Lettering All point, lines, and planes should be labeled using 1/8″ letters with guidelines. Lines should be labeled at each end and planes at each corner. Either letters or numbers can be used.

Points in space should be indicated by two short dashes that are perpendicular to form a cross, *not* a dot. Each dash should be approximately 1/8″ long.

Points on a line should be indicated with a short perpendicular dash on the line, *not* a dot. Label the point with a letter or numeral.

Reference lines are thin black lines that should be labeled like in Fig. 4.1.

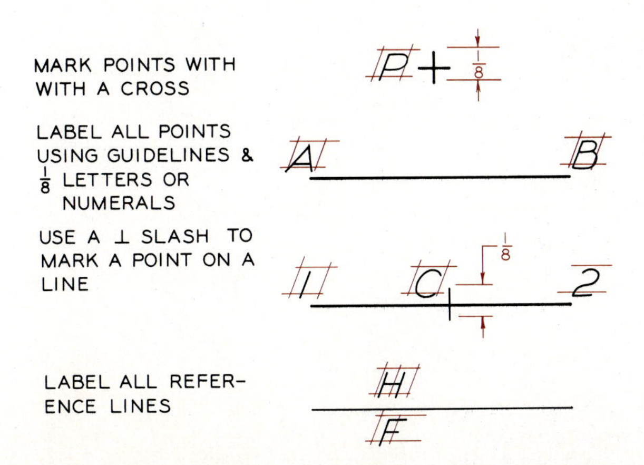

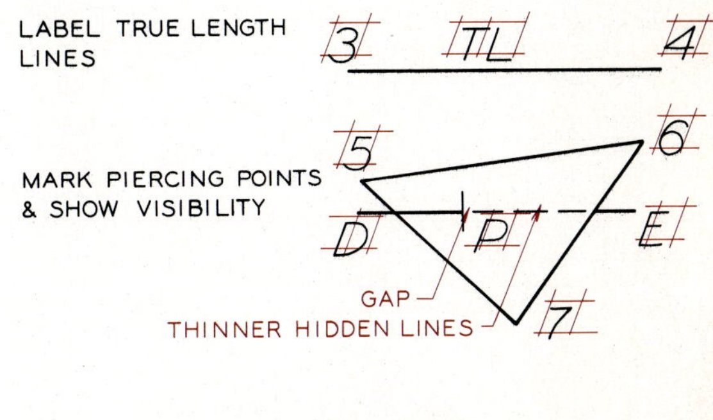

Fig. 4.1 Standard practices for labeling points, lines, and planes.

Object lines used to represent points, lines, and planes should be drawn heavier than reference lines, with an H or F pencil. Hidden lines are drawn thinner than visible lines.

True-length lines should be labeled by the full note, TRUE LENGTH, or by the abbreviation, TL.

True-size planes should be labeled by a note, TRUE SIZE, or by the abbreviation, TS.

Projection lines that are used in constructing the solution to a problem should be precisely drawn with a 4H pencil. These should be thin gray lines, just dark enough to be visible. They need not be erased after the problem is completed.

4.2 ORTHOGRAPHIC PROJECTION OF A POINT

A point is a theoretical location in space and it has no dimension. However, a series of points can establish areas, volumes, and lengths, which are the basis of our physical world.

A point must be projected perpendicularly onto at least two principal planes to establish its true position (Fig. 4.2). Note that when the planes of the projection box at (a) are opened into the plane of the drawing surface in (b), the projectors from each view of point 2 are perpendicular to the reference lines between the planes. The letters *H*, *F*, and *P* are used to represent the horizontal, frontal, and profile planes, respectively, the three principal projection planes.

A point can be located from verbal descriptions with respect to the principal planes. For example, point 2 in Fig. 4.2 can be described as being (1) 4 units left of the profile plane, (2) 3 units below the horizontal plane, and (3) 2 units behind the frontal plane. Each of these measurements must be made in the view where the principal plane being used appears as an edge.

When looking at the front view, the horizontal and profile planes appear as edges. The frontal and profile planes appear as edges in the top view, and the frontal and horizontal planes appear as edges in the side view.

4.3 LINES

A line is a straight path between two points in space. A line can appear in three forms: (1) as a foreshortened line, (2) as a true-length line, or (3) as a point (Fig. 4.3).

Oblique lines are lines that are neither parallel or perpendicular to a principal projection plane, as shown in Fig. 4.4. When line 1–2 is projected onto the horizontal, frontal, and profile planes, it appears foreshortened in each view. This is the general case of a line.

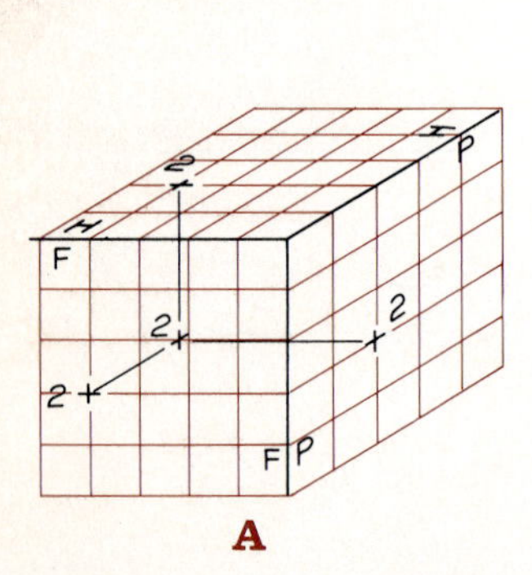

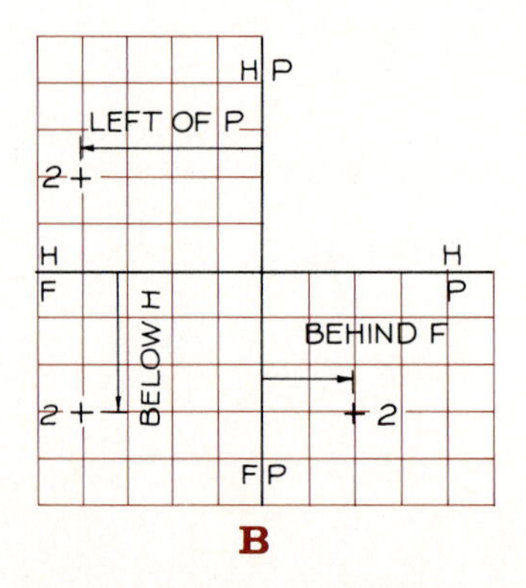

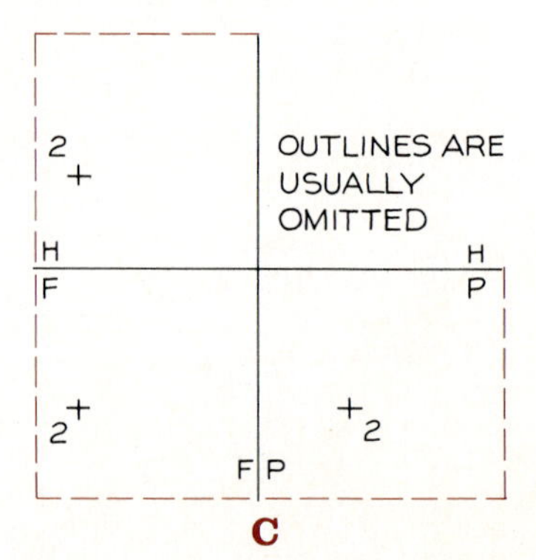

Fig. 4.2. The three projections of point 2 are shown pictorially at A and orthographically at B where the projection planes are opened into the plane of the drawing paper. Point 2 is 4 units to the left of the profile, 3 below the horizontal, and 2 behind the frontal. The outlines of the projection planes are usually omitted in orthographic projection as shown at C.

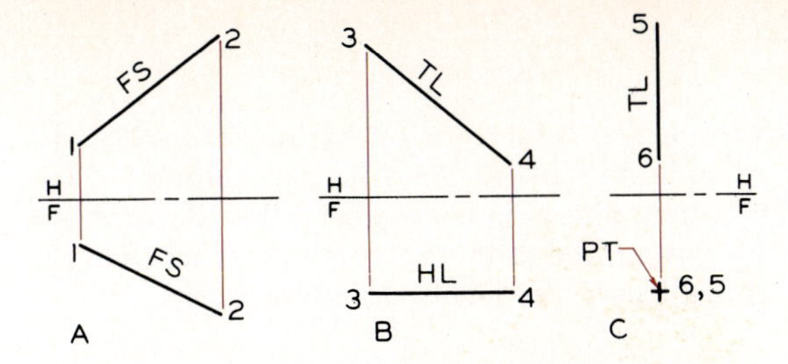

Fig. 4.3 A line in orthographic projection can appear as a point (PT), foreshortened (FS), or true length (TL).

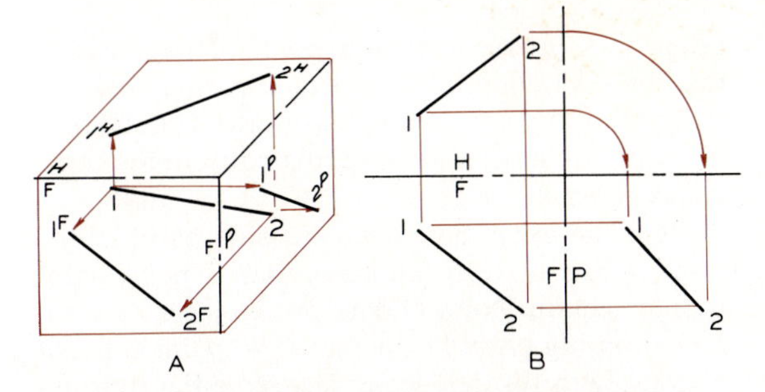

Fig. 4.4 A pictorial of the orthographic projection of an oblique line is shown at A, and as a standard orthographic projection at B.

Principal lines are lines that are parallel to at least one of the principal projection planes.

> The three principal lines are (1) horizontal, (2) frontal, and (3) profile lines, since these are the three principal projection planes.

A principal line is true length in the view where the principal plane to which it is parallel appears true size.

A horizontal line is shown in Fig. 4.5A where it appears true length in the horizontal view, the top view. It may be shown in an infinite number of positions in the top view and still appear true length provided it is parallel to the horizontal plane.

An observer cannot tell whether the line is horizontal or not when looking at the top view. Visibility must be determined by looking at the front or side views. A horizontal line will be parallel to the edge view of the horizontal in the front

Fig. 4.5 Principal lines

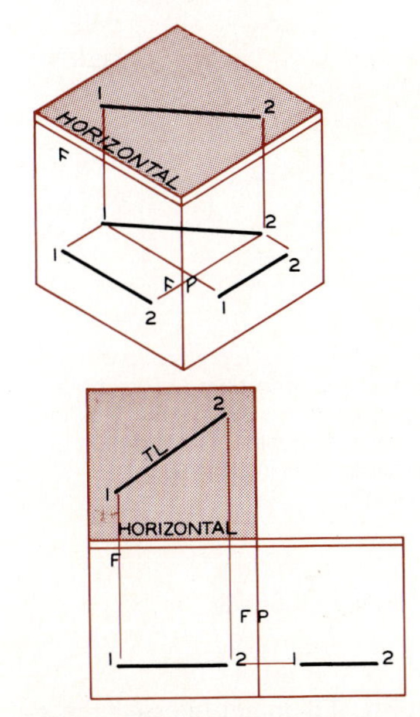

A. Horizontal line The horizontal line is true length in the horizontal view (the top view). The horizontal line will appear parallel to the edge view of the horizontal projection plane in the front and side views.

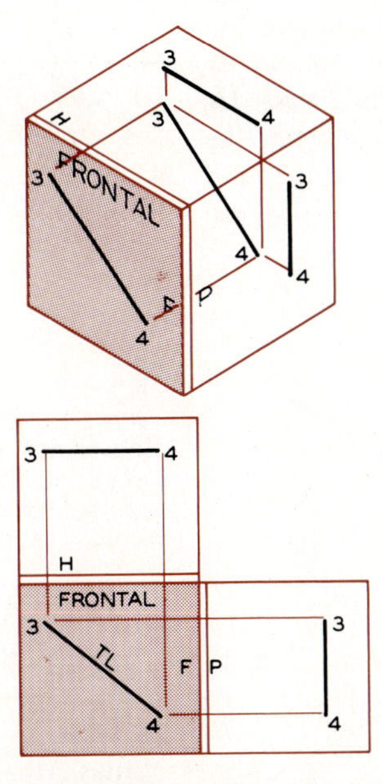

B. Frontal line The frontal line is true length in the front view. It will appear parallel to the edge view of the frontal plane in the top and side views.

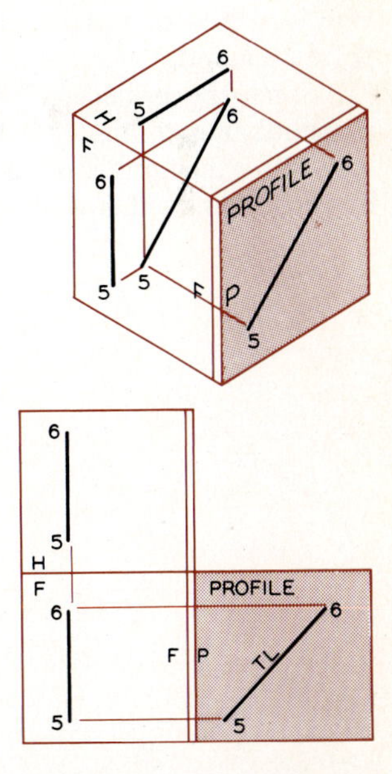

C. Profile line The profile line is true length in the profile view (the side view). It will appear parallel to the edge view of the profile plane in the top and front views.

and side views which is the *H-F* fold line. A line that projects as a point in the front view is a combination horizontal and profile line.

A frontal line is parallel to the frontal projection plane and it appears true length in the front view since the observer's line of sight is perpendicular to it in this view. You cannot see that a line is a frontal line by looking at the front view, but you must observe one of the adjacent views, the top or side views. Line 3–4 in Fig. 4.5B is determined to be a frontal line by observing its top and side view of the frontal plane.

Profile lines are parallel to the profile projection planes and they appear true length in the side views, the profile views. It is necessary to look at a view adjacent to the profile view to tell whether or not a line is a profile line. In Fig. 4.5C line 5–6 is parallel to the edge view of the profile plane in the top and front views.

4.4 LOCATION OF A POINT ON A LINE

The top and front views of line 1–2 are shown in Fig. 4.6. Point *O* is located on the line in the top view and it is required that the front view of the point be found.

Since the projectors between the views are perpendicular to the *H-F* fold line between the views in orthographic projection, point *O* is found by projecting in this same direction from the top view to the front view of the line. Point *O* is located on the line to complete the solution.

If a point is to be located at the midpoint of a line, it will be at the line's midpoint whether the line appears true length or foreshortened.

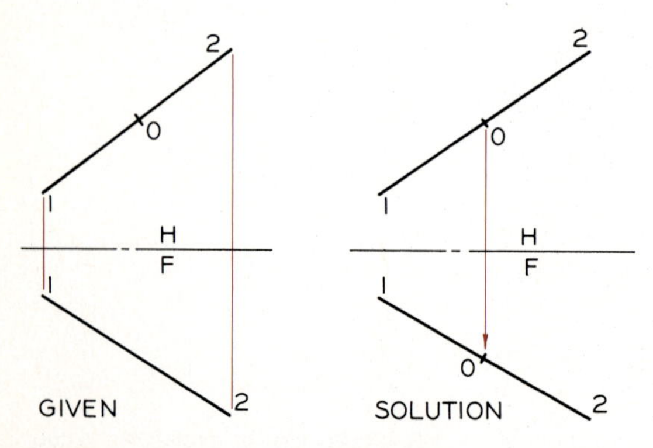

Fig. 4.6 A point on a line that is shown orthographically can be found on the front view by projection. The direction of the projection is perpendicular to the reference line between the two views.

4.5 INTERSECTING AND NONINTERSECTING LINES

Lines that intersect have a point of intersection that lies on both lines and is common to both. Point *O* in Fig. 4.7a is a point of intersection since it projects to a common crossing point in the three views given.

On the other hand, the crossing point of the lines in Fig. 4.7b in the front view is not a point of intersection. Point *O* does not project to a common crossing point in the top view; point *O* is not aligned with the projector. Therefore the lines do not intersect although they do cross. This is verified in the side view where it can be clearly seen that they do not cross.

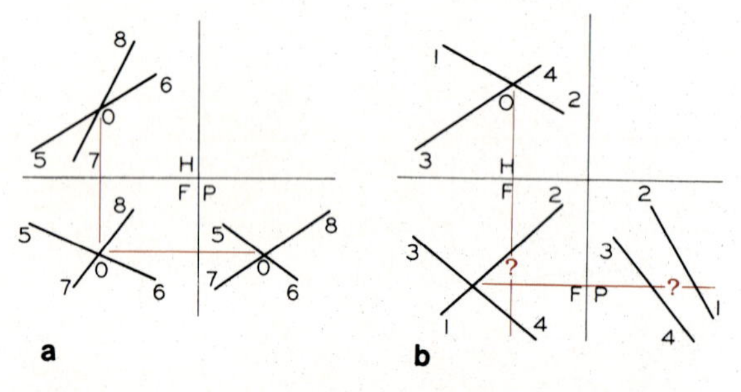

Fig. 4.7 The lines at (b) cross in the top and front views, but they do not intersect. The common point of intersection does not project from view to view. The lines at (a) do intersect because the point of intersection O projects as a point of intersection in all views.

4.6 VISIBILITY OF CROSSING LINES

Lines *AB* and *CD* in Fig. 4.8 do not intersect, however; it is necessary to determine the visibility of the lines by analysis.

Select a crossing point in one of the views, the front view in Step 1, and project it to the top view to determine which line is in front of the other.

Since AB is encountered before CD, then AB is in front of CD and is visible in the front view.

In Step 2 the crossing point in the top view is projected to the front view to find that line CD is higher than AB; therefore CD is visible in the top view.

This process of determining visibility is done by analysis rather than visualization. Two views must be utilized since it would be impossible if only one view were available.

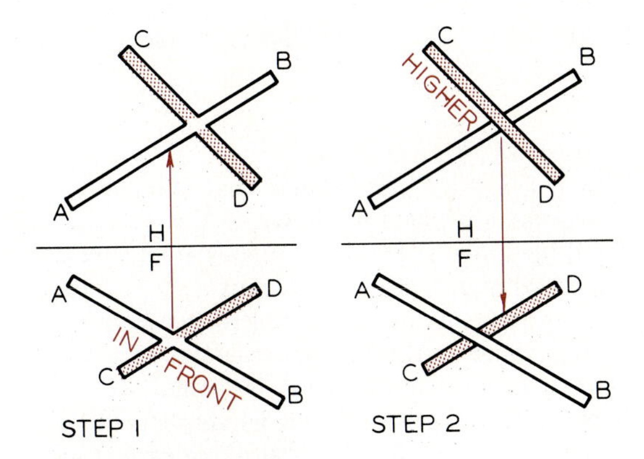

Fig. 4.8 Visibility of lines

Required Find the visibility of the lines in both views.

Step 1 Project the point of crossing from the front to the top view. This projector strikes AB before CD; therefore, line AB is in front and is visible in the front view.

Step 2 Project the point of crossing from the top view to the front view. This projector strikes CD before AB; therefore, line CD is above AB and is visible in the top view.

4.7 VISIBILITY OF A LINE AND A PLANE

The principle of visibility of intersecting lines is used in determining the visibility for a line and a plane. In Step 1 of Fig. 4.9, the intersections of AB and lines 1–3 and 2–3 are projected to the top view to determine that the lines of the plane are in front of AB in the front view. Consequently, the line is shown as a dashed line in the front.

Similarly, the two intersections on AB in the top view are projected to the front view where line

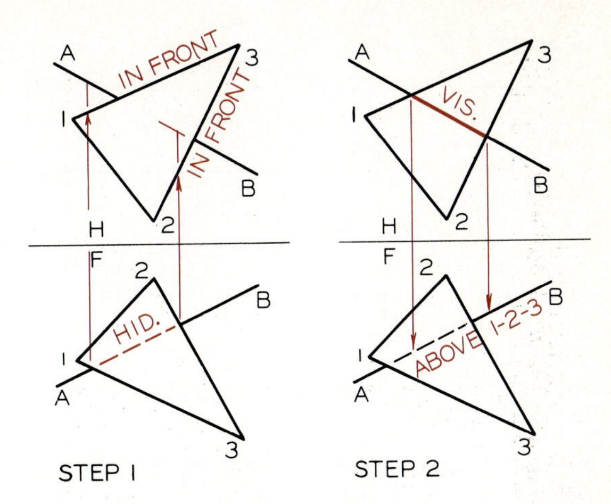

Fig. 4.9 Visibility of line and a plane

Required Find the visibility of the plane and the line in both views.

Step 1 Project the points where AB crosses the plane in the front view to the top view. These projectors encounter lines 1–3 and 2–3 of the plane first; therefore, the plane is in front of the line, making the line invisible in the front view.

Step 2 Project the points where AB crosses the plane in the top view to the front view. These projectors encounter line AB first; therefore, the line is higher than the plane, and the line is visible in the top view.

AB is found to be above the two lines of the plane, 2–3 and 1–3. Therefore AB is drawn as visible in the top view since it is over the plane.

4.8 PLANES

Planes may be considered as infinite in some problems, but it is necessary to determine planes by establishing their limits. A plane can be represented by any of the four methods shown in Fig. 4.10.

Planes in orthographic projection can appear in one of the forms shown in Fig. 4.11: (1) as an edge, (2) as true size, (3) as foreshortened.

Oblique planes are planes that are not parallel or perpendicular to principal projection planes in any view, as shown in Fig. 4.12. This is a general case of a plane.

Fig. 4.10 Representations of a plane

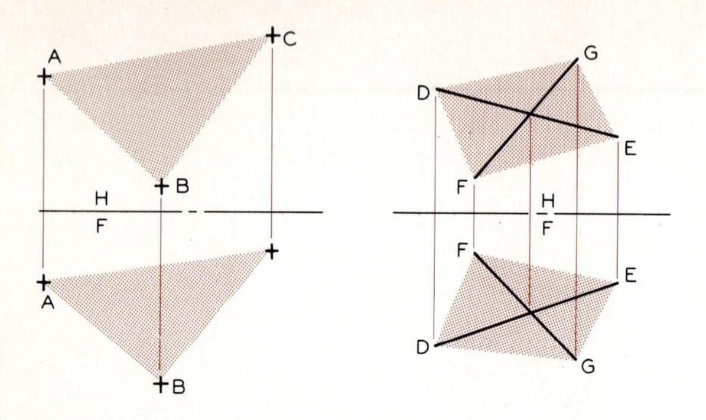

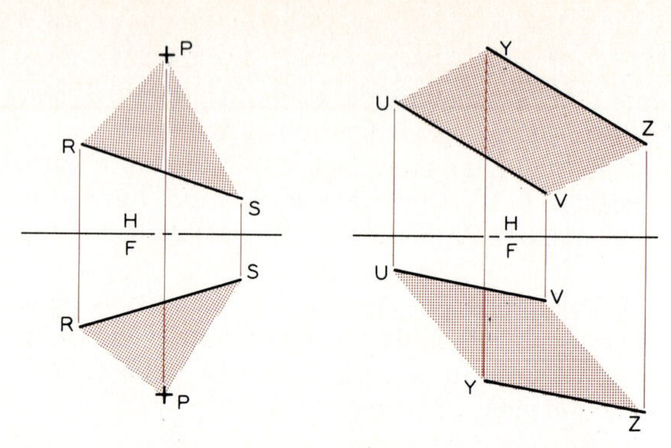

A. Three points not in a straight line can be used to represent a plane.

B. Two intersecting lines can be used to represent a plane.

C. A line and a point not on the line or its extension can be used to represent a plane.

D. Two parallel lines can be used to represent a plane.

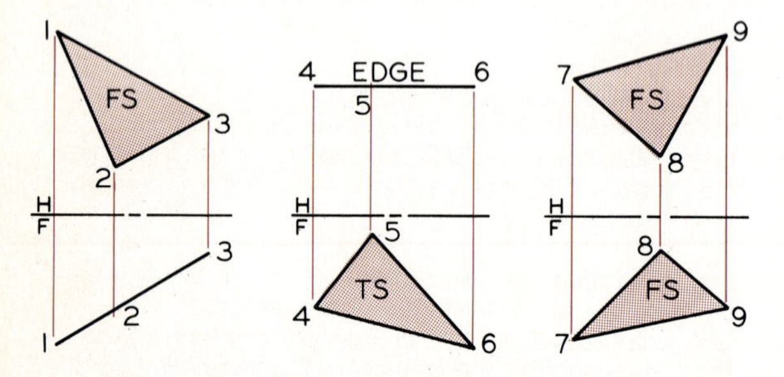

Fig. 4.11 A plane in orthographic projection can appear as an edge, true size (TS), or foreshortened (FS). If a plane is foreshortened in all principal views, it is an oblique plane.

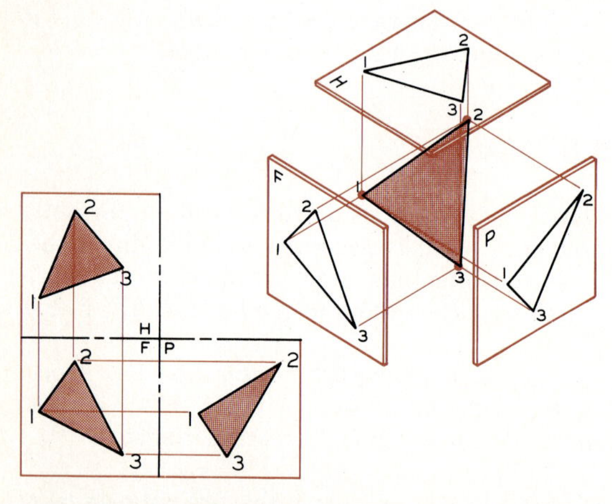

Fig. 4.12 An oblique plane is one that is not parallel or perpendicular to a proejction plane; it can be called a general-case plane. The orthographic projection of an oblique plane is shown here.

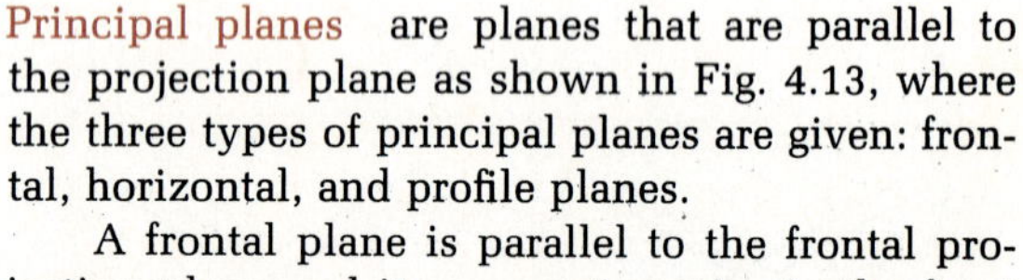

Principal planes are planes that are parallel to the projection plane as shown in Fig. 4.13, where the three types of principal planes are given: frontal, horizontal, and profile planes.

A frontal plane is parallel to the frontal projection plane and it appears true size in the front view. To tell that the plane is frontal, you must observe the top or side views where its parallelism to the edge view of the frontal plane can be seen.

A horizontal plane is parallel to the horizontal projection plane and it is true size in the top view. To tell that the plane is horizontal, you must observe the front or side views where its parallelism to the edge view of the horizontal plane can be seen.

A profile plane is parallel to the profile projection plane and it is true size in the side view. To tell that the plane is a profile plane, you must observe the top or front view where its parallelism to the edge view of the profile plane can be seen.

4.9 A LINE ON A PLANE

Line *AB* is given on the front view of the plane in Fig. 4.14. It is required to find the top view of the line. Points *A* and *B* that lie on lines 1–4 and 2–3 of the plane can be projected to the top view to the same lines of the plane.

Points *A* and *B* are found in the top view and are connected to complete the top view of line *AB*. This is an application of the principle covered in Section 4.4.

Fig. 4.13 Principal planes

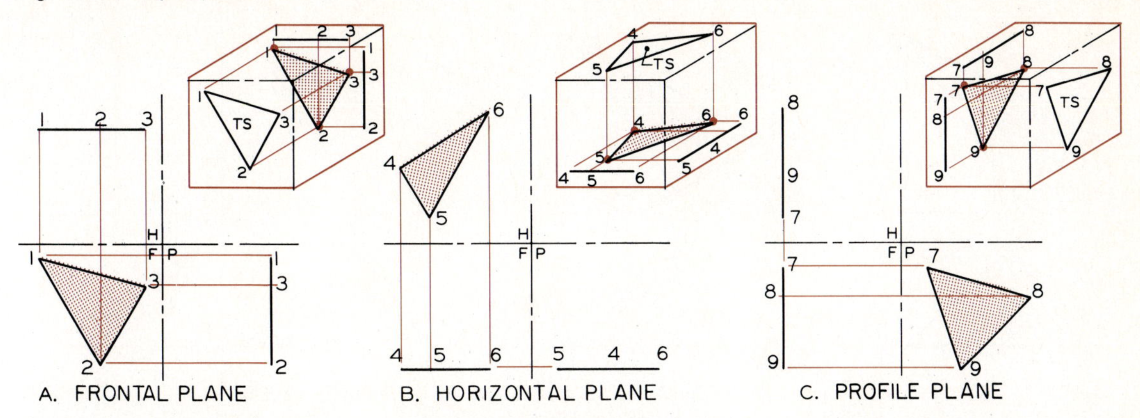

A. Frontal planes are true size and shape (TS) in the front view. They will appear as edges parallel to the frontal plane in the top and side views.

B. Horizontal planes are true size in the horizontal views (top views). They appear as edges parallel to the horizontal plane in the front and side views.

C. Profile planes are true size in the profile views (side views). They appear as edges parallel to the profile plane in the top and front views.

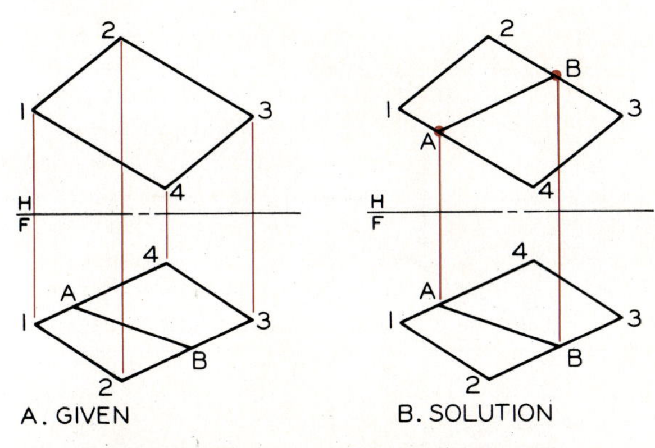

Fig. 4.14 If line *AB* lying on the plane is given, the top view of the line can be found. Points *A* and *B* are projected to lines 1–4 and 2–3, respectively, and are connected to form line *AB.*

4.10 A POINT ON A PLANE

Point *O* is given on the front view of plane 4–5–6 in Fig. 4.15. It is required to locate the point on the plane in the top view.

In Step 1, a line in any direction other than vertical is drawn through the point to establish a line on the plane. The line is projected to the top view in Step 2, and the point is projected from the front view to the top view of the line.

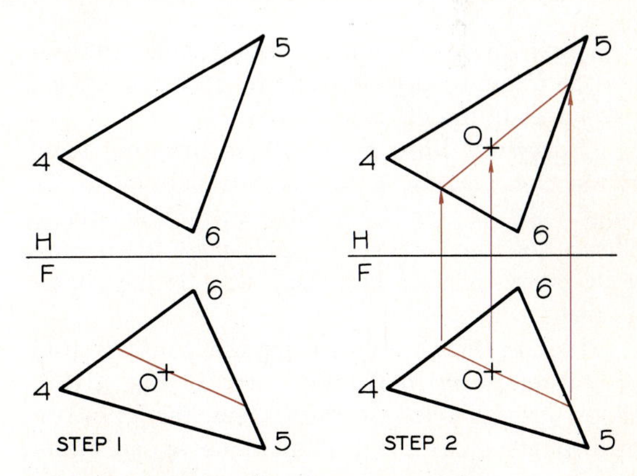

Fig. 4.15 Location of a point on a plane.

Find the top view of point O that lies on the plane.

Step 1 Draw a line through the given view of point O in any convenient direction except vertical.

Step 2 Project the ends of the line to the top view and draw the line. Point O is projected to the line.

Fig. 4.16 Principal lines on a plane.

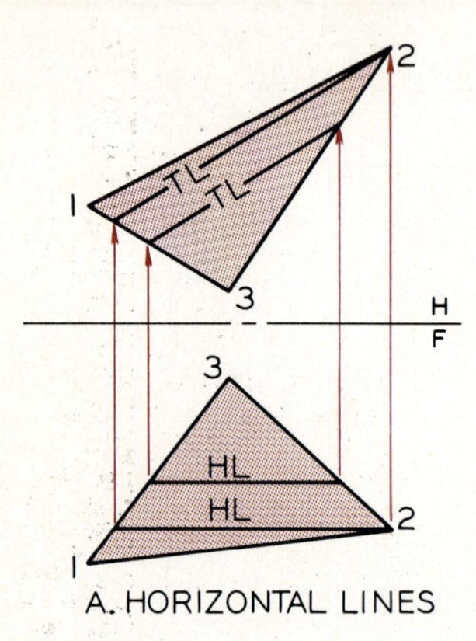

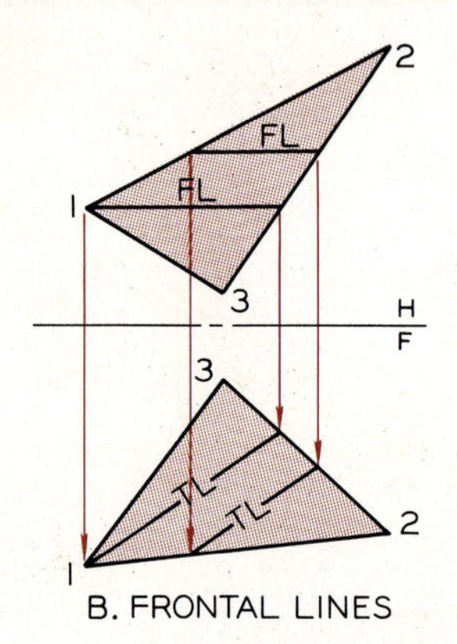

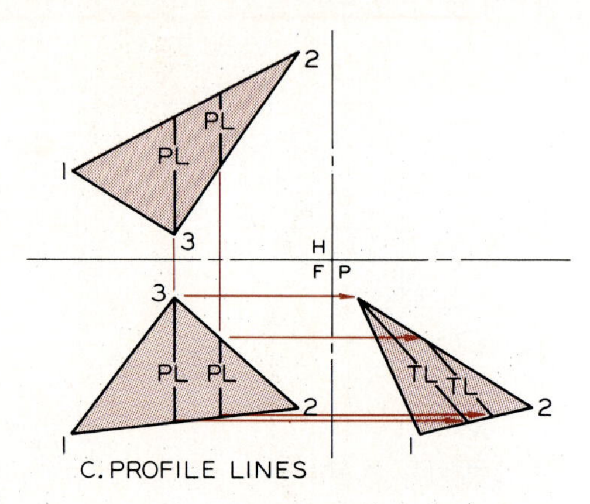

A. Horizontal lines are drawn first in the front view parallel to the edge view of the horizontal plane. These lines are found true length when they are projected to the top view.

B. Frontal lines are drawn first in the top view parallel to the edge view of the frontal plane. These lines are found true length when they are projected to the front view.

C. Profile lines are drawn first in the front view parallel to the edge view of the profile plane. These lines are found true length when they are projected to the profile view (side view)

4.11 PRINCIPAL LINES ON A PLANE

Principal lines—horizontal, frontal, and profile—may be found in any view of a plane when at least two views of the plane are given.

Horizontal lines are drawn in the front view of the plane in Fig. 4.16A that are parallel to the edge view of the horizontal projection plane. These horizontal lines are projected to the top view of the plane where they will be horizontal and true length.

Frontal lines are drawn parallel to the frontal projection plane in the top view in Fig. 4.16B. When projected to the front view, the lines are true length since the frontal plane appears true size in this view.

Profile lines are drawn parallel to the profile projection plane in the top and front views in Fig. 4.16C. When projected to the side view, the lines will appear true length.

An infinite number of principal lines can be drawn on a single plane. Only two have been shown in each case as examples.

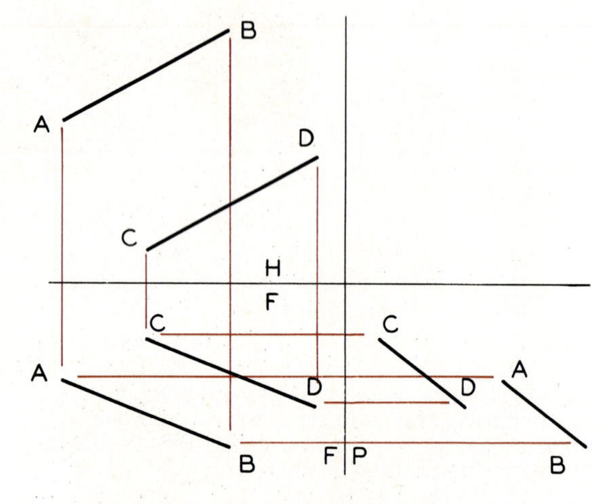

Fig. 4.17 When two lines are parallel, they will project as parallel in all orthographic views.

4.12 PARALLELISM OF LINES

If two lines are parallel, they will appear parallel in all views in which they are seen, except where both appear as points. Lines *AB* and *CD* appear parallel in three views in Fig. 4.17. Parallelism of lines in space cannot be determined if only one view is given; two or more views are required.

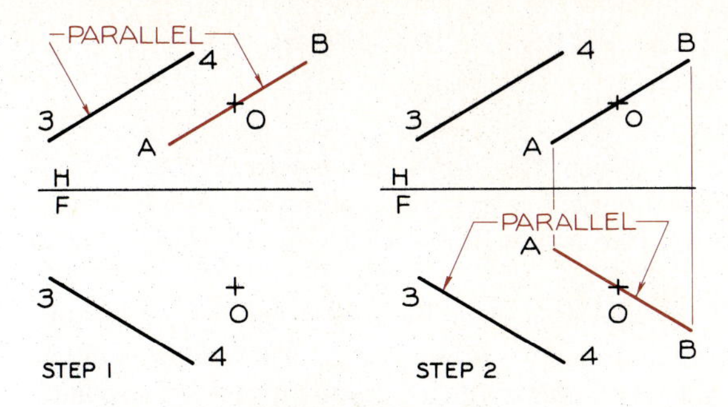

Fig. 4.18 A line parallel to a line.

Draw a line through O that is parallel to the given line.

Step 1 Draw line *AB* parallel to the top view of line 3–4 with its midpoint at O.

Step 2 Draw the front view of line *AB* parallel to the front view of 3–4 through point O.

Using this principle, a line can be drawn parallel to a given line through a specified point as shown in Fig. 4.18. In Step 1, line *AB* is drawn parallel to line 3–4 and through point O, the midpoint of the line.

In Step 2, the front view of *AB* is drawn parallel to the front view of 3–4 and through point O. Line *AB* is parallel to 3–4 since it is parallel to 3–4 in both views.

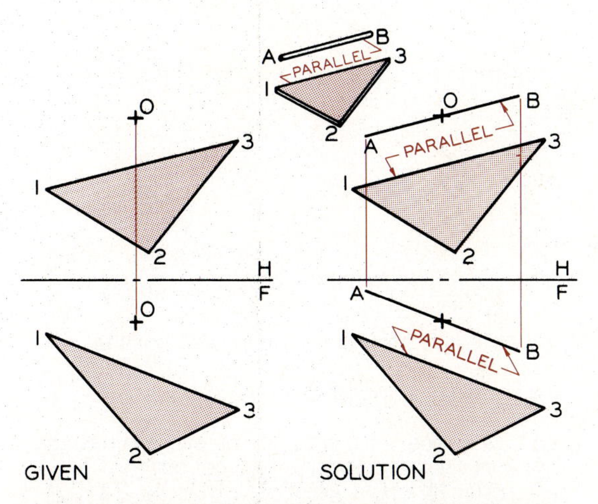

Fig. 4.19 A line can be drawn through point O that is parallel to the given plane if the line is parallel to any line in the plane. Line *AB* is drawn parallel to line 1–3 in the front and top views making it parallel to the plane.

4.13 PARALLELISM OF A LINE AND A PLANE

> A line is parallel to a plane if it is parallel to any line in the plane.

In Fig. 4.19 it is required that a line with its midpoint at point O be drawn that is parallel to plane 1–2–3. This is done by drawing line *AB* parallel to a line in the plane, line 1–3 in this case, in the top and front views.

The line could have been drawn parallel to *any* line in the plane; therefore, there are an infinite number of positions for lines that are parallel to a given plane.

A similar example is shown in Fig. 4.20 where it is required to draw a line with its midpoint at O that is parallel to the plane formed by two intersecting lines. In Step 1, *AB* is drawn parallel to line 1–2 of the plane in the top view. In Step 2, *AB* is drawn parallel to the front view of 1–2, which completes the solution of the problem.

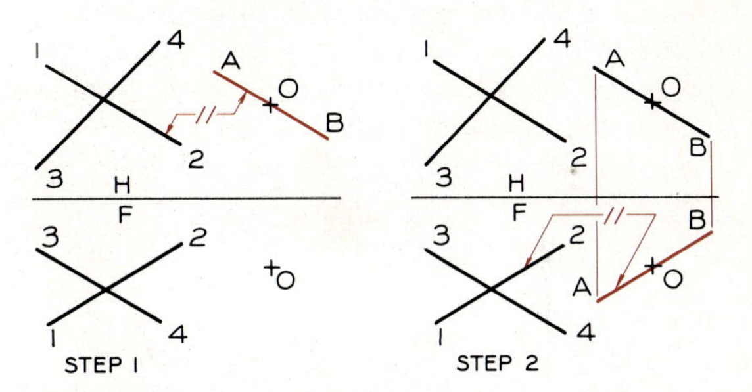

Fig. 4.20 A line parallel to plane

Draw a line parallel to the plane represented by two intersecting lines.

Step 1 Line *AB* is drawn parallel to line 1–2 through point O.

Step 2 Line *AB* is drawn parallel to the same line, line 1–2, in the front view which makes *AB* parallel to the plane.

4.14 PARALLELISM OF PLANES

> Two planes are parallel when intersecting lines in one plane are parallel to intersecting lines in the other, as shown in Fig. 4.21.

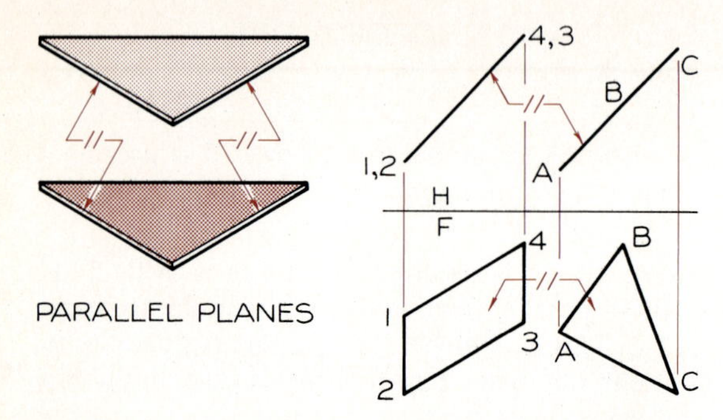

Fig. 4.21 Two planes are parallel when intersecting lines in one are parallel to intersecting lines in the other. When parallel planes appear as edges, the edges will be parallel.

It is easy to determine that planes are parallel when both appear as parallel edges in a view.

It is required that a line be drawn through point *O* and parallel to plane 1–2–3 in Fig. 4.22. In Step 1, *EF* is drawn through point *O* and parallel to line 1–2 in both the top and front views. In Step 2, a second line is drawn through point *O* parallel to line 2–3 of the plane in the front and top views. These two intersecting lines form a plane that is parallel to plane 1–2–3.

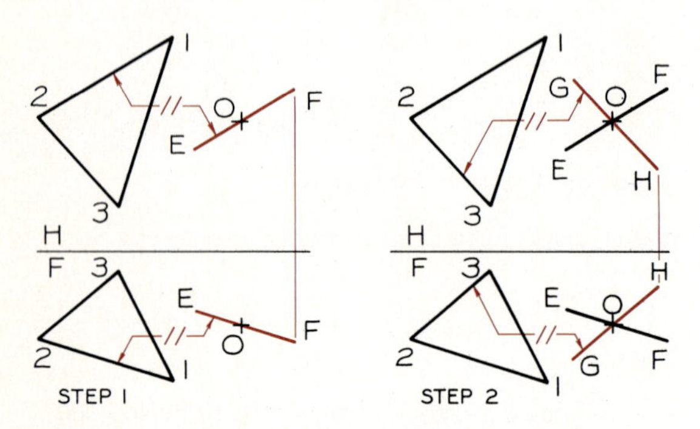

Fig. 4.22 A plane through a point parallel to a plane

Draw a plane through point O that is parallel to the given plane.

Step 1 Draw line *EF* parallel to any line in the plane, line 1–2 in this case. Show the line in both views.

Step 2 Draw a second line parallel to line 2–3 in the top and front views. These two intersecting lines represent a plane parallel to 1–2–3.

4.15 PERPENDICULARITY OF LINES

> When two lines are perpendicular, they will project at true 90° angles when one or both are true length (Fig. 4.23).

It can be seen that the axis is true length in the front view; therefore, any spoke will be shown perpendicular to the axis in the front view. Spokes *OA* and *OB* are examples where one is true length and the other is foreshortened.

When two lines are perpendicular but neither is true length, they will not project with a true 90° angle. This angle will be either less than or greater than 90°.

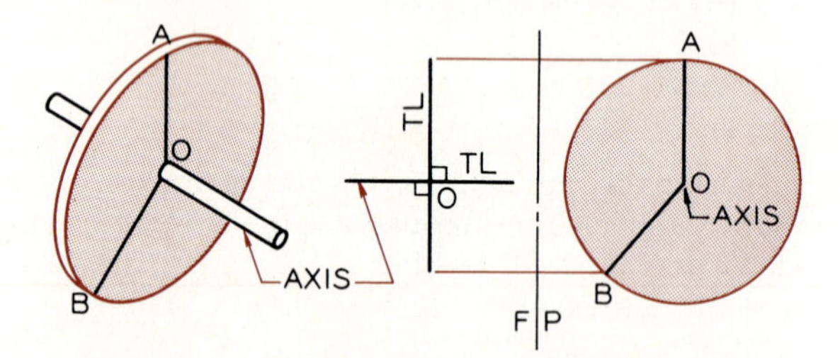

Fig. 4.23 Perpendicular lines will intersect at 90° angles in a view where one or both of the lines appear true length.

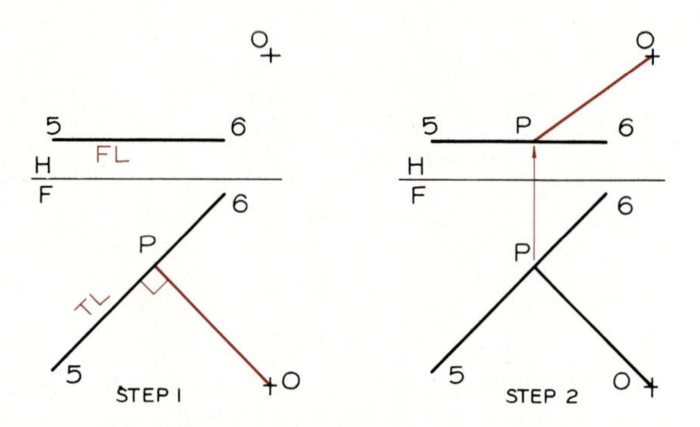

Fig. 4.24 A line perpendicular to a frontal line

Draw a line from point O perpendicular to line 5–6.

Step 1 Since line 5–6 is a frontal line and is true length in the front view, a perpendicular from point O will make a true 90° angle with it.

Step 2 Project point *P* to the top view and connect it to point *O*. Since neither of the lines is true length they will not intersect at 90° in the top view.

4.16 A LINE PERPENDICULAR TO A PRINCIPAL LINE

It is required in Fig. 4.24 to construct a line through point O that is perpendicular to frontal line 5–6 that is true length in the front view. In Step 1, OP is drawn perpendicular to 5–6 since it is true length. In Step 2, point P is projected to the top view of 5–6. Line OP in the top view cannot be drawn as a true 90° angle since neither of the lines is true length in this view.

4.17 A LINE PERPENDICULAR TO AN OBLIQUE LINE

It is required in Fig. 4.25 to construct a line from point O that is perpendicular to oblique line 1–2.

In Step 1, horizontal line OE is drawn at a convenient length. In Step 2, OE can be drawn perpendicular to line 1–2 since OE is true length in the top view.

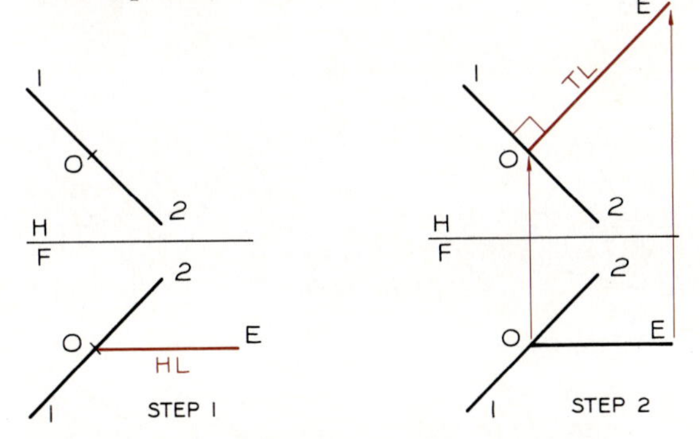

Fig. 4.25 A line perpendicular to an oblique line

Draw a line from point O that is perpendicular to the given line, 1–2.

Step 1 Draw a horizontal line from point O in the front view.

Step 2 Horizontal line *OE* will be true length in the top view; therefore, it can be drawn perpendicular to line 1–2 in this view.

4.18 PERPENDICULARITY INVOLVING PLANES

A line can be drawn perpendicular to a plane if it is drawn perpendicular to any two intersecting lines in the plane, as shown in Fig. 4.26A. Also, a plane is perpendicular to another plane if a line in one is perpendicular to the other. This is illustrated in Fig. 4.26B.

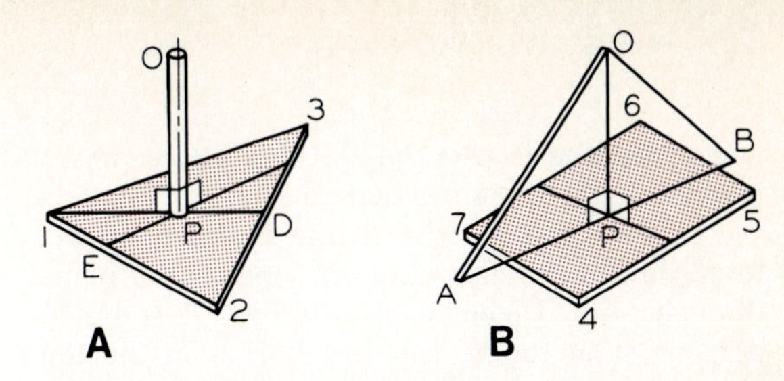

Fig. 4.26 (A) A line is perpendicular to a plane if it is perpendicular to two intersecting lines on the plane. (B) A plane is perpendicular to another plane if the plane contains a line that is perpendicular to the other plane.

4.19 A LINE PERPENDICULAR TO A PLANE

It is required in Fig. 4.27 that a line be drawn from point O on the plane perpendicular to the plane.

In Step 1, a frontal line is drawn on the plane in the top view and is projected to the front view

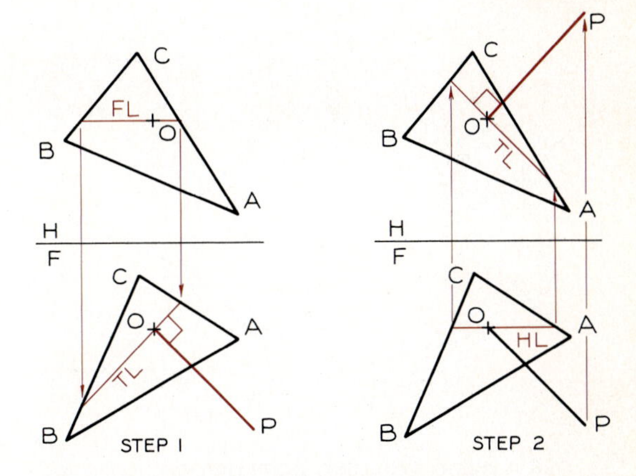

Fig. 4.27 A line perpendicular to a plane

Draw a line from point O that is perpendicular to the plane.

Step 1 Construct a frontal line on the plane through O in the top view. This line is true length in the front view; therefore, line *OP* can be drawn perpendicular to the true length line.

Step 2 Construct a horizontal line through point O in the front view. This line is true length in the top view; therefore, line *OP* can be drawn perpendicular to it.

where the line is true length. Line *OP* is drawn perpendicular to the true-length line.

In Step 2, a horizontal line is drawn in the front view and then in the top view of the plane through point *O*. The top view of line *OP* is drawn perpendicular to the true-length line in the top view. This results in a line perpendicular to the plane since the line is perpendicular to two intersecting lines in the plane, a horizontal and a frontal line.

4.20 A PLANE PERPENDICULAR TO AN OBLIQUE LINE

It is required in Fig. 4.28 to construct a plane through point *O* that is perpendicular to line 1–2.

In Step 1, *AB* is drawn as a frontal line in the top view. Since it will be true length in the front view, *AB* is drawn perpendicular to 1–2 in the front view.

In Step 2, *CD* is drawn as a horizontal line in the front view. Line *CD* is drawn perpendicular to 1–2 in the top view since *CD* is true length in this view. These two intersecting lines, *AB* and *CD*, form a plane that is perpendicular to line 1–2.

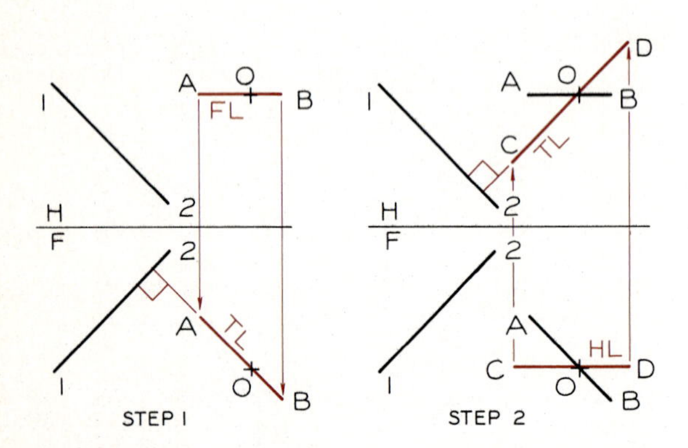

Fig. 4.28 A plane through a point perpendicular to a line

Draw a plane through O that is perpendicular to line 1–2.

Step 1 Draw line *AB* as a frontal line in the top view. Since it will be true length in the front view, it can be drawn perpendicular to line 1–2.

Step 2 Draw line *CD* as a horizontal line in the front view. Since it will be true length in the top view, it can be drawn perpendicular to line 1–2. The intersecting lines form a plane that is perpendicular to the line.

4.21 PERPENDICULARITY OF PLANES

In Fig. 4.29 it is required that a plane be constructed through line *AB* that is perpendicular to plane 1–2–3.

In Step 1, a true-length frontal line is found on the plane. Line *CD* is drawn through *AB* and perpendicular to the frontal line in the plane.

In Step 2, the intersection point between *AB* and *CD* is projected to the top view. A true-length horizontal line is found on the plane. The top view of *CD* is drawn through the point of intersection and perpendicular to the true-length line in the plane. These two intersecting lines, *AB* and *CD*, form a plane that is perpendicular to the plane since a single line in the plane, *CD*, is perpendicular to the plane.

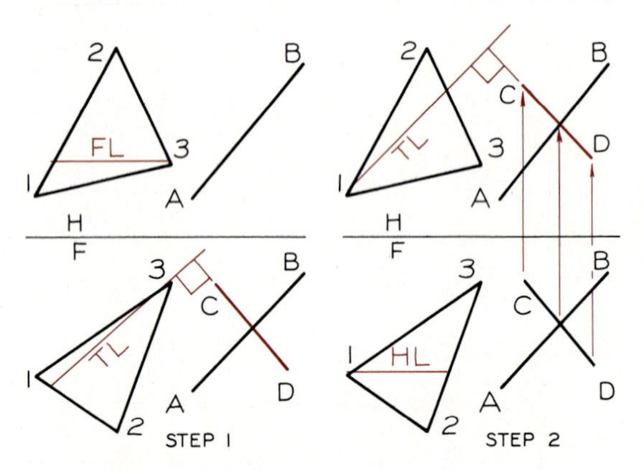

Fig. 4.29 A plane through a line perpendicular to a plane

Draw a plane through the line AB, that is perpendicular to the plane.

Step 1 Draw a frontal line on the plane in the top view and find its true-length view in the front view. Line *CD* is drawn through line *AB* and perpendicular to the true length line.

Step 2 Draw a horizontal line on the plane in the front view and find its true-length view in the top view. Draw line *CD* perpendicular to the true-length line and through line *AB* at the point of intersection projected from the front view.

PROBLEMS

Use Size A (8½″ × 11″) sheets for the following problems, and lay out the problems using instruments. Each square on the grid is equal to 0.20″ or about 5 mm. The problems can be laid out on grid paper or plain paper. Label all reference planes and points in each problem with ⅛″ letters or numbers, using guidelines.

1. (Fig. 4.30) Find the missing third views of the given points in Problems 1A through 1D. Find the missing third views of the lines between the given points in Problems 1E and 1F.

2. (Fig. 4.31) (2A) Find the front and top views of the profile line. (2B) Find the front and side views of the horizontal line. (2C) Find the side view of 5–6. (2D) Find the front and side views of 7–8. (2E and 2F) Find the top and side views of the given lines.

3. (Fig. 4.32) (3A) Find the side view of the line. (3B) Find the front and side views of the horizontal line. (3C) Draw three views of a line from point 2 to point O on the line. (3D) Find the side view of the line and locate the midpoint of the line in each view. (3E and 3F) Find the piercing

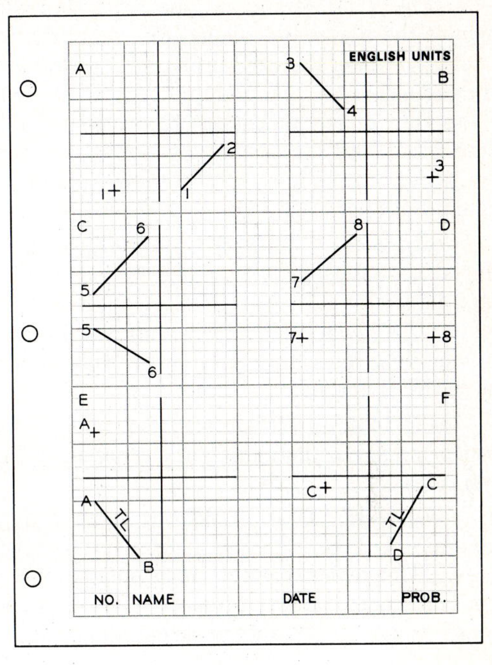

Fig. 4.31 Problems 2A through 2F.

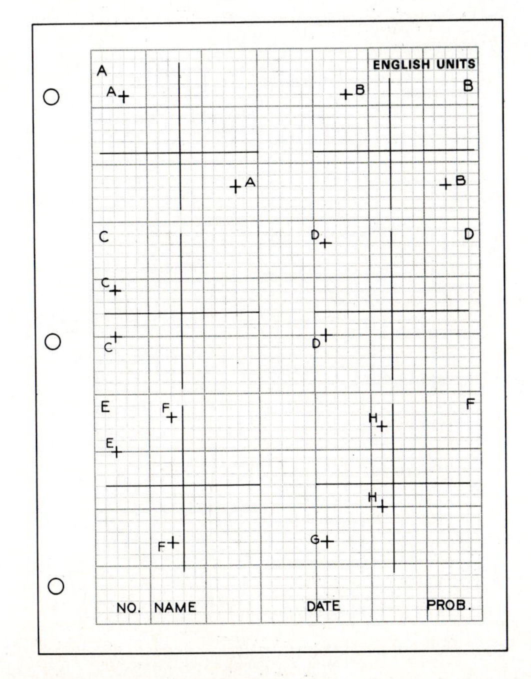

Fig. 4.30 Problems 1A through 1F.

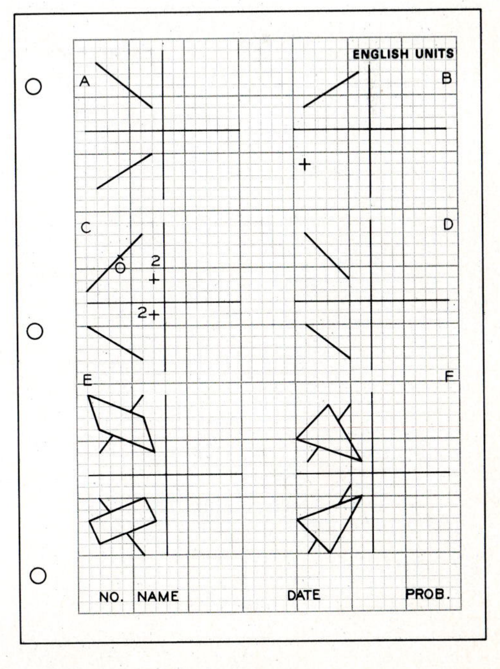

Fig. 4.32 Problems 3A through 3F.

points between the lines and planes, show visibility, and complete the missing third view.

4. (Fig. 4.33) (4A) Find the front and side views of the horizontal plane. (4B) Draw the top and side views of the frontal plane. (4C) Draw the front and top views of the profile plane. (4D) Draw the top view of the plane. (4E) Find the side view of the plane and locate the line upon it in each view. (4F) Find the missing view of the plane and locate point O on each view of the plane.

5. (Fig. 4.34) (5A) Draw a line with its midpoint at O that is parallel to the line. (5B) Construct the front view of a second plane that is parallel to the given plane. (5C and 5D) In each problem, draw lines through point O that are parallel to the given planes.

6. (Fig. 4.35) In each problem, draw lines through point O that are perpendicular to the given lines.

7. (Fig. 4.36) (7A and 7B) Draw a line through each point O that is perpendicular to the given planes. (7C) Draw a plane through point O that is perpendicular to the given line. (7D) Draw a plane through the given line that is perpendicular to the plane.

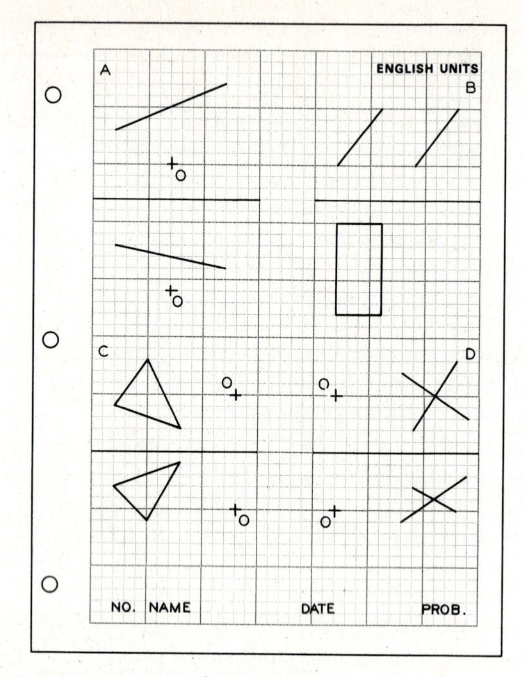

Fig. 4.34 Problems 5A through 5D.

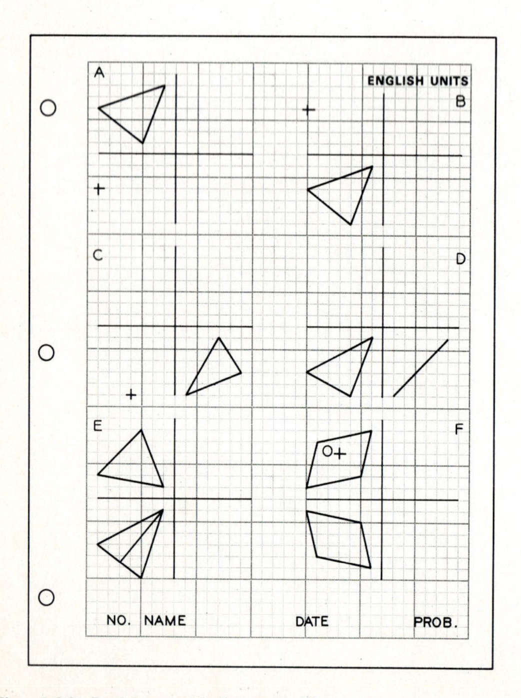

Fig. 4.33 Problems 4A through 4F.

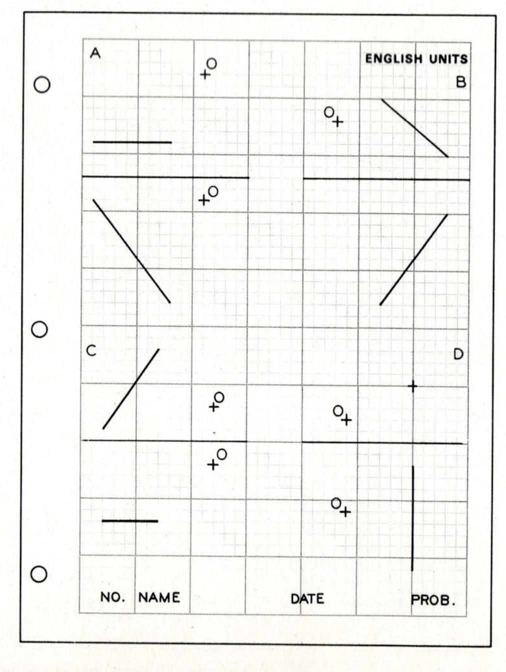

Fig. 4.35 Problems 6A through 6D.

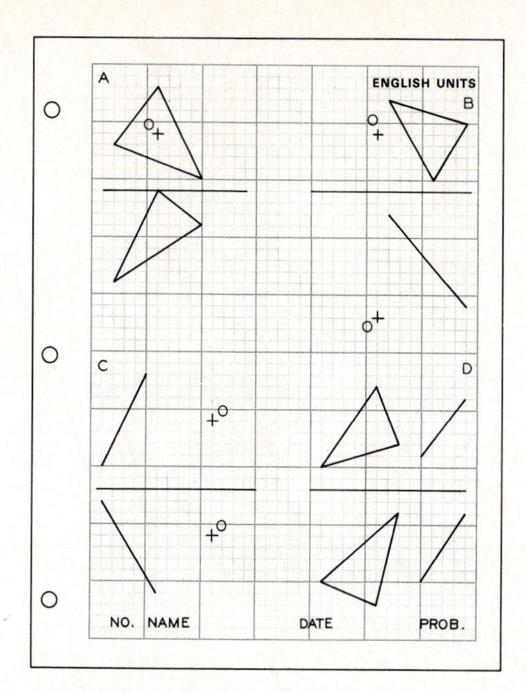

Fig. 4.36 Problems 7A through 7D.

5

PRIMARY AUXILIARY VIEWS

5.1 INTRODUCTION

Descriptive geometry can be defined as the projection of three-dimensional figures onto a two-dimensional plane of paper in such a manner as to allow geometric manipulations to determine lengths, angles, shapes, and other geometric information by means of graphics. Orthographic projection is the system used for laying out descriptive geometry problems.

The primary auxiliary view is a powerful tool of descriptive geometry that permits the analysis of three-dimensional geometry that would be difficult by other means. This area of study permits the measurement and determination of distances, lengths, angles, sizes, and areas that are essential to the solution of technical problems.

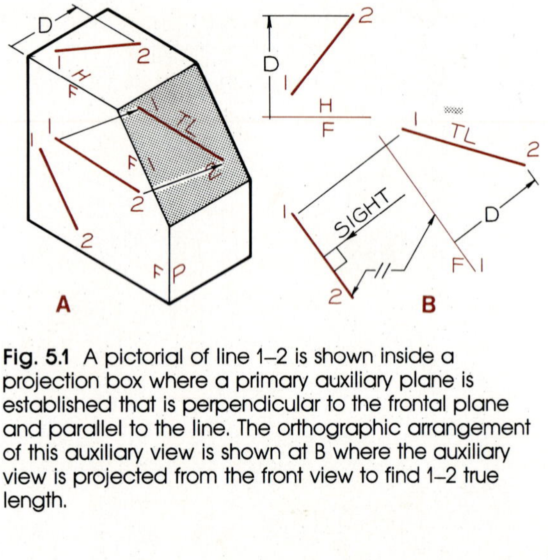

Fig. 5.1 A pictorial of line 1–2 is shown inside a projection box where a primary auxiliary plane is established that is perpendicular to the frontal plane and parallel to the line. The orthographic arrangement of this auxiliary view is shown at B where the auxiliary view is projected from the front view to find 1–2 true length.

5.2 PRIMARY AUXILIARY VIEW OF A LINE

The top and front views of line 1–2 are shown pictorially and orthographically in Fig. 5.1. Since the line is not a principal line, it is not true length in a principal view. To find its true-length view, a primary auxiliary view must be used.

At B the line of sight is drawn perpendicular to the front view of the line and reference line F–1 is drawn parallel to the frontal view. You can see in the pictorial that the auxiliary plane is parallel to the line and perpendicular to the frontal plane, which accounts for it being labeled as F–1.

The auxiliary view is found by projecting parallel to the line of sight and perpendicular to the F–1 reference line. Point 2 is found by transferring distance *D* with your dividers to the auxiliary view, since the frontal plane appears as an edge in both the top and auxiliary views. Point 1 is located in the same manner and the points are connected to find the true-length view of the line. It is labeled TL in this view. The reference lines are labeled as shown in the figure.

Fig. 5.2 True length of a line by a primary auxiliary view

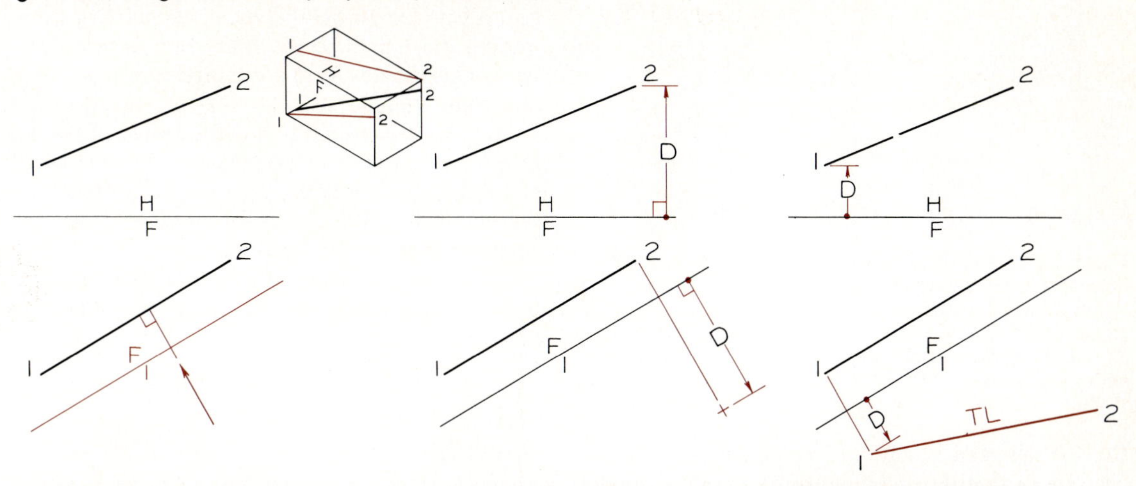

Step 1 To find 1–2 true length, construct the line of sight perpendicular to the line. Draw reference line F–1 parallel to the front view of the line.

Step 2 Project point 2 perpendicularly from the front view. Dimension *D* from the top view locates point 2.

Step 3 Point 1 is located by transferring dimension *D* from the top view. Line 1–2 is true length in the auxiliary view.

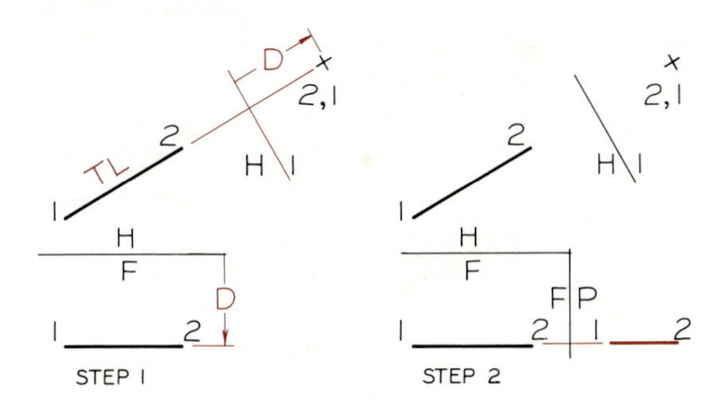

Fig. 5.3 Point view of a line

Step 1 The point view of a line can be found in a primary auxiliary view that is projected from the true-length view of the line.

Step 2 An auxiliary view projected from a foreshortened view of a line will result in a foreshortened view of the line; not a point view.

Figure 5.2 separates the sequential steps required to find the true length of an oblique line. It is beneficial to letter all reference planes using the notation suggested in the various steps with the exception of the dimensions (such as *D*) that are transferred from one view to another with your dividers to locate desired points and lines.

A primary auxiliary view can result in a point view of the line if projected from a true-length view of the line in a principal view (Fig. 5.3). In Step 1, the point view is found by projecting from the horizontal line that is true length in the top view. The auxiliary view that is projected from the front view of the line does not give a point view since the line is foreshortened in the front view.

5.3 TRUE LENGTH BY ANALYTICAL GEOMETRY

You can see in Fig. 5.4 that the length of a frontal line can be found in the front view by the application of analytical geometry (mathematics) and

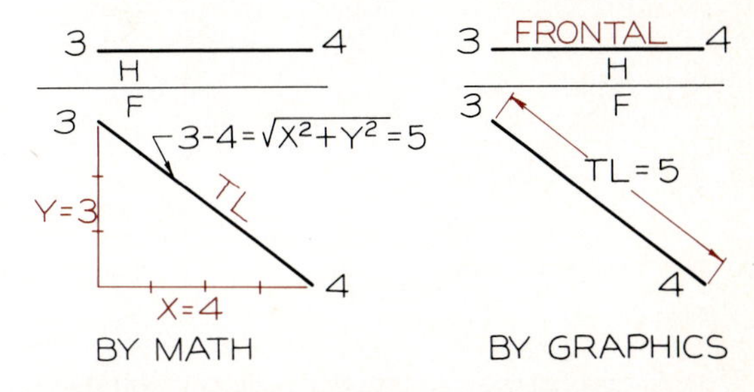

Fig. 5.4 A line that appears true length in a view (the front view in this case) can have its length calculated by application of the Pythagorean theorem and mathematics. Since the line is a frontal line and is true length in the front view, its length can be measured graphically.

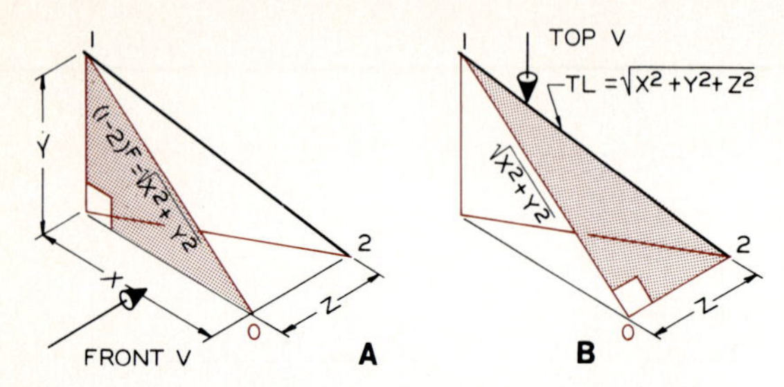

Fig. 5.5 A three-dimensional line that is not true length in a principal view can be found true length by the Pythagorean theorem in two steps. The frontal projection, 1–0, is found using the *x*- and *y*-coordinates. At B, the hypotenuse of right triangle 1–0–2 is found using *x*-, *y*-, and *z*-coordinates.

the Pythagorean theorem. The Pythagorean theorem states that the hypotenuse of a right triangle is equal to the square root of the sum of the squares of the other two sides. The x- and y-coordinates of 4 and 3 result in a length of 5 units for line 3–4.

The length of the line can be graphically measured in the front view since it is true length in this view. The accuracy of the measurement will depend upon the accuracy of your drafting of the views.

The line shown pictorially in Fig. 5.5 can be found true length by analytical geometry by determining the length of the front view where the x- and y-coordinates form a right triangle at A. A second right triangle at B, (1–0–2), is solved to find its hypotenuse, which is the true length of the line in question, 1–2. You can see that the true length of an oblique line is the square root of the sum of the squares of the x-, y- and z-coordinates that correspond to width, height, and depth.

The steps in determining the true length of line 1–2 by analytical geometry are shown in Fig. 5.6.

5.4 THE TRUE-LENGTH DIAGRAM

A true-length diagram is constructed with two perpendicular lines to find a line of true length as shown in Fig. 5.7. This method does not give a direction for the line, but merely its true length.

The two measurements that are laid out on the true-length diagram can be transferred from any two adjacent orthographic views. One measurement is the distance between the endpoints in one of the views. The other measurement, taken

Fig. 5.6 True length of a line—analytical method

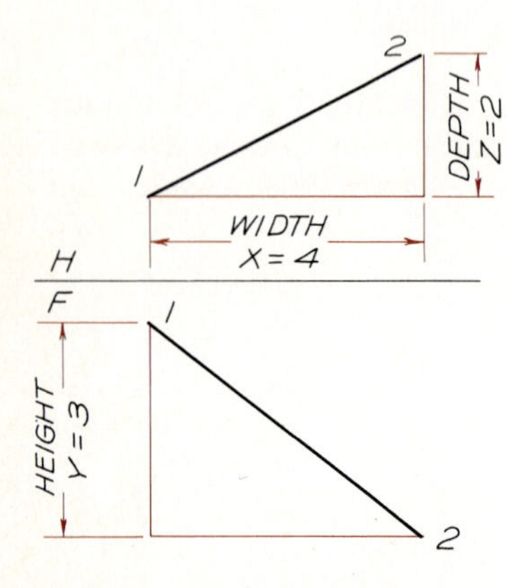

Step 1 Right triangles are drawn with line 1–2 as the hypotenuse in the top and front views. The coordinates, or legs, of the right triangles are drawn parallel and perpendicular to the H-F reference line.

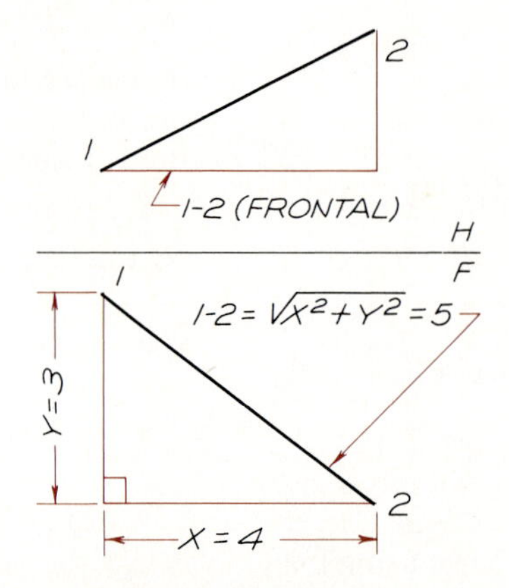

Step 2 The true length of the front view of 1–2 is found by the Pythagorean theorem. The resulting length is found to be 5 units.

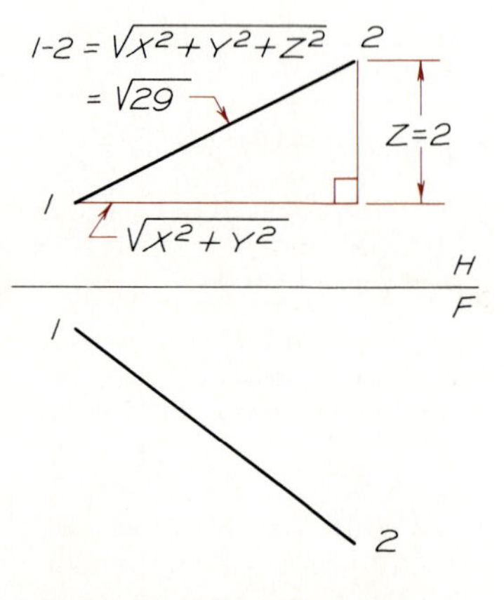

Step 3 The true length of the line is found by combining the length of the front projection with the true length of the *z*-coordinate in the top view. The length is $\sqrt{29}$.

Fig. 5.7 True-length diagram

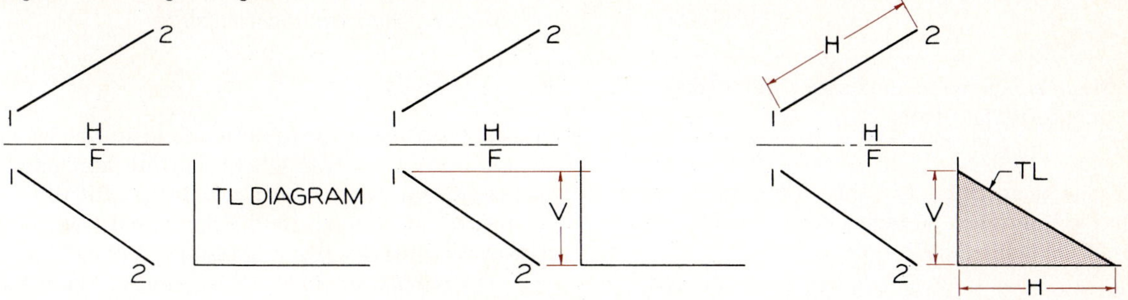

Required Find line 1–2 true length in a TL diagram.

Step 1 Transfer the vertical distance between the ends of 1–2 to the vertical leg of the TL diagram.

Step 2 Transfer the horizontal length of the line in the top view to the horizontal leg of the TL diagram.

Fig. 5.8 Angles between lines and principal planes

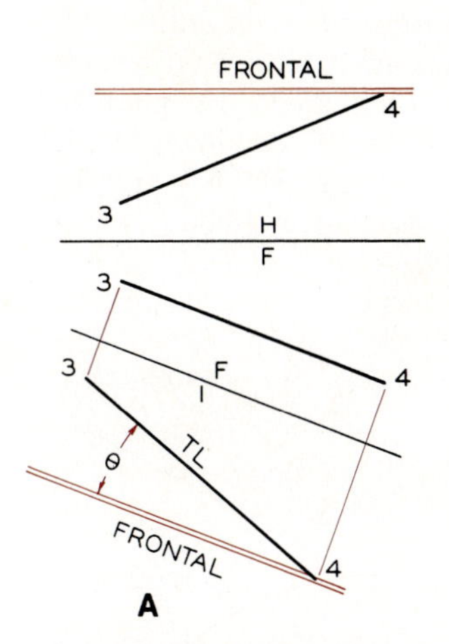

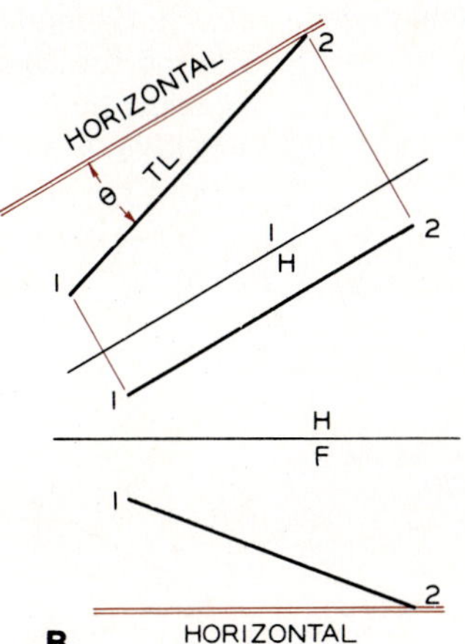

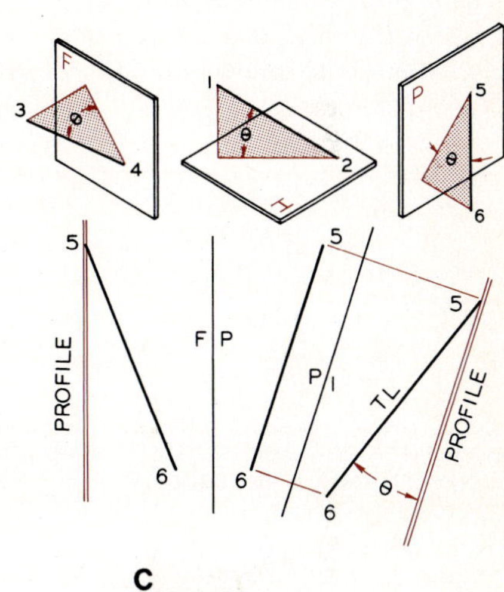

A. When an auxiliary view is projected from the front view, the frontal plane appears as an edge and the line is true length. The angle with the frontal plane can be measured in this view.

B. When an auxiliary view is projected from the top view, the horizontal plane appears as an edge and the line is true length. The angle with the horizontal plane can be measured in this view.

C. When an auxiliary view is projected from the side view, the profile plane appears as an edge and the line is true length. The angle with the profile plane can be measured in this view.

from the adjacent view, is measured between the endpoints in a direction perpendicular to the reference line between the two views.

5.5 ANGLES BETWEEN LINES AND PRINCIPAL PLANES

To measure the angle between a line and a plane, the line must appear true length in the view where the plane appears as an edge. Since a principal plane will appear as an edge in a primary auxiliary view, the angle a line makes with this plane can be measured if the line is found true length in this view.

Line 3–4 in Fig. 5.8A is found true length and the frontal plane as an edge when the auxiliary view is projected from the front view. The angle between 3–4 and the frontal plane can be measured in this view.

Similarly, the angles between 1–2 and the horizontal and the angle between the profile and 5–6 are found in primary auxiliary views where the lines are true length and the planes are edges.

5.6 SLOPE OF A LINE

> **Slope** is defined as the angle a line makes with the horizontal plane.

It may be specified by either of the three methods in Fig. 5.9; it can be indicated as *slope angle, percent grade,* or *slope ratio.*

Slope Angle

The slope of a line can be measured in a view where the line is true length and the horizontal plane appears as an edge. Consequently, the slope of *AB* in Fig. 5.10 can be measured in the front view where Θ is found to be 31°. This angle can also be found by converting its tangent of 0.60 to 31° by using the trigonometric tables.

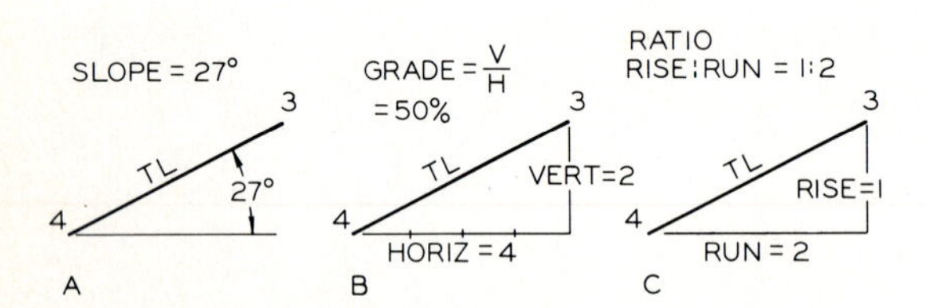

Fig. 5.9 The inclination of a line with the horizontal can be measured and expressed by (A) slope angle, (B) percent grade, or (C) slope ratio.

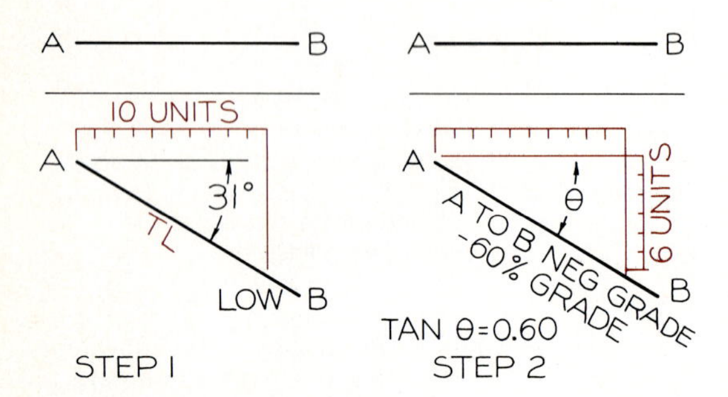

Fig. 5.10 Percent grade of a line

Step 1 The percent grade of a line can be measured in the view where the horizontal appears as an edge and the line is true length (the front view here). Ten units are laid off parallel to the horizontal from the end of the line.

Step 2 A vertical distance from *A* to the line is measured to be 6 units. The percent grade is 6 divided by 10 or 60%. This is negative when the direction is from *A* to *B*. The tangent of this slope angle is 6/10 or 0.60.

Percent Grade

The percent grade of a line is found in the view where the line is true length and the horizontal plane appears as an edge. Grade is the ratio of the vertical (rise) divided by the horizontal (run) between the ends of a line expressed as a percentage.

The percent grade of *AB* is found in Fig. 5.10 by using a combination of mathematics and graphics. Ten units are laid off parallel to the horizontal from *A* using a convenient scale with decimal units in Step 1. In Step 2 the vertical drop of the line after 10 units along the horizontal is measured as 6 units. Since 10 units were used along the horizontal, your arithmetic is simplified in finding the tangent of the angle to be 0.60, which is easily converted into −60% grade. The grade is negative from *A* to *B* since this is downhill. It would be positive from *B* to *A*, which is uphill. Line *CD* has a positive fifty-percent grade (+50%) from C to *D* in Fig. 5.11A.

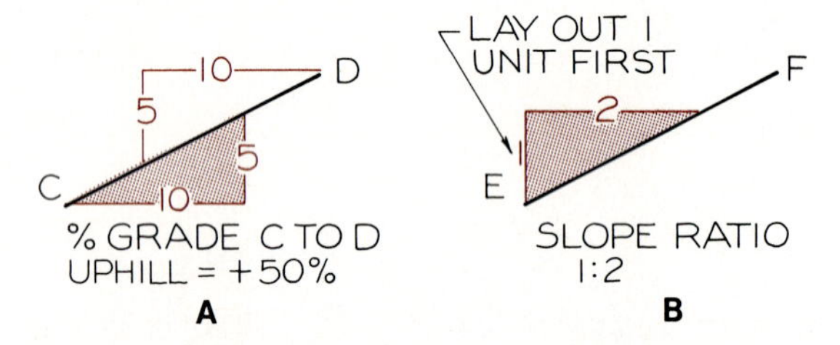

Fig. 5.11 The percent grade of a line is positive if uphill; negative if downhill (part A). The slope ratio is expressed as 1: *xx* where 1 is the rise and *xx* is the horizontal distance. The 1 unit must be drawn first and then the horizontal distance can be drawn (part B).

Slope Ratio

Slope ratio is the same as the percent grade except for the method of expressing the relationship between vertical and horizontal distances. The first number of the ratio is always one, such as 1:10, 1:200, and so on. The first number is the rise (always one) and the second number is horizontal run (see Fig. 5.9).

The graphical method of finding the slope ratio is shown in Fig. 5.11B where the rise of one is laid off on the true-length view of *EF*. The corre-

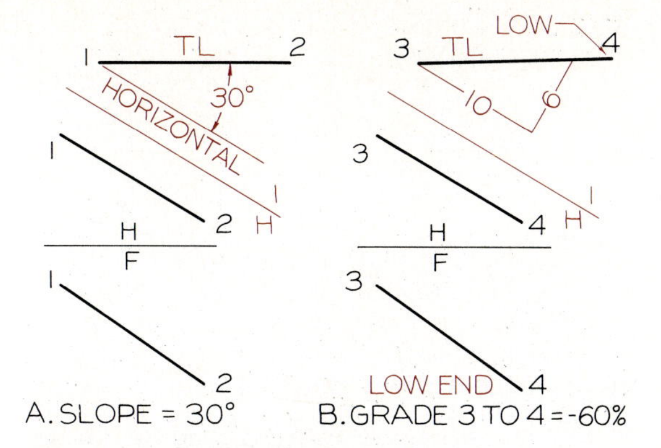

Fig. 5.12 Slope of an oblique line

A. The slope angle can be measured in a view where the horizontal appears as an edge and the line is true length. The slope of 30° is found in an auxiliary view projected from the top view.

B. The percent grade can be measured in a true length view of the line projected from the top view. Line 3–4 has a –60% grade from 3 to the low end at 4.

sponding horizontal is found to be 2, which results in a slope ratio of 1:2. The tangent of the slope angle in this case is ½ or 0.50; therefore this line has a percent grade of 50%

Oblique Lines When a line is oblique and does not appear true length in the front view, it must be found true length by an auxiliary view projected from the top view. This auxiliary view shows the horizontal as an edge and the line true length making it possible to measure the slope angle (Fig. 5.12A).

Similarly, an auxiliary view projected from the top view must be used to find the percent grade of a line of an oblique line (Fig. 5.12B). Ten units are laid off horizontally, parallel to the H–1 reference line and the vertical distance is found to be 6. This gives a −60% grade of the line downhill from 3 to 4.

5.7 COMPASS BEARING OF A LINE

Two types of bearings of a line's direction are (A) compass bearings, and (B) azimuth bearings (Fig. 5.13).

> Compass bearings always begin with the north or south directions and the angles with north and south are measured toward east or west.

The line in part A that makes 30° with north has a bearing of N 30° W. A line making 60° with south toward the east has a compass bearing of South 60° East, or S 60° E. Since a compass can be read only when held horizontally, the compass bearings of a line can be determined only in the top view, the horizontal view.

An azimuth bearing is measured from north in clockwise direction to 360° (Fig. 5.13B). Azimuth bearings are used to avoid confusion that might be caused by reference to the four points of a compass. Azimuth bearings of a line are written as N 120°, N 210°, etc., with this notation indicating that the measurements are made from north.

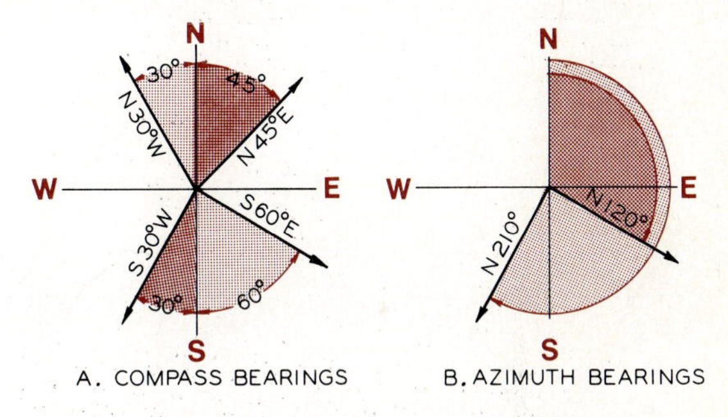

Fig. 5.13 A. Compass bearings are measured with respect to north and south directions on the compass. B. Azimuth bearings are measured with respect to north in a clockwise direction up to 360°.

> The compass bearing (direction) of a line is assumed to be toward the low end of the line unless otherwise specified.

For example, line 2–3 in Fig. 5.14 has a bearing of N 45° E since the line's low end is point 3. It can be seen in the front view that point 3 is the lower end.

The compass bearing and slope of a line are found in Fig. 5.15. In Step 1, the bearing of the line is found in the top view toward point 6, the low end of the line. The line is found true length by projecting from the top view in Step 2 where

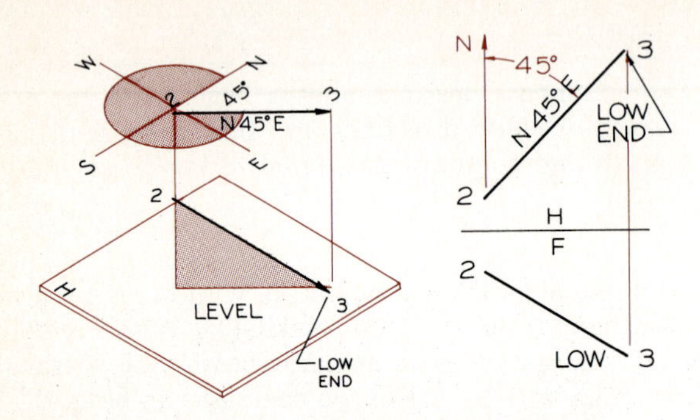

Fig. 5.14 The compass bearing of a line is measured in the top view toward its low end (unless specified toward the high end). Line 2–3 has a bearing of N 45° E toward the low end at 3.

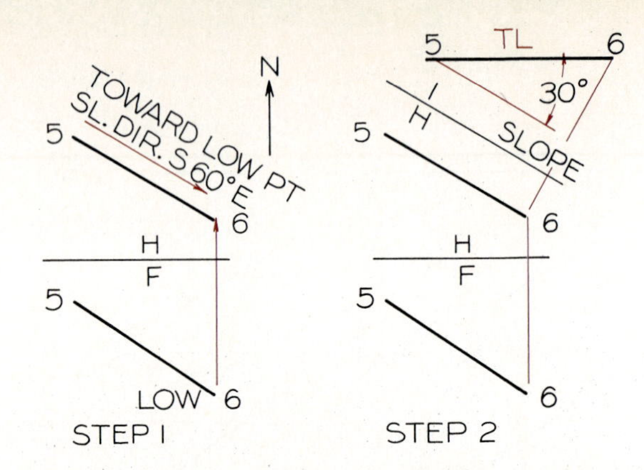

Fig. 5.15 Slope and bearing of a line

Step 1 Slope bearing can be found in top view toward its low end. Direction of slope is S 60° E.

Step 2 The slope angle of 30° is found in an auxiliary view projected from the top view where the line is found true length.

Fig. 5.16 A line from slope specifications

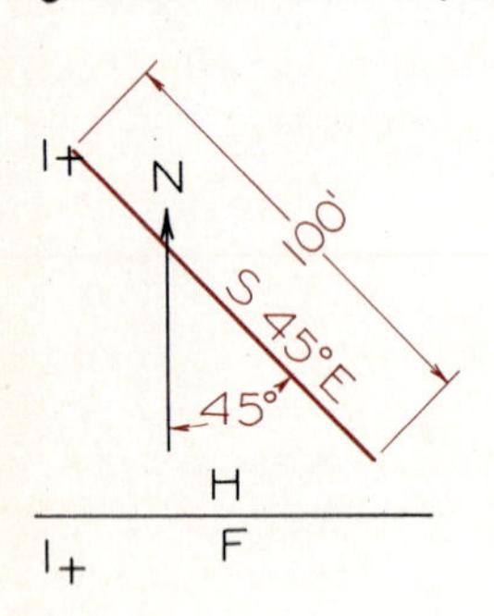

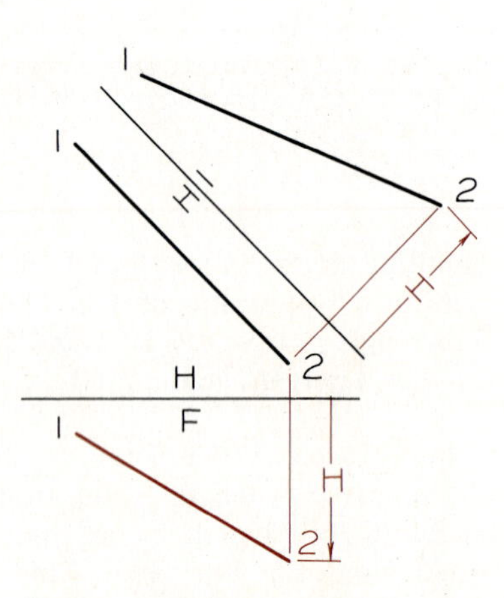

Step 1 It is required to draw a line through point 1 that bears S 45° E for 100′ horizontally and slopes 22°. The bearing and horizontal are drawn in the top view.

Step 2 An auxiliary view is projected from the top view where the 22° slope angle is measured.

Step 3 The front view of 1–2 is found by locating point 2 in the front view.

the slope angle of 30° is found. This information can be used to describe verbally the line as having a compass bearing of S 60° E and a slope of 30° from 5 to 6.

When given verbal information and one point in the top and front views is given, a line can be drawn as shown in Fig. 5.16. If it is known that the line has a bearing of S 45° from point 1 for a horizontal distance of 100′, the top view can be drawn (Step 1). If it is known that the line has a 22° slope, the true length auxiliary view can be constructed (Step 2). The front view can be completed by locating point 2 in Step 3.

5.8 CONTOUR MAPS AND PROFILES

The contour map is the method of representing irregular surfaces of the earth. A pictorial view of a sectional plane and a portion of the earth are shown in Fig. 5.17, along with the conventional

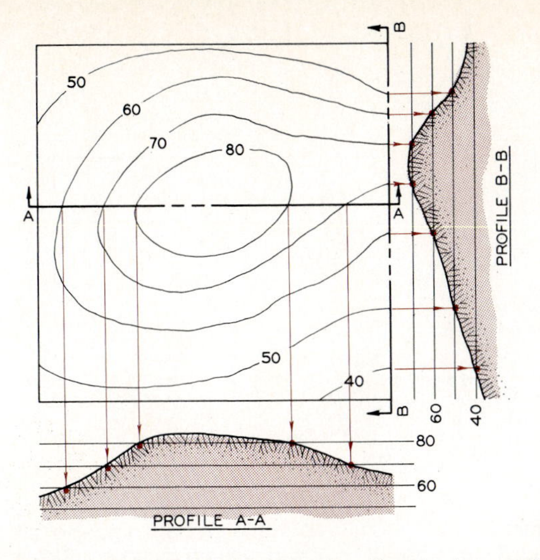

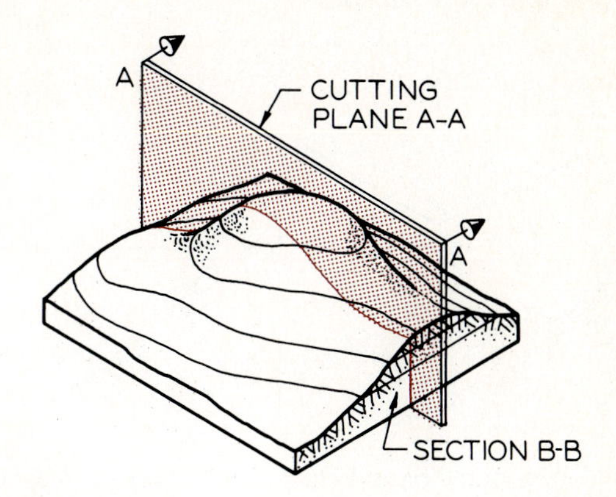

Fig. 5.17 A contour map uses contour lines to show variations in elevation on an irregular surface. Vertical sections taken through a contour map are called profiles.

orthographic views of the contour map and profiles. The following definitions must be understood prior to further discussion.

Contour lines are horizontal lines that represent constant elevations from a horizontal datum such as sea level. Contour lines can be thought of as the intersections of horizontal planes with the surface of the earth. The vertical interval of spacing between the contours in Fig. 5.17 is 10′.

Contour maps are maps on which contour lines are drawn to represent irregularities of the surface (Fig. 5.17). The closer the contour lines are to each other on the map, the steeper the terrain.

Profiles are vertical sections through a contour map that are used to show the earth's surface at any desired location. Two profiles are shown in Fig. 5.17. When applied to topography, a vertical section is called a profile regardless of the direction of the cutting plane in the top view. Contour lines appear as edge views of equally spaced horizontal planes in profiles. The true representation of a profile is drawn with the vertical scale equal to the scale of the contour map; however, this scale is often increased to emphasize changes in elevation that would not otherwise be apparent.

Contoured surfaces such as airfoils, automobile bodies, ship hulls, and household appliances must also be depicted on the drawing board by using contour lines. When applied to objects other than the earth's surface, this technique is representing contours is called **lofting.**

5.9 VERTICAL SECTIONS

In Fig. 5.18, vertical sections (called profiles) are passed through the top view of an underground pipe system that begins at 1 and ends at 3. Auxiliary views are projected from the top view to find the surface of the earth and the pipe under the ground in Steps 1 and 2. The pipe is known to be located 15′ under the ground at points 1, 2, and 3. The percent grade of the pipes is found in Step 3 to complete the problem.

The same scale was used to draw the profiles as was used to construct the contour map. This makes it possible to measure the true lengths and angles of slope in the profiles. The percent grade and the compass bearing of each line is labeled on the contour map.

5.10 PLAN-PROFILES

A plan-profile is a combination drawing that includes a plan with contours and a vertical section called a profile. A plan-profile is used to show an underground drainage system from manhole 1 to manhole 3 in Fig. 5.19 and Fig. 5.20.

Fig. 5.18 Vertical sections

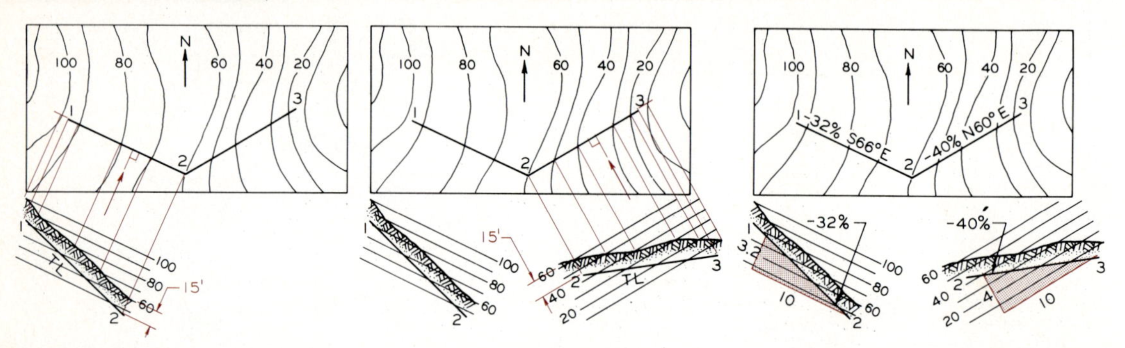

Step 1 To locate pipes 15 feet under the surface from point 1 to 2 to 3, vertical sections are passed through the pipes in the plan view. A profile is taken perpendicularly from pipe 1–2 to find the surface of the ground, the ends of the pipe 15 feet under the surface, and pipe 1–2 true length.

Step 2 A second profile is projected perpendicularly from the plan view of pipe 2–3 to find the surface and pipe 2–3 true length. Notice that the cutting plane is extended beyond point 3 in the top view to provide a more descriptive profile section.

Step 3 The percent grades of pipes 1–2 and 2–3 are found by construction in the profiles. These are negative since the pipes run downhill from 1 to 3. The percent grades and bearings are used to label the top view of the pipes. The elevations of each point can be measured in the profile section.

Fig. 5.19 Plan-profile

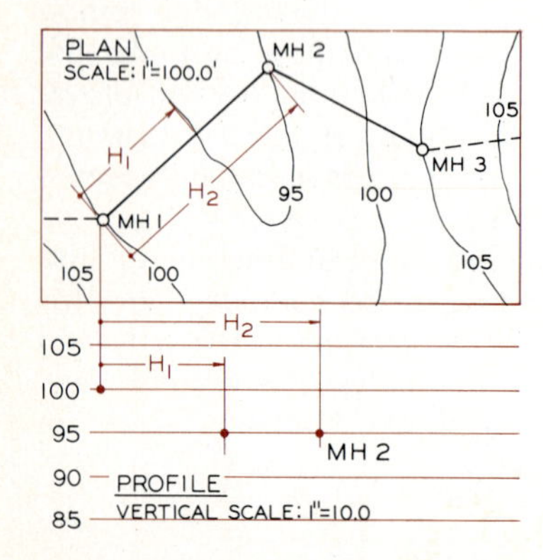

Required Find the profile of the earth over the underground drainage system.

Step 1 Distances H_1 and H_2 from manhole 1 are transferred to their respective elevations in the profile. This is not an orthographic projection.

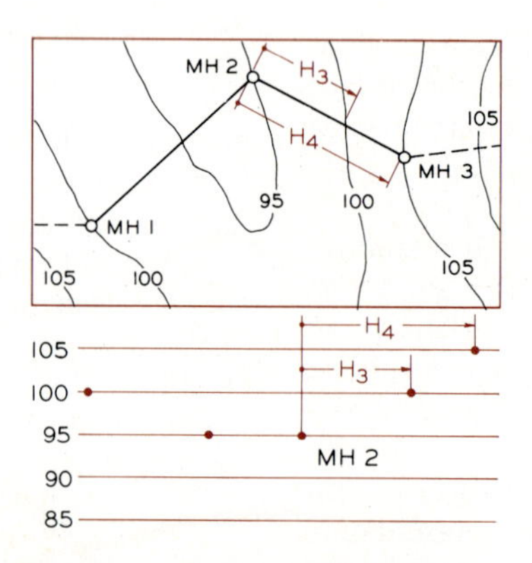

Step 2 Distances H_3 and H_4 are measured from manhole 2 in the plan and are transferred to their respective elevations in the profile. These points represent elevations of points on the earth above the pipe.

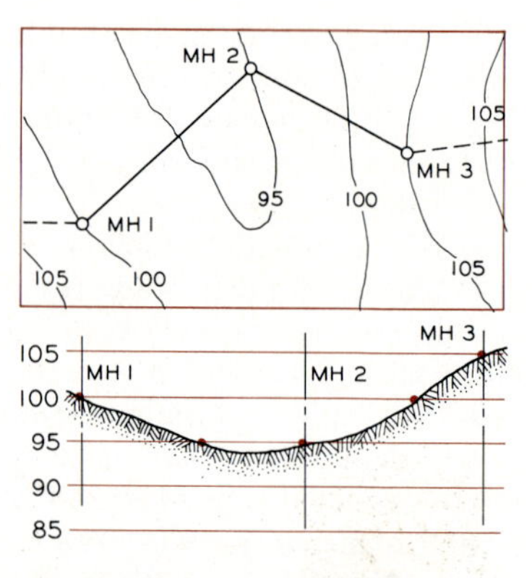

Step 3 The five points are connected with a freehand line and the drawing is crosshatched using earth symbols. Center lines are drawn to show the locations of the three manholes that will be located in Fig. 5.20

Fig. 5.20 Plan-profile, manhole location

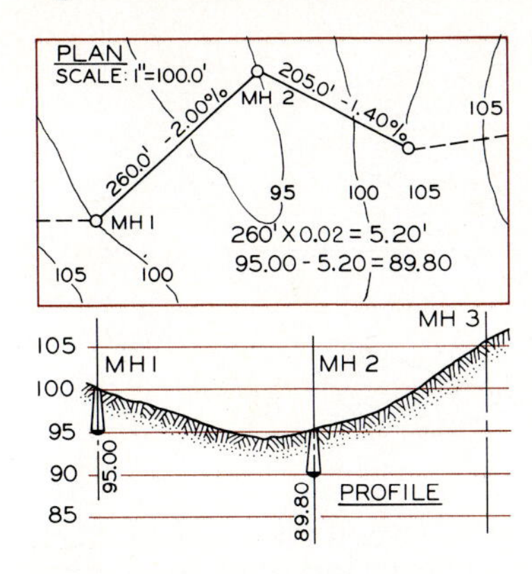

Step 1 The horizontal distance from the MH1 to MH2 is multiplied by the percent grade. The elevation of the bottom of manhole 2 is calculated by subtracting from the elevation of manhole 1.

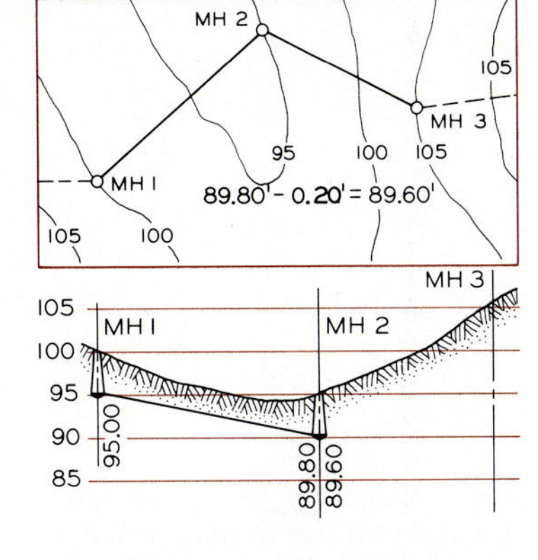

Step 2 The lower side of manhole 2 is 0.20′ lower than the inlet side to compensate for loss of head (pressure) due to the turn in the pipeline. The lower side is found to be 89.60′ and is labeled.

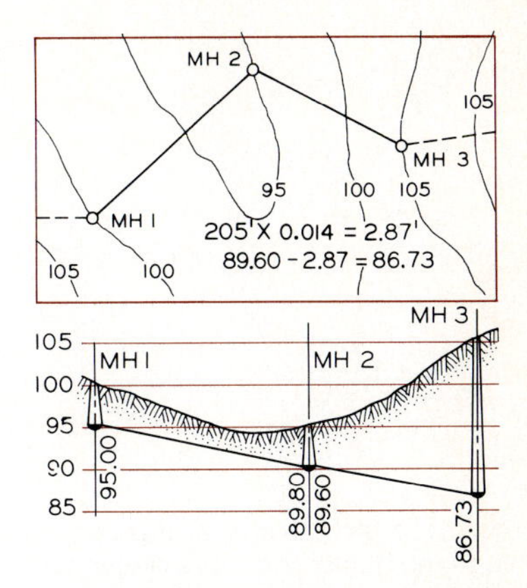

Step 3 The elevation of manhole 3 is calculated to be 86.73′ since the grade is 1.40% from manhole 2 to manhole 3. The flow line of the pipeline is drawn from manhole to manhole and the elevations are labeled.

The profile is drawn with an exaggerated vertical scale to emphasize the variations in the earth's surface and the grade of the pipe. Although the vertical scale is usually increased, it can be drawn at the same scale as is used in the plan if desired.

Manhole 1 is projected to the profile using orthographic projection, but the other points are not orthographic projections (Fig. 5.19). The points where the contour lines cross the top view of the pipe are transferred to their respective elevations in the profile with your dividers. These points are connected to show the surface of the earth over the pipe and the location of manhole center lines.

The vertical drop from one end of a pipe to the other is calculated by multiplying the horizontal distances by the percent grade (Fig. 5.20). The drop from manhole 1 to manhole 2 is found to be 5.20′ by multiplying the horizontal distance of 260.00′ by a −2.00% grade. This drop is subtracted from the depth of manhole 1 to find the elevation of manhole 2; this turns out to be 89.80′.

Since the pipes intersect at manhole 2 at an angle, the flow of the drainage is disrupted at the turn; consequently, a drop of 0.20′ is given from the inlet across the floor of the manhole to compensate for the loss of pressure (head) through the manhole. The elevations on both sides of the manhole are specified as shown in Step 2 of Fig. 5.20.

The true lengths of the pipes cannot be accurately measured in the profile when the vertical scale is different from the horizontal scale. Trigonometry must be used for this computation.

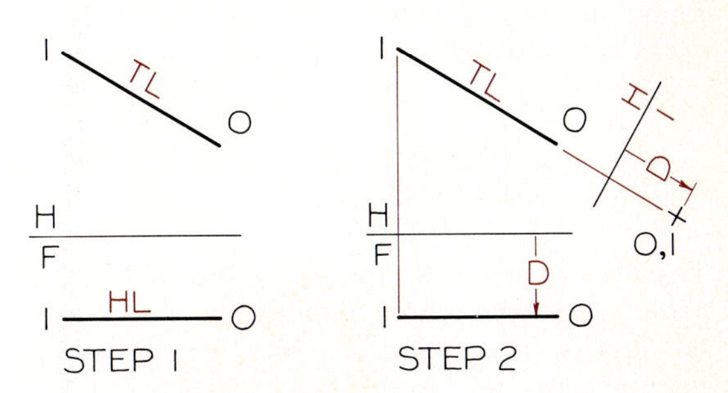

Fig. 5.21 Point view of a line

Step 1 Line 1–0 is horizontal in the front view and is therefore true length in the top view.

Step 2 The point view of 1–0 is found by projecting parallel from the top view to the auxiliary view.

Fig. 5.22 Edge view of a plane

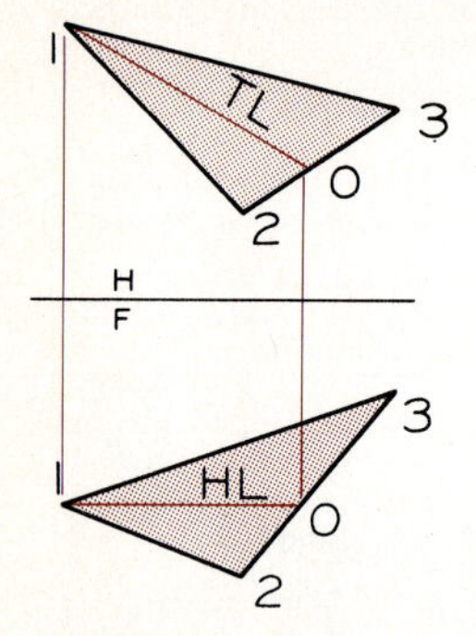

Step 1 To find plane 1–2–3 as an edge, horizontal line 1–0 is drawn on the plane in the front view. Line 1–0 is projected to the top view where it is true length.

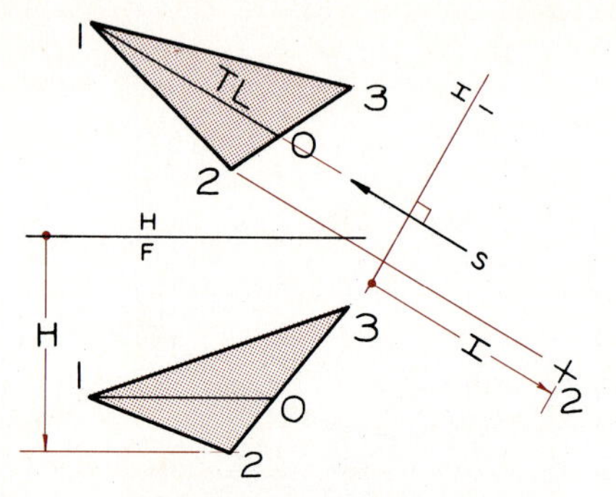

Step 2 A line of sight is constructed parallel to the true-length line 1–0. Reference line H-1 is drawn perpendicular to the line of sight. Point 2 is found in the auxiliary view by transferring dimension *H* from the front view.

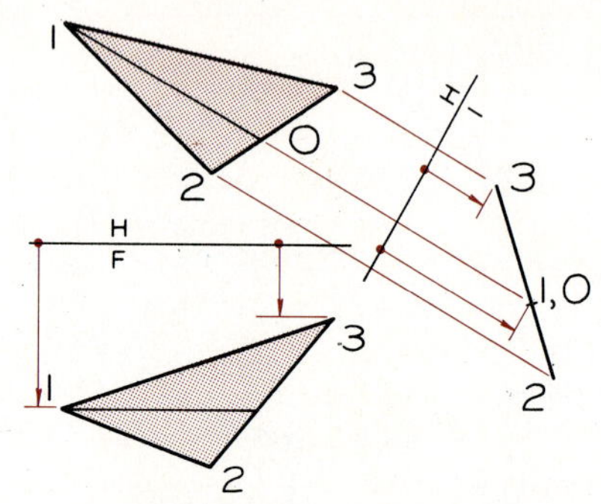

Step 3 Points 1 and 3 are found by transferring their height dimensions from the front view. These points will lie in a straight line which is the edge of the plane. Line 1–0 will appear as a point in this view.

5.11 EDGE VIEW OF A PLANE

The edge view of a plane can be found in any view where a line on the plane appears as a point. A line can be found as a point by projecting from a view where the line is true length (Fig. 5.21).

A true-length line can be drawn on any plane by drawing the line parallel to one of the principal planes and projecting it to the adjacent view, as shown in Step 1 of Fig. 5.22. Since line 1–0 is true length in the top view, its point view may be found in Steps 2 and 3. The remainder of the plane appears as an edge in this auxiliary view.

5.12 DIHEDRAL ANGLES

The angle between two planes is called a *dihedral angle*. This angle can be found in the view where the line of intersection between two planes appears as a point.

The line of intersection, 1–2, between the two planes in Fig. 5.23 is true length in the top view. This makes it possible to find the point view of line 1–2 and the edge view of both planes in a primary auxiliary view that is projected from the true-length view of 1–2.

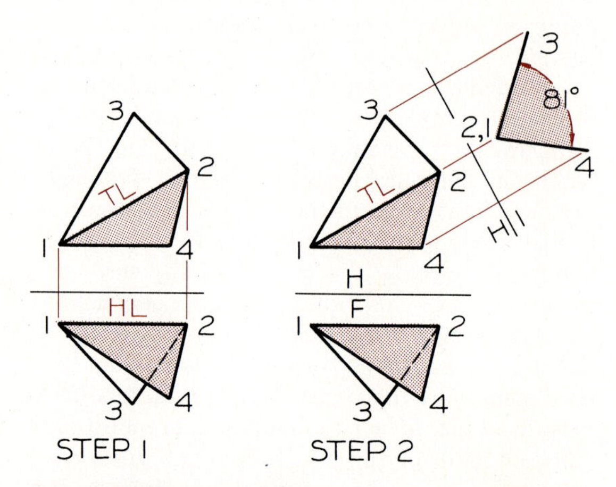

Fig. 5.23 Dihedral angle

Step 1 The line of intersection between the planes, 1–2, is true length in the top view.

Step 2 The angle between the planes (the dihedral angle) can be found in the auxiliary view where the line of intersection appears as a point.

5.13 PIERCING POINTS BY PROJECTION

Figure 5.24 gives the sequential steps necessary to find the piercing point of line 1–2 that passes through the plane. A cutting plane is passed through the line and plane in the top view. The trace of this cutting plane, line *DE*, is then projected to the front view where the piercing point *P* is found in Step 1. The top view of *P* is located in Step 2, and the visibility of the line is found in Step 3 by applying the principles covered in the previous chapter.

5.14 PIERCING POINTS BY AUXILIARY VIEWS

The piercing point of a line and a plane can be found by an auxiliary view as shown in Fig. 5.25. Piercing point *P* can be seen in Step 2 where the plane is found as an edge. Point *P* is projected back to the line in the top and front views from the auxiliary view in Step 3. The location of *P* in the front view is checked by transferring dimension *H* from the auxiliary view with your dividers.

Visibility is easily determined in the top view since it can be seen that *AP* is higher than the plane in the auxiliary view, and is therefore visible in the top view. Analysis of the top view shows that endpoint *A* is the most forward point; therefore this end of the line is visible in the front view.

5.15 PERPENDICULAR TO A PLANE

In Fig. 5.26 it is required that a line be drawn from point *O* that is perpendicular to the plane.

> This line will appear as a true-length perpendicular in the view where the plane appears as an edge.

The true-length perpendicular is drawn in Step 2 to locate piercing point P. Point *P* is found in the top view by drawing line *OP* parallel to the H–1 reference line. It must lie in this direction since *OP* is true length in the auxiliary view. It

Fig. 5.24 Piercing points by projection

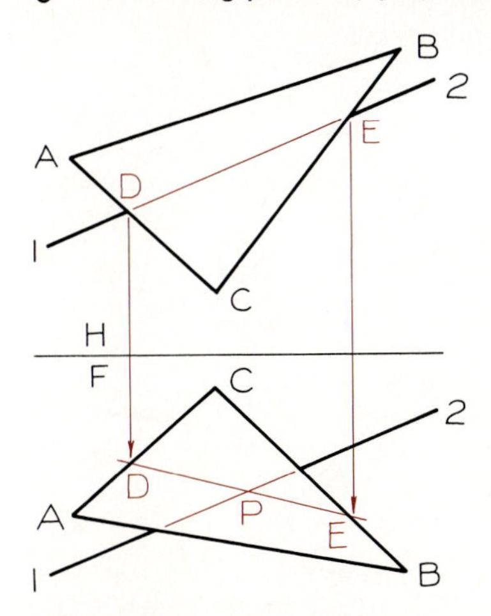

Step 1 Assuming that a vertical cutting plane is passed through the top view of 1–2, the plane intersects *AC* and *BC* at *D* and *E*. The intersection of *DE* with 1–2 in the front view locates the piercing point *P*.

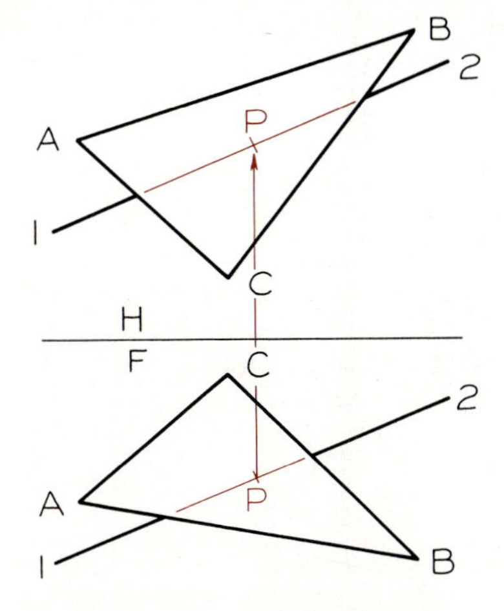

Step 2 Point *P* is projected to the top view of line 1–2 from the front view where it was first located.

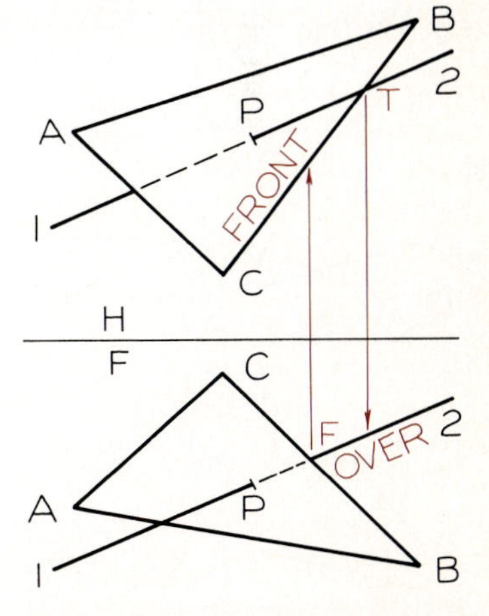

Step 3 Lines *CB* and *P2* cross at point *T* in the top view. By projecting downward from *T*, *P2* is found to be over *CB*; therefore, *PT* is visible in the top view. Intersection *F* can be used to find that line *CB* is in front of *P2*; therefore, *PF* is hidden in the front view.

Fig. 5.25 Piercing points by auxiliary view

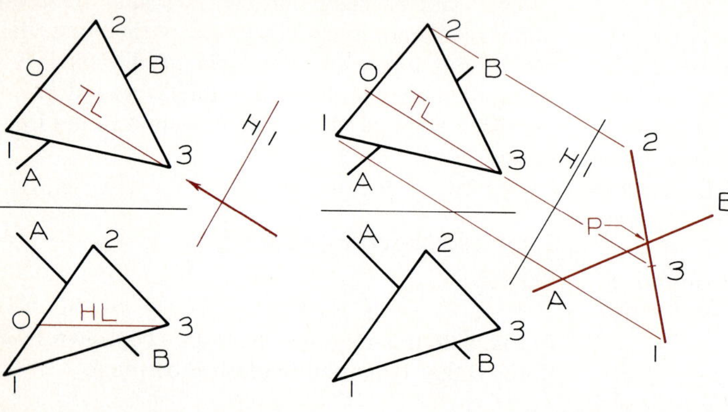

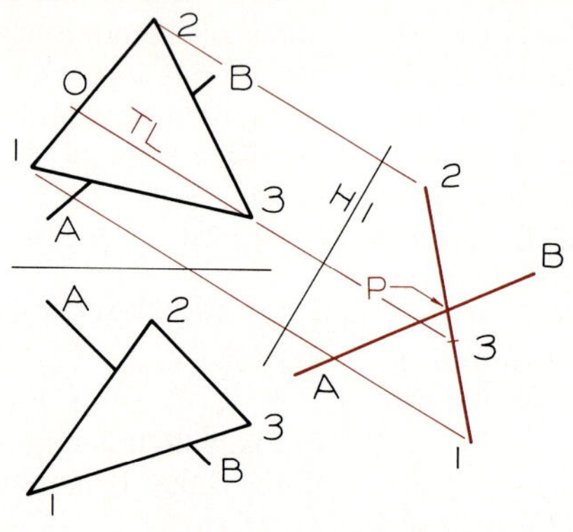

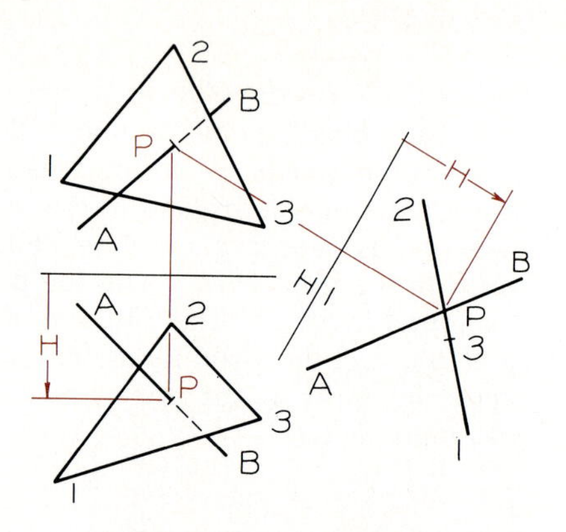

Step 1 Draw a horizontal line on the plane in the front view, project it to the top view to find *TL* line 0–3 on the plane. The line of sight is drawn parallel to the *TL* line.

Step 2 Find the edge view of the plane and project the line *AB* to this view. Point *P* is the piercing point in the auxiliary view.

Step 3 Point *P* is projected to the top and front views to locate the piercing point. Point *A* of the auxiliary view is the highest point and *AP* will be visible in the top view. Point *B* in the top view is the farthest back and therefore *PB* is hidden in the front view.

Fig. 5.26 Line perpendicular to a plane

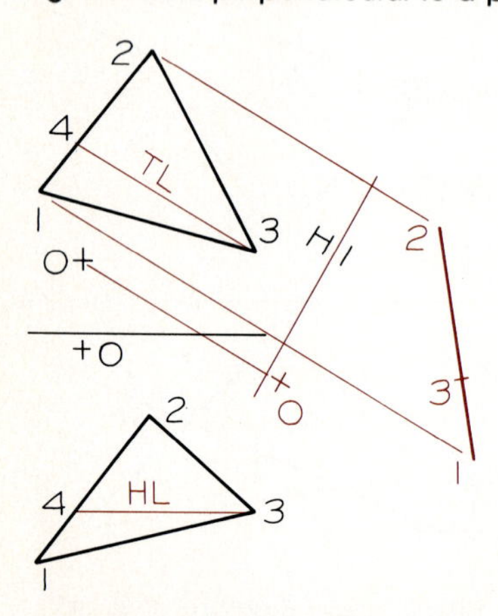

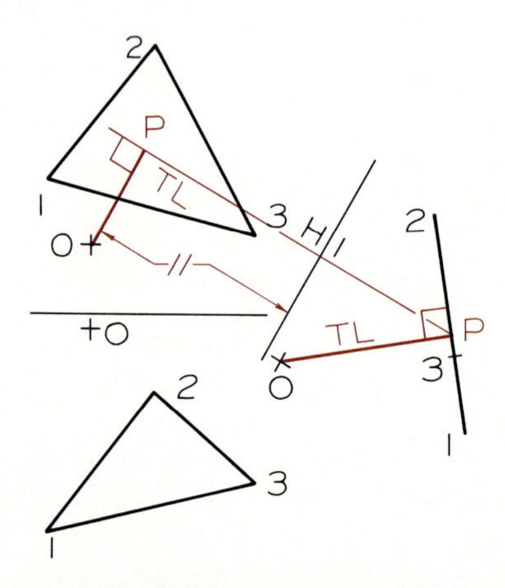

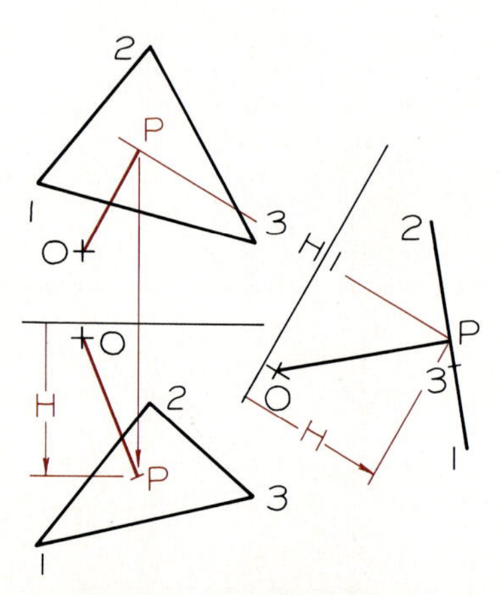

Step 1 Find the edge view of the plane by finding the point view of a line on it. Project from either view. Project point O.

Step 2 Line *OP* is drawn perpendicular to the edge view of the plane. Since *OP* is *TL* in the auxiliary view, it will be parallel to the H-1 reference line in the top view and perpendicular to a *TL* line on the plane.

Step 3 Piercing point *P* is found in the front view by projecting from the top view. Point *P* is accurately located by transferring dimension *H* from the auxiliary view to the front view.

will also be perpendicular to a true-length line in the top view of the plane. The front view of point P is found by projection and measurement in Step 3 along with its visibility.

5.16 INTERSECTIONS BY PROJECTION

The line of interection between two planes can be found by locating the piercing points of two lines on one plane with the other. In Fig. 5.27 a cutting plane is used in Step 1 to find piercing point P by projection. In Step 2, piercing point T is found by the same method. Line PT is the line of intersection, which will always be visible.

The two lines that are selected to be analyzed for their piercing points should be lines of a plane that cross the other plane. Lines AB and 1–2 would be poor selections since they lie outside the other plane. Each line is then analyzed to find its piercing point as if it were a single line.

The visibility is determined by analyzing the points where lines of each plane cross, as covered in the previous chapter.

5.17 INTERSECTIONS BY AUXILIARY VIEW

The intersection between planes can be found by finding the edge view of one of the planes as shown in Fig. 5.28, Step 1. Piercing points L and M are projected from the auxiliary view to their respective lines, 5–6 and 4–6, in the top view in Step 2.

The visibility of plane 4–5–6 in the top view is apparent by inspection of the auxiliary view, where sight line S_1 has an unobstructed view of the 4–L–M portion of the plane. Plane 4–5–L–M is visible in the front view, since sight line S_2 has an unobstructed view of the top view of this portion of the plane.

5.18 SLOPE OF A PLANE

Planes can be established by using verbal specifications of *slope* and *direction of slope* of a plane as defined below.

Slope of a plane is the angle its edge view makes with the edge of the horizontal plane.

Fig. 5.27 Intersection by projection

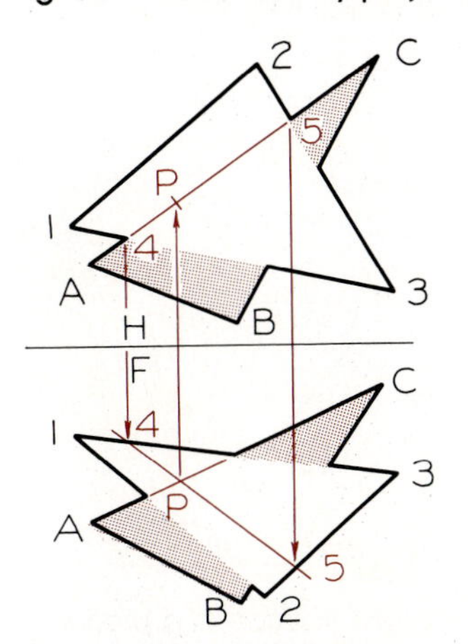

Step 1 Pass a cutting plane through line AC in the top view to locate points 4 and 5 that are projected to the front view. Point P is found on 4–5 in the front view and point P is projected to the top view.

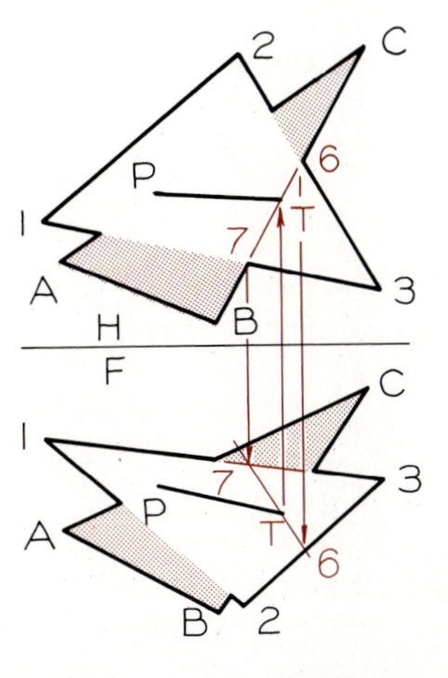

Step 2 Pass a cutting plane through the top view of BC to locate points 6 and 7 that are projected to the front view. Piercing point T is located on 6–7 in the front view and then project point T to the top view.

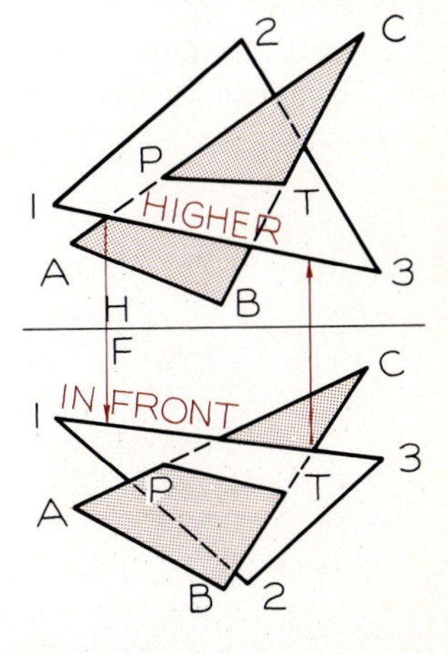

Step 3 Intersection PT is visible. The intersection between AC and 1–3 is projected to the front view to determine that 1–3 is higher in the top view. Line 1–3 is found to be in front of BC in the front view to establish the visibility of the plane.

Fig. 5.28 Intersection by auxiliary view

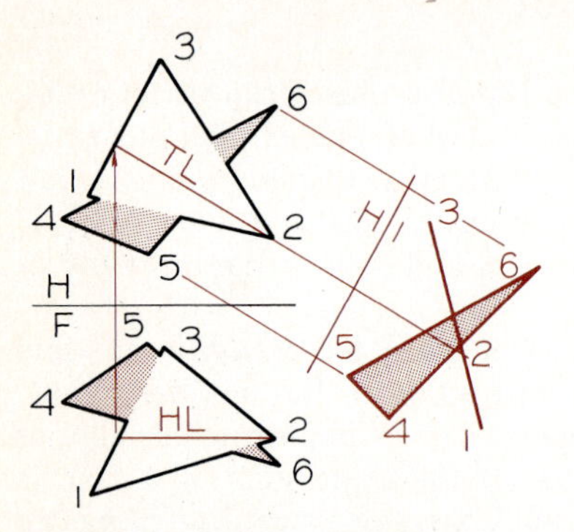

Step 1 Find the edge view of one of the planes and project the other plane to this view also.

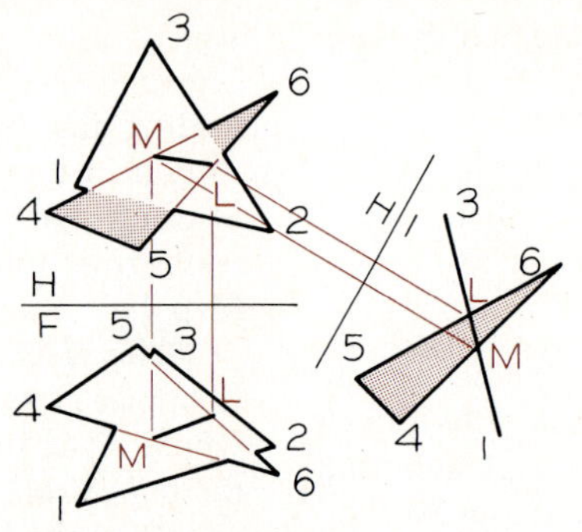

Step 2 Piercing points *L* and *M* can be seen on the edge view of the plane. *LM* is projected to the top and front views.

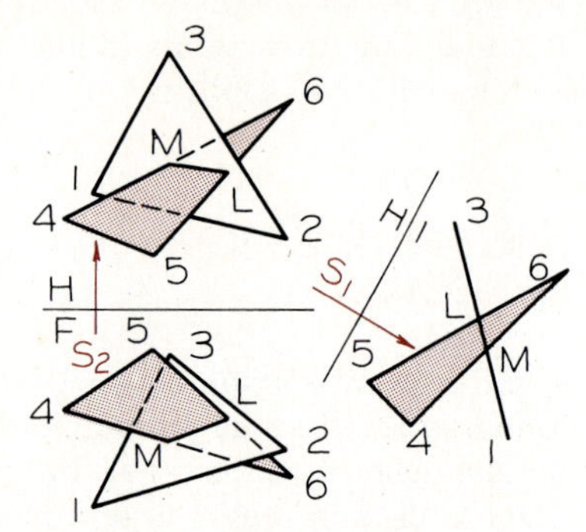

Step 3 The line of sight from the top view strikes L–5 first in the auxiliary view, which makes this portion of the plane visible in the top view. Line 4–5 is farthest forward in the top view and is visible in the front view.

Direction of slope is the compass bearing of a line that is perpendicular to a true-length line in the top view of a plane toward its low side. This is the direction in which a ball would roll on the plane.

It can be seen in Fig. 5.29A that a ball would roll perpendicular to all horizontal lines of the roof toward the low side. This is the direction of slope of the roof. The slope is seen when the roof and the horizontal are edges in a single view.

The direction of slope of a plane is indicated as a compass bearing measured in the top view (Fig. 5.29B).

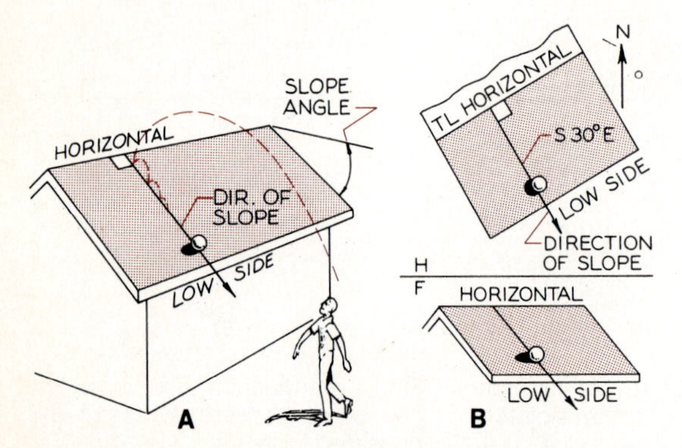

Fig. 5.29 The direction of slope of a plane is the compass bearing of a line on the plane. This is measured in the top view toward the low side of the plane.

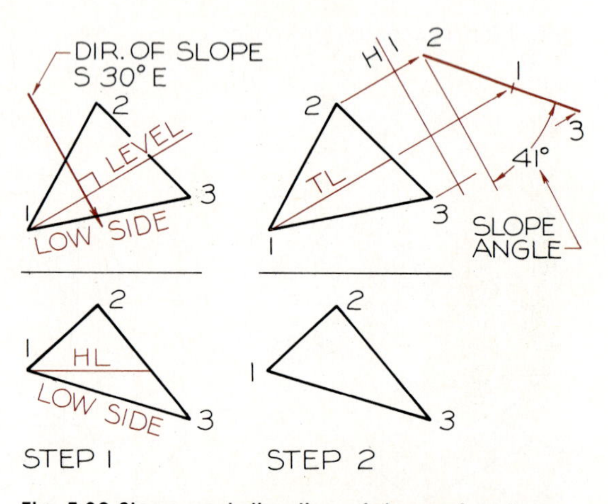

Fig. 5.30 Slope and direction of slope of a plane

Step 1 Slope direction can be found as perpendicular to a true-length level line in the top view toward the low side of the plane, S 30° E in this case.

Step 2 Slope is measured in an auxiliary view where the horizontal is an edge and the plane is an edge, 41° in this case.

Figure 5.30 gives the steps of finding the direction of slope and the slope angle of the plane. An understanding of these terms enables you to verbally describe a sloping plane.

5.19 CUT AND FILL

A level roadway routed through irregular terrain or the embankment of a fill used to build a dam involve the principles of cut-and-fill (Fig. 5.31). This is the process of cutting away equal amounts of the high ground to fill the lower areas to form a nearly level roadway where possible.

In Fig. 5.32, it is required that a level roadway be a given width and an elevation of 60′ be constructed about the given centerline in the contour map using the specified angles of cut and fill.

In Step 1 the roadway is drawn in the top view, and the contour lines in the profile view are

Fig. 5.31 The road across the top of this dam was built by applying the principles of cut and fill. (Courtesy of the Bureau of Reclamation, U.S. Department of the Interior.)

Fig. 5.32 Cut and fill of a level roadway

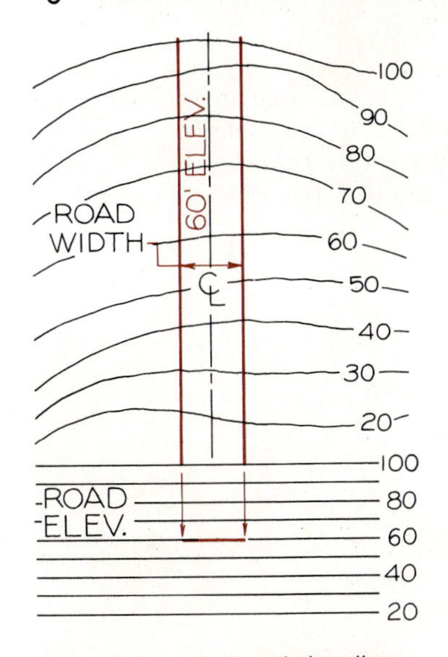

Step 1 Draw a series of elevation planes in the front view at the same scale as the map and label them to correspond to the contours on the map. Draw the width of the roadway in the top view and in the front view at the given elevation, 60′ in this case.

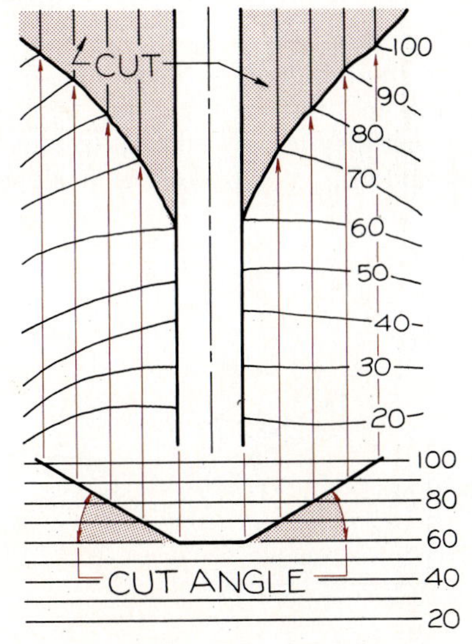

Step 2 Draw the cut angles on the upper sides of the road in the front view according to the given specifications. The points of intersection between the cut angles and the contour planes in the front view are projected to their respective contour lines in the top view to determine the limits of cut.

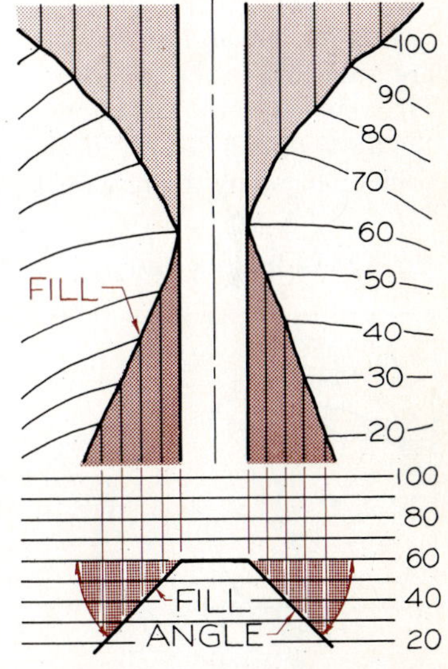

Step 3 Draw the fill angles on the lower sides of the road in the front view. The points in the front view where the fill angles cross the elevation planes are projected to their respective contour lines in the top view to give the limits of the fill. Note that the countour lines have been changed in the cut-and-fill areas to indicate the new contours parallel to the roadway.

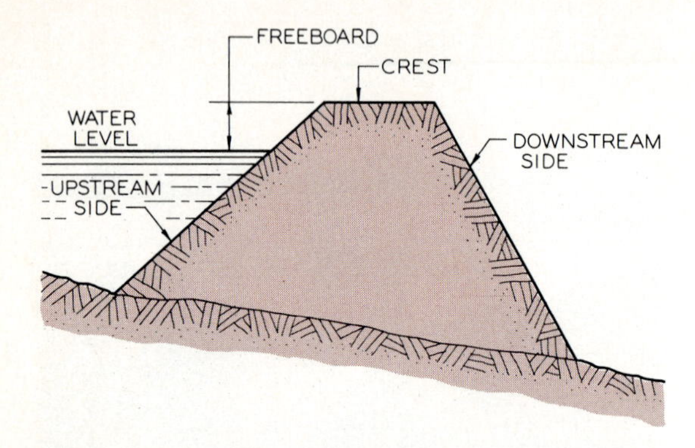

Fig. 5.33 The terms and symbols used in the construction drawing of a dam.

drawn 10′ apart since the contours in the top view are this far apart. The cut angles are measured and drawn on both sides of the roadway on the upper side, Step 2. The points, on the various elevation lines crossed by the cut embankments are projected to the top view to find the limits of cut in this view.

The fill angles are laid off in the profile in Step 3 from given specifications. The crossing points on the profile view of the elevation lines are projected to the top view to find the limits of fill. New contour lines are drawn in Step 3 inside the areas of cut and fill to indicate that they have been changed by the process of cut and fill.

5.20 DESIGN OF A DAM

The design of a dam is a type of cut-and-fill problem. The basic definitions of terms associated with dams are shown in Fig. 5.33. These are: (1) **crest,** the top of the dam, (2) **water level,** the level of water held by the dam, and (3) **freeboard,** the height of the crest above the water level.

An earthen dam is located on the contour map in Fig. 5.34. It makes an arc with its center at point C. The top of the dam is specified to be level to provide a roadway. This method of drawing the top view of the dam and indicating the level of the water held by the dam is shown in the three steps of Fig. 5.34.

These same principles were used in the design and construction of the 726-foot-high Hoover Dam that was built in the 1930s. Since this dam was made of concrete instead of earth, the dam was built in the shape of an arch that is bowed toward the water to take advantage of the compressive strength of concrete (Fig. 5.35).

5.21 STRIKE AND DIP

Strike and **dip** are terms used in geological and mining engineering to describe strata of ore under the surface of the earth. These terms are closely related to slope and direction of slope of a plane.

Fig. 5.34 Graphical design of a dam

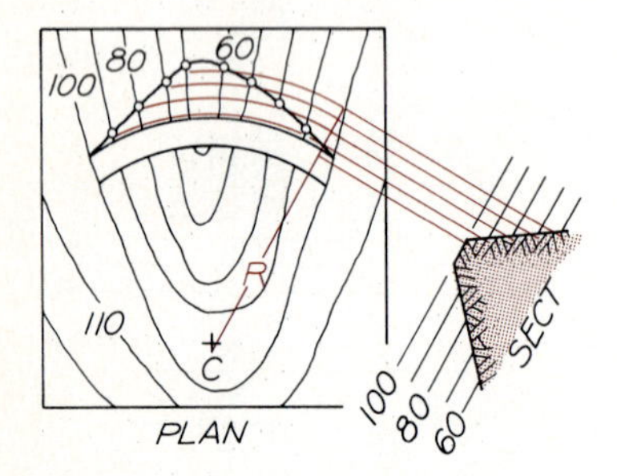

Step 1 A dam in the shape of an arc with its center at C has an elevation of 100′. Draw radius *R* from C and project perpendicularly from this line and draw a section through the dam from specifications. The downstream side of the dam is projected to radial line, *R*. Using the radii from C, use your compass to locate points on their respective contour lines.

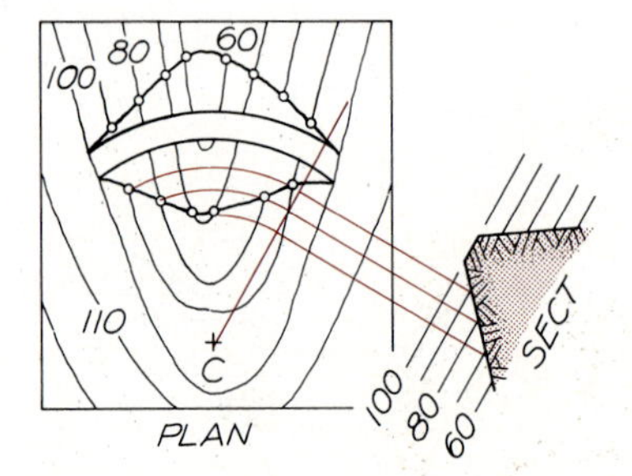

Step 2 The elevations of the dam on the upstream side of the section are projected to the radial line, *R*. Using the center C and your compass, locate points on their respective contour lines in the plan view as they are projected from the section.

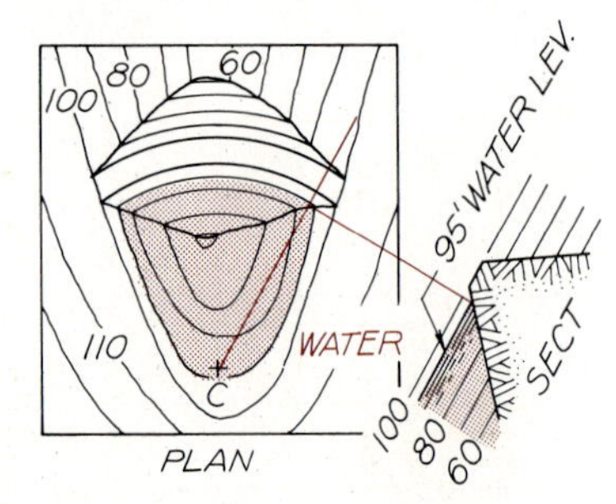

Step 3 The elevation of the water level is 95′ in this case and is drawn in the section. The point where the water intersects the dam is projected to the radial line in the plan view, and is drawn as an arc using center C. The limit of the water is drawn between the 90′ and 100′ contour lines in the top view.

Fig. 5.35 An aerial view of Hoover Dam and Lake Mead, which were built during the period 1931 to 1935. (Courtesy of the Bureau of Reclamation, U.S. Department of the Interior.)

Strike is defined as the compass bearing of a level line in the top view of a plane. Strike has two possible compass bearings since it is the direction of a level line.

Dip is the angle the edge view of a plane makes with the horizontal plus its general compass direction, such as NW or SW. The dip angle is found in the primary auxiliary view projected from the top view, and its general direction measured in the top view. Dip direction is measured perpendicular to a level line in the top view of the plane toward the low side.

The steps of finding the strike and dip of a plane are given in Fig. 5.36. Strike can be measured in the top view by finding a true-length line on the plane in this view. The dip angle requires an auxiliary view that must be projected from the top view in order for the horizontal to appear as an edge and the plane as an edge.

5.22 DISTANCES FROM A POINT TO A PLANE

Descriptive geometry principles can be used to find various distances from a point to a plane. An example of this is shown in Fig. 5.37 where the distance from point *O* on the ground to an ore vein under the ground is found.

Fig. 5.36 Strike and dip of a plane

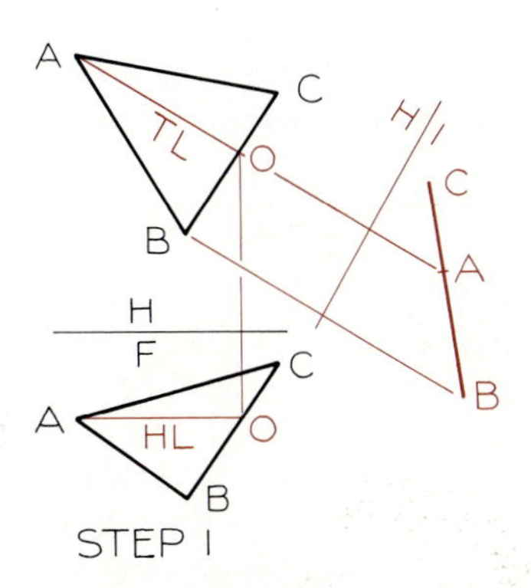

Step 1 Find the edge view of the plane by projecting from the top view.

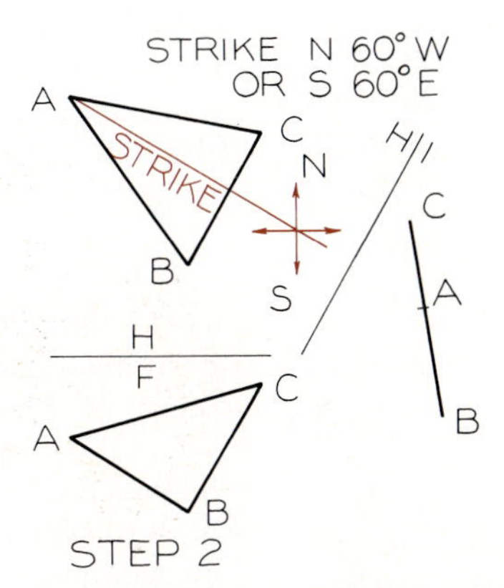

Step 2 Strike is the compass direction of a level line on the plane and is measured in the top view. The strike of the plane is either N 60° W or S 60° E.

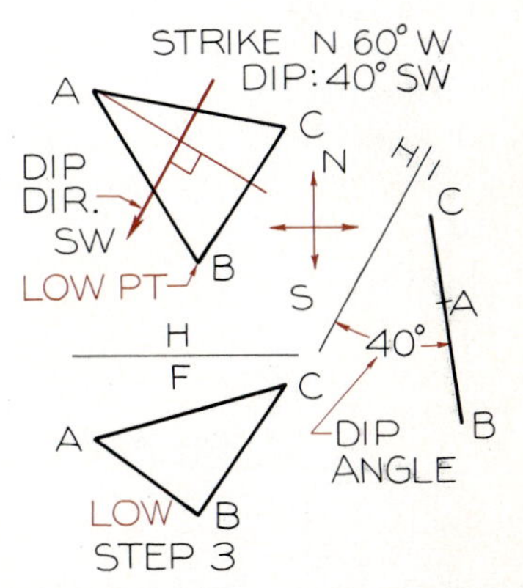

Step 3 The dip of a plane is the angle it makes with the horizontal (40° in the auxiliary view) plus the general compass direction toward the low side in the top view (SW). Dip direction is perpendicular to a *TL* line in the top view. Dip is written 40° SW.

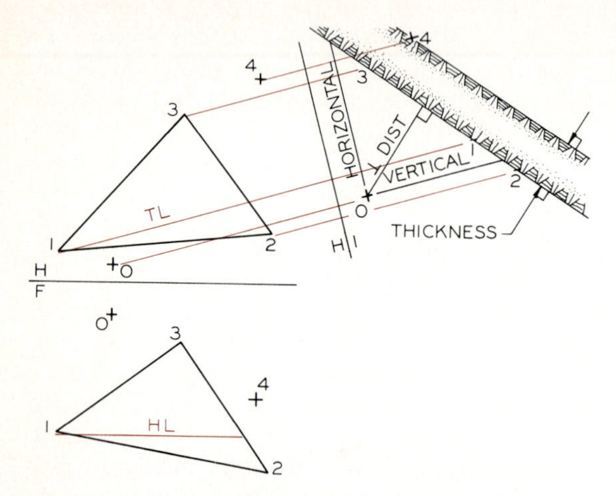

Fig. 5.37 The vertical, horizontal, and perpendicular distances from a point to an ore vein can be found in an auxiliary view projected from the top view. The thickness of an ore vein is perpendicular to the upper and lower planes of the vein.

Three points are located on the top plane of an ore vein. Point O is the point on the earth from which the tunnels are to be drilled to the vein for mining. Point 4 is a point on the lower plane of the vein.

The edge view of plane 1–2–3 is found by projecting from the top view. The lower plane is drawn parallel to the upper plane through point 4. The horizontal distance from point O to the plane is drawn parallel to the H–1 reference line. The vertical distance is perpendicular to the H–1 line. The shortest distance to the plane is perpendicular to the plane. Each of these lines is true length in this view where the ore vein appears as an edge.

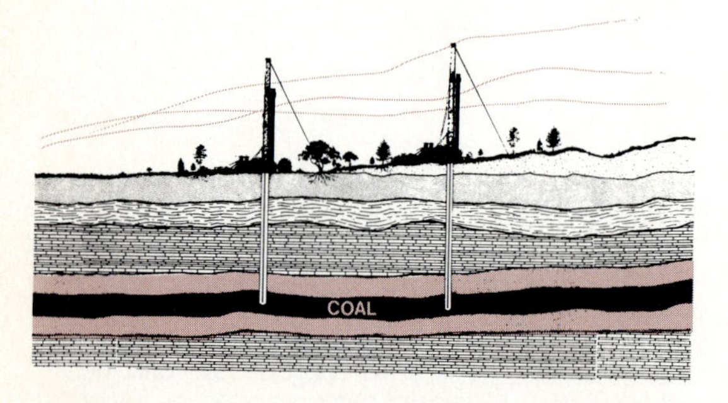

Fig. 5.38 Test wells are drilled into coal zones to determine which coal seams will contribute significantly to the production of gas. (Courtesy of Texas Eastern News.)

The process of finding the distance from a point to a plane or a vein is a technique often used in solving mining and geological problems. Figure 5.38 illustrates test wells that are drilled into coal zones to learn more about them.

5.23 OUTCROP

Strata of ore or rock formations usually approximate planes of a uniform thickness. This assumption is employed in analyzing data concerning the orientation of ore veins that are underground. A vein of ore may be inclined to the surface of the earth and may actually outcrop on its surface. Outcrops permit open-surface mining operations at a minimum of mining expense.

The steps of finding the outcrop of an ore vein are given in Fig. 5.39. The locations of sample drillings, A, B, and C are shown on the contour map and their elevations are located on the contours of the profile. These points are known to lie on the upper surface of the vein. Point D is known to lie on the lower plane of the vein.

The edge view of the ore vein can be found in an auxiliary view projected from the top view (Step 1). The points on the upper surface are projected back to their respective contour lines in the top view in Step 2. The points on the lower surface of the vein are projected to the top view in Step 3. These two lines are drawn to show the limits of the outcrop in the top view. If the ore vein does continue uniformly at its angle of inclination to the surface, the space between these two lines will be the edge of the vein on the surface of the earth.

5.24 INTERSECTIONS BETWEEN PLANES—CUTTING PLANE METHOD

The top and front views of two planes are given in Fig. 5.40 where it is required to find the line of intersection between them if the planes are infinite in size. Cutting planes are passed through either view at any angle and projected to the adjacent view. The two points L and M that are found in the top view are connected to form the top view of the line of intersection. The compass direction of this line can be used to describe its direction of slope toward its low end.

Fig. 5.39 Ore vein outcrop

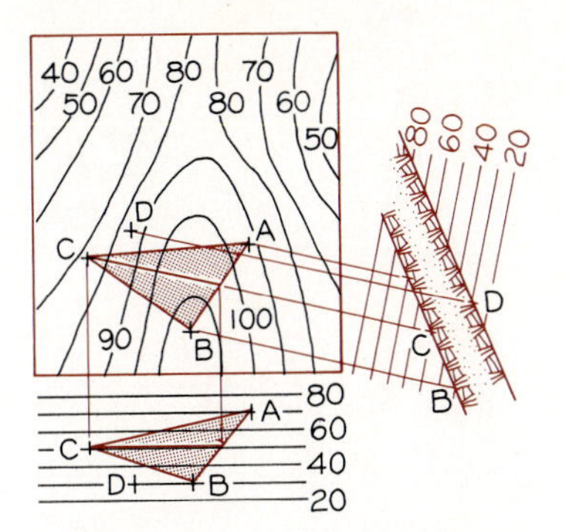

Step 1 Using points *A, B,* and *C* on the upper surface of the plane, find its edge view by projecting an auxiliary off the top view. The lower surface of the plane is found by drawing it parallel to the upper surface through point *D,* a known point on the lower surface.

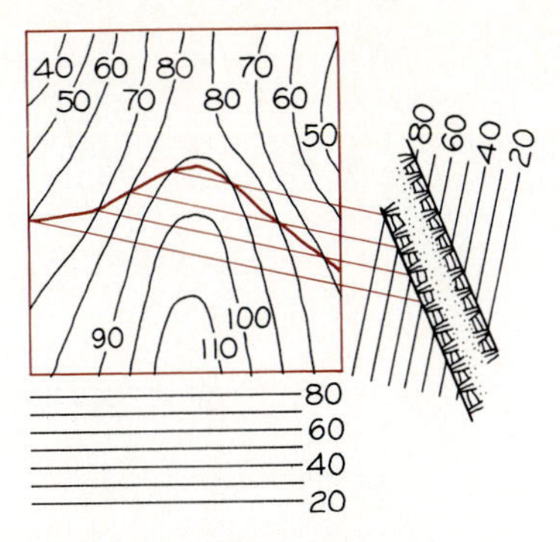

Step 2 Points of intersection between the upper surface and the contour lines in the auxiliary view are projected to their respective contour lines in the top view to find one line of the outcrop.

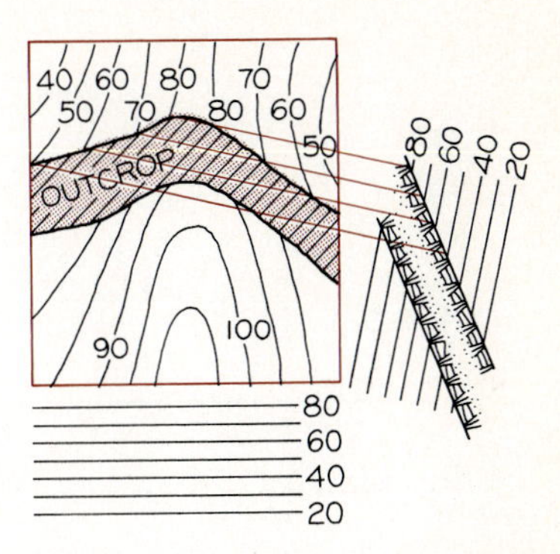

Step 3 Points from the lower surface in the auxiliary view are projected to their respective contour lines in the top view to find the second line of outcrop. Cross-hatch this area to indicate the outcrop area.

Fig. 5.40 Intersection of planes by cutting-plane method

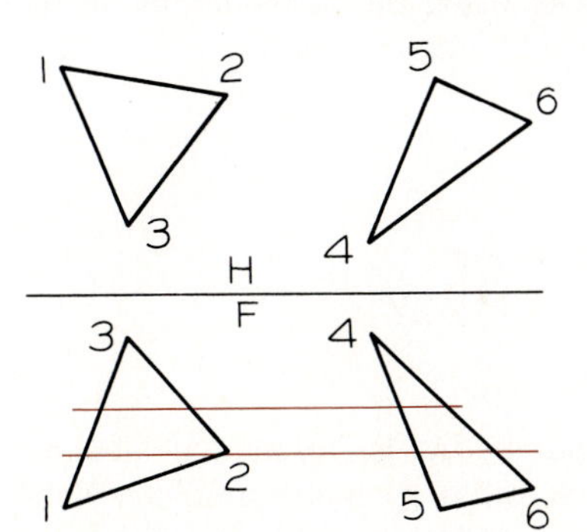

Step 1 Pass cutting planes through the front view of the planes. These planes can be drawn in any direction.

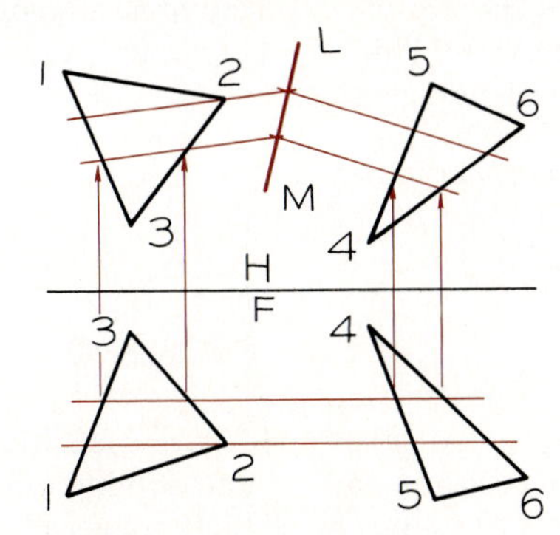

Step 2 Project the intersections of the cutting planes in the front view to the top view of the planes. Line of intersection, *LM,* is found in the top view.

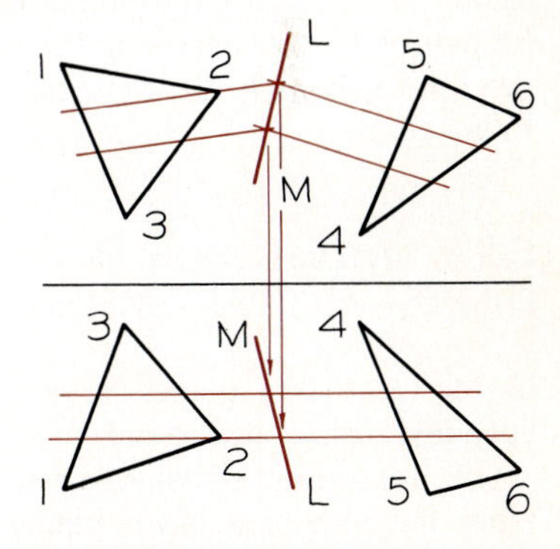

Step 3 Points *L* and *M* are projected to their respective cutting planes in the front view to complete the solution.

Fig. 5.41 Intersection between ore veins by auxiliary view

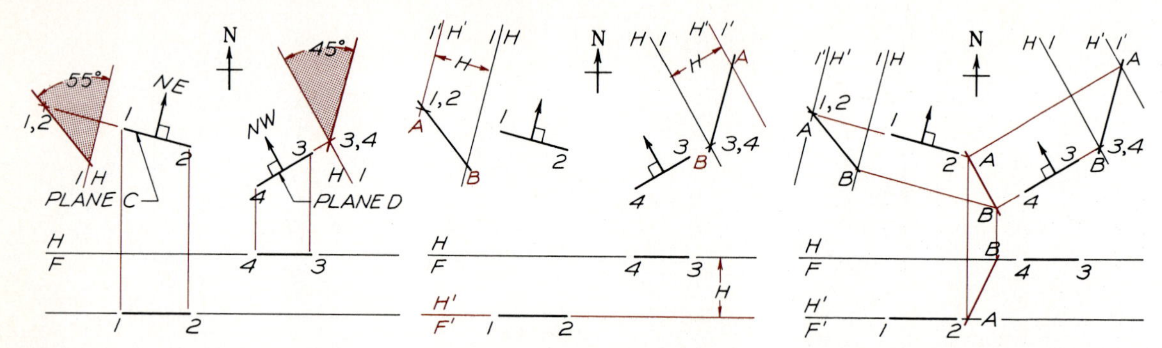

Step 1 Lines 1–2 and 3–4 are strike lines and are true length in the horizontal view. The point view of each strike line is found by auxiliary views, using a common reference plane. The edge views of the ore veins can be found by constructing the dip angles with the H-1 line through the point views. The low side is the side of the dip arrow.

Step 2 A supplementary horizontal plane, H′-F′, is constructed at a convenient location in the front view. This plane is shown in both auxiliary views located *H* distance from the H-1 reference line. The H′-1′ plane cuts through each ore vein edge in the auxiliary views to locate point *A* on each plane.

Step 3 Points *A*, which were established on each auxiliary view by the H′-1′ plane, are projected to the top view, and they intersect at point *A*. Points *B* on the H-1 plane are projected to their intersection in the top view at point *B*. Points *A* and *B* are projected to their respective planes in the front view. Line *AB* is the line of intersection between the two planes.

The front view of the line of intersection is found by projecting the points from the top view to their respective planes in the front view. This is the line on which all lines on the planes would intersect.

This technique of geometry has application in the area of geology when the intersections of underground veins are being analyzed.

5.25 INTERSECTION BETWEEN PLANES—AUXILIARY METHOD

In Fig. 5.41 two planes have been located and specified using strike and dip. Since the given strike lines are true-length level lines in the top view, the edge view of the planes can be found in the view where the strike appears as a point, Step 1. The edge views are drawn using the given dip angles and directions.

Horizontal datum planes H-F and H′-F′, are used to find lines on each plane that will intersect when projected from the auxiliary views to the top view. Points *A* and *B* are connected to determine the line of intersection between the two planes in the top view. These points are projected to the front view, where line *AB* is found.

5.26 SOLUTION OF DESCRIPTIVE GEOMETRY PROBLEMS

Figure 5.42 illustrates the techniques of labeling and solving a descriptive geometry problem. Note that some of the lettering and numbering is aligned with inclined lines and reference lines to which the labeling applies, and other lettering is not aligned but is parallel to the edge of the paper. You may use either technique or a combination of the two.

Always use guidelines and ⅛″ lettering for best results. Observe the difference in line qualities that are used in the problem solution. Guidelines and projection lines need be only dark enough to be seen and used as guides.

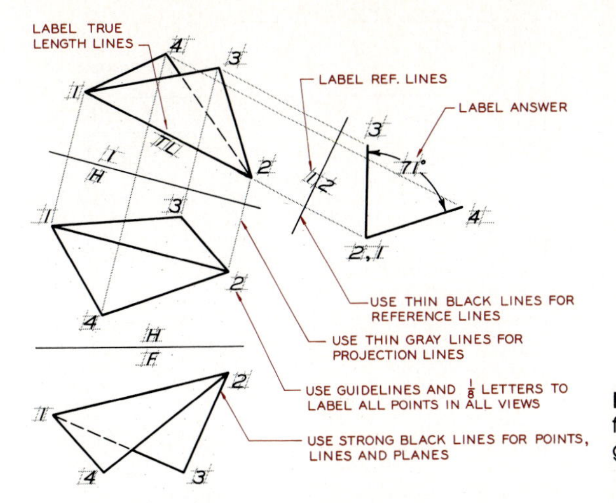

Fig. 5.42 Rules that should be followed in solving descriptive geometry problems.

PROBLEMS

Use Size A (8½″ × 11″) sheets for the following problems, and lay out the problems using instruments. Each square on the grid is equal to 0.20″ or about 5 mm. The problems can be laid out on grid or plain paper. Label all reference planes and points in each problem with ⅛″ letters and/or numbers, using guidelines.

1. Fig. 5.43. (1A–1D) Find the true length views of the lines as indicated by the given lines of sight by an auxiliary view.

2. Fig. 5.44. (2A–2D) Find the angles that these lines make with the respective principal planes indicated by the given auxiliary reference lines.

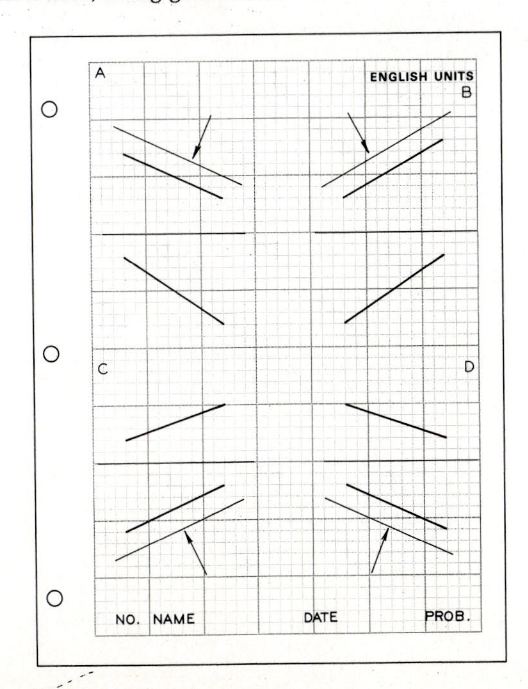

Fig. 5.43 Problems 1A through 1D.

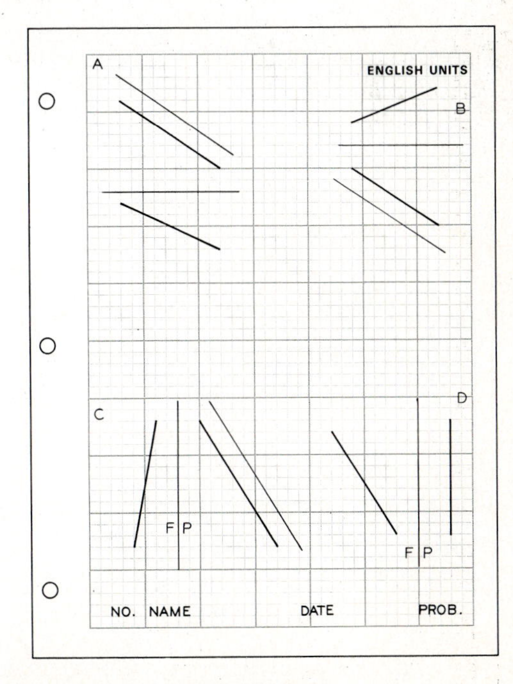

Fig. 5.44 Problems 2A through 2D.

3. Fig. 5.45. (3A and 3B) Find the lines' true length by the true-length diagram method, using the same diagram for both lines. (3C and 3D) Find the point views of the lines.

4. Fig. 5.46. (4A–4D) Find the slope angle, tangent of the slope angle, and the percent grade of the four lines.

5. Fig. 5.47. (5A and 5B) Find the edge views of the two planes. (5C) Find the angle between the line and plane.

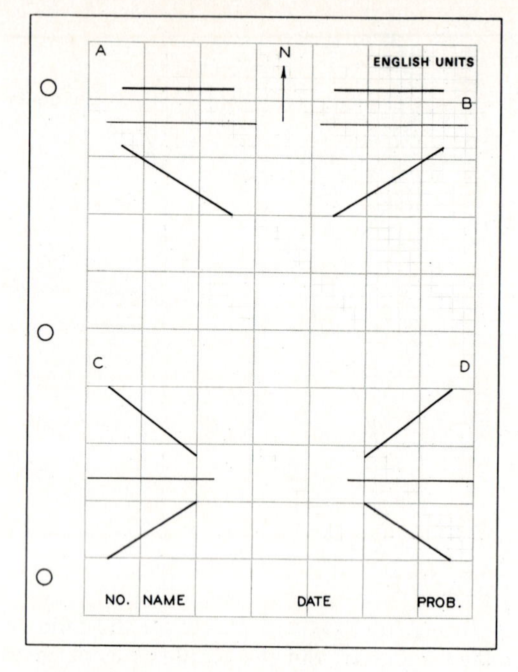

Fig. 5.46 Problems 4A through 4D.

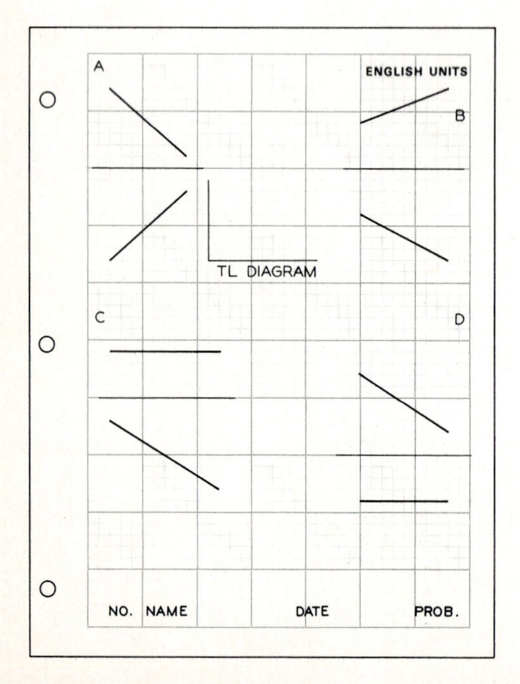

Fig. 5.45 Problems 3A through 3D.

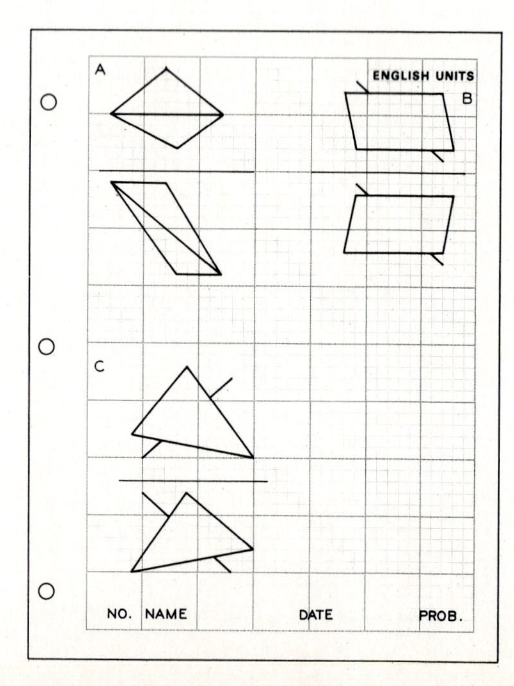

Fig. 5.47 Problems 5A through 5C.

6. Fig. 5.48. (6A) Find the angle between the planes. (6B) Find the piercing point by projection. (6C) Find the piercing point by an auxiliary view.

7. Fig. 5.49. (7A) Construct a line perpendicular to the plane and through point O on the plane by an auxiliary view. (7B) Construct a line perpendicular to the plane from point O by an auxiliary view.

8. Fig. 5.50. Find the line of intersection between the planes by projection in part A and by an auxiliary view in part B. Show visibility.

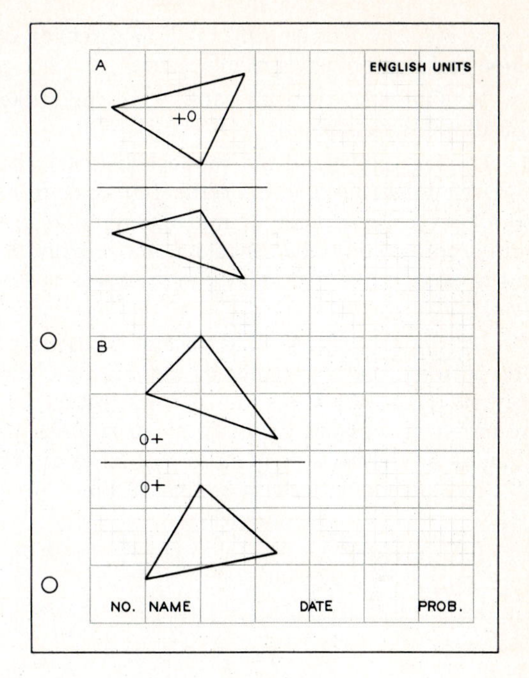

Fig. 5.49 Problems 7A and 7B.

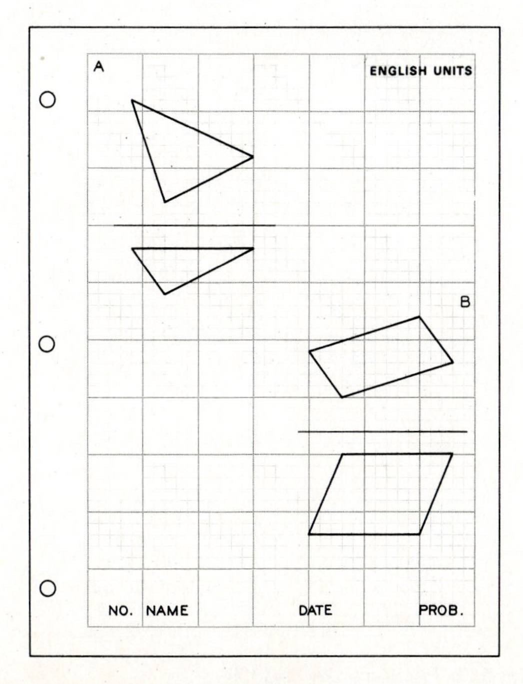

Fig. 5.48 Problems 6A through 6C.

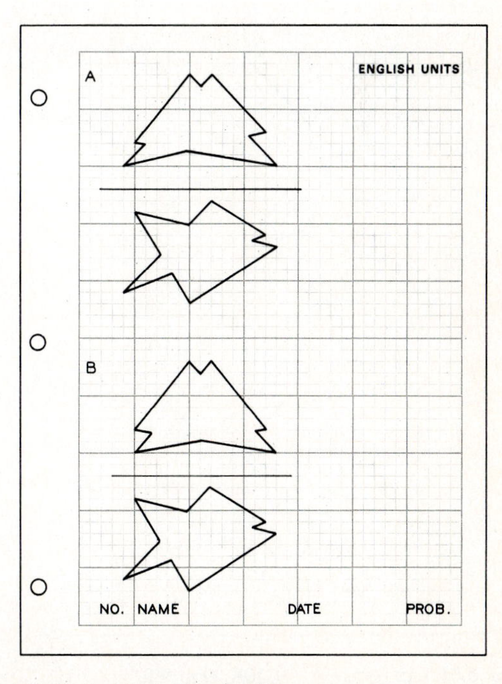

Fig. 5.50 Problems 8A and 8B.

9. Fig. 5.51. (9A and 9B) Find the direction of slope and slope of angle of the planes.

10. Fig. 5.51. (10A and 10B) Find the strike and dip of the planes.

11. Fig. 5.52. Find the shortest distance, the horizontal distance, and the vertical distance from point *O* on the ground to the underground ore vein represented by the triangle. Point *B* is on the lower plane of the vein. Find the thickness of the vein.

12. Fig. 5.53. (12A) Find the line of intersection between the two planes by the cutting-plane method. (12B) Find the line of intersection between the two planes indicated by strike lines 1–2 and 3–4. The plane with strike 1–2 has a dip of 30°, and the one with strike 3–4 has a dip of 60°.

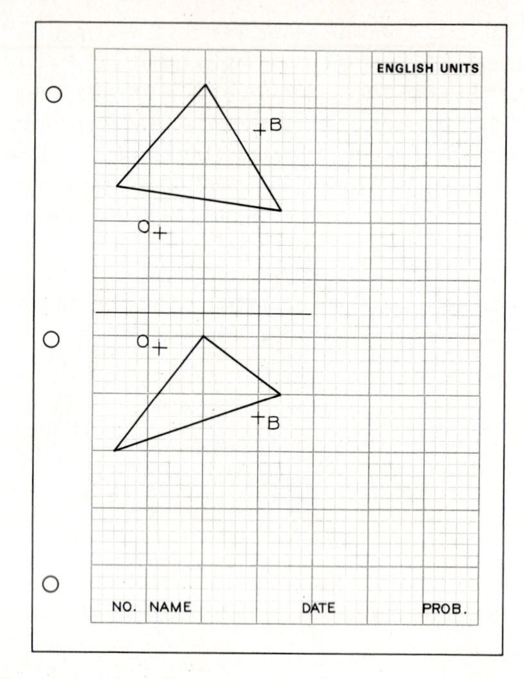

Fig. 5.52 Problem 11.

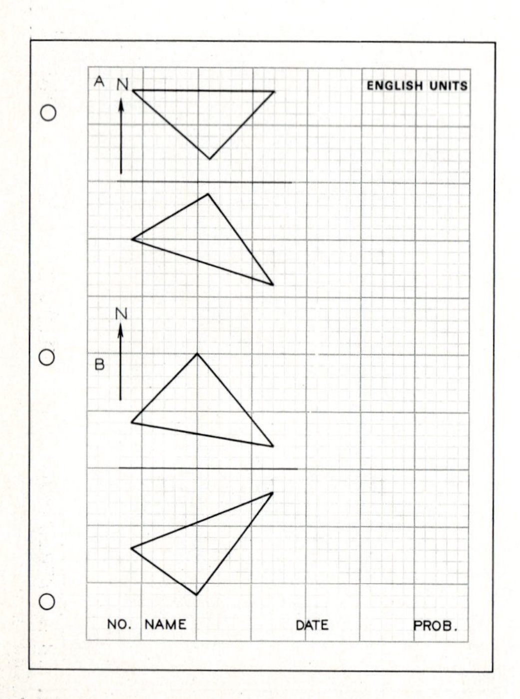

Fig. 5.51 Problems 9A, 9B, 10A, and 10B.

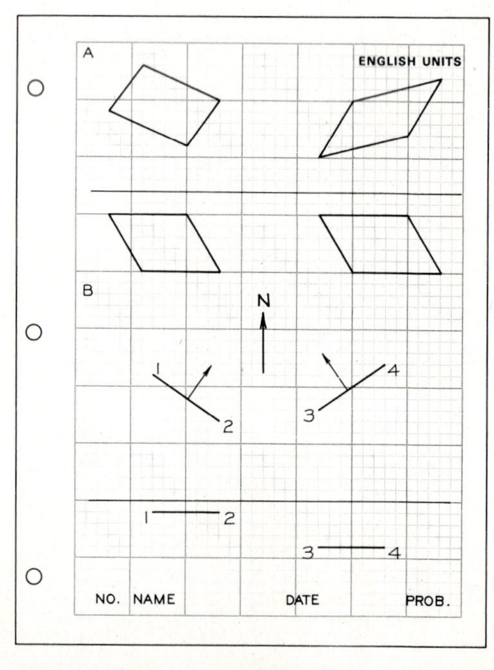

Fig. 5.53 Problems 12A and 12B.

13. Fig. 5.54. Find the limits of cut and fill in the plan view of the roadway. The roadway has a cut angle of 35° and a fill angle of 40°.

14. Fig. 5.55. Find the outcrop of the ore vein represented by the triangle (upper surface). Point *B* is on the lower surface.

15. Fig. 5.56. Complete the plan-profile drawing of the drainage system from manhole 1, through manhole 2, to manhole 3, using the grades indicated. Allow a drop of 0.20′ across each manhole to compensate for loss of pressure.

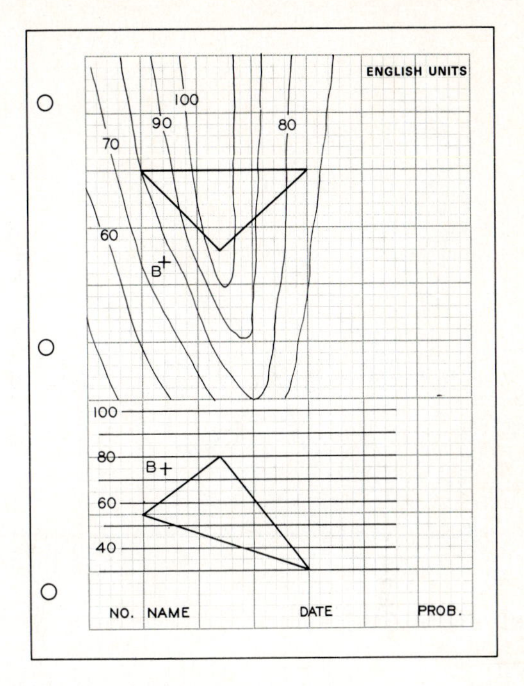

Fig. 5.55 Problem 14.

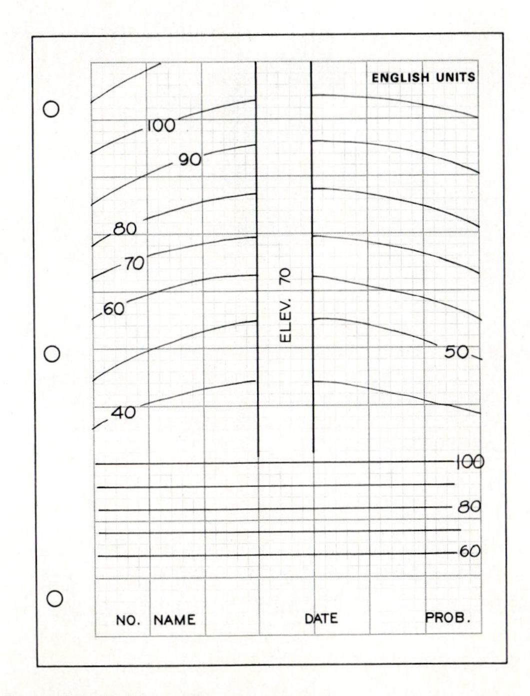

Fig. 5.54 Problem 13.

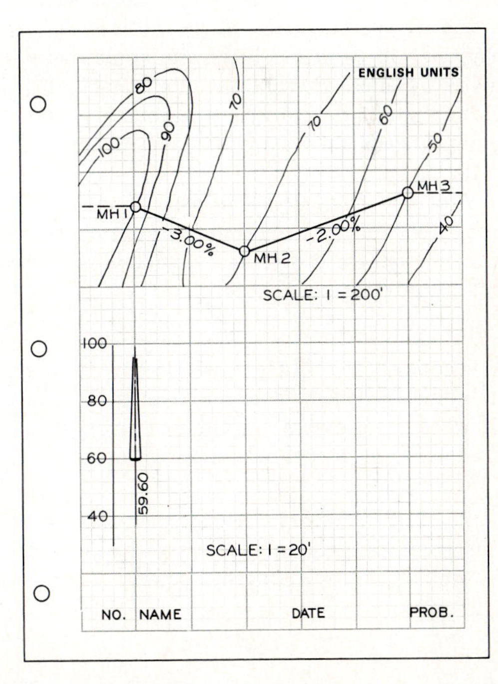

Fig. 5.56 Problem 15.

SUCCESSIVE AUXILIARY VIEWS

6.1 INTRODUCTION

A design cannot be detailed with complete specifications necessary for construction unless its complete geometry has been determined. This usually requires the application of descriptive geometry principles. Typical details that are needed are true shapes of planes, angles between planes, distances from points to lines, and angles between lines and planes. The Comsat satellite (Fig. 6.1) is an example of a design where various problems of geometry were solved by the use of successive auxiliary views.

You will recall that a primary auxiliary view is a supplementary view that is found by projecting orthographically from a primary view–a horizontal, frontal, or profile view.

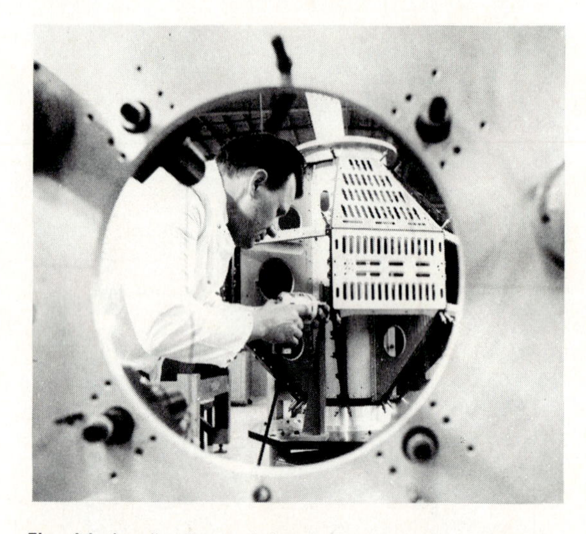

Fig. 6.1 Applications of descriptive geometry requiring successive auxiliary views are seen in the structure of this Comsat satellite. The angles between the planes and the true-size views of the surfaces were determined by applying the principles of descriptive geometry. (Courtesy of TRW Systems.)

> A **secondary auxiliary view** is an auxiliary view that is projected from a primary auxiliary view. A **successive auxiliary view** is an auxiliary view of a secondary auxiliary view.

Usually three auxiliary views projected from the primary view are adequate to solve the more complex problems of descriptive geometry.

6.2 POINT VIEW OF A LINE

When a line appears true length, its point view can be found by projecting an auxiliary view with parallel projectors from it. In Fig. 6.2, line 3–4 is true length in the top view since it is horizontal in the front view. The point view is found in the primary auxiliary view by constructing reference line H–1 perpendicular to the true-length line. The height dimension, H, is transferred to the auxiliary view to locate the point view of 4–3.

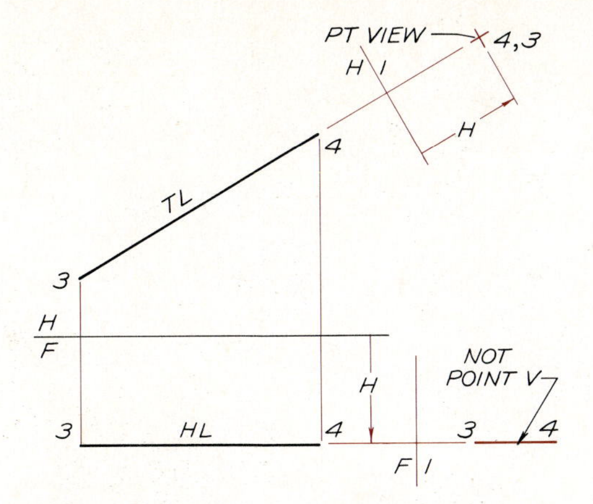

Fig. 6.2 The point view of a line can be found by projecting an auxiliary view from the true-length view of the line.

The line in Fig. 6.3 is not true length in either view, which requires that the line be found true length by a primary auxiliary view. In this example, the line is found true length by projecting from the front view, but this view could have been projected from the top as well. The point view of the line is found by projecting from the true-length line to a secondary auxiliary view. This point is labeled 4, 3 since point 4 is seen first.

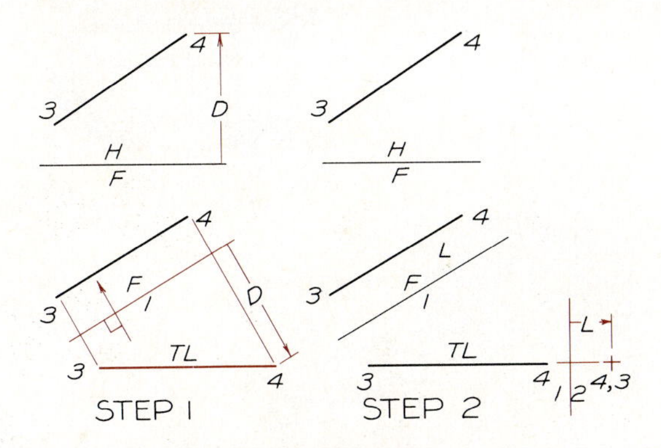

Fig. 6.3 Point view of an oblique line

Step 1 A line of sight is drawn perpendicular to one of the views, the front view in this example. Line 3–4 is found true length by projecting perpendicularly from the front view.

Step 2 A secondary reference line 1–2 is drawn perpendicular to the true-length view of 3–4. The point view is found by transferring dimension *L* from the front view to the secondary auxiliary view.

6.3 ANGLE BETWEEN PLANES

The angle between two planes is called a **dihedral angle.** This angle can be found in the view where the line of intersection appears as a point.

Since this results in the point view of a line that lies on both planes, the planes will appear as edges in this view.

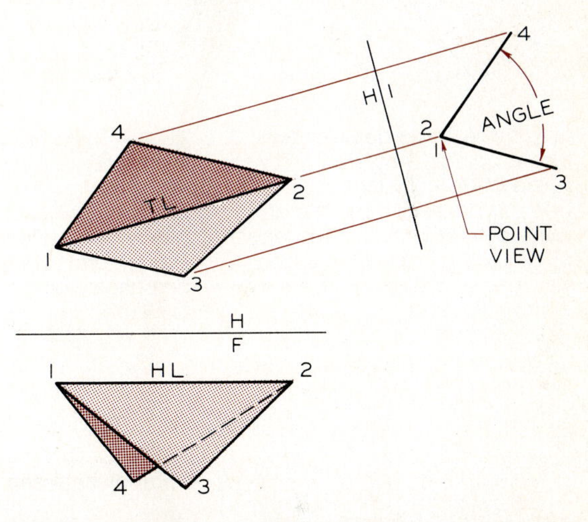

Fig. 6.4 The angle between two planes can be found in the view where the line of intersection between them projects as a point. Since the line of intersection, 1–2, is true length in the top view, it can be found as a point in a view that is projected from the top view.

The two planes in Fig. 6.4 represent a special case since the line of intersection, 1–2, is true length in the top view. This permits you to find its point view in a primary auxiliary view where the true angle can be measured.

A general case is given in Fig. 6.5 where the line of intersection between the two planes is not true length in either of the principal views.

The line of intersection, 1–2, is found true length in a primary auxiliary view, and the point view of the line is then found in the secondary auxiliary view. The dihedral angle is measured in the secondary auxiliary view.

It is apparent that this principle must be used to determine the angles between intersecting planes such as those shown in Fig. 6.6, where the

Fig. 6.5 Angle between two planes

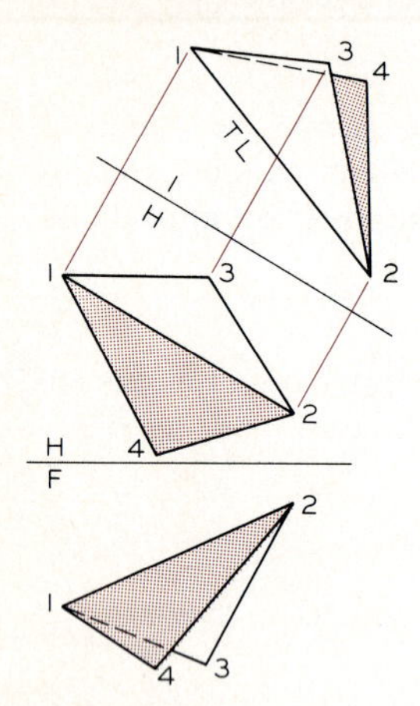

Step 1 The angle between two planes can be measured in a view where the line of intersection appears as a point. The line of intersection is first found true length by projecting a primary auxiliary view perpendicularly from the top view, in this case.

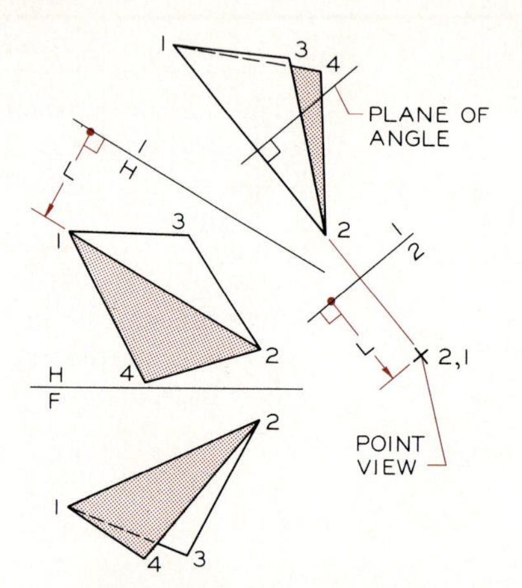

Step 2 The point view of the line of intersection is found in the secondary auxiliary view by projecting parallel to the true length view of 1–2. Note that the plane of the dihedral angle is an edge and perpendicular to the true-length line of intersection.

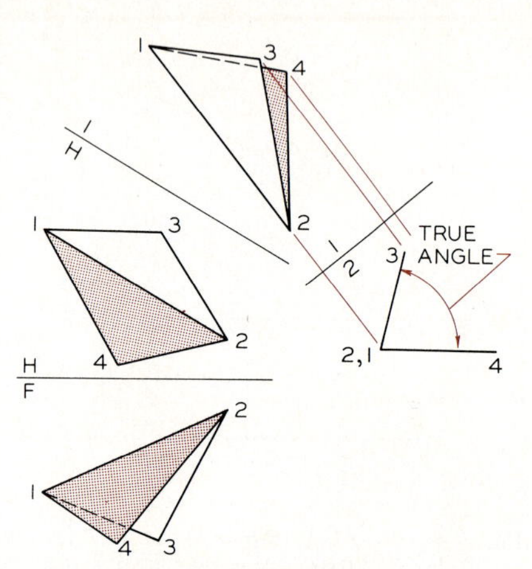

Step 3 The edge views of the planes are completed in the secondary auxiliary view by locating points 3 and 4. The angle between the planes, the dihedral angle, can be measured in this view.

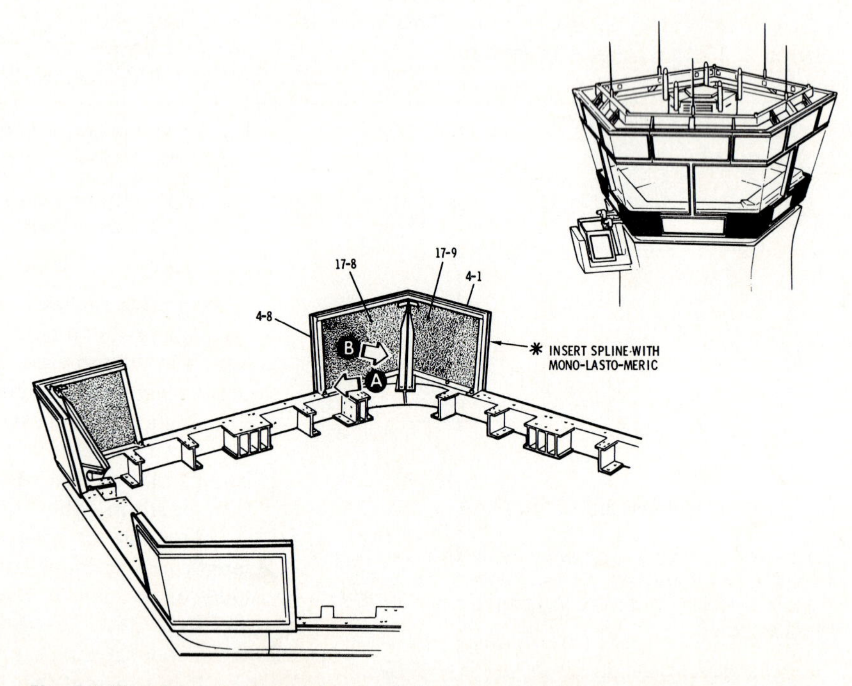

Fig. 6.6 The determination of the angle between the planes of the corner panels of the control tower utilized principles of descriptive geometry. (Courtesy of the Federal Aviation Agency.)

Fig. 6.7 True size of a plane

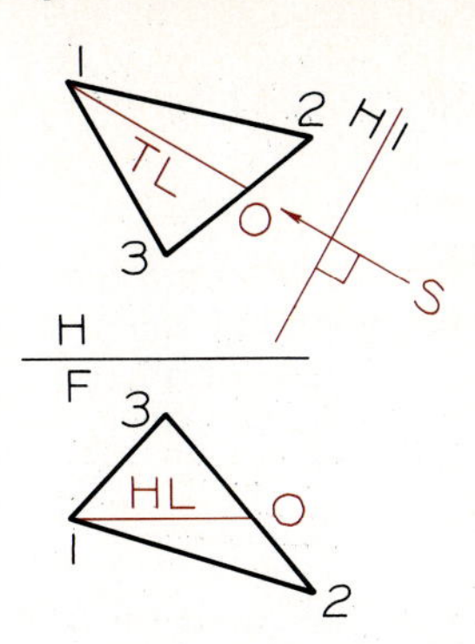

Step 1 The edge view of a plane can be found by finding the point view of any line that lies on it. Horizontal line 1-O is drawn on the front view and is then projected to the top view. A line of sight is drawn parallel to 1-O

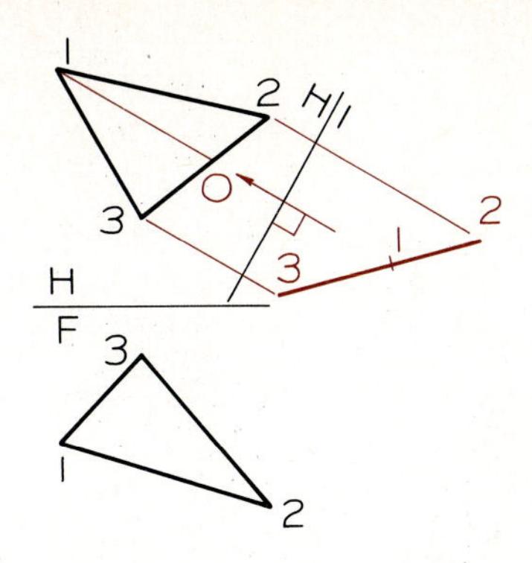

Step 2 Line 1-O is found as a point in the primary auxiliary view and the plane appears as an edge.

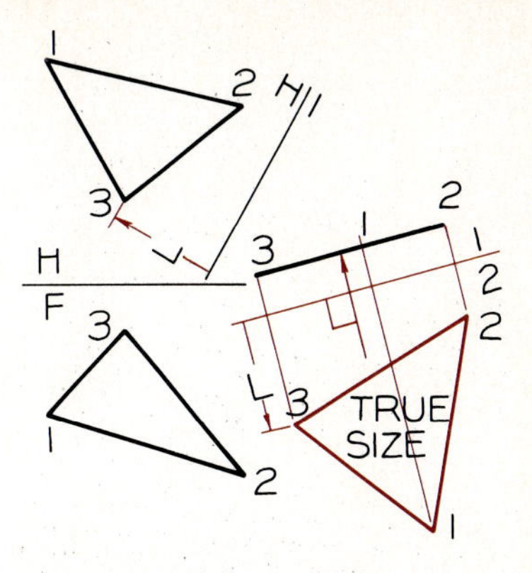

Step 3 The true-size view of a plane can be found by locating a line of sight that is perpendicular to an edge view of the plane. A secondary auxiliary view is projected perpendicularly from the edge of plane 1–2–3 to find its true-size view.

side panels of a control tower join. This is necessary in order to design and fabricate corner braces to hold the structure together.

6.4 TRUE SIZE OF A PLANE

A plane can be found true size in a view that is projected perpendicularly from an edge view of a plane.

In Fig. 6.7, the true size of a plane 1–2–3 is found by first finding the edge view of the plane in Steps 1 and 2. In Step 3, the secondary auxiliary view is projected perpendicularly from the edge view. The result is a true-size view of the plane where each angle is true size.

This principle can be used to find the angle between lines such as bends in a fuel line of an aircraft engine (Fig. 6.8). A problem of this type is shown in Fig. 6.9 where the top and front views of intersecting lines are given. It is required that

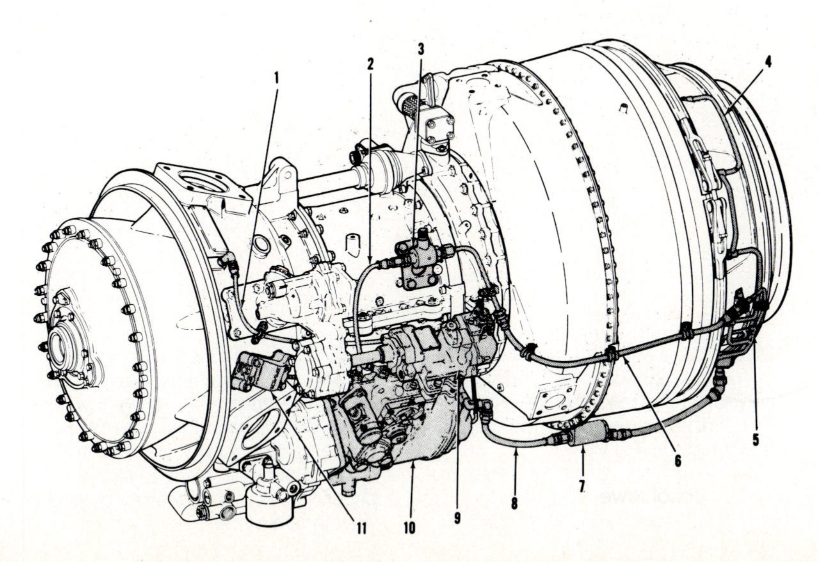

Fig. 6.8 The angles of bend in the fuel line were found by the application of the principle of finding the angle between two lines. (Courtesy of Avco Lycoming.)

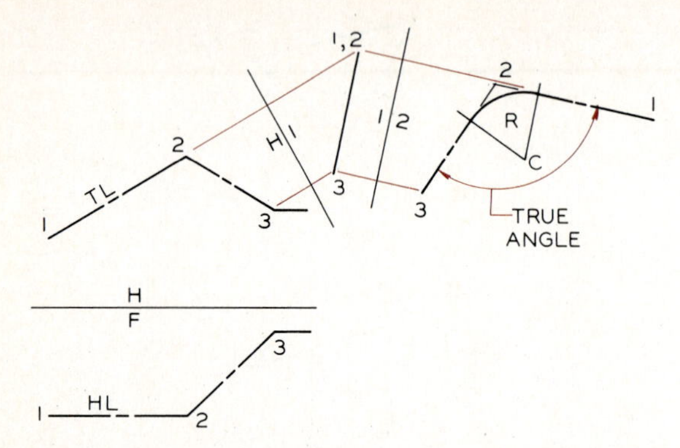

Fig. 6.9 The angle between two lines can be found by finding the plane of the lines' true size.

the angles be determined at each bend, and that a given radius of curvature be shown.

The angle 1–2–3 is found as an edge in the primary auxiliary view and as a true-size angle in the secondary view. The angle can be measured in this view and the radius of curvature drawn using principles of geometric construction.

6.5 SHORTEST DISTANCE FROM A POINT TO A LINE

The shortest distance from a point to a line can be measured in the view where the line appears as a point.

This distance will appear perpendicular to the line whenever the line appears true length.

This type of problem is solved in Fig. 6.10 by finding the line 1–2 true length in a primary auxiliary view along with point 3. The line is found as a point in the secondary auxiliary view, where the distance from point 3 is true length. Since line *O*–3 is true length in this view, it must be parallel to the 1–2 reference line in the preceding view, the primary auxiliary view. It is also perpendicular to the true length view of line 1–2 in the primary auxiliary view. Point *O* is projected back to the principal views to complete the solution.

Fig. 6.10 Shortest distance from a point to a line

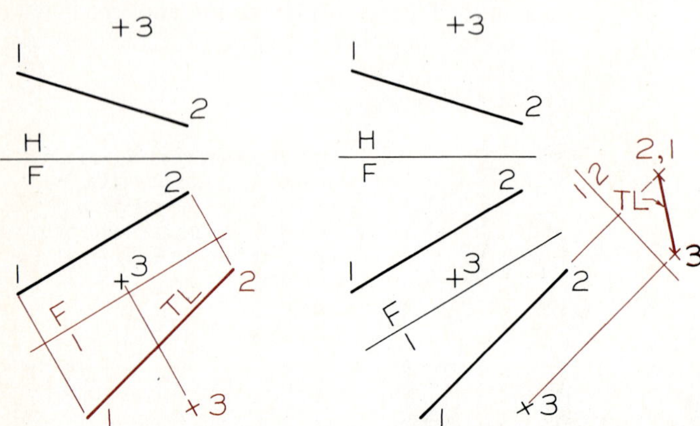

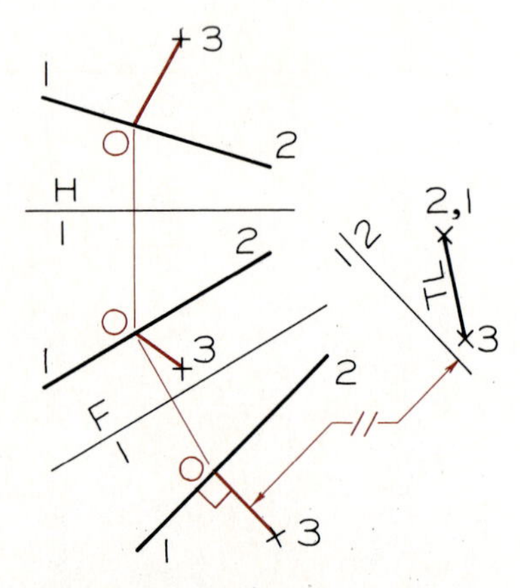

Step 1 The shortest distance from a point to a line can be found in the view where the line appears as a point. Line 1–2 is found true length by projecting from the front view.

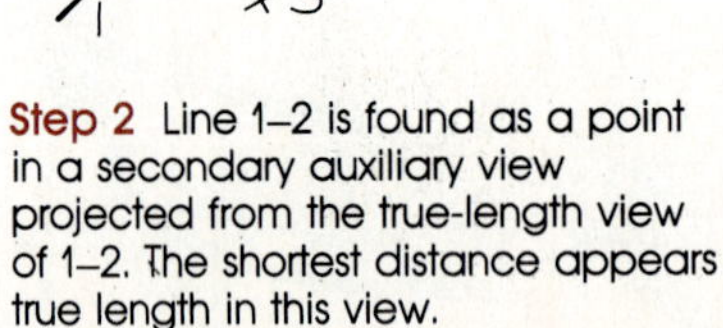

Step 2 Line 1–2 is found as a point in a secondary auxiliary view projected from the true-length view of 1–2. The shortest distance appears true length in this view.

Step 3 Since 3-O is true length in the secondary auxiliary view, it must be parallel to the 1–2 reference line in the primary auxiliary view and perpendicular to the line. The front and top views of 3-O are found by projecting from the primary auxiliary view in sequence.

6.6 SHORTEST DISTANCE BETWEEN SKEWED LINES—LINE METHOD

> The shortest distance between two skewed lines (randomly positioned lines) can be measured in the view where one of the lines appears as a point.

The shortest distance between two lines is perpendicular to both lines. The location of the shortest distance is both functional and economical, as demonstrated by the connector between two pipes in Fig. 6.11, since a standard connector is a 90° Tee.

A problem of this type is solved by the line method in Fig. 6.12. Line 3–4 is found as a point in the secondary auxiliary view where the shortest distance is drawn perpendicular to line 1–2. Since the distance is true length in the secondary auxiliary view, it must be parallel to the 1–2 reference line in the primary auxiliary view. Point *O* is found by projection and *OP* is drawn perpendicular to line 3–4. The line is projected back to the given principal views.

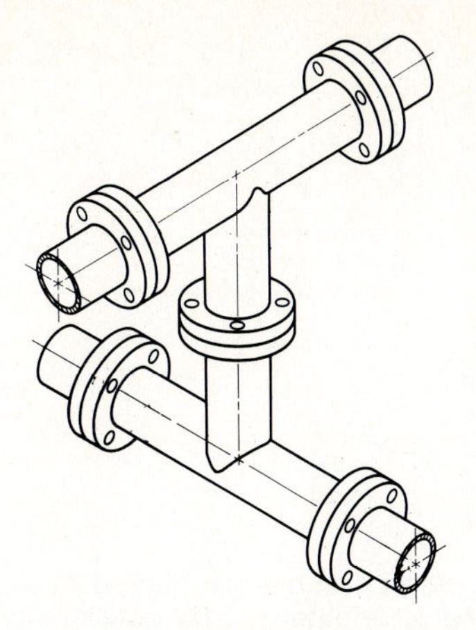

Fig. 6.11 The shortest distance between two lines, or pipes, is a line that is perpendicular to both. This is the most economical connection and the most functional since perpendicular connectors are standard.

6.7 SHORTEST DISTANCE BETWEEN SKEWED LINES—PLANE METHOD

The distance between skewed lines can be solved using the alternative plane method. This involves the construction of a plane through one of the

Fig. 6.12 Shortest distance between skewed lines—line method

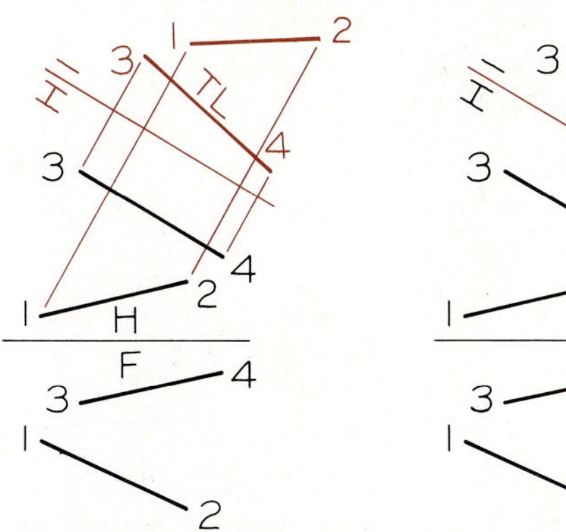

Step 1 The shortest distance between two skewed lines can be found in the view where one of the lines appears as a point. Line 3–4 is found true length by projecting from the top view along with line 1–2.

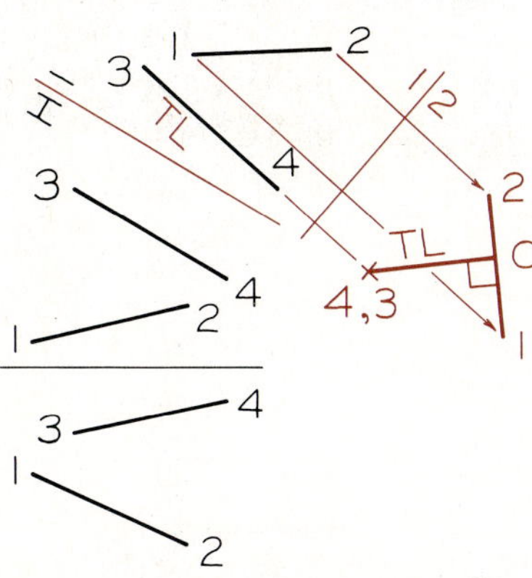

Step 2 The point view of line 3–4 is found in a secondary auxiliary view projected from the true-length view of 3–4. The shortest distance between the lines is drawn perpendicular to line 1–2.

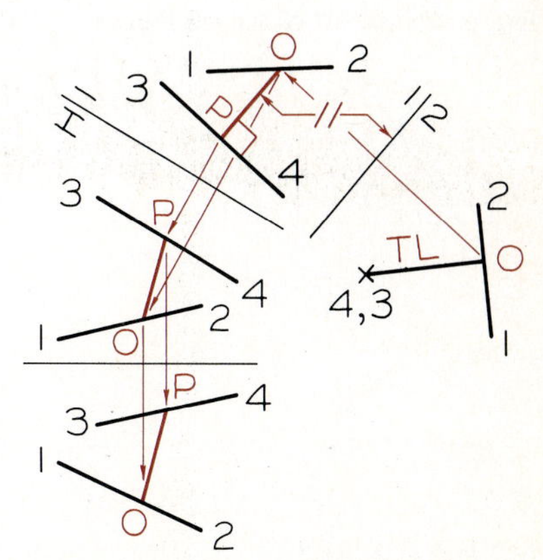

Step 3 Since the shortest distance is true length in the secondary auxiliary view, it must be parallel to the reference line in the preceding view. Points *O* and *P* are projected back to the given view to show the shortest distance in all views.

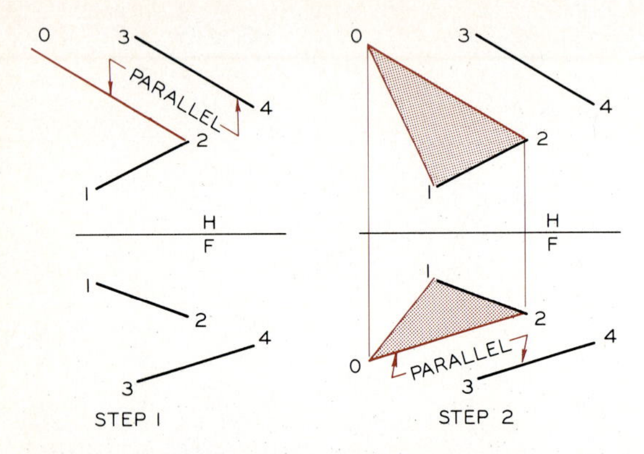

Fig. 6.13 A plane can be constructed through a line and parallel to another line by construction.

Step 1 Line O-2 is drawn parallel to line 3–4 to a convenient length.

Step 2 The front view of line O-2 is drawn parallel to the front view of line 3–4. The length of the front view of O-2 is found by projecting from the top view of O. Plane 1–2–O is parallel to line 3–4.

lines parallel to the other, as shown in Fig. 6.13. The top and front views of 0–2 are drawn parallel to their respective views of line 3–4. The resulting plane, 1–2–0, is parallel to line 3–4. Both lines will appear parallel in a view where the plane appears as an edge.

In Fig. 6.14, plane 3–4–0 is constructed and its edge view is found and both lines appear parallel (Step 1). A secondary auxiliary view is projected perpendicularly from these parallel lines to find the view where the lines cross (Step 2). This crossing point represents the point view of the shortest distance between the two lines. It will appear true length and perpendicular to the two lines when it is projected to the primary auxiliary view, where it is labeled as line *LM*. It is projected back to the given views to complete the solution of locating the shortest distance between two lines (Step 3).

This principle was applied to the design of the separation of power lines shown in Fig. 6.15 where the clearance is critical.

6.8 SHORTEST LEVEL DISTANCE BETWEEN SKEWED LINES

When it is required to find the shortest level (horizontal) distance between two skewed lines, the plane method must be used instead of the line method. Also, the primary auxiliary view must be projected from the top view in order to find a view where the horizontal plane appears as an edge.

Fig. 6.14 Shortest distance between skewed lines—plane method

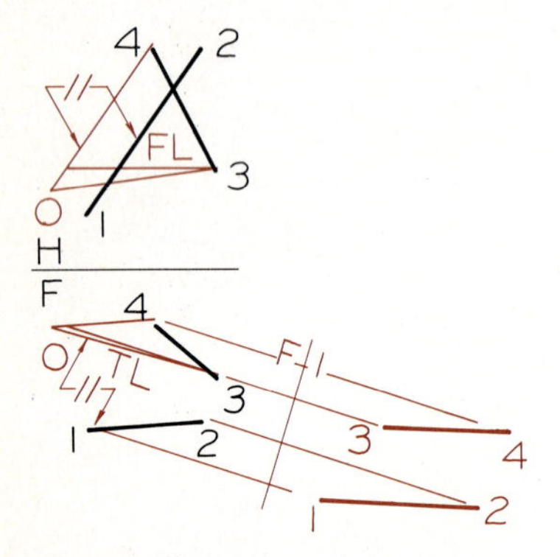

Step 1 Construct a plane through line 3–4 that is parallel to line 1–2. When this plane is found as an edge by projecting from the front view, the two lines will appear parallel in this auxiliary view.

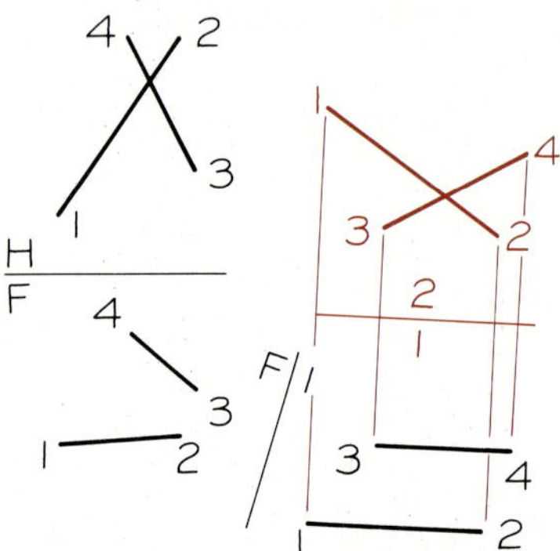

Step 2 The shortest distance will appear true length in the primary auxiliary view, where it will be perpendicular to both lines. Draw a secondary auxiliary view by projecting perpendicularly from the lines in the primary auxiliary view.

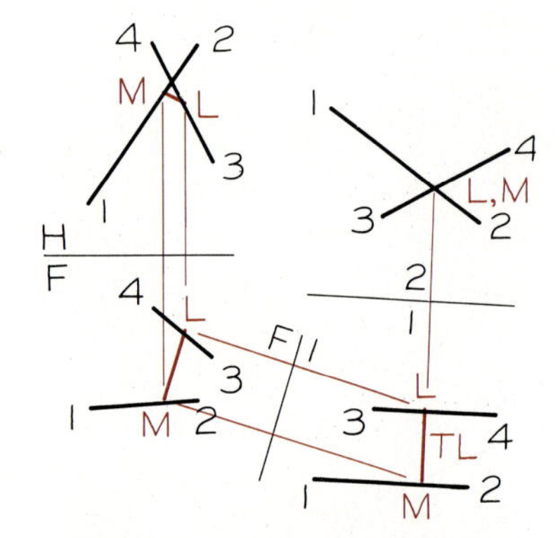

Step 3 The crossing point of the two lines is the point view of the perpendicular distance, *LM*, between the lines. This distance is projected to the primary auxiliary view, where it is true length, and back to the given views.

Fig. 6.15 The determination of clearance between power lines is a critical problem for the electrical engineer. This is an application of the shortest distance between two lines. (Courtesy of the Tennessee Valley Authority.)

In Fig. 6.16, plane 3–4–*O* is drawn parallel to line 1–2 and an edge view of the plane is found in the primary auxiliary view. The lines appear parallel in this view and the horizontal (H–1) appears as an edge. A line of sight is drawn parallel to the H–1, and the secondary reference line, 1–2, is drawn perpendicular to the H–1 (Step 2). The crossing point of the lines found in the secondary auxiliary views locates the point view of the shortest horizontal distance between the lines (Step 3). This line, *LM*, is true length in the primary auxiliary view and parallel to the H–1 plane.

Line *LM* is projected back to the given views. As a check, this line must be parallel to the H-line in the front view since it is a level or horizontal line.

6.9 SHORTEST GRADE DISTANCE BETWEEN SKEWED LINES

Many lines representing highways, power lines, or conveyors are connected by lines at a specified grade other than horizontal or perpendicular. Conveyors, such as the one shown in Fig. 6.17, are used to transport aggregates or grain for mixing. The slopes of such a system cannot exceed specified grades for efficient operation.

If a 50% grade connector between two lines must be found between the two lines in Fig. 6.18, the plane method is used as in the two previous examples. A view of the lines where they appear

Fig. 6.16 Shortest level distance between skewed lines—plane method

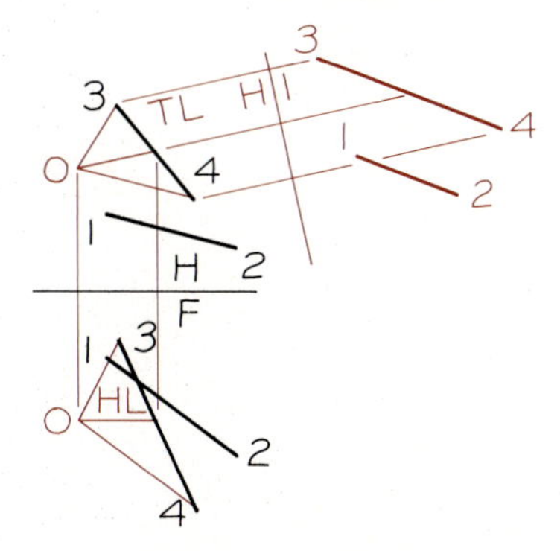

Step 1 Construct plane *O*–3–4 parallel to line 1–2 by drawing line *O*–4 parallel to 1–2. Find the edge view of *O*–3–4 by projecting off the top view. The lines will appear parallel in this view. *Note:* The auxiliary view *must* be projected from the *top view* to find the horizontal plane as an edge.

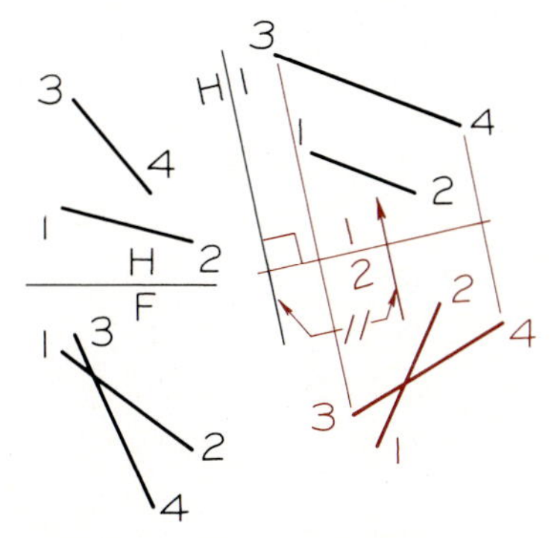

Step 2 An infinite number of horizontal (level) lines can be drawn parallel to H-1 between the lines in the auxiliary view, but only the shortest level line will appear true length in the primary auxiliary view. Construct the secondary auxiliary view by projecting parallel to the horizontal (H-1) to find the point view of the shortest level line.

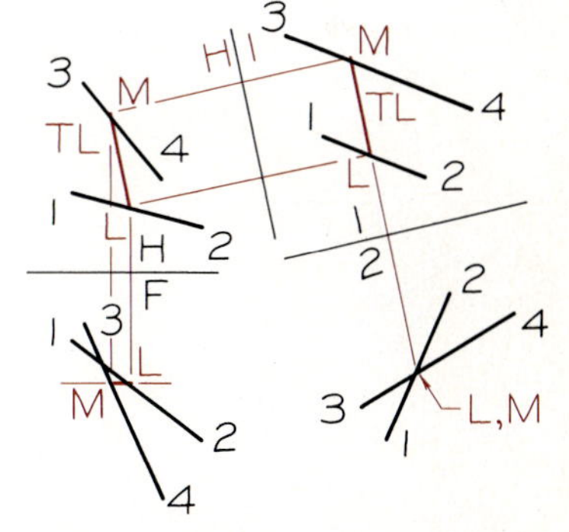

Step 3 The crossing point of the two lines in the secondary auxiliary view establishes the point view of the level connector, *LM.* Project *LM* back to the given views. *LM* is parallel to the H-plane in the front view, which verifies that it is a level line.

Fig. 6.17 These conveyors represent the application of skewed-line problems that must be solved using descriptive geometry principles.

parallel is constructed and a 50% grade line is constructed from the edge view of the horizontal (H-1). It should be noted that the auxiliary views must be projected from the top view in order to have an edge view of the horizontal from which the 50% grade is constructed.

The grade line can be constructed in two directions from the H-1 line, but the shortest distance will be the most nearly perpendicular to both lines, which is the one shown in Step 2. The secondary auxiliary view is projected parallel with this 50% grade line to find the crossing point of the lines to locate the shortest connector, *LM*, that is at a 50% grade.

Line *LM* is projected back to all views. This line appears true length in the primary auxiliary view where the lines appear parallel.

By now, it should be apparent that any connector between skewed lines will appear true length in the view where the lines appear parallel. Perpendicular lines, horizontal lines, and grade lines are true length in this view.

6.10 ANGULAR DISTANCE TO A LINE

Standard connectors used to connect pipes and structural members are available in two standard angles—90° and 45°. Consequently, it is economical to incorporate these into a design rather than having to design specially made connectors.

In Fig. 6.19 it is required to locate the point of intersection on line 1–2 of a line drawn from

Fig. 6.18 Grade distance between skewed lines

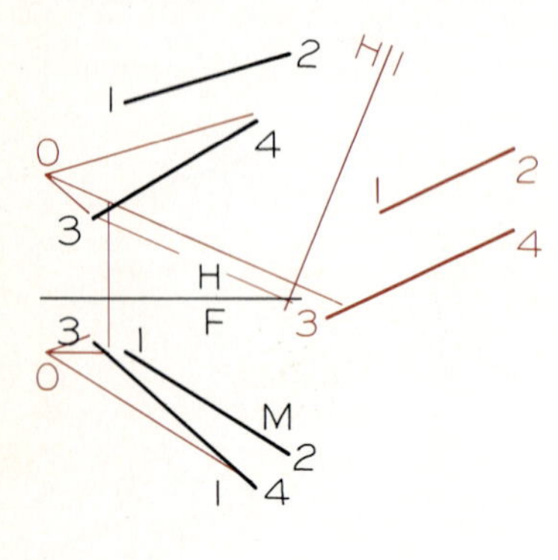

Step 1 To find a level line or a line on a grade between two skewed lines, the primary auxiliary must be projected from the top view. Plane 3–4–O is constructed parallel to line 1–2. The edge view of the plane is found where the lines appear parallel.

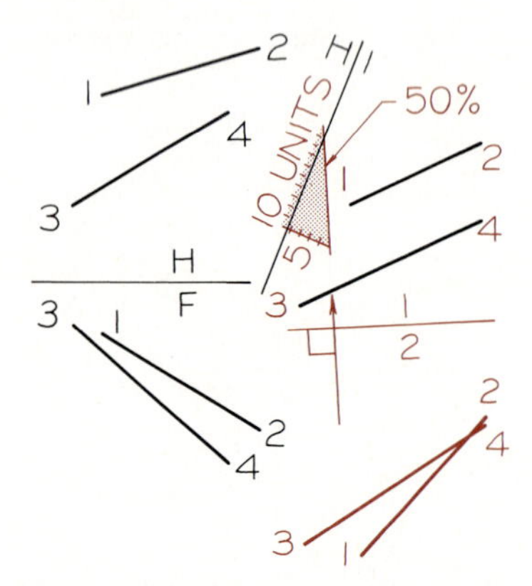

Step 2 Construct a 50-percent grade line with respect to the edge view of the H-1 line in the primary auxiliary view that is as nearly perpendicular to the lines as possible. Project the secondary auxiliary view parallel to the grade line. The shortest grade distance will appear true length in the primary auxiliary view.

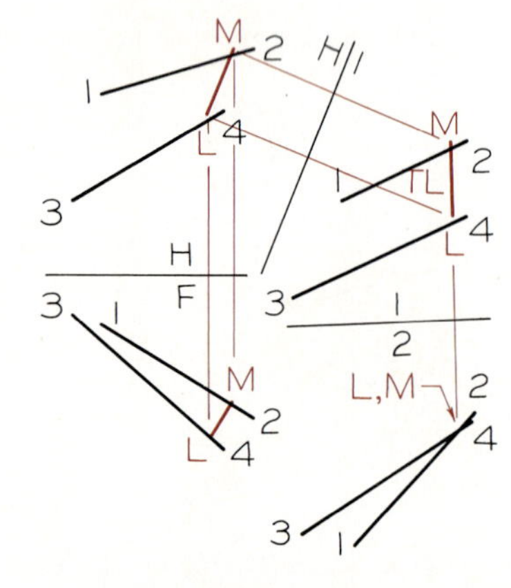

Step 3 The point of crossing of the two lines in the secondary auxiliary view establishes the point view of the 50-percent grade line, *LM*. This line is projected back to the previous views in sequence.

Fig. 6.19 Line through a point with a given angle to a line

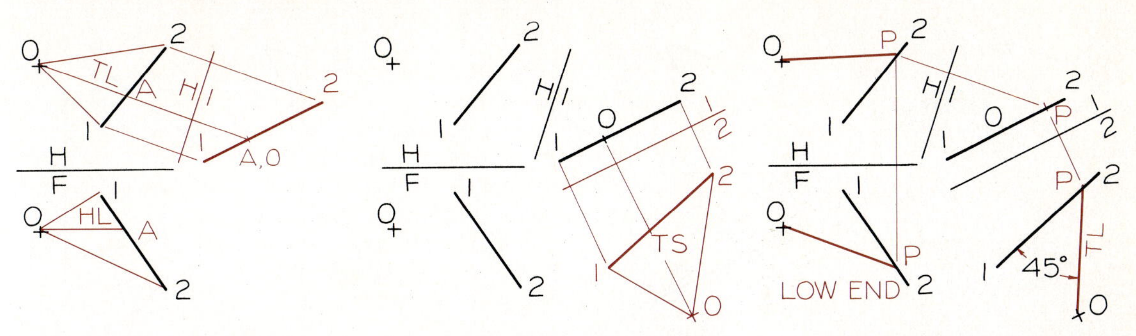

Step 1 Connect point O to each end of the line to form a plane 1–2–O in both views. Draw a horizontal line in the front view of the plane and project it to the top view, where it is true length. Find the edge view of the plane by obtaining the point view of line *A-O.*

Step 2 Find the true size of plane 1–2–O projecting perpendicularly from the edge view of the plane in the primary auxiliary view. The plane can be omitted in this view and only line 1–2 and point O are shown.

Step 3 Line *OP* is constructed at the specific angle with line 1–2, 45° in this case. Note that, if the angle is toward point 2 the line slopes downhill, and if toward point 1 it slopes uphill. Point *P* is projected back to the other views in sequence to show the line, *OP,* in all views.

point *O* at a 45° angle to the line. The plane of the line and point, 1–2–*O,* is found as an edge in the primary auxiliary view. The angle can be measured in this view where the plane of the line and point is true size.

The 45° connector is drawn from *O* to the line toward point 2 to slope downhill, or toward point 1 if it is to slope uphill. This can be determined by referring to the front view where height can be easily seen. Line *OP* is projected back to the given views.

6.11 ANGLE BETWEEN A LINE AND A PLANE—PLANE METHOD

> The angle between a line and plane can be measured in the view where the plane appears as an edge and the line true length.

In Fig. 6.20, the edge view of the plane is found in a primary auxiliary view projected from any primary view. The plane is then found true size in Step 2 where the line is foreshortened. The line, *AB,* can be found true length in a third auxiliary view projected perpendicularly from the secondary auxiliary view of *AB.* The line appears true length and the plane as an edge in the third successive auxiliary view.

The piercing point is projected back to the views in sequence and the visibility is determined for each view.

6.12 ANGLE BETWEEN A LINE AND A PLANE—LINE METHOD

To find the angle between a line and a plane by the line method, the line is first found as a point, and then true length as shown in Fig. 6.21. The plane is foreshortened in Step 2 where the line *AB* appears as a point.

The plane is found as an edge in a third auxiliary view by finding the point view of a line on the plane in this view. Since the view is projected from the point view of line *AB,* the line will appear true length. This view satisfies the condition that the line be true length and the plane an edge. The angle is measured in the third view. The piercing point is projected back to previous views and the visibility is determined to complete the solution.

Fig. 6.20 Angle between a line and a plane—plane method

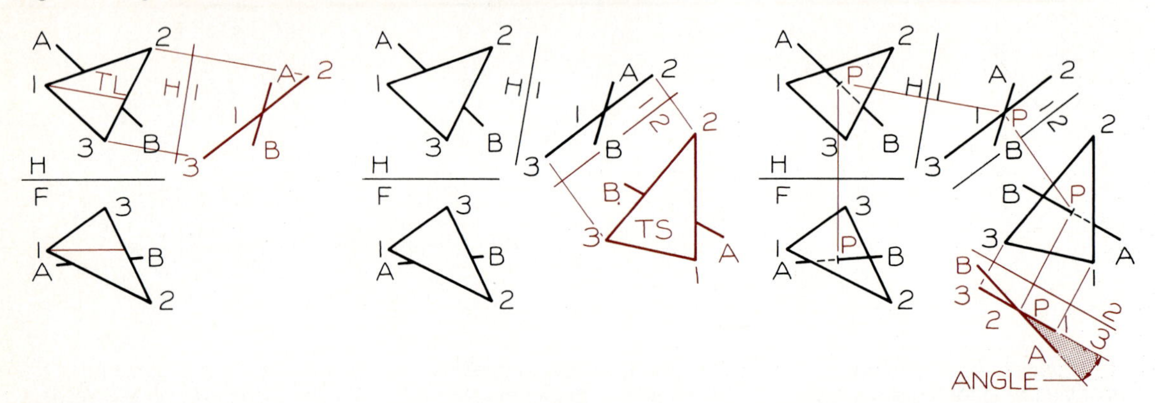

Step 1 The angle between a line and a plane can be measured in the view where the plane is an edge and the line is true length. The plane is found as an edge by projecting off the top view. The line is not true length in this view.

Step 2 The plane is found true size by projecting perpendicularly from the edge view of the plane.

Step 3 A view projected in any direction from a true-size view of a plane will show the plane as an edge. A third successive auxiliary view is projected perpendicularly from line *AB*. The line appears true length and the plane as an edge in this view. The angle is measured in this view. The piercing points and visibility are shown in the views by projecting back in sequence.

Fig. 6.21 Angle between a line and a plane—line method

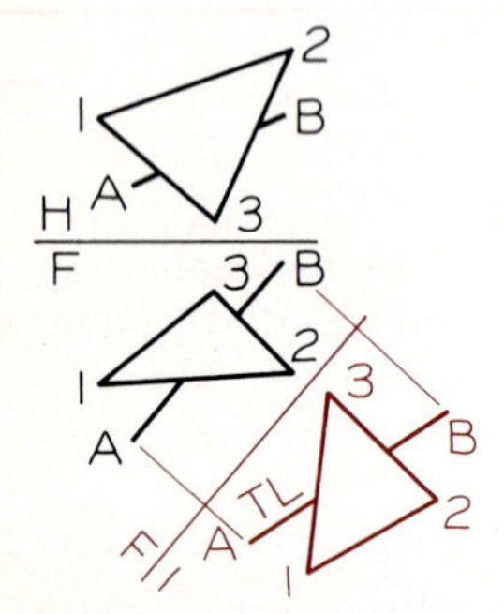

Step 1 The angle between a line and a plane can be measured in the view where the plane appears as an edge and the line is true length. Find a view where line *AB* is true length and project the plane to this view also.

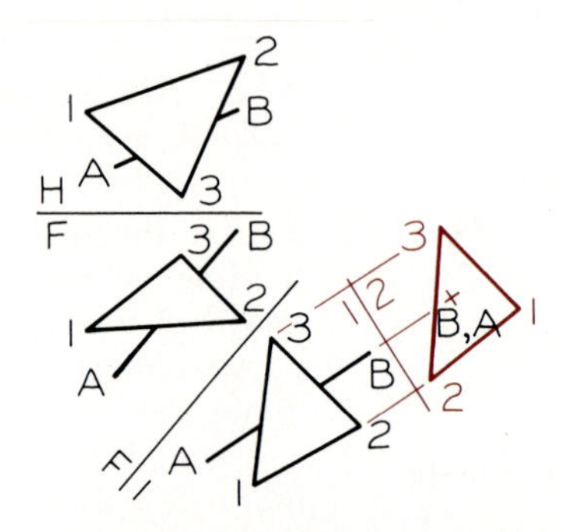

Step 2 Construct a point view of line *AB* in the secondary auxiliary view. Plane 1–2–3 does not appear true size in this view unless the line is perpendicular to the plane. The point view of the line in this view is the piercing point on the plane.

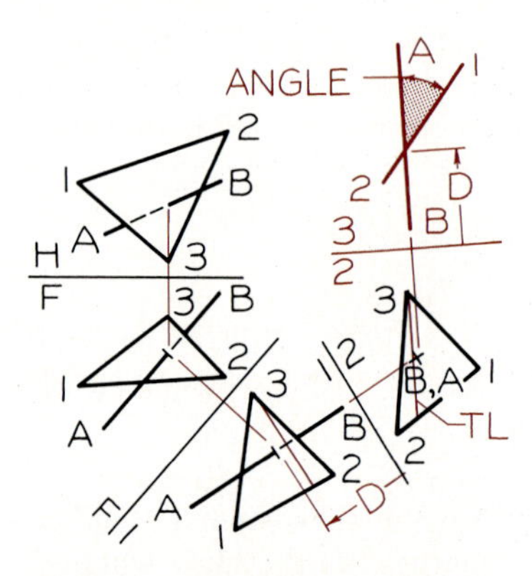

Step 3 An edge view of the plane is found by finding the point view of a line on the plane. Line *AB* will appear true length in this view since it was a point in the secondary auxiliary view. Measure the angle, project back to each view to locate the piercing point, and determine the visibility of the line.

6.13 ELLIPTICAL VIEWS OF A CIRCLE

Circles appear as circles only when your line of sight is perpendicular to the plane of the circle. When this line of sight is oblique to the plane, the circle will appear as an ellipse. The following definitions are given to explain the terminology of ellipses (Fig. 6.22).

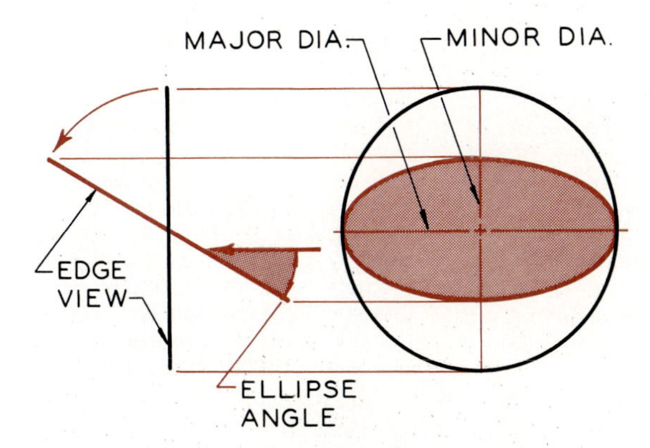

Fig. 6.22 The ellipse angle is the angle the line of sight makes with the edge view of a circle. An ellipse template can be used to construct the ellipse by aligning it with the major and minor axes.

Major diameter is the largest diameter measured across an ellipse. It is always true length.

Minor diameter is the shortest diameter across an ellipse and it is perpendicular to the major diameter at its midpoint.

Ellipse angle is the angle between the line of sight and the edge view of the plane of the circle.

Cylindrical axis is the centerline of a cylinder that connects the centers of the circular ends of a right cylinder.

Ellipse template is a template of various sizes of ellipses used to draw the ellipses by aligning the major and minor diameters. The templates are available in 5° intervals (Fig. 6.23).

It is required in Fig. 6.24 to draw a circle through the three vertexes of the triangular plane, and to show the circle in all views. This is done by finding the true size of the triangle and then finding the center of the circle where the perpendicular bisectors of the sides intersect. The circle is drawn in this view.

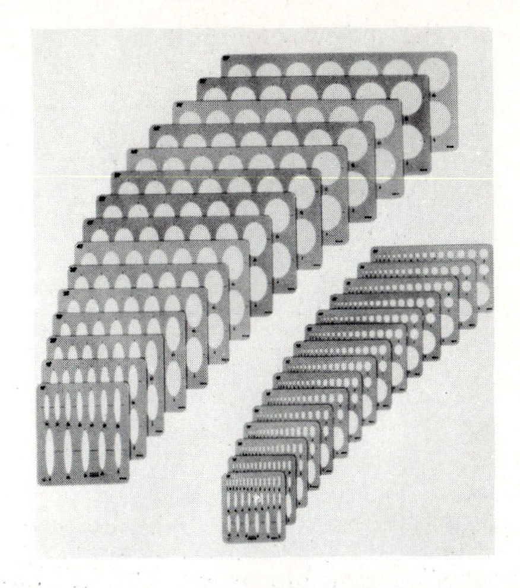

Fig. 6.23 Typical ellipse templates used for ellipse representation. (Courtesy of the A. Lietz Company.)

The circle is projected to the primary auxiliary view where it appears as an edge. The major and minor diameters are projected back to the top view where they are parallel and perpendicular to the H-1 reference line. The ellipse guide angle is found to be 45° since this is the angle the line of sight makes with the edge view of the circle.

The elliptical view of the circle is found in the front view projecting an edge view of the plane from the front view. The major and minor diameters are drawn parallel and perpendicular to the F-1 reference line.

> The angle of the ellipse template is the angle the line of sight makes with the edge view of the circle, 40°.

The elliptical ends of right cylinders will be perpendicular to the cylinder's axis (Fig. 6.25A). Consequently, the major diameters will be perpendicular to the axis. When the major diameter is not drawn perpendicular to the axis (Fig. 6.25B), it is apparent to even the untrained eye that the cylinder is not a right cylinder. That is, the ends of the cylinder are not perpendicular to the cylinder's axis.

Fig. 6.24 Elliptical views of a circle

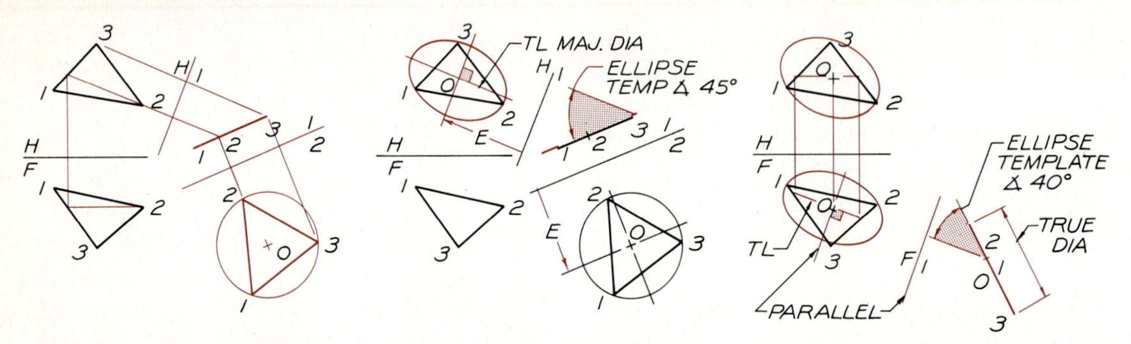

Step 1 To construct a circle that will pass through each vertex of a triangle, find plane 1–2–3 true size. The center of the circle is found at the intersection of the perpendicular bisectors of each side of the triangle.

Step 2 Draw diameters parallel and perpendicular to the 1–2 line in the secondary auxiliary view. Project these lines to the primary auxiliary and top views, where they will represent the major and minor diameters of an ellipse. Select the ellipse template for drawing the top view by measuring the angle between the line of sight and the edge view of the plane.

Step 3 Determine the ellipse template for drawing the ellipse in the front view by finding the edge view of the plane in an auxiliary view projected from the front view. The ellipse angle is measured in the auxiliary view as shown. The major diameter is true length and is parallel to a true-length line on the plane in the front view. The minor diameter is perpendicular to it.

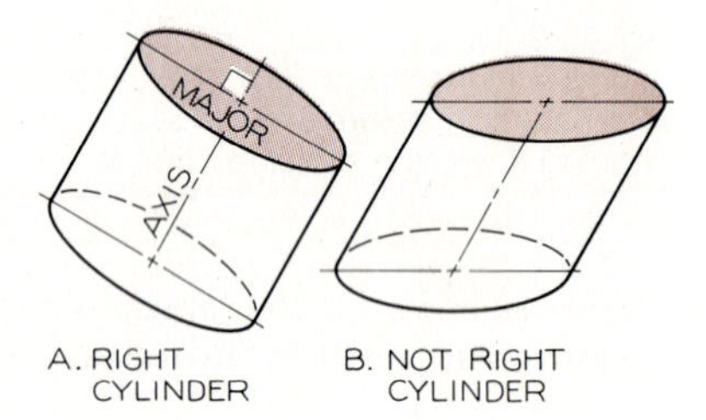

Fig. 6.25 The axis of the cylinder (the centerline) is drawn perpendicular to the major diameter of its elliptical end, if it is a right cylinder. It should be apparent to you by inspection that the cylinder at B is not a right cylinder.

PROBLEMS

Use Size A (8½″ × 11″) sheets for the following problems and lay out the problems using instruments. Each square on the grid is equal to 0.20″ or about 5 mm. The problems can be laid out on grid or plain paper. Label all reference planes and points in each problem with ⅛″ letters and/or numbers, using guidelines.

The crosses marked "1" and "2" are to be used for placing the primary and secondary reference lines. The primary reference line should pass through "1" and the secondary through "2."

1 and 2. Find the point views of the line in Fig. 6.26.

3 and 4. Find the angles between the planes in Fig. 6.26.

5 and 6. Find the true size views of the planes in Fig. 6.27. Project from the front view in Problem 5 and from the top view in Problem 6.

7 and 8. Find the angles between the lines in Fig. 6.28. Project from the top views of both problems.

9. Find the shortest distance from the point to the line in Fig. 6.29 and show the distance in all views. Use the plane method and project from the left side view. Scale: full size.

10. Find the shortest distance from point O to the line in Fig. 6.29 and show the distance in all

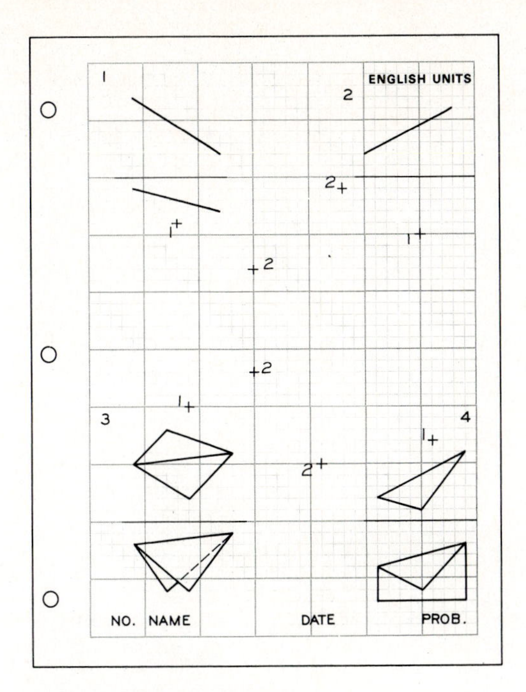

Fig. 6.26 Problems 1 through 4.

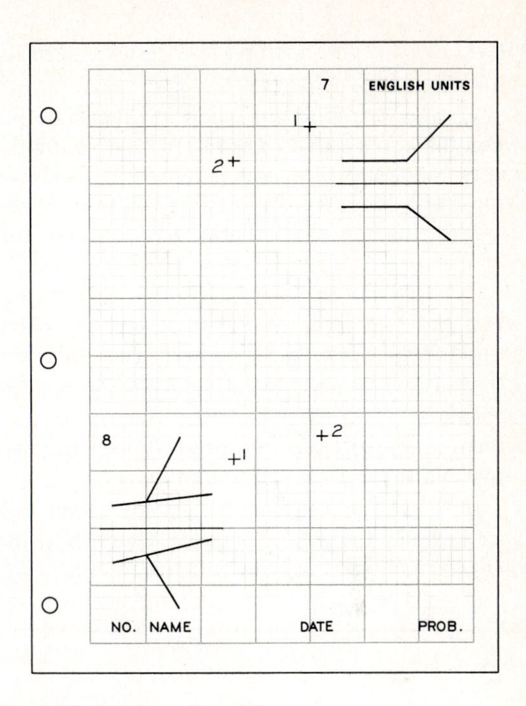

Fig. 6.28 Problems 7 and 8.

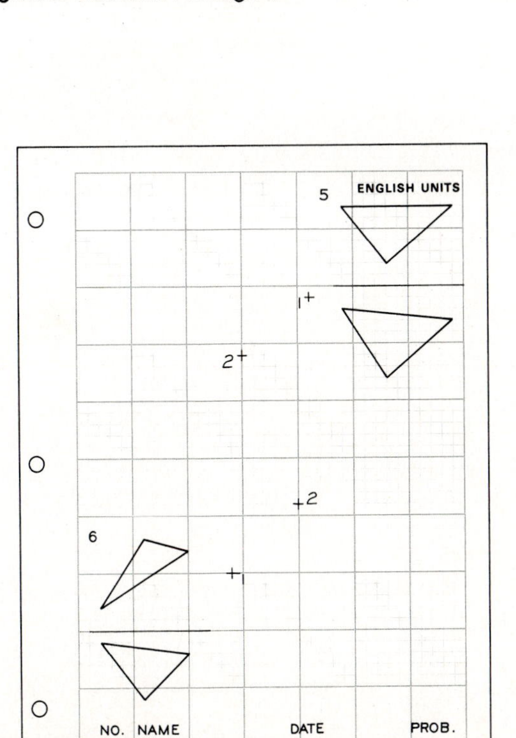

Fig. 6.27 Problems 5 and 6.

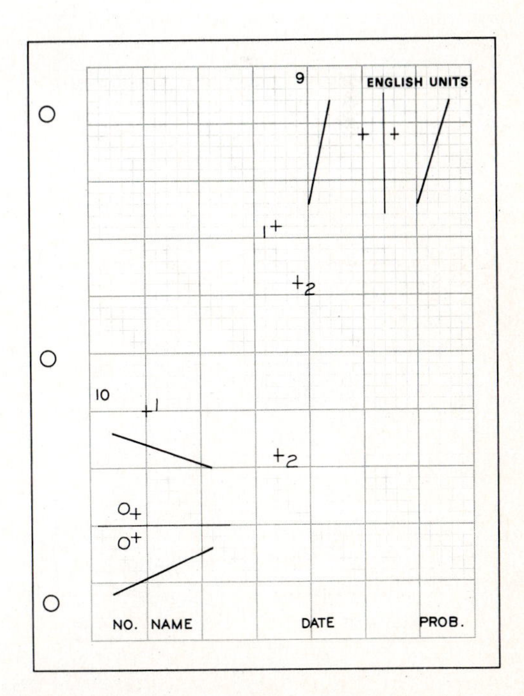

Fig. 6.29 Problems 9 and 10.

views. Use the line method and project from the top view. Scale: full size.

11 and 12. Find the shortest distances between the two skewed lines in Fig. 6.30 using the line method, and show the distances in all views. Begin each problem by finding line 3–4 true length, using the cross marks given. Scale: full size.

13. Find the shortest horizontal distance between the two lines in Fig. 6.31 by the plane method. Project from the top view. Scale: full size.

14. On a separate sheet of paper, redraw Problem 11 (Fig. 6.31) and find the shortest 20 percent grade between two lines. Project the first view from the top view. Scale: full size.

15. Find the shortest 25-percent grade distance between the two lines in Fig. 6.32. Show the distance in all views. Scale: full size.

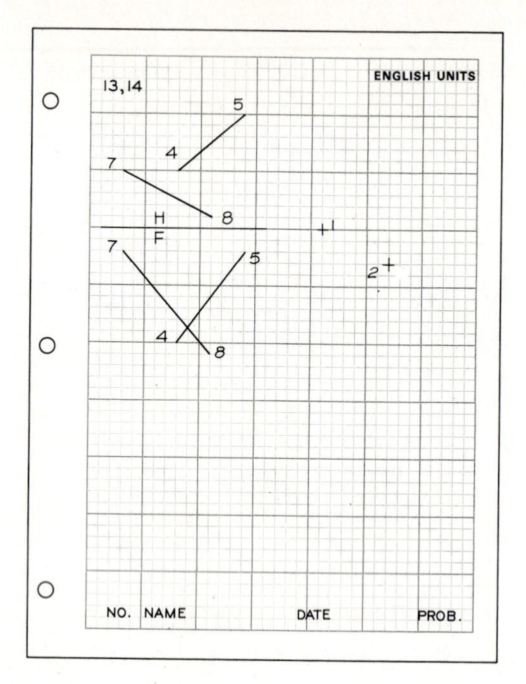

Fig. 6.31 Problems 13 and 14.

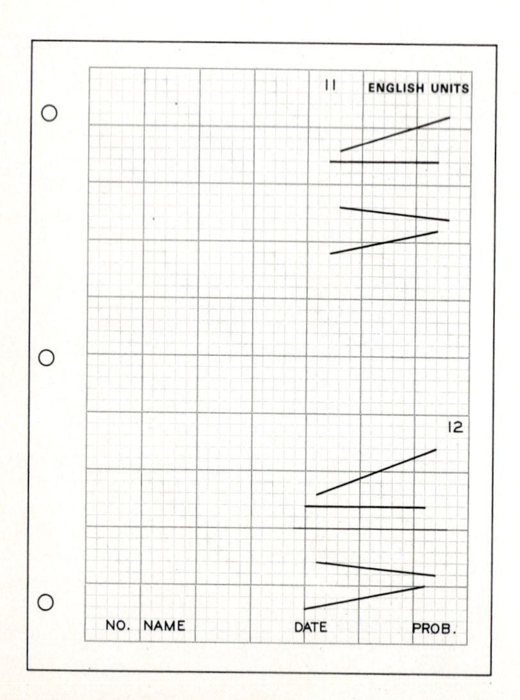

Fig. 6.30 Problems 11 and 12.

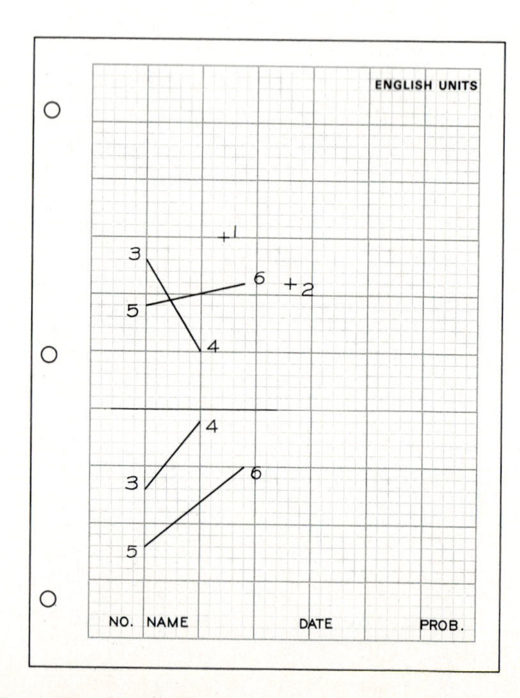

Fig. 6.32 Problem 15.

16. Find the connector from point O that will intersect line 1–2 at a 60° angle (Fig. 6.33). Show this line in all views. Project from the top view. Scale: full size.

17. Find the angle between the line and the plane in Fig. 6.34 by using the plane method. Project from the front view and show the visibility in all views. Scale: full size.

18. Same as Problem 17, except use the line method.

19. Construct a circle that will pass through each vertex of the triangle in Fig. 6.35. Project from the top view and show the elliptical views of the circle in all views.

20. Find the front view of the elliptical path of a circular section through the sphere in Fig. 6.35. The edge view of the section is shown in the top view.

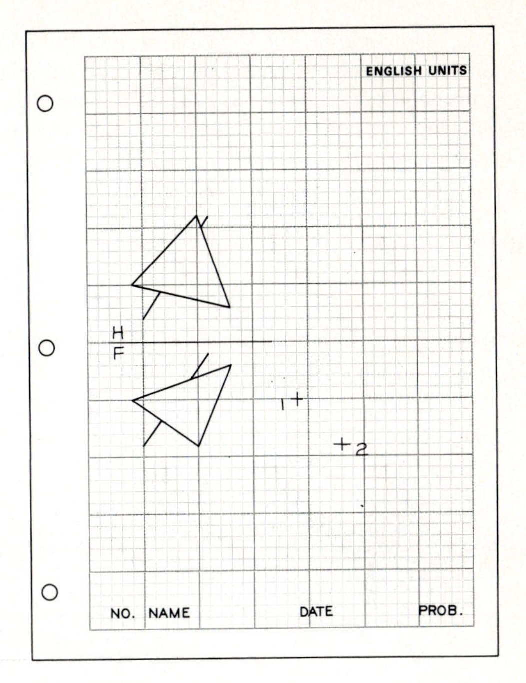

Fig. 6.34 Problems 17 and 18.

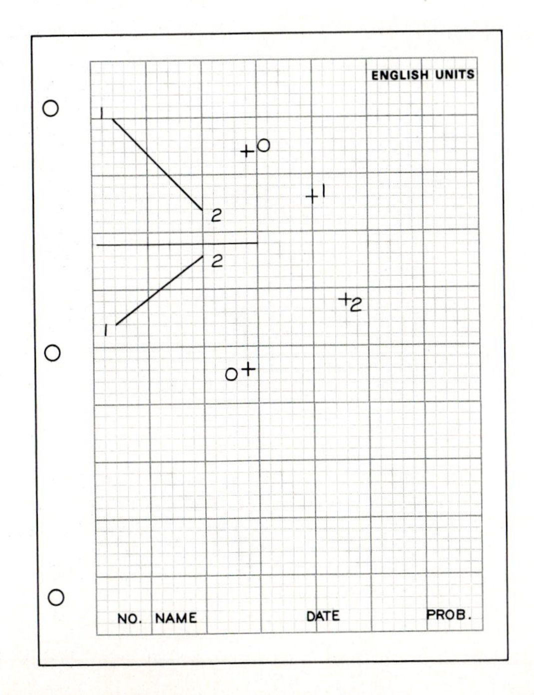

Fig. 6.33 Problem 16.

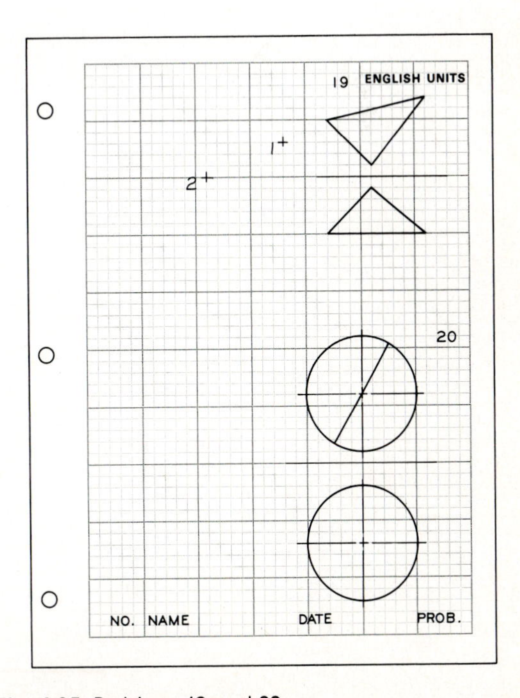

Fig. 6.35 Problems 19 and 20.

21. The line in Fig. 6.36 represents the centerline of a right cylinder that has each circular end perpendicular to the axis. Complete the views of the cylinder and show the ends, which will appear as ellipses.

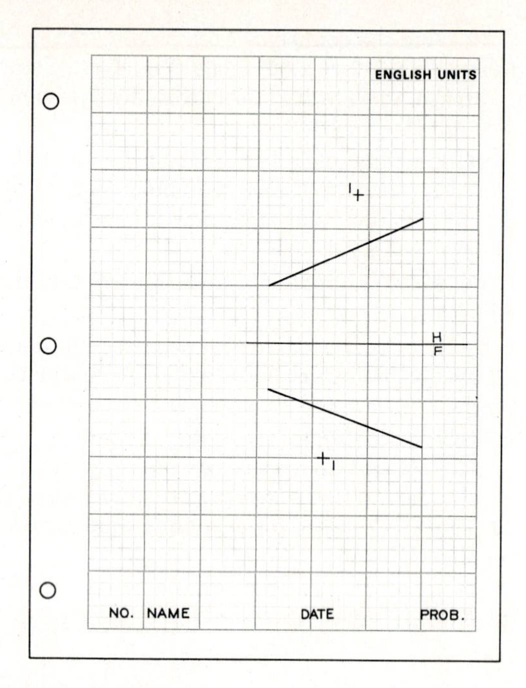

Fig. 6.36 Problem 21.

7

REVOLUTION

7.1 INTRODUCTION

The orthicon camera in Fig. 7.1 was designed to permit the camera to be revolved about three axes; thus it is possible to aim it in any direction for tracking space vehicles. This is just one of many designs that was based on the principles of revolution.

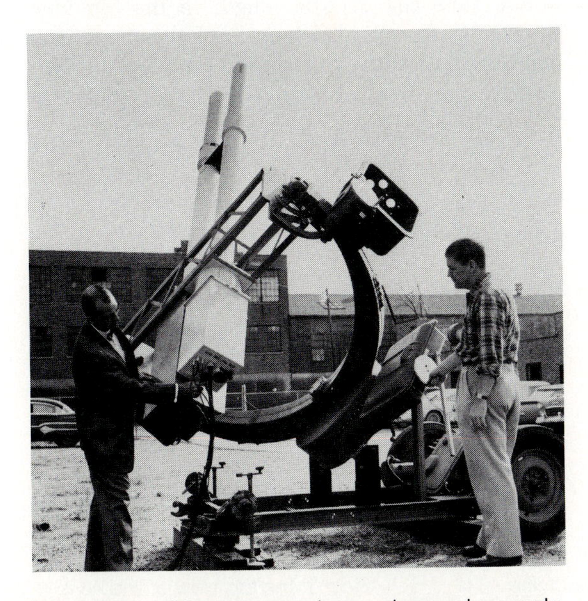

Fig. 7.1 This orthicon camera is an advanced example of a design that utilizes principles of revolution. Its cradle was designed to permit the camera to be revolved into any position by revolving it about three axes. (Courtesy of ITT Industrial Laboratories.)

Revolution is another method of solving problems that can be solved by auxiliary views. In fact, revolution techniques were developed and used before auxiliary views came into use. The understanding of revolution will reinforce your understanding of auxiliary-view principles, which is necessary for the solution of spatial problems.

7.2 TRUE LENGTH OF A LINE IN THE FRONT VIEW

The simple object in Fig. 7.2 is used to demonstrate how an inclined surface can be found true size by auxiliary view and by revolution. When the auxiliary view method is used, the observer changes position to an auxiliary vantage point where he or she can look perpendicularly at the inclined surface.

When the revolution method is used, the top view of the object is revolved about an axis until the edge view of the inclined plane is perpendicular to the standard line of sight from the front view. In other words, the observer's line of sight does not change, and the conventional lines of sight between adjacent orthographic views are used. Note that the height dimension, *H*, in Fig. 7.2B does not change when the object is revolved.

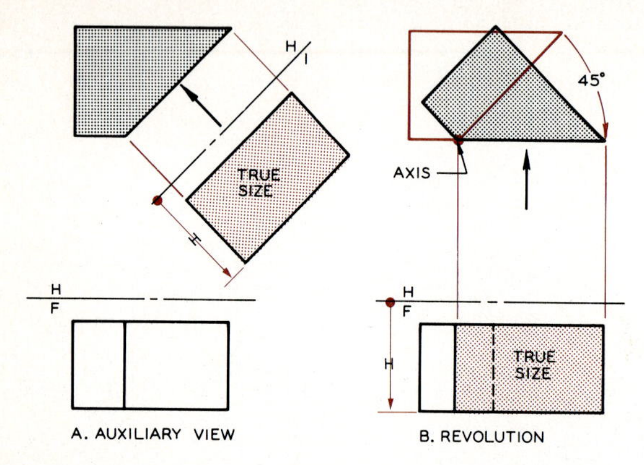

Fig. 7.2 The surface is found true size by an auxiliary view in Part A. At B, the surface is found true size by revolving the top view.

A single line can be found true length in the front view by revolution as shown in Fig. 7.3. By establishing the point view of an axis in the top view, line *AB* is revolved into a position that is parallel to the frontal plane. Therefore, the line will appear true length in the front view.

Note that the top view represents the circular base of a right cone and the front view is the triangular view of a cone. Line *AB′* is the outside element of the cone and is therefore true length.

Figure 7.4 illustrates the technique of finding line 1–2 true length in the front view. When in its first position, the observer's line of sight is not perpendicular to the triangular plane or line 1–2. But when it is revolved to be perpendicular to the line of sight, the triangle appears true size and line 1–2′ is true length.

7.3 TRUE LENGTH OF A LINE IN THE TOP VIEW

A surface that appears as an edge in the front view can be found true size in the top view by a primary auxiliary view or by a single revolution (Fig. 7.5). When revolution is used, you need not change your position, but the standard line of sight that gives the top view can be used.

The axis of revolution is located as a point in the front view, and is true length in the top view. The edge view of the plane is revolved until it is a horizontal edge in the front view (Fig. 7.5A). It is projected to the top view to find the surface true size. As in the auxiliary-view method, the depth dimension *(D)* does not change.

Line *CD* is found true length in the top view by revolving the line into a horizontal position in Step 2 of Fig. 7.6. The arc of revolution in the front view represents the base of a cone of revolution. Line *CD′* is true length in the top view since it is an outside element of the cone. Note that the depth dimension in the top view does not change.

Fig. 7.3 True length in the front view

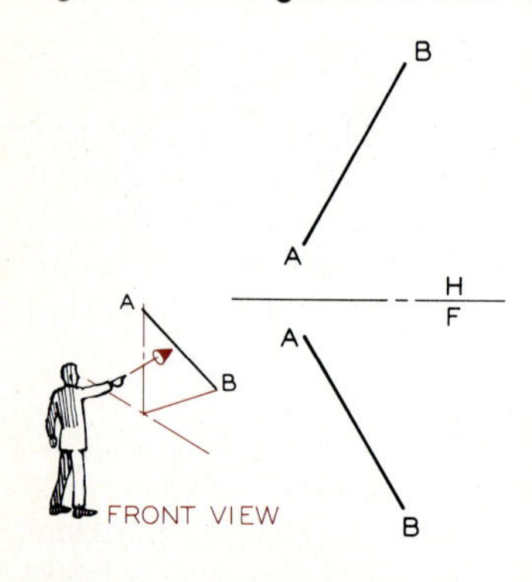

Given The top and front views of line *AB*.

Required Find the true-length view of line *AB* in the front view by revolution.

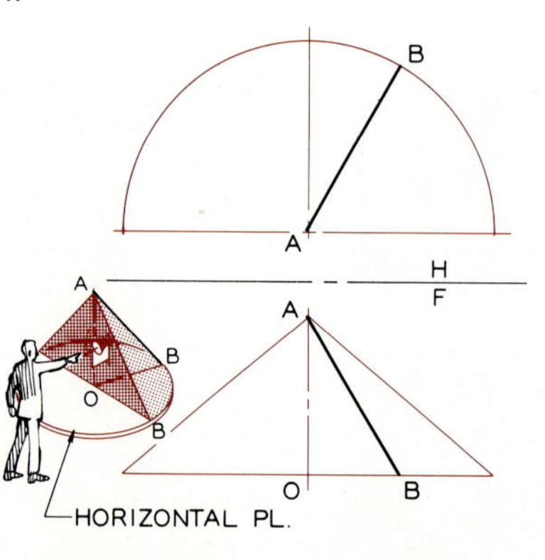

Step 1 The top view of line *AB* is used as a radius to draw the base of a cone with point *A* as the apex. The front view of the cone is drawn with a horizontal base through point *B*. Line *AO* is the axis of the cone.

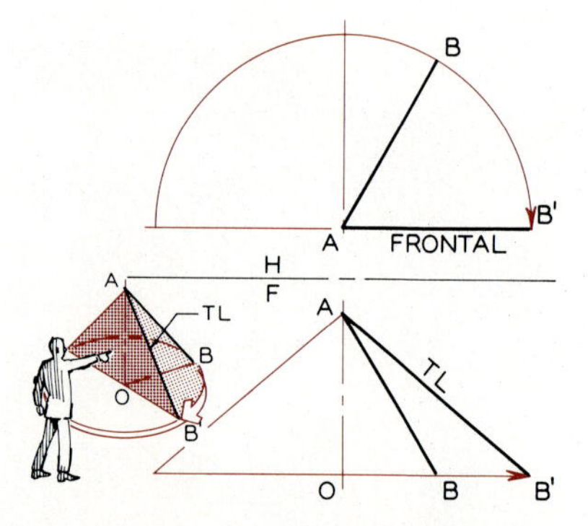

Step 2 The top view of line *AB* is revolved to be parallel to the frontal plane *AB′*. When projected to the front view, frontal line *AB′* is the outside element of the cone and is true length.

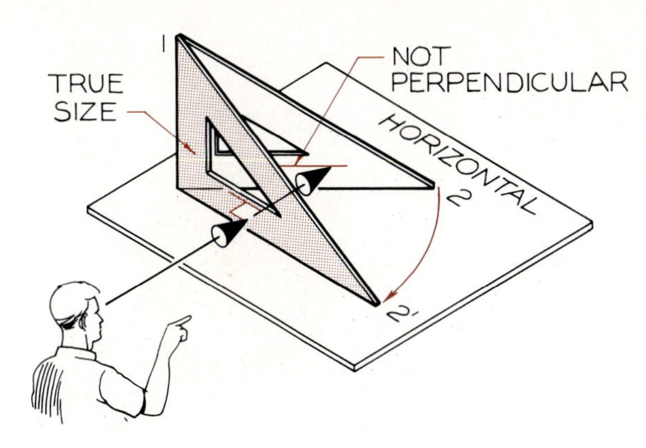

Fig. 7.4 Line 1–2 of the triangle does not appear true length in the front view because your line of sight is not perpendicular to it. When the triangle is revolved to a position where your line of sight is perpendicular to it, line 1–2′ can be seen true length.

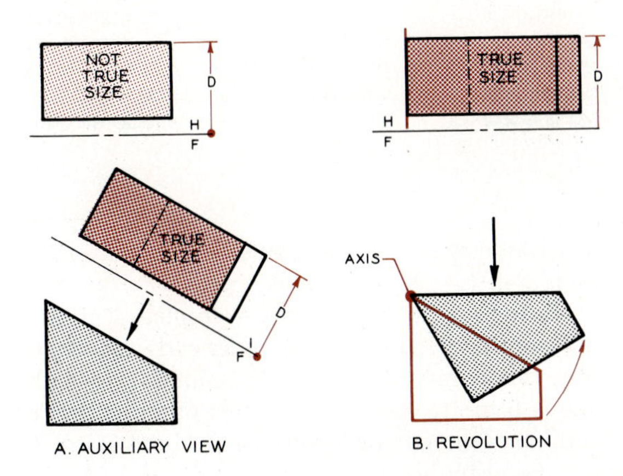

Fig. 7.5 At A, the inclined plane is found true size by an auxiliary with a line of sight that is perpendicular to the surface. At B, the surface is found true size by revolving the front view until it is perpendicular to the standard line of sight from the top view.

Fig. 7.6 True length of a line in the top view

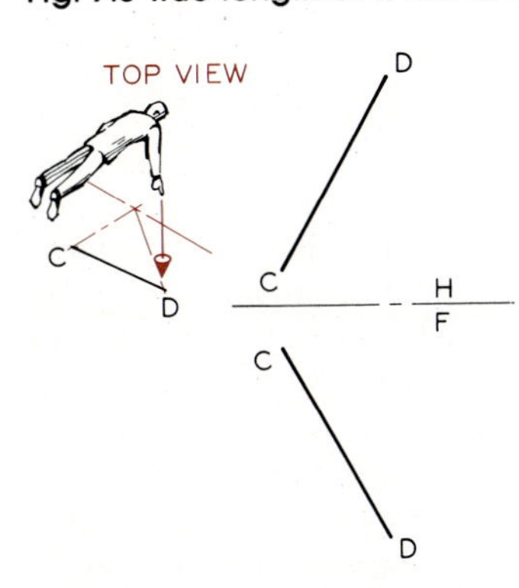

Given The top and front views of line *CD*.

Required Find the true-length view of line *CD* in the top view by revolution.

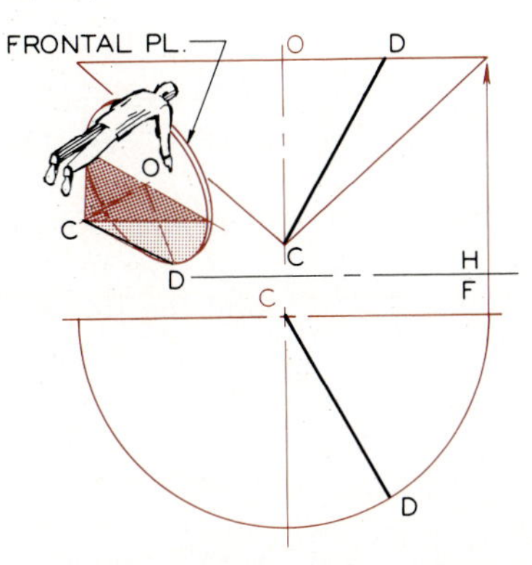

Step 1 The front view of line *CD* is used as a radius to draw the base of a cone with point C as the apex. The top view of the cone is drawn with the base shown as a frontal plane. The axis, *CO*, is perpendicular to the frontal base.

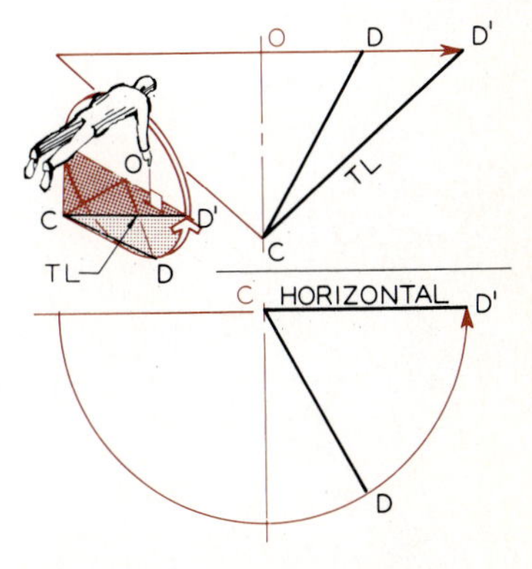

Step 2 The front view of line *CD* is revolved into position *CD′* where it is horizontal. When projected to the top view, *CD′* is the outside element of the cone and is true length.

Fig. 7.7 True length of a line in the side view

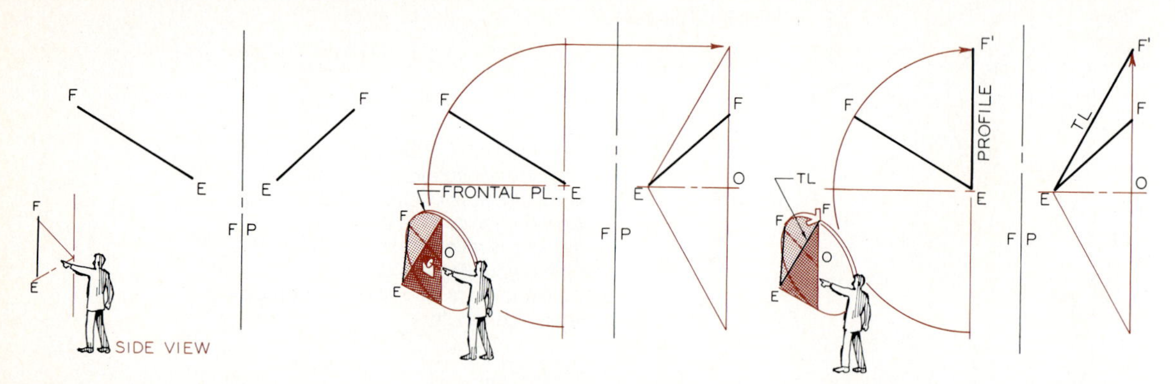

Given The front and side views of line *EF*.

Required Find the true-length view of line *EF* in the profile review by revolution.

Step 1 The front view of line *EF* is used as a radius to draw the circular view of the base of a cone. The side view of the cone is drawn with a base through point *F* that is a frontal edge.

Step 2 Line *EF* in the frontal view is revolved to position *EF'* where it is a profile line. Line *EF'* in the profile view is true length, since it is a profile line and the outside element of the cone.

7.4 TRUE LENGTH OF A LINE IN THE PROFILE VIEW

Line *EF* in Fig. 7.7 is found true length by revolving it in the front view until it is parallel to the edge view of the profile plane, Step 1. The circular view of the cone is projected to the side view, where the triangular shape of the cone is seen. Since *EF'* is a profile in line in Step 2, *EF'* is true length in the side view, where it is the outside element of the cone.

In the previously covered examples, each line has been revolved about one of its ends. However, a line can be revolved about any point on its length. Line 5–6 in Fig. 7.8 is an example of a line that is found true length by revolving it about point *O* near its midpoint.

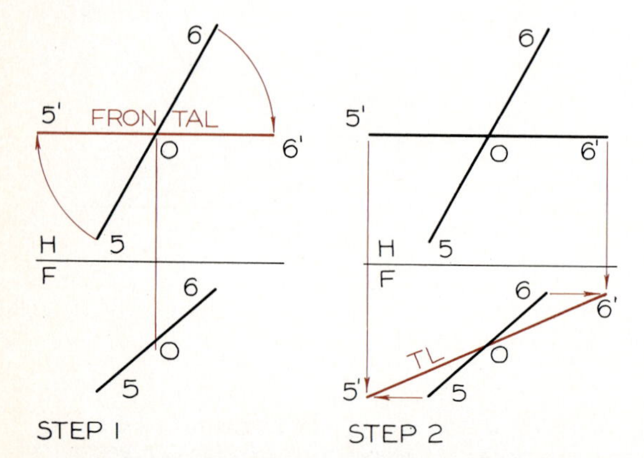

Fig. 7.8 In the preceding examples, the lines have been revolved about their ends, but they can be found true length by revolving them about any point on them. Line 5–6 is revolved into a frontal position in the top view and is found true length in the frontal view.

7.5 ANGLES WITH A LINE AND PRINCIPAL PLANES

The angle between a line and a plane will appear true size in the view where the plane is an edge and the line is true length. Two principal planes appear as edges in all principal views. Consequently, when a line appears true length in a principal view the angle between the line and two principal planes can be measured.

Two examples of finding lines true length by revolution are shown in Fig. 7.9. The angles with the principal planes can be measured in these views where the lines are true length.

The angle between the horizontal and the profile planes can be measured in part A in the front view. The angle with frontal and profile planes can be measured in the top view, part B.

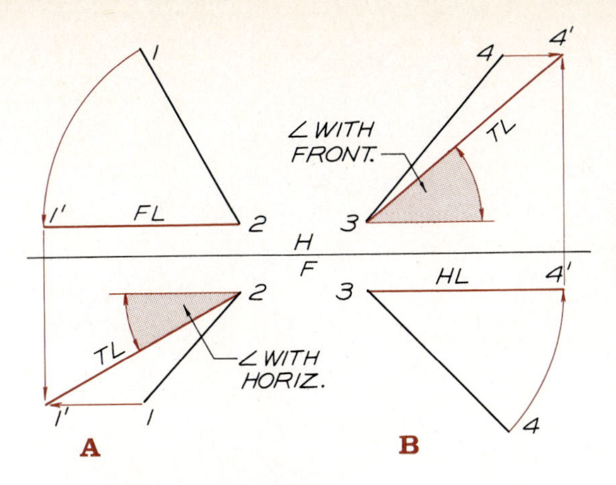

Fig. 7.9 Angles with principal planes

A. The angle with the horizontal plane can be measured in the front view if the line appears true length.

B. The angle with the frontal plane can be measured in the top view if the line appears true length.

Fig. 7.10 True size of a plane

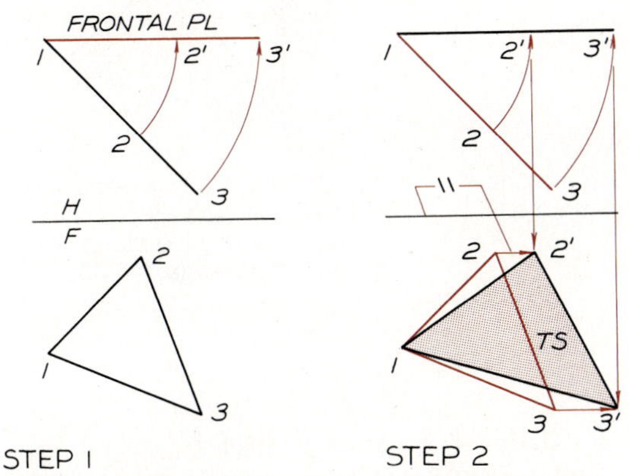

Step 1 The edge view of the plane is revolved to be parallel to the frontal plane.

Step 2 Points 2′ and 3′ are projected to the horizontal projectors from 2 and 3 in the front view.

7.6 TRUE SIZE OF A PLANE

A plane that appears as an edge in an orthographic view can be found true size by revolving the edge until it is parallel to a reference line between it and the adjacent view. In Fig. 7.10, the edge view in the top view is revolved until it is parallel to the frontal plane, Step 1.

In Step 2, the true-size view is found by projecting down from the revolved view and horizontally across the given front view.

A plane that does not appear as an edge in a given view is found true size by an auxiliary view and one revolution in Fig. 7.11. The plane is found as an edge (Step 1) and the edge is revolved to be parallel to the F–1 reference line (Step 2). The revolved view is found true size in Step 3.

Fig. 7.11 True size of a plane

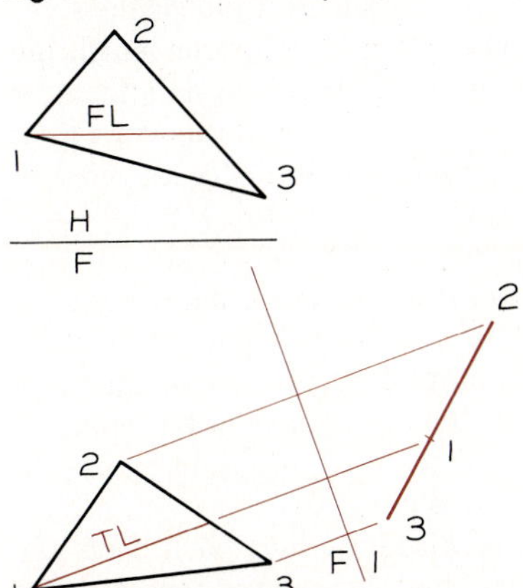

Step 1 An edge view of the plane is found by projecting from the front view.

Step 2 The plane is revolved to a position parallel to the F–1 reference line.

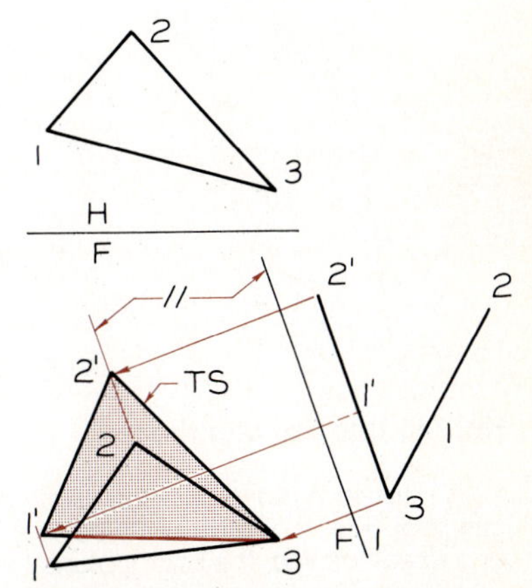

Step 3 Points 1′ and 2′ are found in the front view by projecting from the auxiliary view to find the true size of the plane.

Fig. 7.12 Edge view of a plane

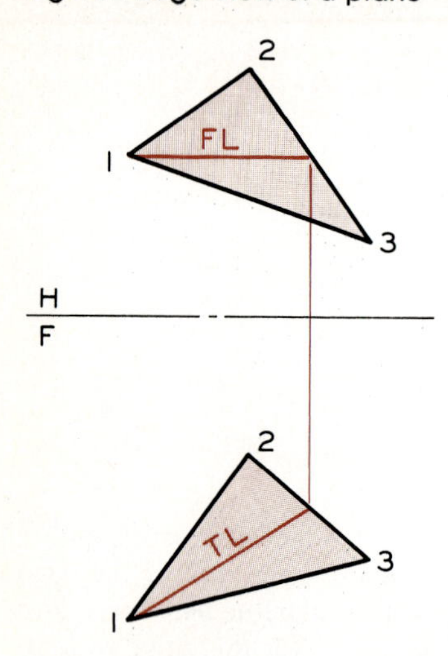

Step 1 It is required that we find the edge view of plane 1–2–3. A frontal line is found true length on the front view of the plane.

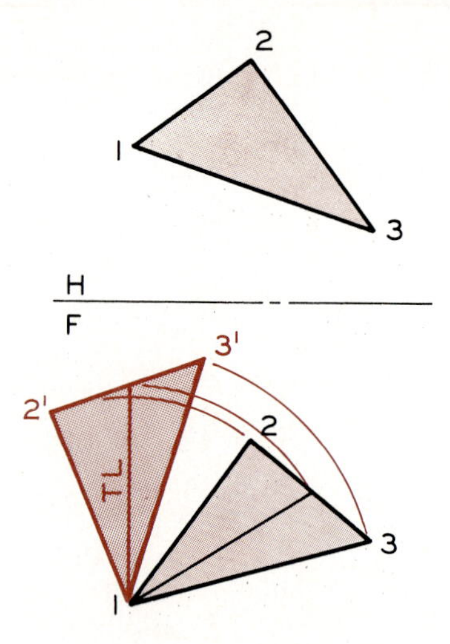

Step 2 The front view of the plane is revolved until the true-length line is vertical.

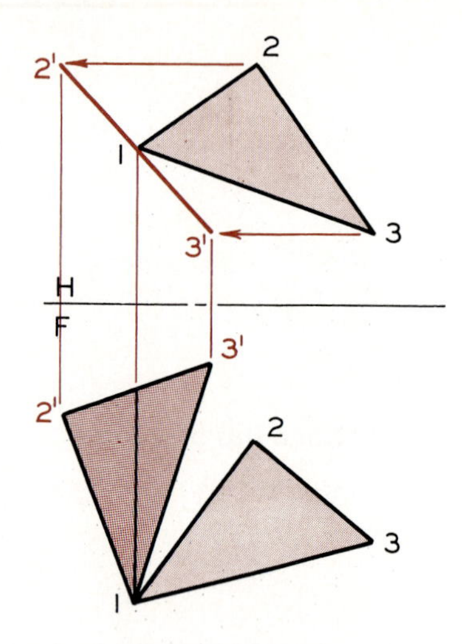

Step 3 Since the true-length line is vertical, it will appear as a point in the top view and the plane will appear as an edge, 1–2′–3′.

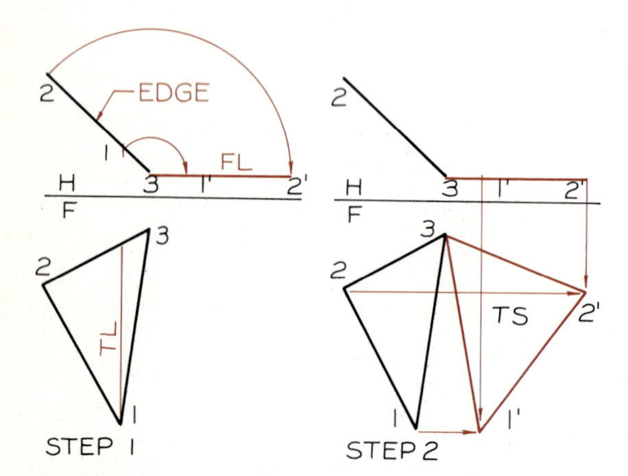

Fig. 7.13 True size of a plane

Step 1 When a plane appears as an edge in a principal view, it can be revolved to a position parallel to a reference line, the frontal plane in this case.

Step 2 Points 1′ and 2′ are projected to the front view to intersect with the horizontal projectors from the original points 1 and 2. The plane is true size in this view.

7.7 TRUE SIZE OF A PLANE BY DOUBLE REVOLUTION

The edge view of a plane can be found by revolution without the use of auxiliary views (Fig. 7.12). A frontal line is drawn on plane 1–2–3 and the line appears true length in the front view. The plane is revolved until the true-length line becomes vertical in the front view (Step 2). The true-length line will project as a point in the top view, and therefore, the plane will appear as an edge in this view (Step 3). Note that the projectors from the top view of points 2 and 3 are parallel to the *H-F* reference line.

A second revolution, called a **double revolution,** can be made to revolve this edge view of the plane until it is parallel to the frontal plane as shown in Fig. 7.13. The top views of the points 1′ and 2′ are projected to the front view where the plane 1′–2′–3 is true size.

This second revolution could have been performed in Fig. 7.12, but this would have resulted in an overlapping of view that would have made it difficult to observe the separate steps.

Double revolution is used in Fig. 7.14 to find the oblique plane of the object, plane 1–2–3, true

Fig. 7.14 True size by double revolution

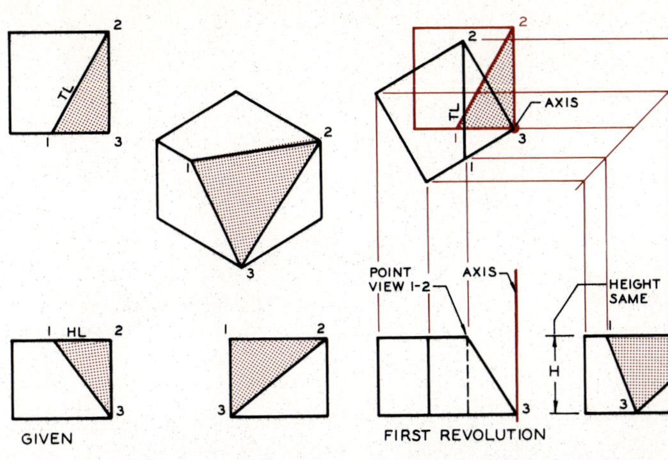

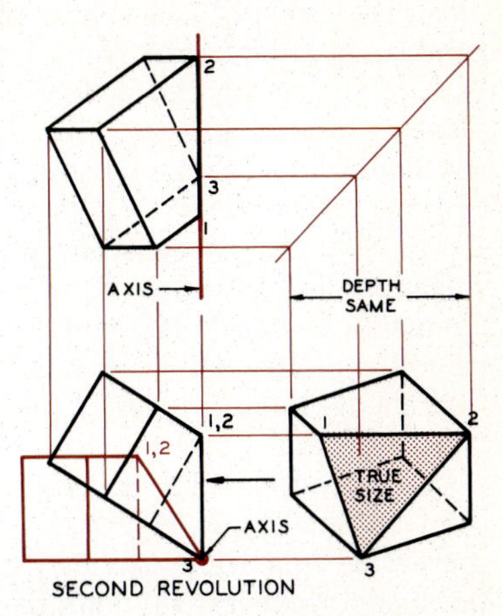

Given Three views of a block with an oblique plane across one corner.

Required Find the plane true size by revolution.

Step 1 Since line 1–2 is horizontal in the frontal view, it is true length in the top view. The top view is revolved into a position where line 1–2 can be seen as a point in the front view.

Step 2 Since plane 1–2–3 was found as an edge in step 1, this plane can be revolved into a vertical position in the front view, so that it will appear true size in the side view. The depth dimension does not change, since it is parallel to the axis of revolution.

Fig. 7.15 Double revolution

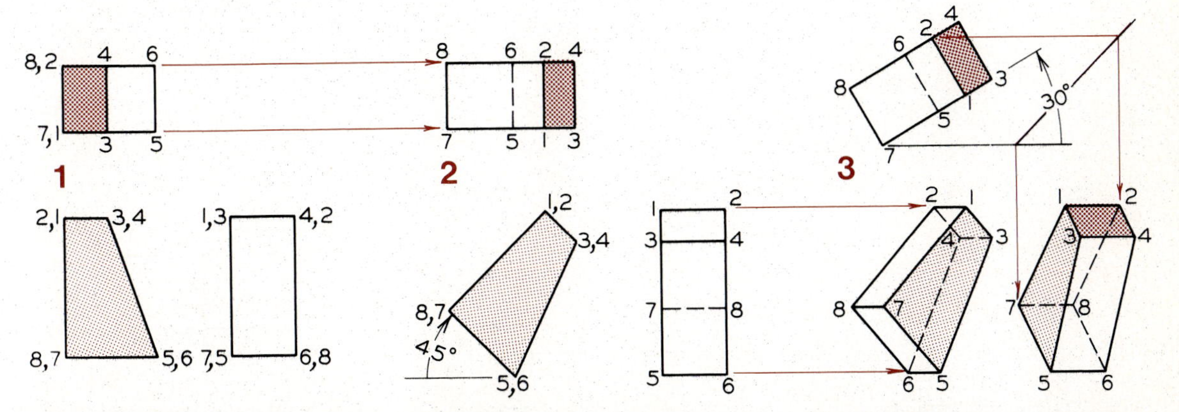

Given Three views of a surface with an inclined plane.

Step 1 The front view is revolved 45°, and the new width is projected to the top view where the depth is unchanged.

Step 2 The top view is revolved 30° to change the width and depth, but the height remains the same. The resulting front view is an axonometric pictorial.

size. In Step 1, the true-length line 1–2 on the plane is revolved in the top view until it is perpendicular to the frontal plane. Consequently, line 1–2 appears as a point in the front view, and the plane appears as an edge. This changes the width and depth dimensions, but the height dimension does not change.

In Step 2, the edge view of the plane is revolved into a vertical position parallel to the profile plane. The plane is found true size by projecting to the profile view where the depth dimension is unchanged and the height dimension has been increased.

A second example of double revolution of a solid is shown in Fig. 7.15 with dimensions transferred from view to view. The front view is revolved clockwise 45° and new top and side views are drawn with the depth dimension remaining constant. The top view is revolved 30° counterclockwise in Step 2. New front and side views are constructed by using projectors from the side view in Step 1 and the top view in Step 2. The resulting views are axonometric pictorials and none of the surfaces is true size.

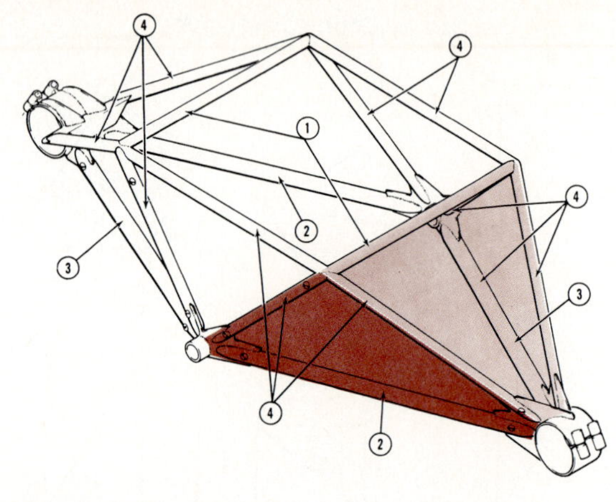

Fig. 7.16 The angle between any two planes of the helicopter engine mount can be found by using revolution principles. (Courtesy of Bell Helicopter Corporation.)

7.8 ANGLE BETWEEN PLANES

The engine mount frame of a helicopter is an application where the angle between two intersecting planes must be found, in order to provide its design specifications (Fig. 7.16). If orthographic views of this frame were given, the angles could be found by revolution.

In Fig. 7.17, the angle between two planes is found by drawing the edge view of the dihedral angle (the angle between the planes) perpendicular to the line of intersection and the plane of the angle is projected to front view (Step 1). The edge

Fig. 7.17 Angle between planes

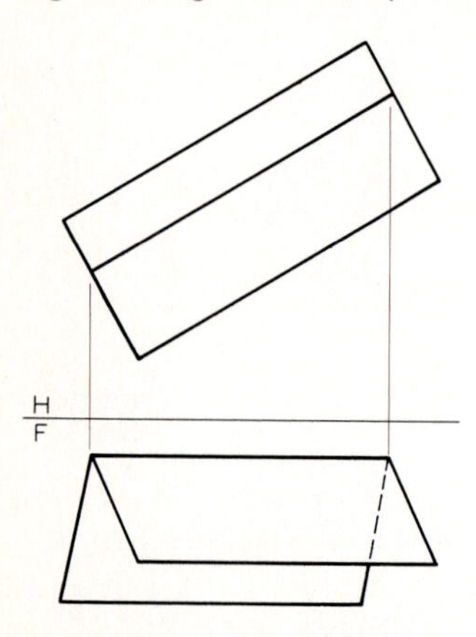

Given The top and front views of two intersecting planes.

Required Find the angle between the planes by revolution

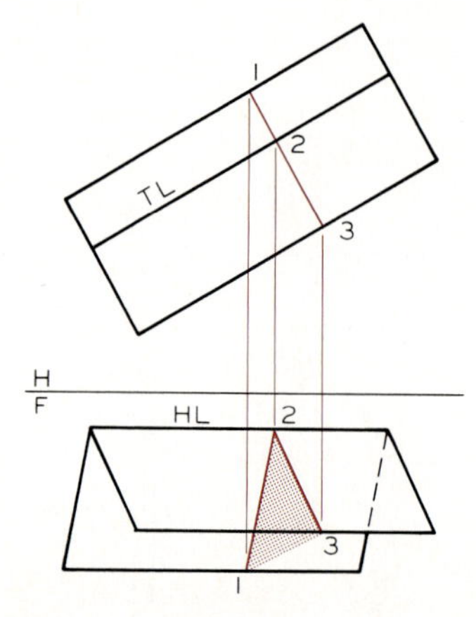

Step 1 A right section is drawn perpendicular to the true-length line of intersection between the planes in the top view and is projected to the front view. The section is not true size in the front view.

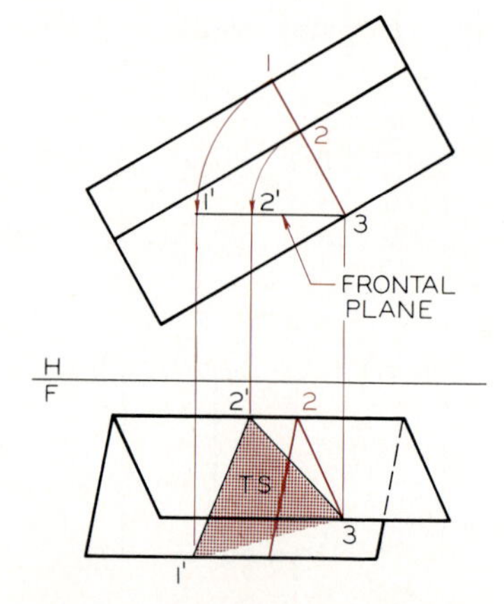

Step 2 The edge view of the right section is revolved to position 1′–2′–3 in the top view to be parallel to the frontal plane. This section is projected to the front view, where it is true size since it is a frontal plane.

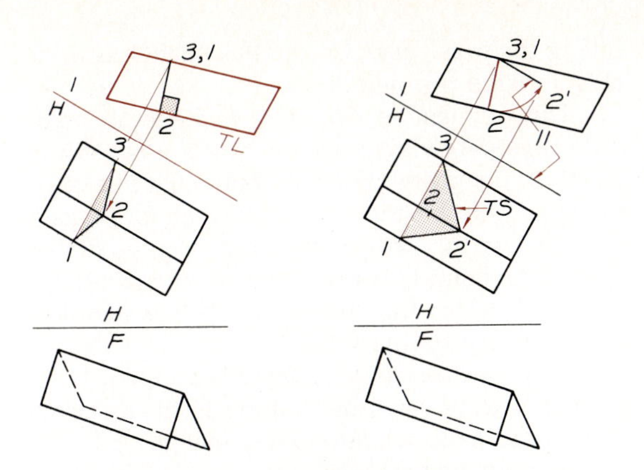

Fig. 7.18 Angle between oblique planes

Step 1 A true-length view of the line of intersection is found in an auxiliary view projected from the top view. The right section is constructed perpendicular to the true length of the line of intersection and is projected to the top view.

Step 2 The edge view of the right section is revolved to be parallel to the H-1 reference line so the plane will appear true size in the top view after being revolved. The angle between the planes can be found by measuring angle 1–2′–3.

view of the plane in the top view is revolved until it is a frontal plane in the top view, and it is then projected to the front view where its true-size view is found (Step 2).

A similar problem is solved in Fig. 7.18. In this example the line of intersection does not appear true length in one of the given views; consequently an auxiliary view is used in Step 1 to find the line of intersection true length. The plane of the angle between the planes can be drawn as an edge perpendicular to the true-length line of intersection (Step 1). The foreshortened view of plane 1–2–3 is projected to the top view in Step 1. The edge view of plane 1–2–3 is then revolved in the primary auxiliary view until it is parallel to the H-1 line (Step 2). When it is projected back to the top view, angle 1–2′–3 is true size.

7.9 LOCATION OF DIRECTIONS

It is necessary that you be able to locate the basic directions of up, down, forward, and back in any view that you are given in order to solve more ad-

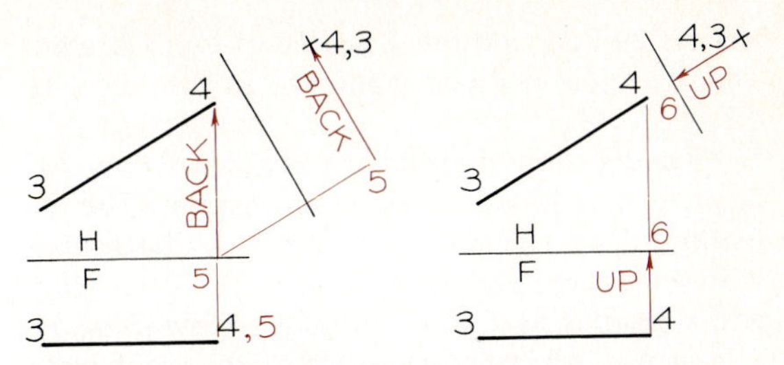

Fig. 7.19 To find the direction of back, forward, up, and down in an auxiliary view, construct an arrow pointing in the desired direction in the given principal views and project this arrow to the auxiliary view. The directions of back and up are shown here.

vanced problems of revolution. In Fig. 7.19, the directions of back and up are located by first drawing directional arrows in the given top and front views.

Line 4–5 is drawn pointing back in the top view, and its front view appears as a point. This directional arrow, 4–5, is projected to the auxiliary view as any other line would be drawn to locate the direction of back. By drawing the arrow on the other end of the line, you would find the direction of forward.

The direction of up is located in Fig. 7.19 by drawing line 4–6 in the direction of up in the front view and in the top view as a point. The arrow is found in the primary auxiliary by the usual pro-

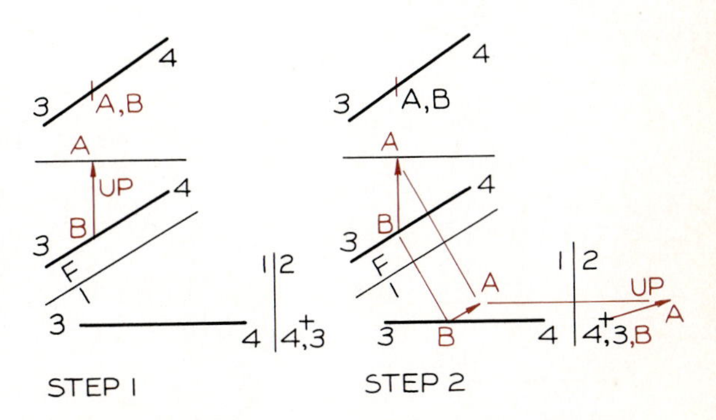

Fig. 7.20 Direction in a secondary auxiliary view

Step 1 To find the direction of up in the secondary auxiliary view, arrow *AB* is drawn pointing upward in the front view. It appears as a point in the top view.

Step 2 Arrow *AB* is projected to the primary and secondary auxiliary views like any other line. The direction of up is located in the secondary auxiliary view.

jection method, and the direction of up is located. The direction of down would be in the opposite direction.

The location of directions in secondary auxiliary views is found in the same manner. The direction of up is found in Fig. 7.20 by beginning with an arrow pointing upward in the front view and appearing as a point in the top view (Step 1). The arrow, *AB*, is projected from the front view to the primary and then to a secondary auxiliary view to give the direction of up in all views. The other directions can be found in the same manner by beginning with the two given principal views of a known directional arrow.

7.10 REVOLUTION OF A POINT ABOUT AN AXIS

In Fig. 7.21 it is required that point O be revolved about axis 3–4 into its most forward position. The circular path of revolution is drawn in the primary auxiliary view where the axis is a point (Step 1). The direction of forward is drawn in Step 2 and the new location of point O is found at O′. By projecting back through the successive views, point O′ is found in each view. Note that O′ lies on the line in the front view; this verifies that O′ is in its most forward position.

The problem in Fig. 7.22 requires an additional auxiliary view since the axis 3–4 is not true length in the given views. Therefore the line must be found true length before it can be found as a point where the path of revolution can be drawn as a circle. Point O is revolved into its highest position, O′, where the "up" arrow, 3–5, is found in the secondary auxiliary view.

By projecting back to the given views, O′ is located in each view. Its position in the top view is over the axis, which verifies that the point is located at its highest position.

The paths of revolution will appear as edges when the axis is true length, and as ellipses when the axis is not true length. The angle of the ellipse guide for drawing the ellipse in the front view is the angle the projectors from the front view make with the edge view of the revolution in the primary auxiliary view. To find the ellipse in the top view, an auxiliary view must be used to find the path of revolution as an edge perpendicular to the true-length axis projected from the top view.

The handcrank of a casement window (Fig. 7.23) is an example of a problem that must be solved using this principle to determine the clearances between the sill and the window frame.

Fig. 7.21 Revolution about an axis

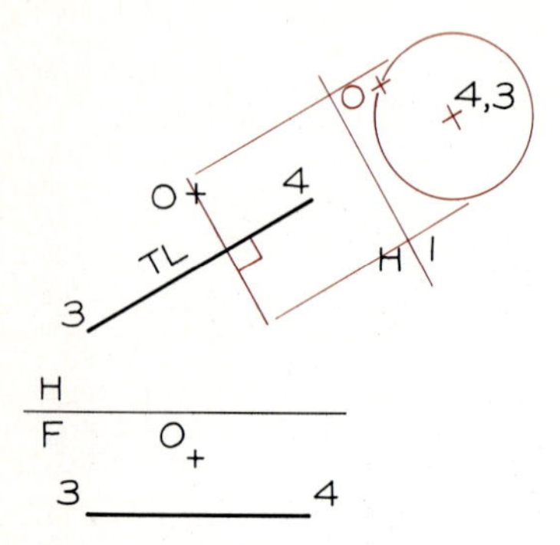

Step 1 To rotate point O about axis 3–4, it is necessary to find the point view of the axis in a primary auxiliary view. The circular path is drawn and the path of revolution is shown in the top view as an edge that is perpendicular to the axis.

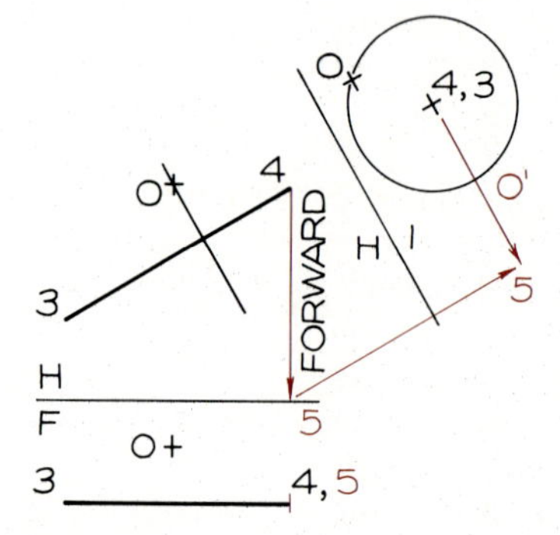

Step 2 If it is required to rotate point O to its most forward position, draw an arrow pointing forward in the top view. It will appear as a point in the front view. The arrow, 4–5, is found in the auxiliary view to locate point O′.

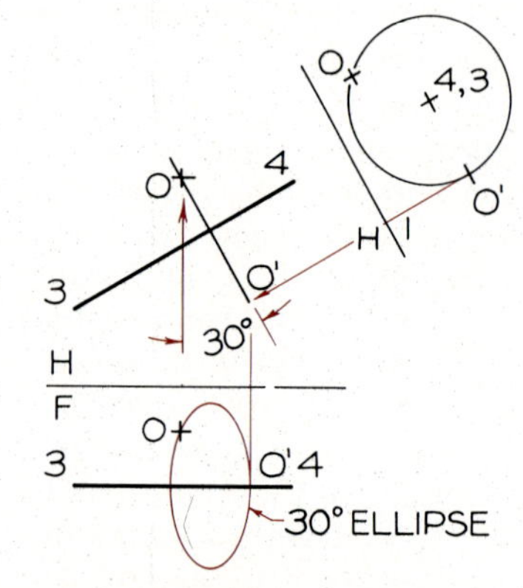

Step 3 Point O′ is projected back to the given views. The path of revolution appears as an ellipse in the front view since the axis is not true length in this view. A 30° ellipse is drawn since this is the angle your line of sight makes with the circular path in the front view.

Fig. 7.22 Revolution of a point about an axis

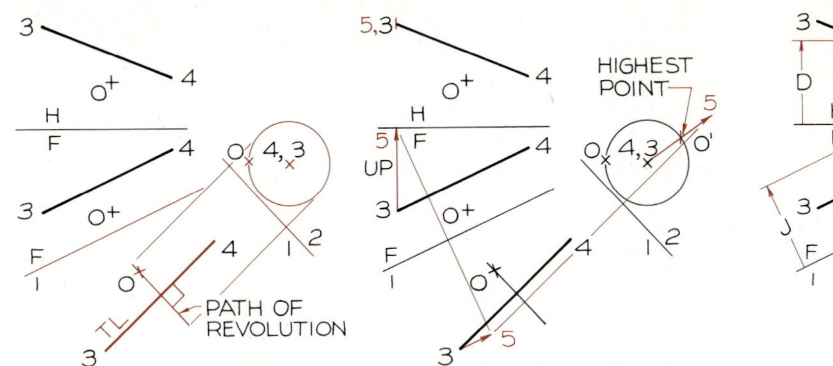

Step 1 To rotate O about axis 3–4, the axis is found as a point in a secondary auxiliary view where the circular path is drawn. The path appears as an edge in the primary auxiliary view where the axis is true length.

Step 2 If it is required to rotate O to its highest position, construct an arrow 3–5 in the front and top views that points upward. The direction of this arrow in the secondary auxiliary view locates the highest position, O′.

Step 3 Point O′ is projected back to the given views by transferring the dimensions *J* and *D* using your dividers. The highest point lies over the line in the top view to verify its position. The path of revolution is elliptical wherever the axis is not true length.

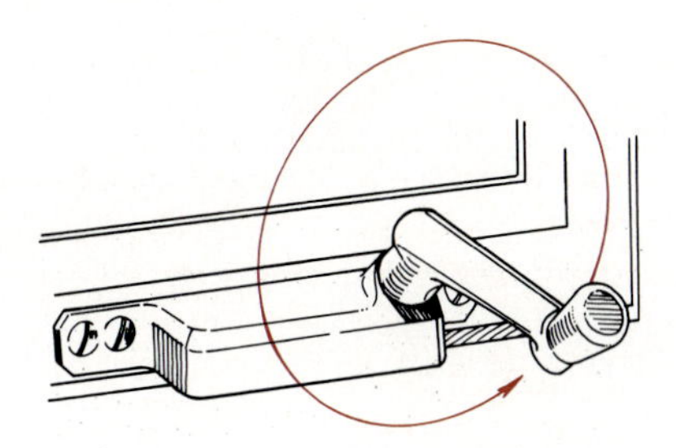

Fig. 7.23 This handcrank that is used on casement windows is an example of a problem that must be solved by using revolution principles. The handle must be properly positoned so to not interfere with the window sill or wall.

7.11 REVOLUTION OF A LINE ABOUT AN AXIS

Line 3–4 is revolved about axis 1–2 in Fig. 7.24. The point view of the axis 1–2 is found as a point in Step 1. A circle is drawn tangent to line 3–4 with its center at the point view of 1–2, and arcs are drawn through each end of the line as well. The perpendicular is revolved into the desired position and the new endpoints are found, 3′ and 4′.

The top vew of line 3′–4′ is found by projecting parallel to the H-1 line from the original points of 3 and 4 in the top view, as shown in Step 2. These projectors intersect those from the auxiliary view. The front view is obtained by projecting from the top view and transferring the height dimensions from the primary auxiliary view (Step 2).

7.12 REVOLUTION OF A RIGHT PRISM ABOUT ITS AXIS

Conveyors, ducts, and stairways are used to connect various parts of industrial installations. Such an example is the coal chute between two buildings shown in Fig. 7.25, which is used to convey coal at a continuous rate. You can easily understand why it is necessary to have the sides of the enclosed chute vertical and the bottom of the chute's right section horizontal in order for the coal to be transported efficiently.

In Fig. 7.26, it is required that the right section be positioned about centerline *AB* so that two of its sides will be vertical. This is done by finding the point view of the axis in Step 1, and the direction of up is projected to this view. In Step 2, the right section is found in the other views in Step 2. In Step 3, the sides of the chute are constructed parallel to the axis.

The bottom of this chute will be horizontal, and will be properly positioned for conveying materials such as coal.

Fig. 7.24 Revolution of a line about an axis

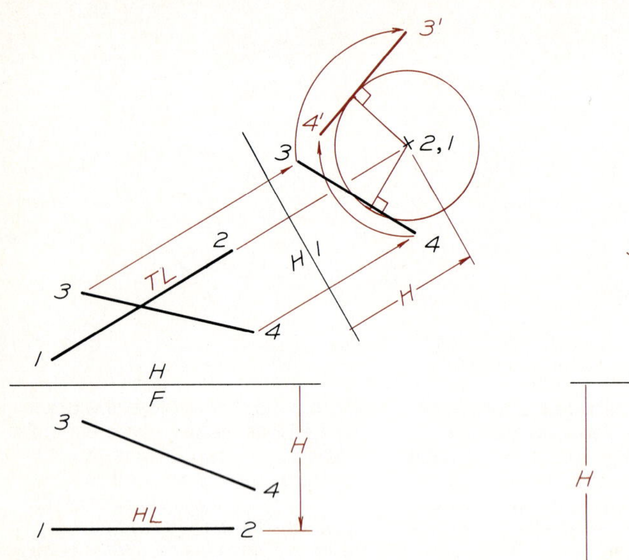

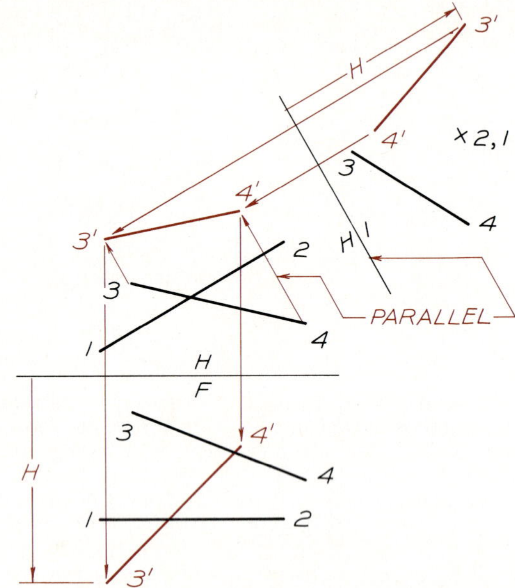

Step 1 The axis, 1–2, is found as a point in the auxiliary view. Line 3–4 is revolved to its specified position in this view.

Step 2 The new position of line 3–4 is projected back to the top view where projectors parallel to the top view of the H-1 reference line intersect those from the auxiliary view to find line 3′–4′. Line 3′–4′ is located in the front view.

Fig. 7.25 A conveyor chute must be properly installed so that two edges of its right section are vertical so the conveyors will function properly. This requires the application of the revolution of a prism about its axis. (Courtesy of Stephens-Adamson Manufacturing Company.)

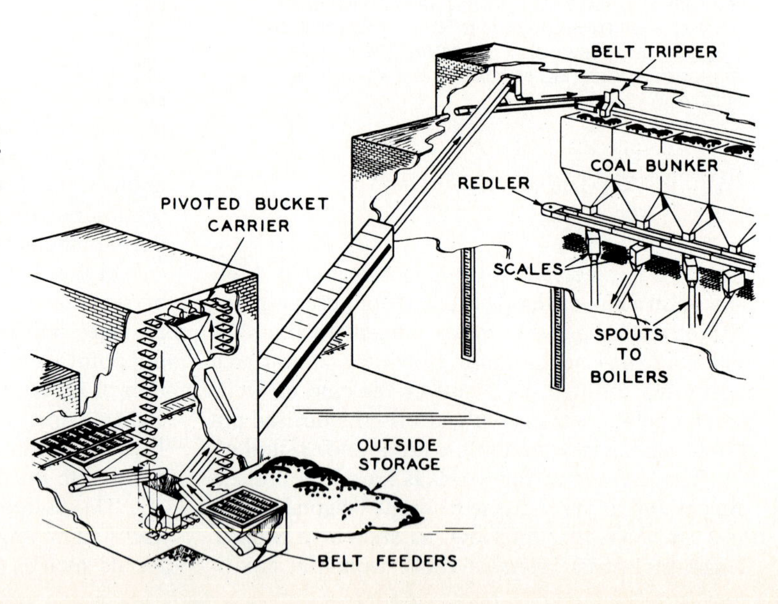

Fig. 7.26 Revolution of a prism about its axis

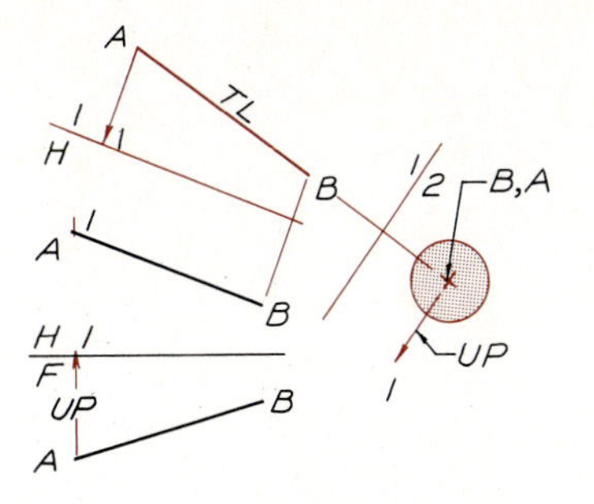

Step 1 Locate the point view of centerline AB in the secondary auxiliary view by drawing a circle about the axis with a diameter equal to one side of the square right section. Draw a vertical arrow in the front and top views and project it to the secondary auxiliary view to indicate the direction of vertical in this view.

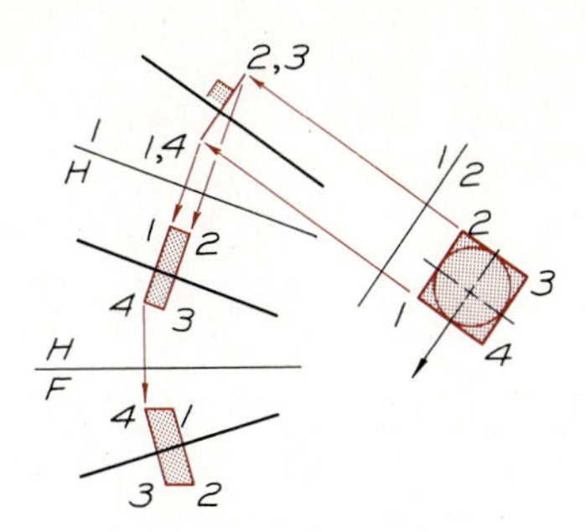

Step 2 Draw the right section, 1–2–3–4, in the secondary auxiliary view with two sides parallel to the vertical directional arrow. Project this section back to the successive views by transferring measurements with dividers. The edge view of the section could have been located in any position along centerline *AB* in the primary auxiliary view, so long as it was perpendicular to the centerline.

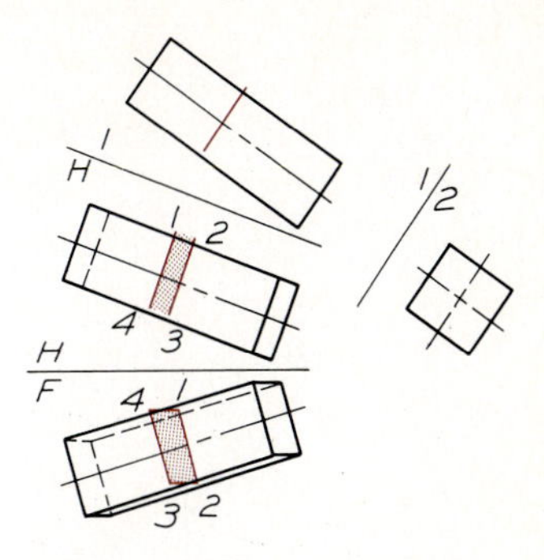

Step 3 Draw the lateral edges of the prism through the corners of the right section so that they are parallel to the centerline in all views. Terminate the ends of the prism in the primary auxiliary view where they appear as edges that are perpendicular to the centerline. Project the corner points of the ends to the top and front views to establish the ends in these views.

7.13 ANGLE BETWEEN A LINE AND PLANE

The angle between the line and plane is found by a combination of auxiliary views and revolution in Fig. 7.27. The plane is found true size in a secondary auxiliary view in Step 1.

The line is revolved in Step 2 until it is parallel to the 1–2 reference line. The line can then be found true length in the primary auxiliary view in Step 3. Since the line appears true length and the plane as an edge in this view, the true angle can be measured here.

7.14 A LINE AT A SPECIFIED ANGLE WITH TWO PRINCIPAL PLANES

In Fig. 7.28 it is required that a line be drawn through point O that will make angles of 35° with the frontal plane and 44° with the horizontal plane, and slopes forward and downward.

The cone containing elements making 35° with the frontal plane is drawn in Step 1. In Step 2, the cone with elements making 44° with the horizontal plane is drawn. In Step 2 the length of the elements of the cone must be equal to elements of the first cone so that the cones will intersect with equal elements.

Lines 0–1 and 0–2 are found in Step 3 where these lines are elements that lie on each cone and make the specified angle with the principal planes.

These principles can be applied to the determination of intersections between piping systems that are joined by standard connectors.

7.15 REVOLUTION OF PARTS ON DETAIL DRAWINGS

It is standard practice to revolve features such as those shown in Fig. 7.29 to make the views more descriptive. The front view of the part at A is true size because the top view has been revolved. Similarly, the front view of the part at B is more descriptive with the revolution in the top view.

These are called conventional practices, and they should be used to conserve effort while gaining additional clarity in the preparation of working drawings.

Fig. 7.27 Angle between a line and plane

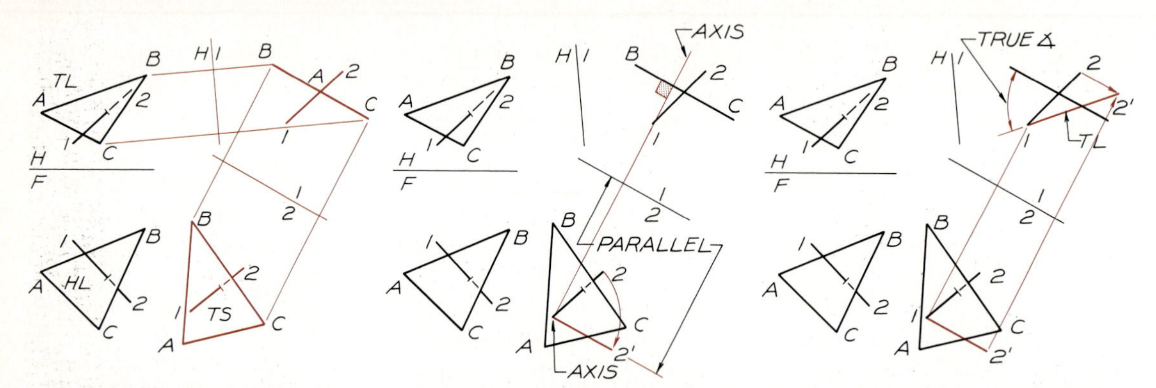

Step 1 Construct plane *ABC* as an edge in a primary auxiliary view, which can be projected from either view. Determine the true size of the plane in the secondary auxiliary view, and project line 1–2 to each view.

Step 2 Revolve the secondary auxiliary view of the line until it is parallel to the 1–2 reference line. The axis of revolution appears as a point through point 1 in the secondary auxiliary view. The axis appears true length and is perpendicular to the 1–2 line and plane *ABC* in the primary auxiliary view.

Step 3 Point 2′ is projected to the primary auxiliary view where the true length of line 1–2′ is found by projecting the primary auxiliary view of point 2 parallel to the 1–2 line as shown. Since the plane appears as an edge and the line appears true length in this view, the true angle between the line and the plane is found.

Fig. 7.28 A line at specified angles

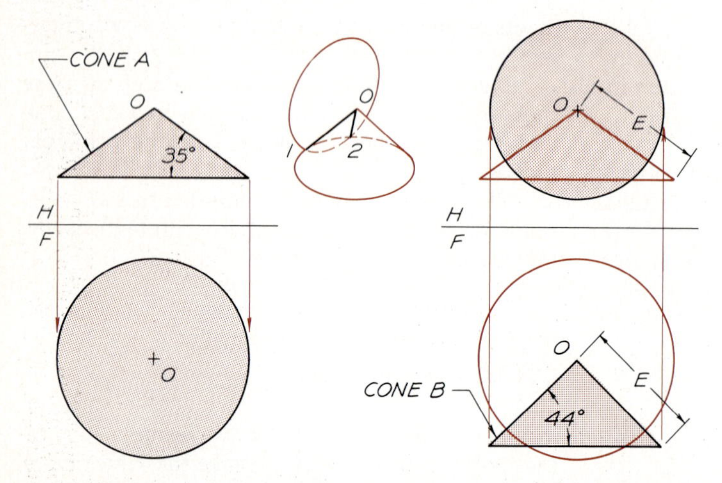

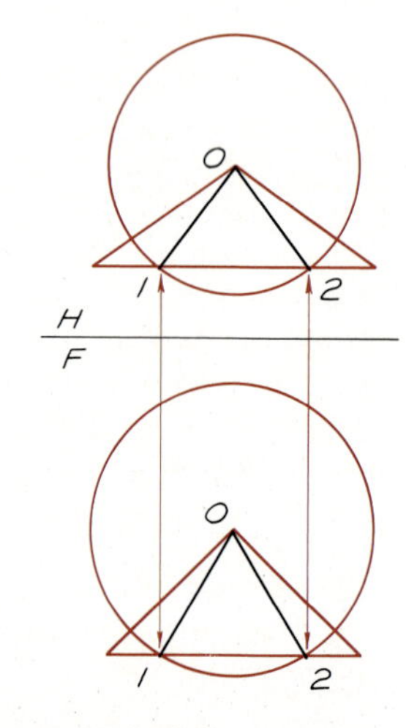

Step 1 Draw a triangular view of a cone in the top view such that the extreme elements make an angle of 35° with the frontal plane. Construct the circular view of the cone in the front view, using point *O* as the apex. All elements of this cone make an angle of 35° with the frontal plane.

Step 2 Draw a triangular view of a cone in the front view such that the elements make an angle of 44° with the horizontal plane. Draw the elements of this cone equal in length to element *E* of cone *A*. All elements of cone *B* make an angle of 44° with the horizontal plane.

Step 3 Since the elements of cones *A* and *B* are equal in length, there will be two common elements that lie on the surface of each cone, elements *O*–1 and *O*–2. Locate points 1 and 2 at the point where the bases of the cone intersect in both views. Either of these lines will satisfy the problem requirements.

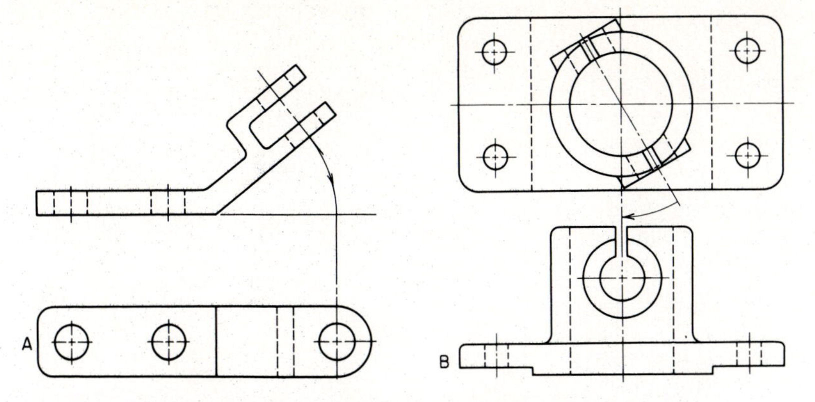

Fig. 7.29 It is conventional practice to revolve features of parts in order to show the features true size in the adjacent orthographic views. When this is done, it is unnecessary to show the arrows of rotation since this is understood as standard practice.

PROBLEMS

Use Size A (8½" × 11") sheets for the following problems and lay out the problems using instruments. Each square on the grid is equal to 0.20" or about 5 mm. The problems can be laid out on grid or plain paper. Label all reference planes and points in each problem with ⅛" letters and/or numbers, using guidelines.

The crosses marked "1" and "2" are to be used for placing primary and secondary reference lines. The primary reference line should pass through "1" and the secondary through "2".

1–4. Find the true-length views of the lines by revolution in Fig. 7.30.

5. Find the true-size view of the plane by an auxiliary view and a single revolution in Fig. 7.31.

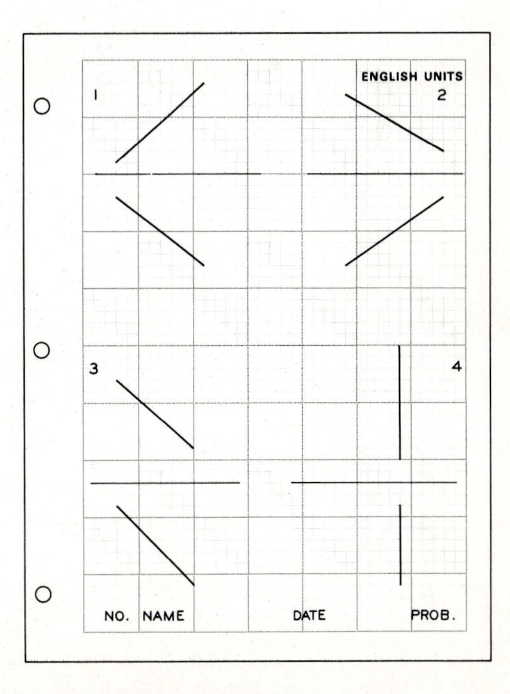

Fig. 7.30 Problems 1 through 4.

6. Find the true-size view of the plane by revolution in Fig. 7.31.

7 and 8. Find the true-size views of the planes by double revolution in Fig. 7.32.

9 and 10. Find the dihedral angles between the planes in Fig. 7.33.

11 and 12. Revolve the points about the given axes in Fig. 7.34 and show the points in all views. In Problem 11, revolve the point into its most forward position, and in Problem 12 into its highest position.

13. The centerline of a conveyor chute is given in the top and front views of Fig. 7.35. The chute has a 10-foot-square cross section. Construct the necessary views to revolve the 10-foot-square into a position where two sides of the right section will be vertical planes. Show the chute in all views. Scale: 1″ = 10′.

14–19. Lay out these problems on a size B sheet (11″ × 17″) using the horizontal format. Position the crossing division lines at the center of the sheet. The grid is spaced at 0.25″ or approximately 6 mm apart. Problem 14 requires that you lay out the prism in area 1 (Fig. 7.36) and rotate it as specified about its corner point 0 in the remaining areas of the sheet.

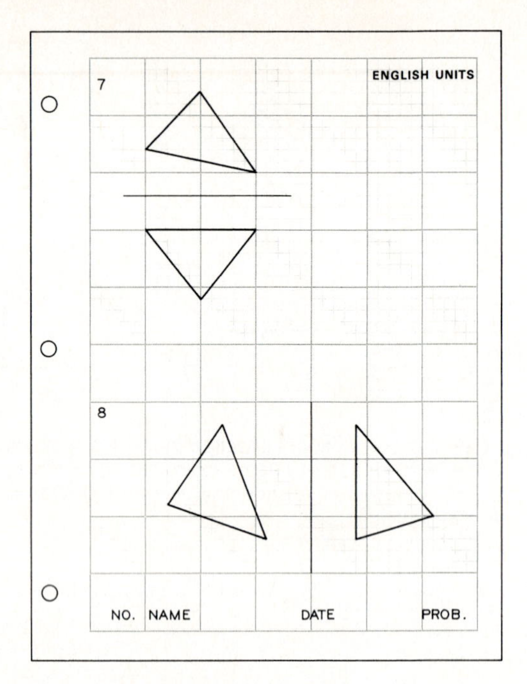

Fig. 7.32 Problems 7 and 8.

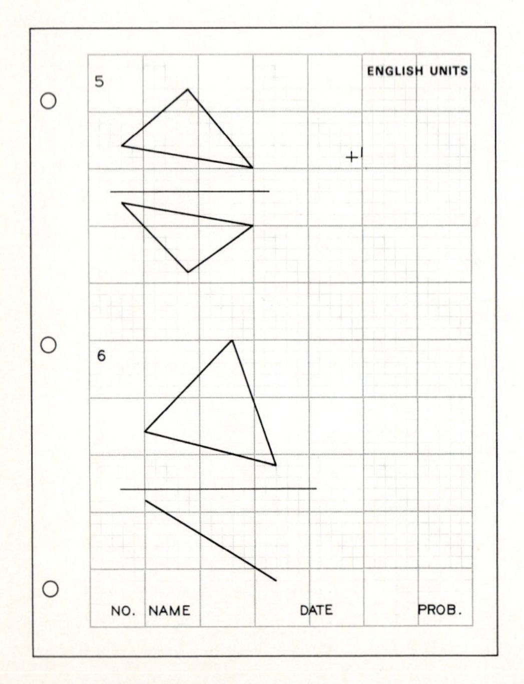

Fig. 7.31 Problems 5 and 6.

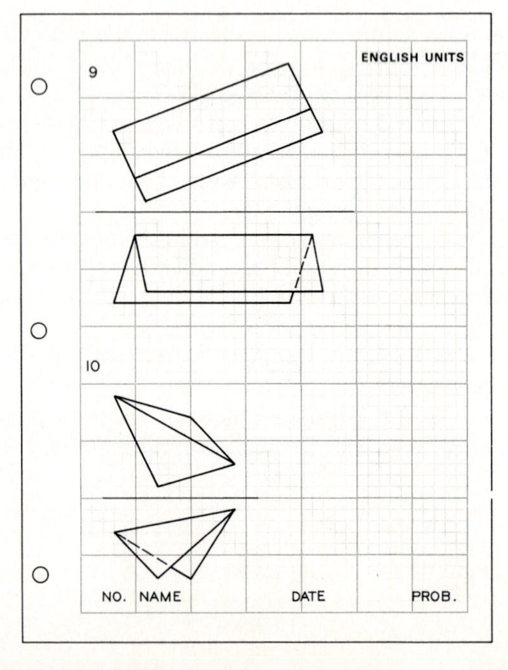

Fig. 7.33 Problems 9 and 10.

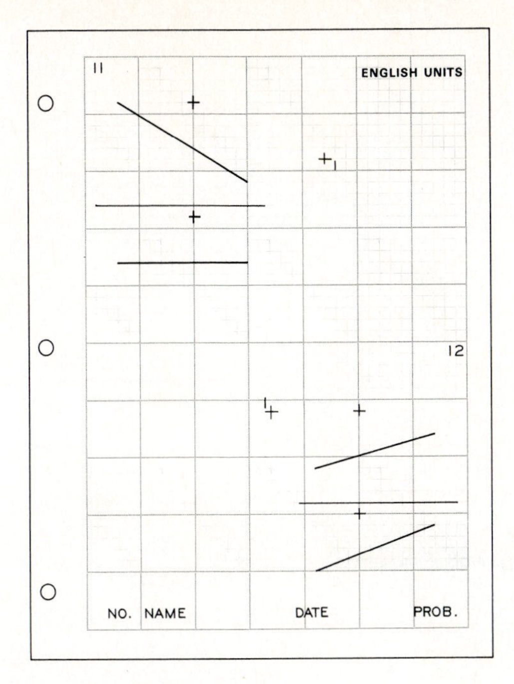

Fig. 7.34 Problems 11 and 12.

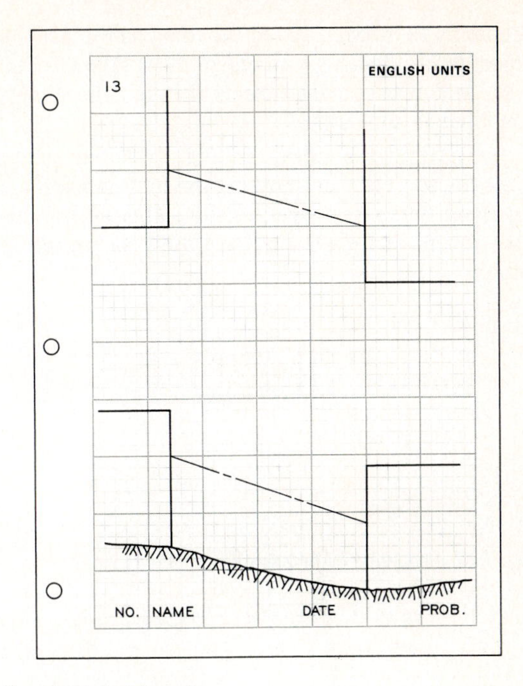

Fig. 7.35 Problem 13.

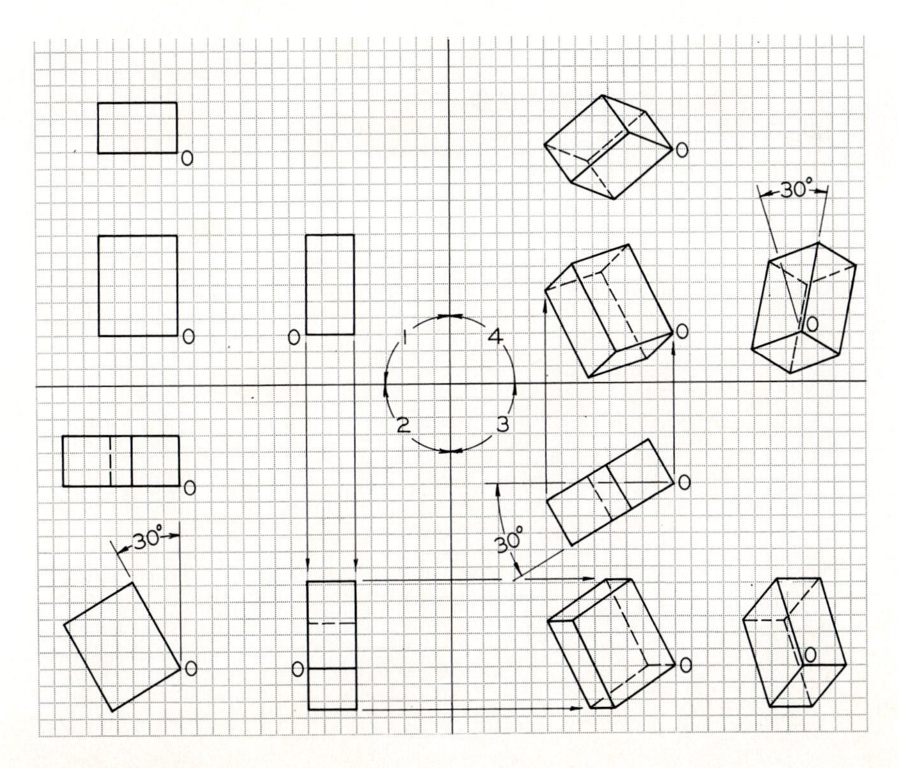

Fig. 7.36 Problem 14.

Problems 15 through 19 require that you replace the object in Problem 14 with one of those given in Fig. 7.37, and rotate these objects through the angles specified in each step.

20. Draw the object in Fig. 7.38, but complete the top view by showing the inclined surface revolved into a true-size position. This will eliminate the need for an auxiliary view as presently shown.

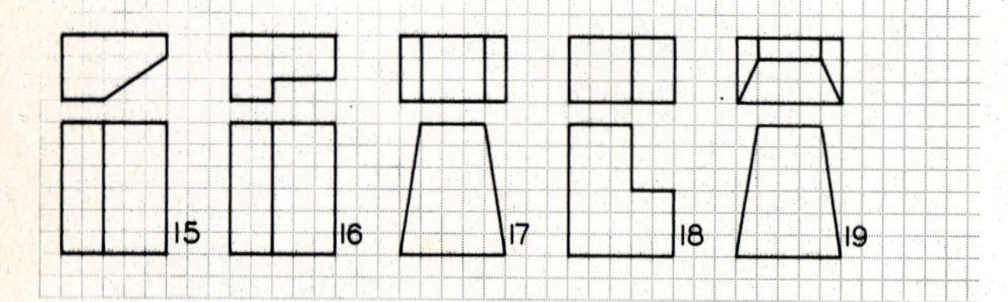

Fig. 7.37 Problems 15 through 19.

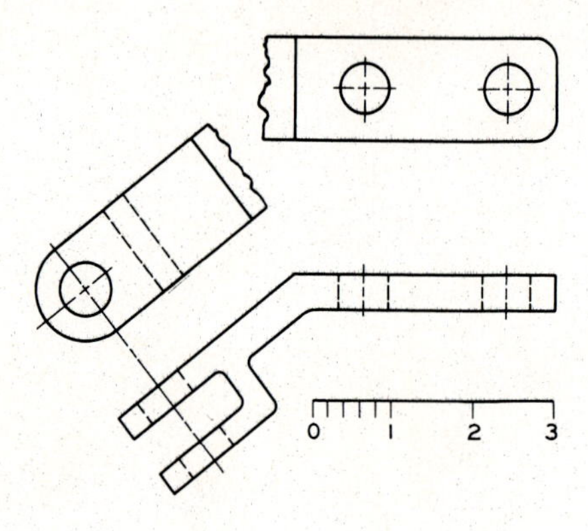

Fig. 7.38 Problem 20.

8

VECTOR GRAPHICS

8.1 INTRODUCTION

When analyzing a system for strength, it is necessary to consider the forces of tension and compression within the system. These forces are represented by vectors. Vectors may also be used to represent other quantities such as distance, velocity, and electrical properties.

Graphical methods are useful in the solution of vector problems, which are often very complicated to solve by conventional trigonometric and algebraic methods. Each method can serve as an effective check on the solutions found by the other methods.

8.2 BASIC DEFINITIONS

A knowledge of the terminology of graphical vectors is necessary to understand the techniques of problem-solving with vectors. The following definitions will be used throughout this chapter.

Force A push or a pull that tends to produce motion. All forces have (1) magnitude, (2) direction, (3) a point of application, and (4) sense. A force is represented by the rope being pulled in Fig. 8.1A.

Vector A graphical representation of a quantity of force that is drawn to scale to indicate magnitude, direction, sense, and point of application. The vector shown in Fig. 8.1B represents the force of the rope pulling the weight, W.

Magnitude The amount of push or pull. In drawings, this is represented by the length of the vector line. Magnitude is usually measured in pounds or kilograms of force.

Direction The inclination of a force (with respect to a reference coordinate system).

Point of application The point through which the force is applied on the object or member, point A in Fig. 8.1A.

Sense Either of the two opposite ways in which a force may be directed, i.e., toward or away from the point of application. The sense is shown by an

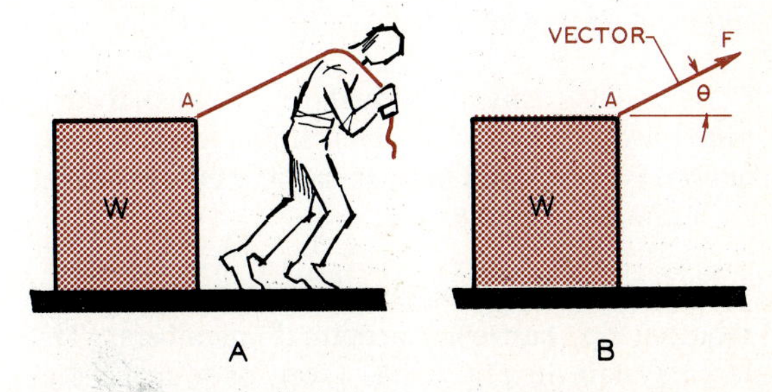

Fig. 8.1 Representation of a force by a vector.

arrowhead attached to one end of the vector line. It is shown in part B of Fig. 8.1 by the arrowhead at F.

Compression The state created in a member by subjecting it to opposite pushing forces. A member tends to be shortened by compression (Fig. 8.2A). Compression is represented by a plus sign (+) or the letter C.

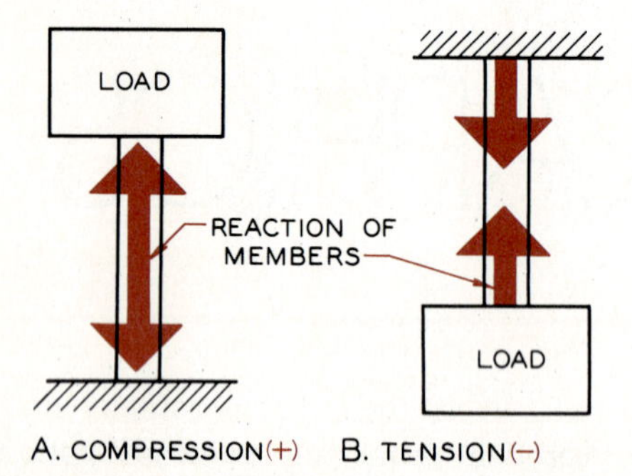

Fig. 8.2 Comparison of tension and compression in a member.

Tension The state created in a member by subjecting it to opposite pulling forces. A member tends to be stretched by tension, as shown in Fig. 8.2B. Tension is represented by a minus sign (−) or the letter T.

Force system The combination of all forces acting on a given object. Figure 8.3 shows a force system.

Resultant A single force that can replace all the forces of a force system and have the same effect as the combined forces.

Equilibrant The opposite of a resultant; it is the single force that can be used to counterbalance all forces of a force system.

Components Any individual forces which, if combined, would result in a given single force. For example, Forces A and B are components of resultant R_1 in Step 1 of Fig. 8.3.

Space diagram A diagram depicting the physical relationship between structural members. The force system in Fig. 8.3 is given as a space diagram.

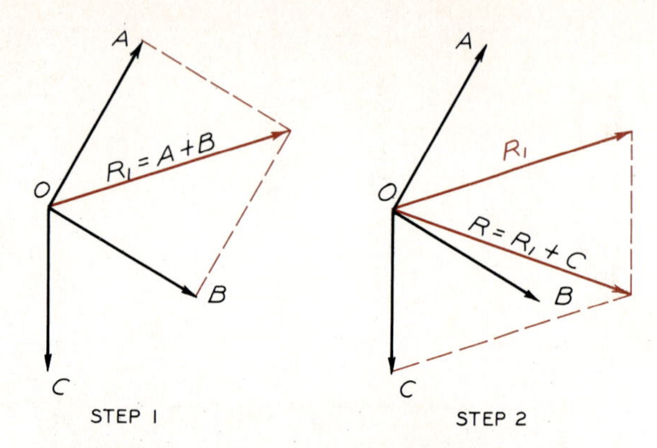

Fig. 8.3 Resultant by the parallelogram method

Step 1 Draw a parallelogram with its sides parallel to vectors A and B. The diagonal R_1, drawn from point P to point O, is the resultant of forces A and B.

Step 2 Draw a parallelogram using vectors R_1 and C to find diagonal R. This resultant can replace forces A, B, and C.

Vector diagram A diagram composed of vectors that are scaled to their appropriate lengths to represent the forces within a given system. The vector diagram is used to solve for unknowns.

Statics The study of forces and force systems that are in equilibrium.

Metric units The kilogram (kg) is the standard unit for indicating mass (loads). A comparison of kilograms with pounds is shown in Fig. 8.4. The metric ton is 1000 kilograms. One kilogram = 2.2 pounds.

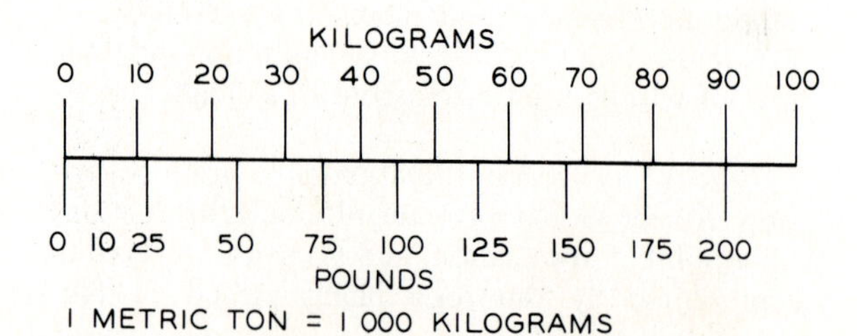

Fig. 8.4 The kilogram (kg) is the standard metric unit for measuring forces, which are represented by pounds in the English system: 1 kilogram = 2.2 pounds.

8.3 COPLANAR, CONCURRENT FORCE SYSTEMS

When several forces, represented by vectors, act through a common point of application, the system is said to be **concurrent.** Vectors *A*, *B*, and *C* act through a single point in Fig. 8.3; therefore this is a concurrent system. When only one view is necessary to show the true length of all vectors, as in Fig. 8.3, the system is **coplanar.**

Engineering designs are analyzed to determine the total effect of the forces applied in a system. Such an analysis requires that the known forces be resolved into a single force—the **resultant**—that will represent the composite effect of all forces on the point of application. The resultant is found graphically by two methods: (1) the parallelogram method and (2) the polygon method.

8.4 RESULTANT OF A COPLANAR, CONCURRENT SYSTEM—PARALLELOGRAM METHOD

In the system of vectors shown in Fig. 8.3, all the vectors lie in the same plane and act through a common point. The vectors are scaled to a known magnitude.

The vectors for a force system must be known and drawn to scale in order to apply the parallelogram method to determine the resultant. Two vectors are used to find a parallelogram; the diagonal of the parallelogram is the resultant of these two vectors and has its point of origin at point *P* (Fig. 8.3). Resultant R_1 can be called the *vector sum* of vectors *A* and *B*.

Since vectors *A* and *B* have been replaced by R_1, they can be disregarded in the next step of the solution. Again, resultant R_1 and vector *C* are resolved by completing a parallelogram, i.e., by drawing a line parallel to each vector. The diagonal of this parallelogram, *R*, is the resultant of the entire system and is the vector sum of R_1 and *C*. This resultant, *R*, can be analyzed as though it were the only force acting on the point; therefore the analysis of a particular point-of-force application is simplified by finding the resultant.

8.5 RESULTANT OF A COPLANAR, CONCURRENT SYSTEM—POLYGON METHOD

The system of forces shown in Fig. 8.3 is shown again in Fig. 8.5, but in this case the resultant is found by the polygon method. The forces are drawn to scale and in their true directions, with each force being drawn head-to-tail to form the polygon. In this example, the vectors are drawn in a counterclockwise sequence, beginning with vector *A*. Note that the polygon does not close; this means that the system is not in *equilibrium*. In other words, it would tend to be in motion, since the forces are not balanced in all directions. The resultant *R* is drawn from the tail of vector *A* to the head of vector *C* to close the polygon. The resultant is equal in length, direction, and sense to the resultant found by the parallelogram method of the previous article.

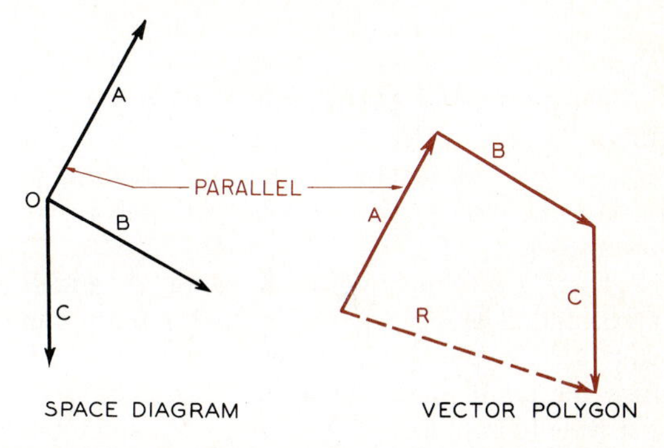

Fig. 8.5 Resultant of a coplanar, concurrent system as determined by the polygon method, in which the vectors are drawn head-to-tail.

8.6 RESULTANT OF A COPLANAR, CONCURRENT SYSTEM—ANALYTICAL METHOD

Vectors can be solved analytically by application of algebra and trigonometry. The analytical example in Fig. 8.6 is given to afford a comparison between the graphical and analytical methods.

In Step 1, the vertical components, which are parallel to the *Y*-axis, are drawn from the ends of both vectors to form right triangles, and their lengths are found through the use of the trigonometric functions of the angles the vectors make with the *X*-axis.

Fig. 8.6 Resultant by the analytical method

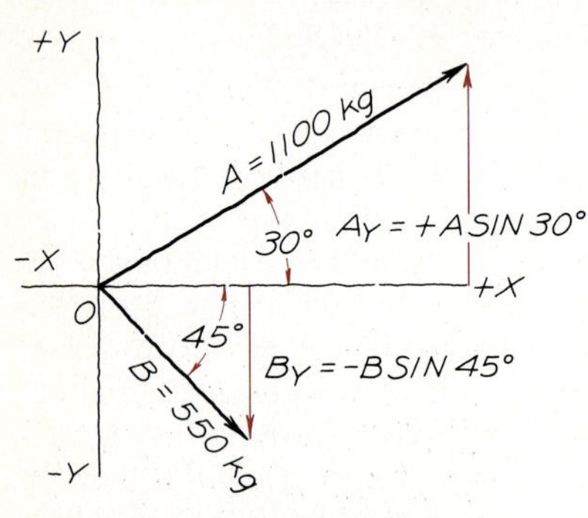

Step 1 The Y-components (vertical components) are found to be the sine functions of the angles the vectors make with the X-axis. The Y-component of A is positive and the Y-component of B is negative.

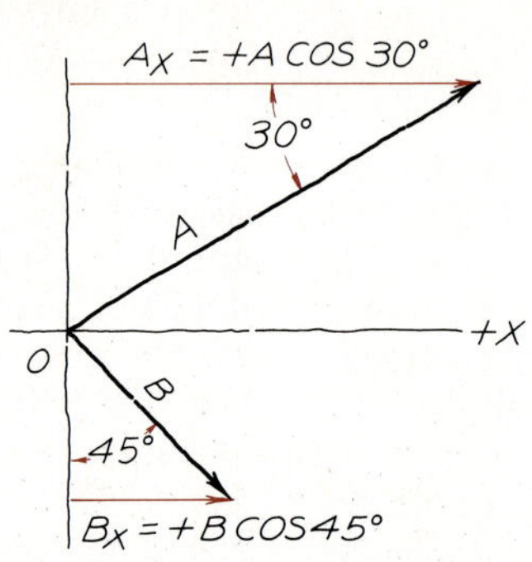

Step 2 The X-components (horizontal components) are the cosine functions of 30° and 45° in this case, both in the positive direction.

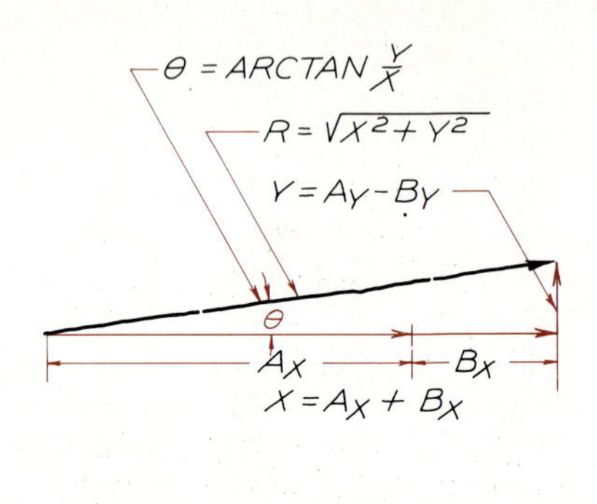

Step 3 The Y-components and X-components are summed to find the components of the resultant, X and Y. The Pythagorean theorem is applied to find the magnitude of the resultant. Its angle with the X-axis is the arctangent of Y/X.

The horizontal component of each vector is drawn parallel to the X-axis through the end of each vector. The lengths of these components are found to be the cosine functions of the given vectors in Step 2.

The Y-components of each vector, A_y and B_y, can be added, since each lies in the same direction (Step 3). The resulting value is $Y = A_y - B_y$, since the components have opposite senses. The horizontal component is $X = A_x + B_x$, since both components have equal directions and senses.

A right triangle is sketched using the X- and Y-components that were found trigonometrically. The vertical and horizontal components are laid off head-to-tail and the head of the horizontal component is connected to the tail of the vertical to form a right triangle of forces. The magnitude of the resultant is found by the Pythagorean theorem,

$$R = \sqrt{X^2 + Y^2}$$

The direction of the resultant is

$$\text{angle } \theta = \arctan Y/X$$

and it is measured from the horizontal X-axis.

Law of Sines

The law of sines is illustrated in Fig. 8.7A. When any three values are known, the remaining un-

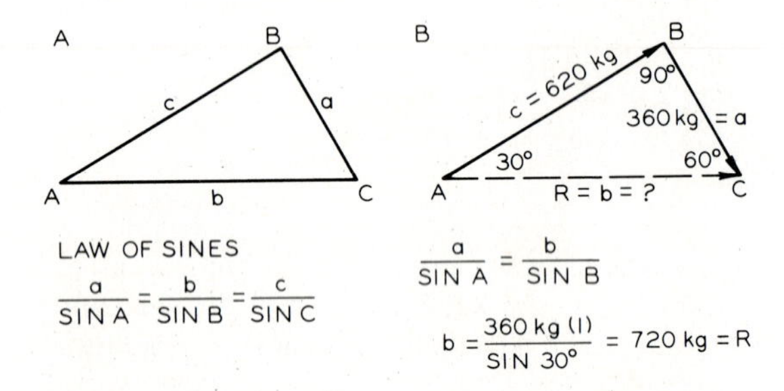

Fig. 8.7 The law of sines is illustrated in part A. This principle is used in part B to solve for resultant *R (b)* when two angles and vectors are known.

knowns of a triangle can be computed. An example is given (Fig. 8.7B) where two sides of a triangle are vectors of known magnitude and direction. This enables you to find the resultant mathematically, as shown.

> An *equilibrant* has the same magnitude, direction, and point of application as the *resultant* in any system of forces.

The difference is the sense. Note that the resultant of the system of forces shown in Fig. 8.8 is solved for through the parallelogram method. The equili-

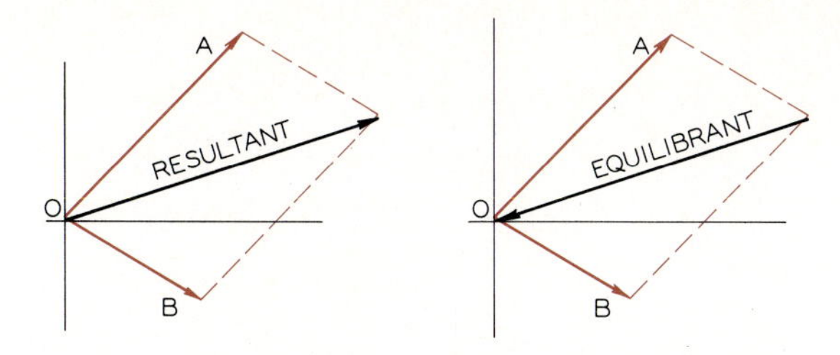

Fig. 8.8 The resultant and equilibriant are equal in all respects except in sense (position of arrowhead).

brant can be applied at point O to balance the forces A and B and thereby cause the system to be in a state of equilibrium.

8.7 RESULTANT OF NONCOPLANAR, CONCURRENT FORCES—PARALLELOGRAM METHOD

When vectors lie in more than one plane of projection, they are said to be **noncoplanar;** therefore more than one view is necessary to analyze their spatial relationships. The resultant of a system of noncoplanar forces can be found by the parallelogram method, regardless of their number, if their true projections are given in two adjacent orthographic views.

Vectors 1 and 2 in Fig. 8.9 are used to construct the top and front views of a parallelogram. The diagonal of the parallelogram, R_1, is found in both views. As a check, the front view of R_1 must be an orthographic projection of its top view; if it is not, there is an error in construction.

In Step 2, resultant R_1 and vector 3 are resolved to form resultant R in both views. The top and front views of R must project orthographically if there is no error in construction. Resultant R can be used to replace vectors 1, 2, and 3. Since R is an oblique line, its true length can be found by auxiliary view, as shown in Fig. 8.10 or by revolution.

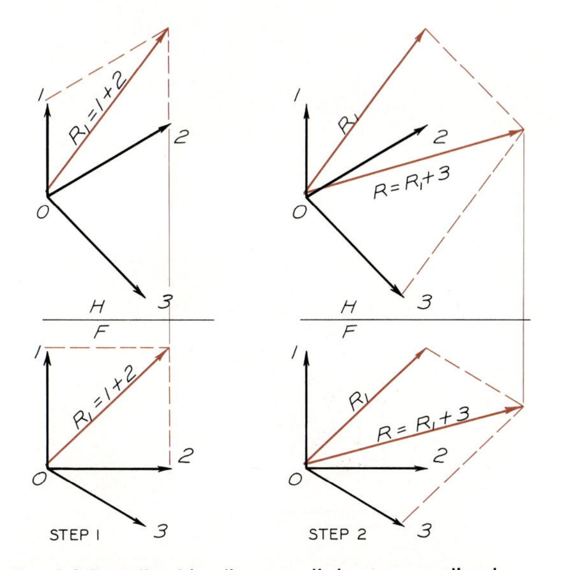

Fig. 8.9 Resultant by the parallelogram method

Step 1 Vectors 1 and 2 are used to construct a parallelogram in the top and front views. The diagonal, R_1, is the resultant of these two vectors.

Step 2 Vectors 3 and R_1 are used to construct a second parallelogram to find the views of the overall resultant, R.

8.8 RESULTANT OF NONCOPLANAR, CONCURRENT FORCES—POLYGON METHOD

The same system of forces that was given in Fig. 8.9 is solved in Fig. 8.10 for the resultant of the system by the polygon method.

In Step 1, each vector is laid head-to-tail in a clockwise direction, beginning with vector 1. The vectors are drawn in each view to be orthographic projections at all times (Step 2). Since the vector polygon did not close, the system is not in equilibrium. The resultant R is constructed from the tail of vector 1 to the head of vector 3 in both views.

Resultant R is an oblique line and requires an auxiliary view to find its true length. The magnitude of the resultant can be measured in the true-length auxiliary view by using the same scale as was used to draw the original views.

8.9 RESULTANT OF NONCOPLANAR, CONCURRENT FORCES—ANALYTICAL METHOD

We are required to solve for the resultant of the system in Fig. 8.11 by the analytical method, using trigonometry and algebraic equations. The projected lengths of the vectors are known in both views.

In Step 1, the summation of the forces in the X-direction is found in the front view. Since this

Fig. 8.10 Resultant by the polygon method

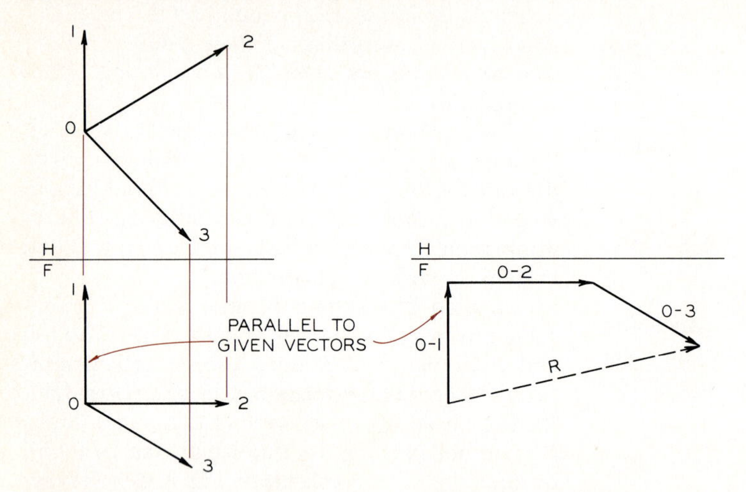

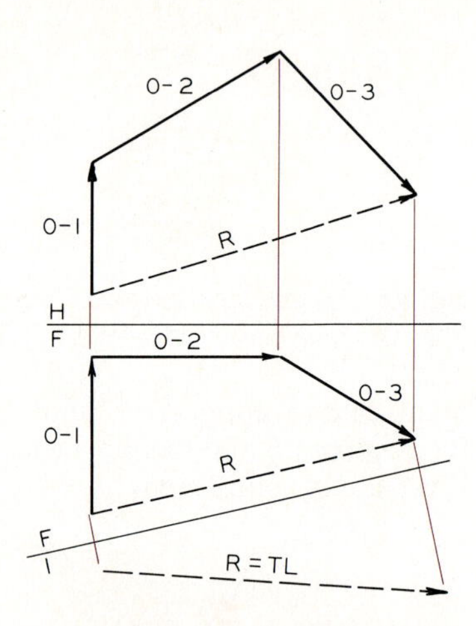

Required Find the resultant of this system of concurrent, noncoplanar forces by the polygon method.

Step 1 Each vector is laid off head-to-tail in the front view. The front view of the resultant is the vector found.

Step 2 The same vectors are laid off head-to-tail in the top view to complete the three-dimensional polygon. The resultant is found true length by an auxiliary view.

Fig. 8.11 Resultant by the analytical method

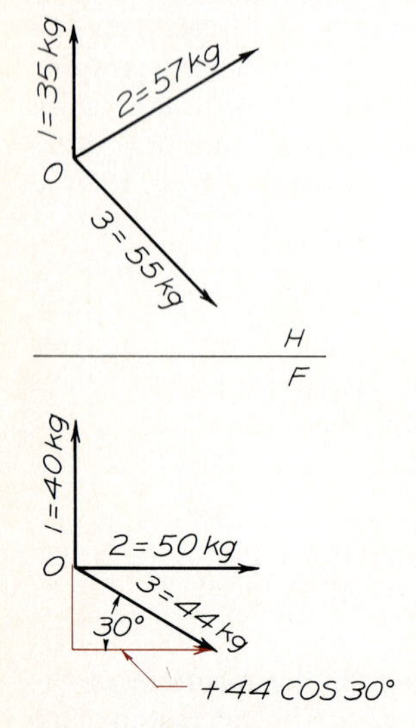

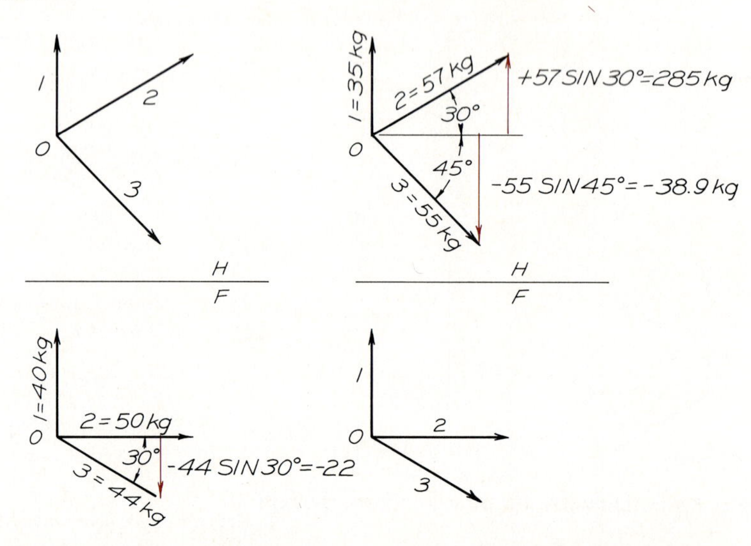

Step 1 The *X*-component is found in the front or top view. The *X*-components are found to be: force 1 = 0 kg; force 2 = 50 kg; force 3 = 44 kg cos 30°. These values are positive.

Step 2 The *Y*-component must be found in the front view. The *Y*-components are found to be: force 1 = 40 kg; force 2 = 0 kg; force 3 = 44 kg sin 30°.

Step 3 The *Z*-component must be found in the top view. The *Z*-components are found to be: force 1 = 35 kg; force 2 = 57 kg sin 30°; force 3 = 55 kg sin 45°. The resultant is found in Fig. 8.12

left and right direction can be seen in both the top and front views, either view can be used for finding the X-component of the system. The summation in the X-direction is expressed in the following equation:

$$\sum F_x = (2) + (3)\cos 30°$$
$$= 50 + 44\cos 30° = 88.2 \text{ kg}(+).$$

The X-component is found to be 88.2 lb in the positive direction, which is toward the right. Vector 1 is vertical and consequently has no component in the X-direction.

Summation of forces in the Y-direction is found in the front view. This summation is expressed in the following equation:

$$\sum F_y = (1) - (3)\sin 30°$$
$$= 40 - 44\sin 30° = 18 \text{ kg}(+).$$

Vector 2 is horizontal and has no vertical component.

The summation of forces in the Z-direction is found in the top view. Positive direction is considered to be backward, and negative to be forward. This summation is expressed in the following equation:

$$\sum F_z = (1) + (2)\sin 30° - (3)\sin 45°$$
$$= 35 + 57\sin 30° - 55\sin 45°$$
$$= 24.61 \text{ lb}(+).$$

The resultant that can be used to replace vectors 1, 2, and 3 can be found from these three components by the following equation:

$$R = \sqrt{X^2 + Y^2 + Z^2}.$$

By substitution of the X-, Y-, and Z-components found in the three previous summations, the equation can be solved as follows:

$$R = \sqrt{88.2^2 + 18^2 + 24.6^2} = 93.3 \text{ kg}.$$

The resultant force of 93.3 kg is of no value unless its direction and sense are known. To find this information, we must refer to the two orthographic views of the force system, as shown in Fig. 8.12. The X- and Z-components, 88.2 kg and 24.6 kg, are drawn to form a right triangle in the top view. The hypotenuse of this triangle depicts the direction and sense of the resultant in the top view. The angular direction of the top view of re-

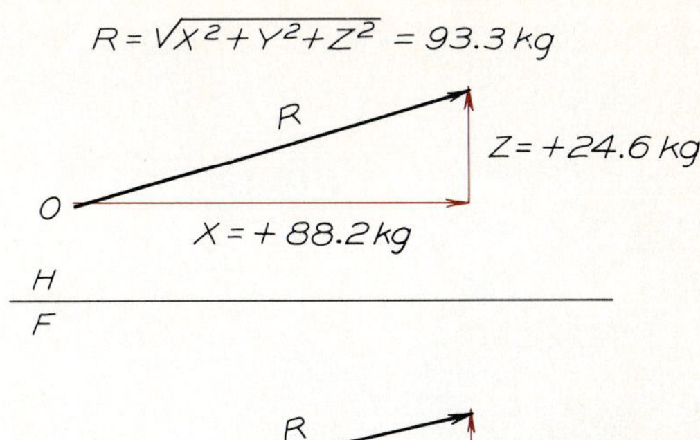

Fig. 8.12 The three components X, Y, and Z, found in Fig. 8.11 are used to find the resultant $R = X^2 + Y^2 + Z^2 = 93.3$ kg.

sultant is found in the following equation:

$$\tan\theta = \frac{24.6}{88.2} = 0.279; \quad \theta = 15.6°.$$

The angular direction of the resultant is found in the front view by constructing a triangle with the X- and Y-components, 88.2 kg and 18 kg. The hypotenuse of this right triangle is the direction of the resultant. The direction of the resultant in the front view is expressed in the following equation:

$$\tan\phi = \frac{18}{88.2} = 0.204; \quad \phi = 11.5°.$$

These two angles, found in the top and front views, establish the direction of the resultant vector, whose sense can be described as upward, to the right, and back.

8.10 FORCES IN EQUILIBRIUM

In the previous examples, the vectors were drawn from given or known magnitudes and directions. The same principles can be applied to structural systems in which the magnitudes and senses are not given.

An example of a coplanar, concurrent structure in equilibrium can be seen in the loading cranes in Fig. 8.13, which are used for the handling of cargo on board ship.

The coplanar, concurrent structure given in Fig. 8.14 is designed to support a load of $W = 1000$ kg. The maximum loading in each is used to de-

Fig. 8.13 The cargo cranes on the cruise ship *Santa Rosa* are examples of coplanar, concurrent force systems that are designed to remain in equilibrium. (Courtesy of Exxon Corporation.)

termine the type and size of structural members used in the structural design.

In Step 1, the only known force, $W = 1000$ kg, is laid off parallel to the given direction. Unknown forces A and B are drawn as vectors to close the force polygon. Each vector must be drawn head-to-tail.

In Step 2, vectors A and B can be analyzed to determine whether they are in tension or compression. Vector B points upward to the left, which is toward point O when transferred to the structural diagram. A vector that acts toward a point of application is in compression. Vector A points away from point A when transferred to the structural diagram and is, therefore, in tension.

The length versus the cross section of a member will be considered when selecting a member, but the determination of force in the member is found in the same manner in the vector polygon regardless of member length.

A similar example of a force system involving a pulley is solved in Fig. 8.15 to determine the loads in the structural members caused by the weight of 100 lb. The only difference between this solution and the previous one is the construction of two equal vectors to represent the loads in the cable on both sides of the pulley.

8.11 TRUSS ANALYSIS

Vector polygons can be used to analyze structural trusses to determine the loads in each member by two graphical methods: (1) joint-by-joint analysis, and (2) Maxwell diagrams.

Joint-by-Joint Analysis

The truss shown in Fig. 8.16 is called a Fink truss, and is loaded with forces of 3000 lb that are concentrated at joints of the structural members.

Fig. 8.14 Coplanar forces in equilibrium

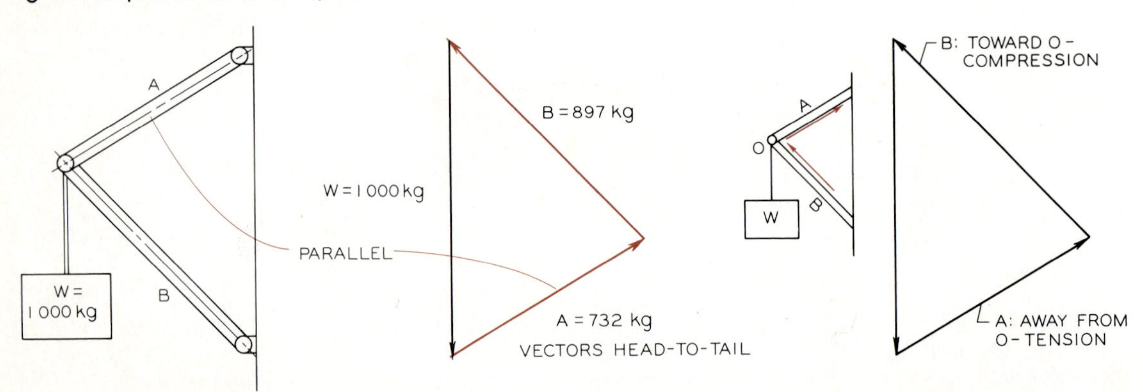

Required Find the forces in the two structural members caused by the load of 1000 kg.

Step 1 Draw the known load of 1000 kg as a vector. Draw the vectors *A* and *B* to the same scale and parallel to their directions. Arrowheads are drawn head-to-tail.

Step 2 Vector *A* points away from the point O when transferred to the structural diagram. Therefore, vector *A* is in tension. Vector *B* points toward point O and is in compression.

Fig. 8.15 Determination of forces in equilibrium

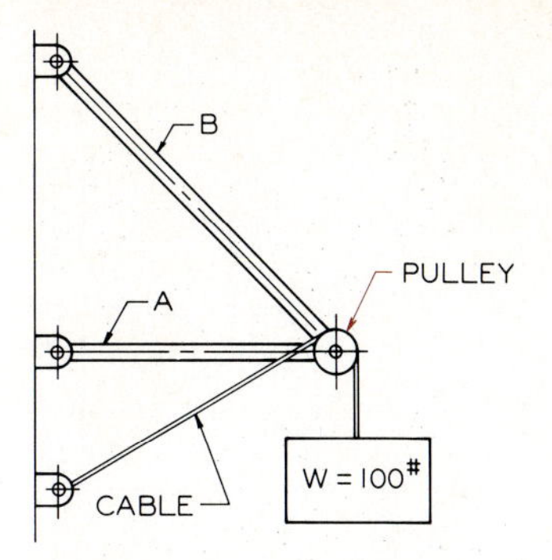

Required Find the forces in the members caused by the load of 100 lb supported by the pulley.

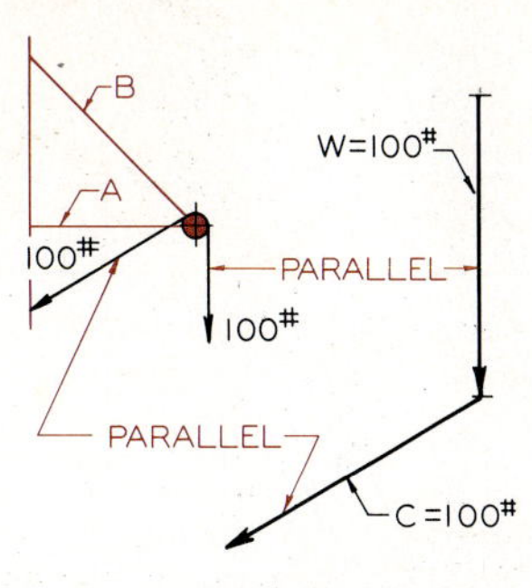

Step 1 The force in the cable is equal to 100 lb on both sides of the pulley. These two forces are drawn as vectors head-to-tail parallel to their directions in the space diagram.

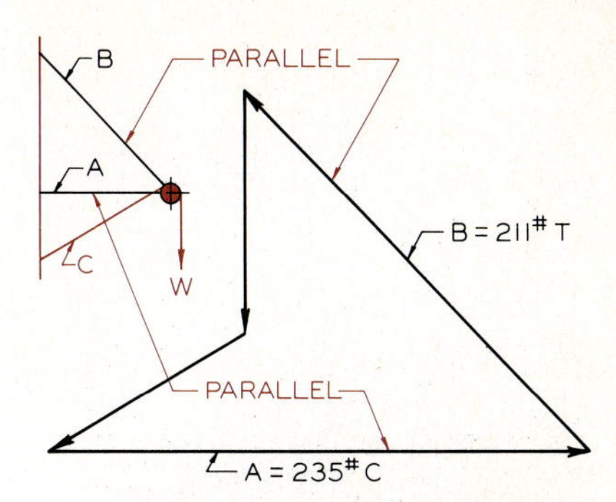

Step 2 *A* and *B* are drawn to close the polygon, and arrowheads are placed to give a head-to-tail arrangement. The sense of *A* is toward the point of application, and in compression; *B* is away from the point and is in tension.

Fig. 8.16 Joint analysis of a truss

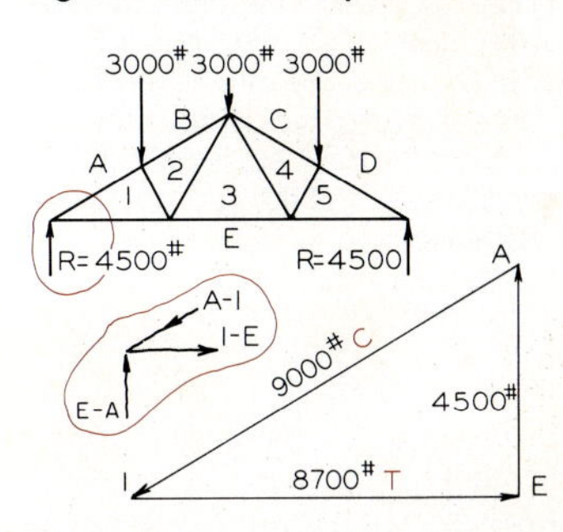

Step 1 The truss is labeled using Bow's notation, with letters between the exterior loads and numbers between interior members. The lower left joint can be analyzed, since it has only two unknowns, *A*–1 and 1–*E*. These vectors are found by drawing them parallel to their directions from both ends of the reaction of 4500 lb. The vectors are laid off in a head-to-tail order.

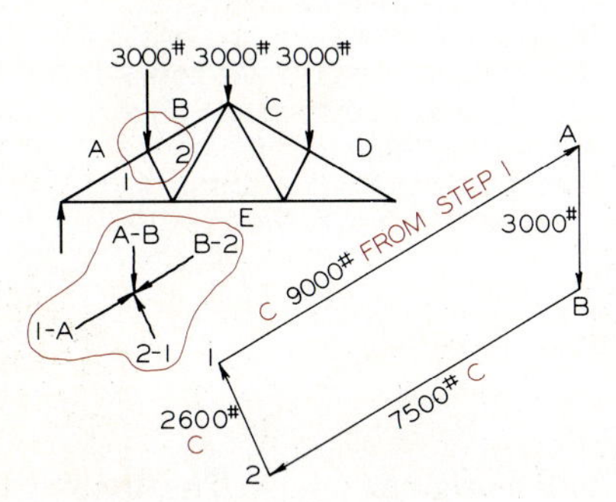

Step 2 Using the vector 1–*A* found in step 1 and load *AB*, the two unknowns *B*–2 and 2–1 can be found. The known vectors are laid out beginning with vector 1–*A* and moving clockwise about the joint. Vectors *B*–2 and 2–1 close the polygon. If the sense of a vector is toward the point of application, it is in compression; if away from the point, it is in tension.

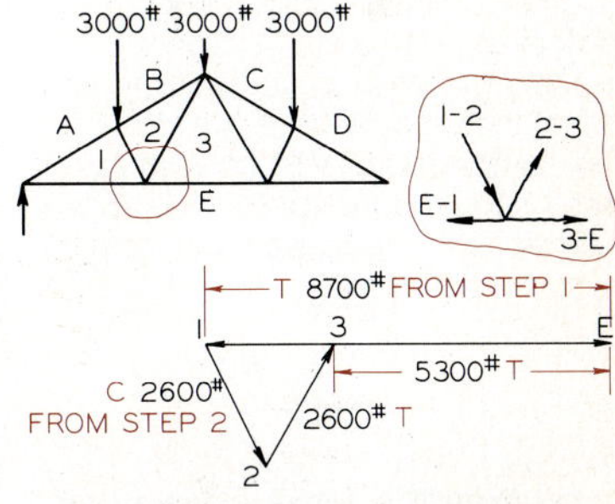

Step 3 The third joint can be analyzed by laying out the vectors *E*–1 and 1–2 from the previous steps. Vectors 2–3 and 3–*E* close the polygon and are parallel to their directions in the space diagram. The senses of 2–3 and 3–*E* are away from the point of application; these vectors are in tension.

A special method of designating forces, called **Bow's notation,** is used. The exterior forces applied to the truss are labeled with letters placed between the forces. Numerals are placed between the interior members.

Each vector used to represent the load in each member is referred to by the number on each of its sides by reading in a clockwise direction. For example, the first vertical load at the left is called *AB*, with *A* at the tail and *B* at the head of the vector.

We first analyze the joint at the left where the reaction of 4500 pounds (denoted by #) is known. This force, reading in a clockwise direction about the joint, is called *EA* with an upward sense. The tail is labeled *E* and the head *A*. Continuing in a clockwise direction, the next force is *A*-1 and the next 1–*E*, which closes the polygon and ends with the beginning letter, *E*. The arrows are placed, beginning with the known vector *EA*, in a head-to-tail arrangement.

Tension and compression can be determined

Fig. 8.17 Truss analysis

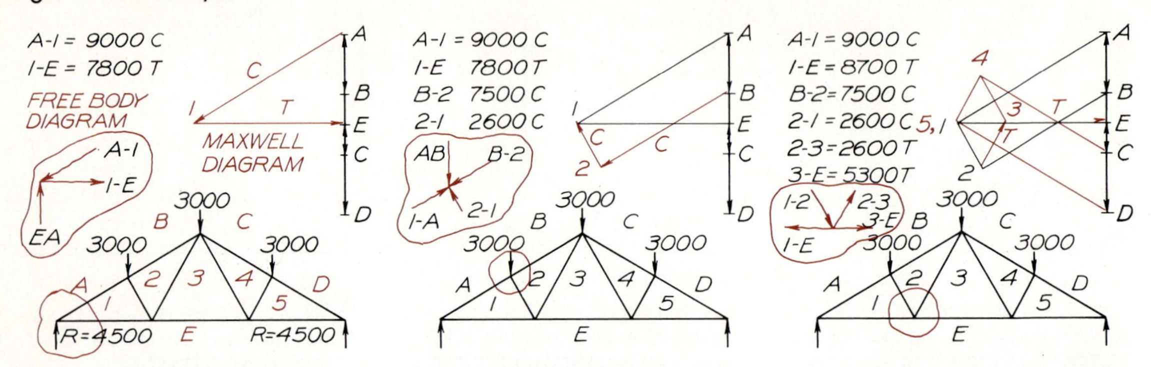

Step 1 Label the spaces between the outer forces of the truss with letters and the internal spaces with numbers, using Bow's notation. Add the given load vectors in a Maxwell diagram, and sketch a free-body diagram of the first joint. Using vectors *EA*, *A*-1, and 1-*E* drawn head-to-tail, draw a vector diagram to find their magnitudes. Vector *A*-1 is in compression (+) because its sense is toward the joint, and 1-*E* is in tension (−) because its sense is away from the joint.

Step 2 Draw a sketch of the next joint to be analyzed. Since *AB* and *A*-1 are known, we have to determine only two unknowns, 2–1 and *B*–2. Draw these parallel to their direction, head-to-tail, in the Maxwell diagram using the existing vectors found in step 1. Vectors *B*–2 and 2–1 are in compression since each has a sense toward the joint. Note that vector *A*–1 becomes 1–*A* when read in a clockwise direction.

Step 3 Sketch a free-body diagram of the next joint to be analyzed. The unknowns in this case are 2–3 and 3–*E*. Determine the true length of these members in the Maxwell diagram by drawing vectors parallel to given members to find point 3. Vectors 2–3 and 3–*E* are in tension because they act away from the joint. This same process is repeated to find the loads of the members on the opposite side.

by relating the sense of each vector to the original joint. For example, *A*–1 has a sense toward the joint and is in compression, while 1–*E* is away and in tension.

Since the truss is symmetrical and equally loaded, the loads in the members on the right will be equal to those on the left.

The other joints are analyzed in the same manner in Steps 2 and 3; the procedure is to begin with known vectors found in the previous polygons and then solve for the unknowns. Note that the sense of the vectors is opposite at each end. Vector *A*–1 has a sense toward the left in Step 1, and toward the right in Step 2.

Maxwell Diagrams

The Maxwell diagram is exactly the same as the joint-by-joint analysis except that the polygons are positioned to overlap, with some vectors common to more than one polygon. Again, Bow's notation is used to good advantage.

The first step (Fig. 8.17) is to lay out the exterior loads beginning clockwise about the truss—*AB*, *BC*, *CD*, *DE*, and *EA*—head-to-tail. A letter is placed at each end of the vectors. Since they are parallel, this polygon will be a straight line.

The structural analysis begins at the joint through which reaction *EA* acts. A free-body diagram is drawn to isolate this joint for easier analysis. The two unknowns are members *A*–1 and 1–*E*. These vectors are drawn parallel to their direction in the truss in Step 1 of Fig. 8.17 with *A*–1 beginning at point *A* and 1–*E* beginning at point *E*. These directions are extended to a point intersection, which locates point 1. Since this joint is in equilibrium, as are all joints of a system in equilibrium, the vectors must be drawn head-to-tail. Because resultant *EA* has an upward sense, vector *A*–1 must have its tail at *A*, giving it a sense toward point 1. By relating this sense to the free-body diagram, we can see that the sense is toward the point of application, which means that *A*–1 is a compression member. Vector 1–*E* has a sense away from the joint, which means that it is a tension member. The vectors are coplanar and can be scaled to determine their loads.

In Step 2, vectors 1–*A* and *AB* are known, while vectors *B*–2 and 2–1 are unknown. Since there are only two unknowns it is possible to solve for them. A free-body diagram showing the joint to be analyzed is sketched. Vector *B*–2 is drawn parallel to the structural member through point *B* in the Maxwell diagram and the line of vector 2–1 is extended from point 1 until it intersects with *B*–2, where point 2 is located. The sense of each vector is found by laying off each vector head-to-tail. Both vectors *B*–2 and 2–1 have a sense toward the joint in the free-body diagram; therefore, they are in compression.

The next joint is analyzed in sequence to find the stresses in 2–3 and 3–*E* (Step 3). The truss will have equal forces on each side, since it is symmetrical and is loaded symmetrically. The total Maxwell diagram is drawn to illustrate the completed work in Step 3.

If all the polygons in the series do not close at every point with perfect symmetry, there is an error in construction. If the error of closure is very slight, it can be disregarded, since safety factors are generally applied in derivation of working stresses of structural systems to assure safe construction. Arrowheads are usually omitted on Maxwell diagrams, since each vector will have opposite senses when applied to different joints.

Fig. 8.18 The structural members of this tripod support for a moon vehicle can be analyzed graphically to determine design loads. (Courtesy of NASA.)

8.12 NONCOPLANAR STRUCTURAL ANALYSIS—SPECIAL CASE

Structural systems that are three-dimensional require the use of descriptive geometry, since it is necessary to analyze the system in more than one plane. The manned flying system (MFS) in Fig. 8.18 can be analyzed to determine the forces in the support members (Fig. 8.19). Weight on the moon can be found by multiplying earth weight by a factor of 0.165. A tripod that must support 182 lb on earth has to support only 30 lb on the moon. This is a special case; since members *B* and *C* lie in the same plane and appear as an edge in the front view, we need to determine only two unknowns.

A vector polygon is constructed in the front view in Step 1 of Fig. 8.19 by drawing force *F* as a vector and using the other vectors as the other sides of the polygon. One of these vectors is actually a summation of vectors *B* and *C*. The top view is drawn using the vectors *B* and *C* to close the polygon from each end of vector *A*. In Step 2, the front view of vectors *B* and *C* is found.

The true lengths of the vector are found in a true-length diagram in Step 3. The vectors are measured to determine their loads. Vector *A* is found to be in compression because its sense is toward the point of concurrency. Vectors *B* and *C* are in tension.

8.13 NONCOPLANAR STRUCTURAL ANALYSIS—GENERAL CASE

The structural frame shown in Fig. 8.20 is attached to a vertical wall to support a load of $W = 600$ lb. Since there are three unknowns in each of the views, we are required to construct an auxiliary view that will give the edge view of a plane containing two of the vectors, thereby reducing the number of unknowns to two (Step 1). We no longer need to refer to the front view.

A vector polygon is drawn by constructing vectors parallel to the members in the auxiliary view (Step 1). An adjacent orthographic view of the vector polygon is also drawn by constructing its vectors parallel to the members in the top view (Step 2). A true-length diagram is used in Step 3 to find the true length of the vectors to determine their magnitudes.

Fig. 8.19 Noncoplanar structural analysis—a special case

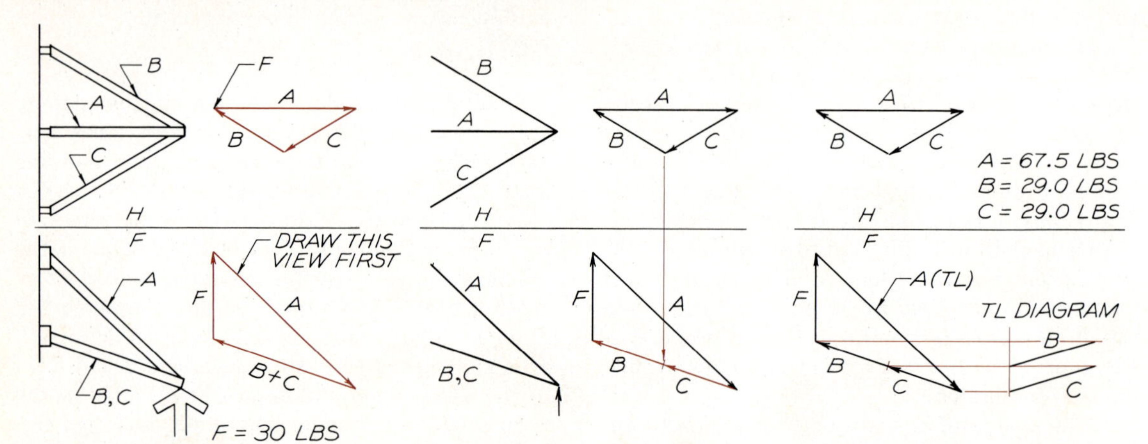

Step 1 Two forces, *B* and *C*, coincide in the front view, resulting in only two unknowns in this view. Vector *F* (30 lb) is drawn, and the other two unknowns are drawn parallel to their front view to complete the front view of the vector polygon. The top view of *A* can be found by projection, from which vectors *B* and *C* can be found.

Step 2 The point of intersection of vectors *B* and *C* in the top view is projected to the front view to separate these vectors. All vectors are drawn head-to-tail. Vectors *B* and *C* are in tension because their vectors are acting away from the point in the space diagram, while *A* is in compression.

Step 3 The completed top and front views found in step 2 do not give the true lengths of vectors *B* and *C*, since they are oblique. The true lengths of these lines are determined by a true-length diagram where they are scaled to find the forces in each member.

Fig. 8.20 Noncoplanar structural analysis—general case

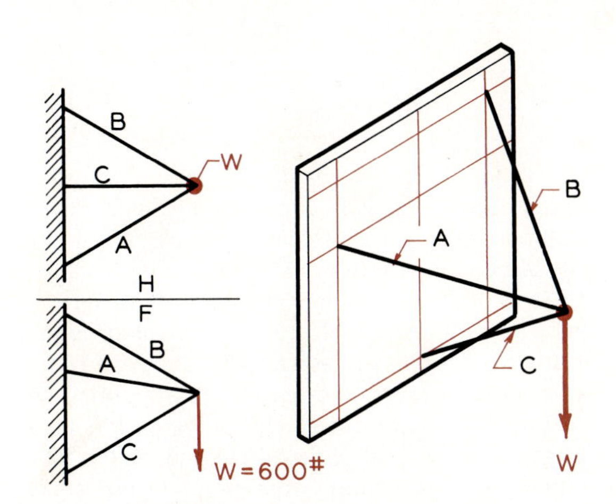

Given The top and front views of a three-member frame which is attached to a vertical wall and supports a weight of 600 lb.

Required Find the loads in the structural members.

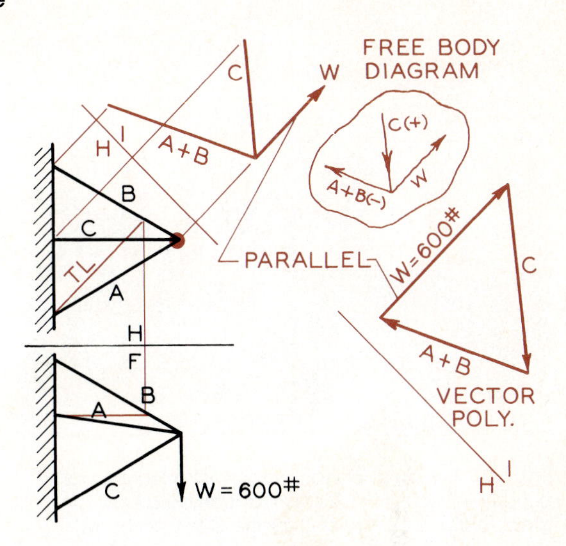

Step 1 To limit the unknowns to two, construct an auxiliary view to find two vectors lying in the edge of a plane. Use the auxiliary view and top view in the remainder of the problem. Draw a vector polygon parallel to the members in the auxiliary view in which $W = 600$ lb is the only known vector. Sketch a free-body diagram for preliminary analysis.

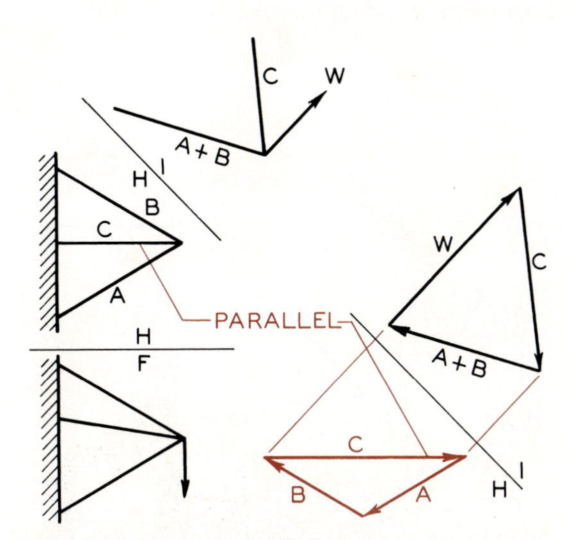

Step 2 Construct an orthographic projection of the view of the vector polygon found in Step 1 so that its vectors are parallel to the members in the top view. The reference plane between the two views is parallel to the *H*–1 plane. This portion of the problem is closely related to the problem in Fig. 8.19.

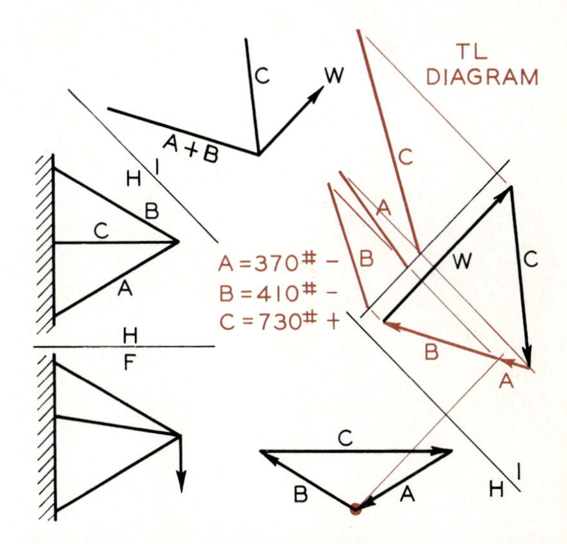

Step 3 Project the intersection of vectors *A* and *B* in the horizontal view of the vector polygon to the auxiliary view polygon to establish the lengths of vectors *A* and *B*. Determine the true lengths of all vectors in a true-length diagram to determine their magnitudes. Analyze for tension or compression, as covered in Section 8.11.

Fig. 8.21 Tractor sidebooms represent noncoplanar, concurrent systems of forces that can be solved graphically. (Courtesy of Trunkline Gas Company.)

A three-dimensional system is the side-boom tractors used for lowering pipe into a ditch during pipeline construction (Fig. 8.21).

8.14 NONCONCURRENT, COPLANAR VECTORS

Forces may be applied in such a manner that they are not concurrent, as illustrated in Fig. 8.22. Bow's notation can be used to locate the resultant of this type of nonconcurrent system.

In Step 1, the vectors are laid off to form a vector diagram in which the closing vector is the resultant, $R = 68$ lb. Each vector is resolved into two components by randomly locating point O on the interior or exterior of the polygon and connecting point O with the end of each vector. The components, or strings, from point O are equal and opposite components of adjacent vectors. For example, component o-b is common to vectors AB and BC. Since the strings from point O are equal and opposite, the system has not changed statically.

In Step 2, each string is transferred to the space diagram of the vectors where it is drawn between the respective vectors to which it applies. (The figure thus produced is called a *funicular diagram.*) For instance, string o-b is drawn in the

Fig. 8.22 Resultant of nonconcurrent forces

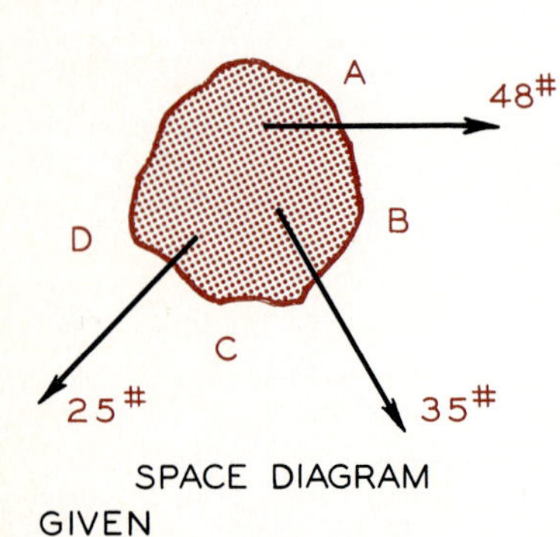

Required Find the resultant of the known forces applied to the above object. The forces are nonconcurrent.

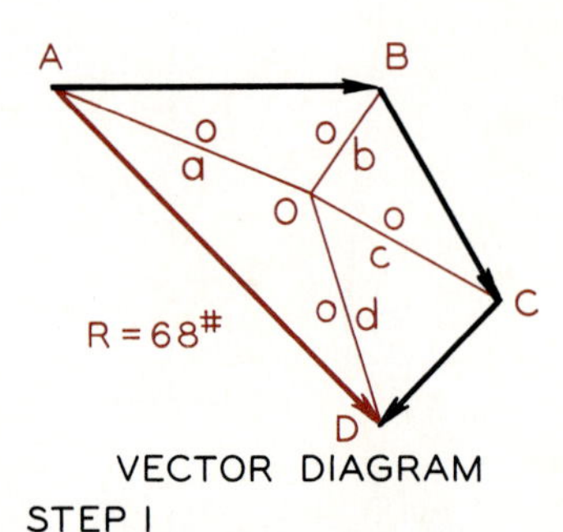

Step 1 The vectors are drawn head-to-tail to find resultant *R*. Point *O* is conveniently located for the construction of strings to the ends of each vector.

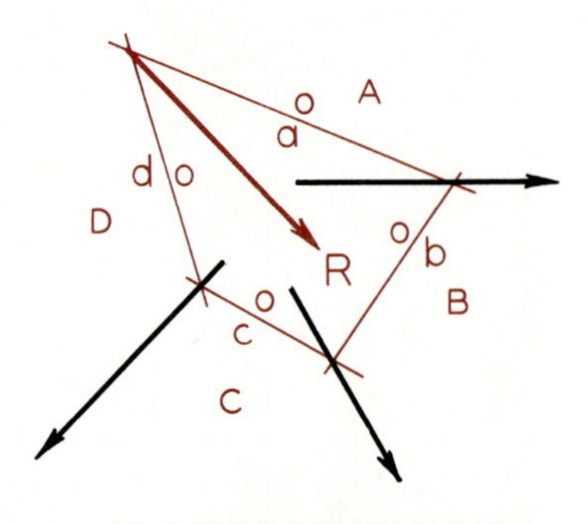

Step 2 Each string is drawn between the two vectors to which it applies in Step 1. *Example: o–c* between *BC* and *CD*. These strings are connected in sequence until the strings *o–a* and *o–d* establish the position of *R*, found in Step 1.

area between the vectors *AB* and *BC*. String *o-c* is drawn in the *C*-area to connect at the intersection of *o-b* and vector *BC*. The point of intersection of the last strings, *o-a* and *o-d*, locates a point through which the resultant *R* will pass. The resultant has now been determined with respect to magnitude, sense, direction, and point of application.

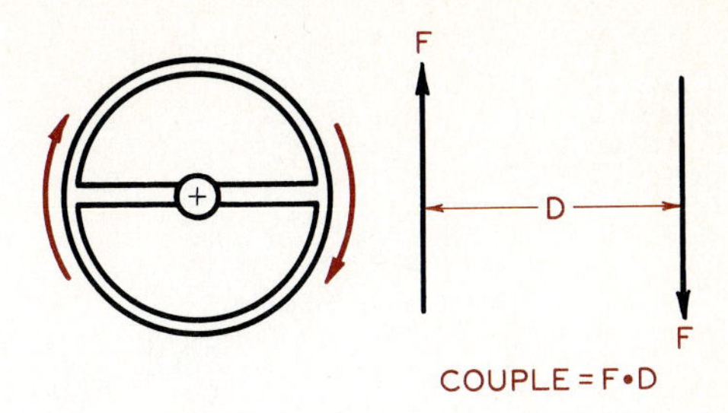

Fig. 8.23 Representation of a couple or moment.

8.15 NONCONCURRENT SYSTEMS RESULTING IN COUPLES

A *couple* is the descriptive name given to two parallel, equal, and opposite forces that are separated by a distance and are applied to a member in such a manner that they cause the member to rotate.

An important quantity associated with a couple is its *moment*.The moment of any force is a measure of its rotational effect. An example is shown in Fig. 8.23, in which two equal and opposite forces are applied to a wheel. The forces are separated by the distance *D*. The moment of the couple is found by multiplying one of the forces by the perpendicular distance between it and a point on the line of action of the other: $F \times D$. If the force is 20 lb and the distance is 3 ft, the moment of the couple would be given as 60 ft-lb.

A series of parallel forces is applied to a beam in Fig. 8.24. The spaces between the vectors are labeled with letters that follow Bow's notation. We are required to determine the resultant.

After constructing a vector diagram, we have a straight line that is parallel to the direction of the forces and that closes at point *A*. We then locate pole point *O* and draw the strings of a funicular diagram.

The strings are transferred to the space diagram and are drawn in their respective spaces. For example, *o–c* is drawn in the *C*-space between vectors *BC* and *CD*. The last two strings, *o–d* and *o–a*, do not close at a common point, but are found to be parallel; the result is therefore a cou-

Fig. 8.24 Couple resultants

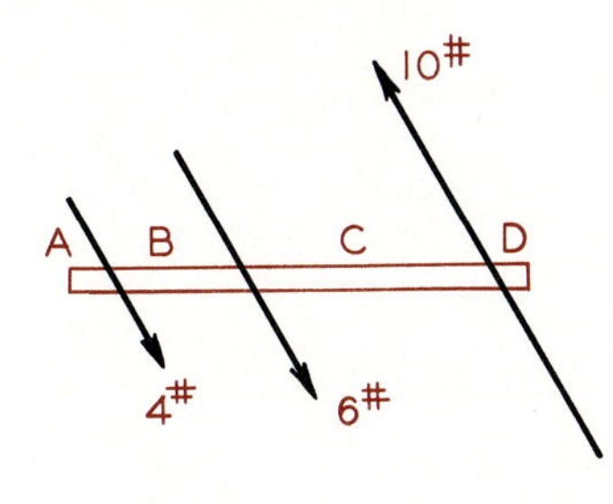

Required Find the resultant of these nonconcurrent forces applied to this beam.

Step 1 The spaces between each force are labeled in Bow's notation.

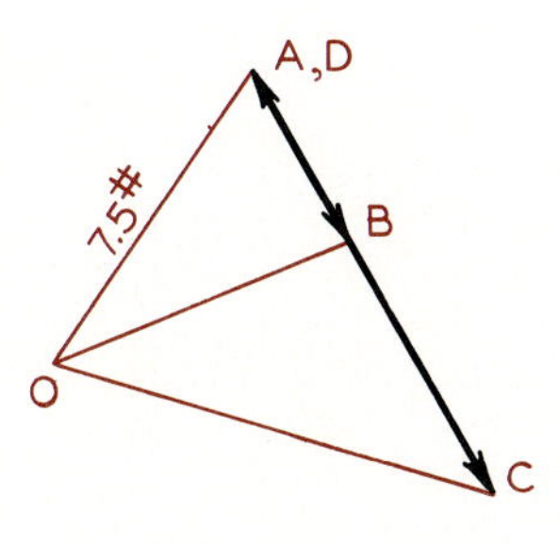

Step 2 The vectors are laid out head-to-tail; they will lie in a straight line since they are parallel. Pole point O is located in a convenient location and the ends of each vector are connected with point O.

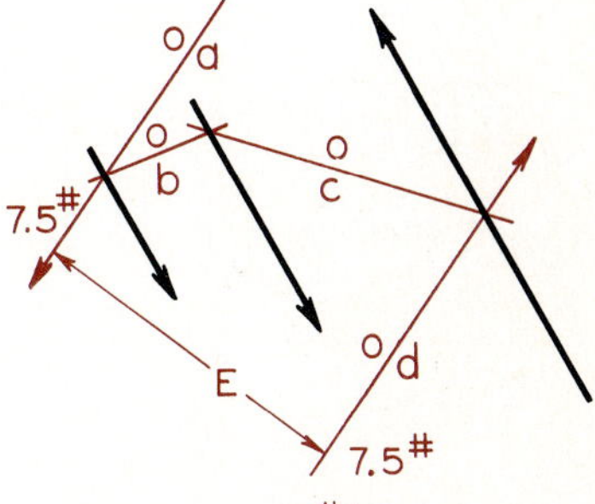

Step 3 Strings *o–a*, *o–b*, *o–c*, and *o–d* are successively drawn between the vectors to which they apply. Since strings *o–a* and *o–d* are parallel, the resultant will be a couple equal to 7.5 lb × *E*, where *E* is the distance between *o–a* and *o–d*.

Fig. 8.25 The boom of this crane can be analyzed for its resultant as a parallel, nonconcurrent system of forces when the cables have been disregarded.

ple. The distance between the forces of the couple is the perpendicular distance, E, between strings o–a and o–d in the space diagram, using the scale of the space diagram. The magnitude of the force is the scaled distance from point O to A and D in the vector diagram, using the scale of the vector diagram. The moment of the couple is equal to 7.5 lb × E in a counterclockwise direction.

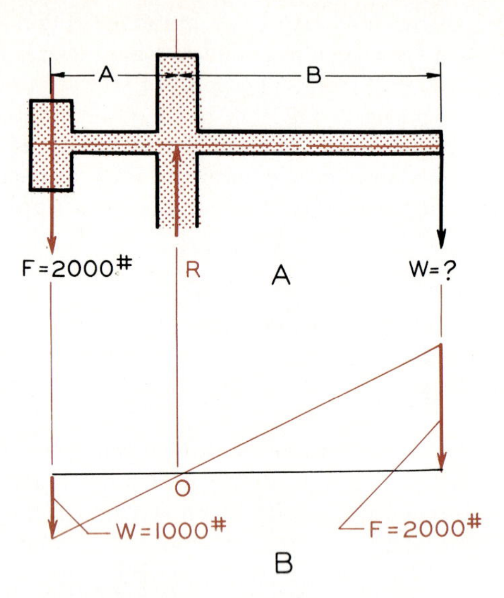

Fig. 8.26 Determining the resultant of parallel, nonconcurrent forces.

Fig. 8.27 Beam analysis with parallel loads

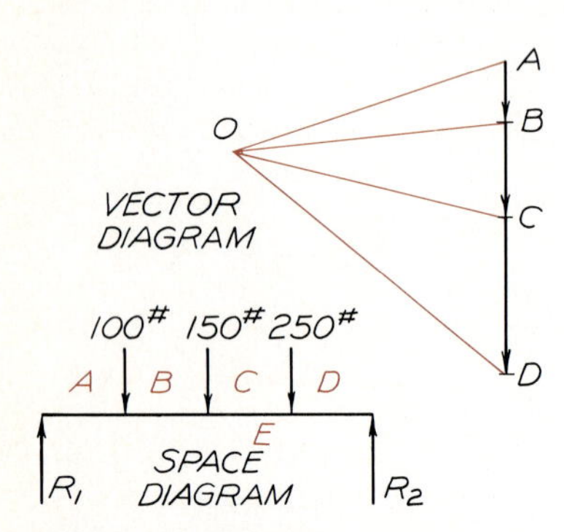

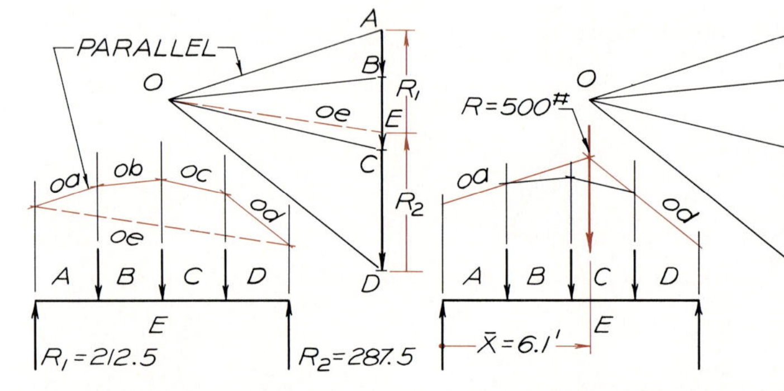

Step 1 Letter the spaces between the loads with Bow's notation. Find the graphical summation of the vectors by drawing them head-to-tail in a vector diagram at a convenient scale. Locate pole point O at a convenient location and draw strings from point O to each end of the vectors.

Step 2 Extend the lines of force in the space diagram, and draw a funicular diagram with string o–a in the A-space, o–b in the B-space, o–c in the C-space, etc. The last string, which is drawn to close the diagram, is o–e. Transfer this string to the vector polygon and use it to locate point E, thus establishing the lengths of R_1 and R_2, which are EA and DE, respectively.

Step 3 The resultant of the three downward forces will be equal to their graphical summation, line AD. Locate the resultant by extending strings o–a and o–d in the funicular diagram to a point of intersection. The resultant R = 500 lb will act through this point in a downward direction. $\bar{X}$ is a locating dimension.

8.16 RESULTANT OF PARALLEL, NONCONCURRENT FORCES

Forces applied to beams, such as those shown in Fig. 8.25, are parallel and nonconcurrent in many instances, and they may have the effect of a couple, tending to cause a rotational motion.

The beam in Fig. 8.26 is on a rotational crane that is used to move building materials in a limited area. The magnitude of the weight W is unknown, but the counterbalance weight is known to be 2000 lb; column R supports the beam as shown. Assuming that the support cables have been omitted, we desire to find the weight W that would balance the beam.

This problem can be solved by the application of the law of moments, i.e., the force is multiplied by the perpendicular distance to its line of action from a given point, or $F \times A$. If the beam is to be in balance, the total effect of the moments must be equal to zero, or $F \times A = W \times B$.

The graphical solution (Fig. 8.26B) is found by constructing a line to represent the total distance between the forces F and W. Point O is projected from the space diagram to this line. Point O is the point of balance where the summation of the moments will be equal to zero. Vectors F and W are drawn to scale at each end of the line by transposing them to the opposite ends of the beam. A line is drawn from the end of vector F through point O and extended to intersect the direction of vector W. This point represents the end of vector W, which can be scaled, resulting in a magnitude of 1000 lb.

8.17 RESULTANT OF PARALLEL, NONCONCURRENT FORCES ON A BEAM

The beam given in Fig. 8.27 is supported at each end and must in turn support three given loads. We are required to determine the magnitude of each support, R_1 and R_2, along with the resultant of the loads and its location. The spaces between all vectors are labeled in a clockwise direction with Bow's notation in Step 1, and a force diagram is drawn.

In Step 2 the lines of force in the space diagram are extended and the strings from the vector diagram are drawn in their respective spaces, parallel to their original direction. *Example:* String oa is drawn parallel to string OA in space A between forces EA and AB, and string ob is drawn in space B beginning at the intersection of oa with vector AB. The last string, oe, is drawn to close the funicular diagram. The direction of string oe is transferred to the force diagram, where it is laid off through point O to intersect the load line at point E. Vector DE represents support R_2 (refer to Bow's notation as it was applied in Step 1). Vector EA represents support R_1.

The magnitude of the resultant of the loads (Step 3) is the summation of the vertical downward forces, or the distance from A to D, or 500 lb. The location of the resultant is found by extending the extreme outside strings in the funicular diagram, oa and od, to their point of intersection. The resultant is found to have a magnitude of 500 lb, a vertical direction, a downward sense, and a point of application established by $\overline{X}$.

PROBLEMS

Problems should be presented in instrument drawings on Size A (8½″ × 11″) paper, grid or plain. Each grid square represents .20″. All notes, sketches, drawings, and graphical work should be neatly prepared in keeping with good practices. Written matter should be legibly lettered using ⅛″ guidelines.

1. In Fig. 8.28A, determine the resultant of the force system by the parallelogram method at the left of the sheet. Solve the same system using the vector polygon method at the right of the sheet. Scale: 1″ = 100 lb (note that each grid square equals .20″). (B) In part B of the figure, determine the resultant of the concurrent, coplanar force system shown at the left of the sheet by the parallelogram method. Solve the same system using the polygon method at the right of the sheet. Scale 1″ = 100 lb.

2. (A and B) In Fig. 8.29 solve for the resultant of each of the concurrent, noncoplanar force systems by the parallelogram method at the left of the sheet. Solve for the resultant of the same systems by the vector polygon method at the right of the sheet. Find the true length of the resultant in both problems. Letter all construction. Scale: 1″ = 600 lb.

3. (A and B) In Fig. 8.30, the concurrent, coplanar force systems are in equilibrium. Find the loads in each structural member. Use a scale of 1″ = 300 lb in part A and a scale of 1″ = 200 lb in part B. Show and label all construction.

4. In Fig. 8.31, solve for the loads in the structural members of the truss. Vector polygon scale: 1″ = 2000 lb. Label all construction.

5. In Fig. 8.32, solve for the loads in the structural members of the concurrent, noncoplanar force system. Find the true length of all vectors. Scale: 1″ = 300 lb.

6. In Fig. 8.33, solve for the loads in the structural members of the concurrent, noncoplanar force system. Find the true length of all vectors. Scale: 1″ = 400 lb.

7. (A) In Fig. 8.34, find the resultant of the coplanar, nonconcurrent force system. The vectors are drawn to a scale of 1″ = 100 lb. (B) In part B of the figure, solve for the resultant of the coplanar,

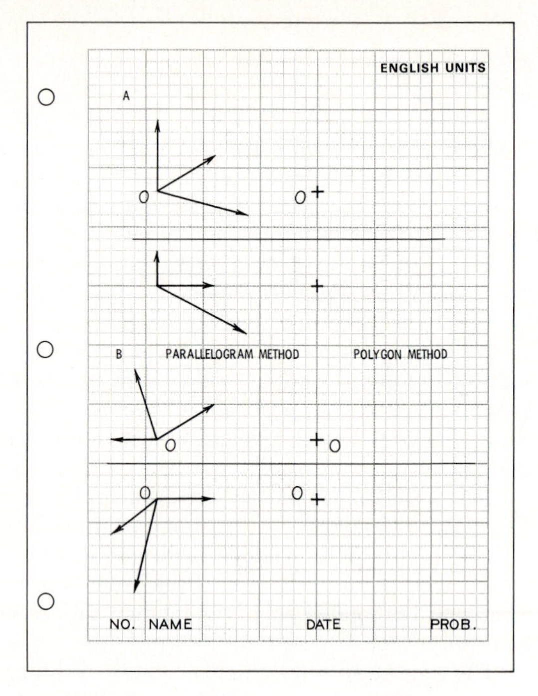

Fig. 8.29 Resultant of concurrent, noncoplanar vectors.

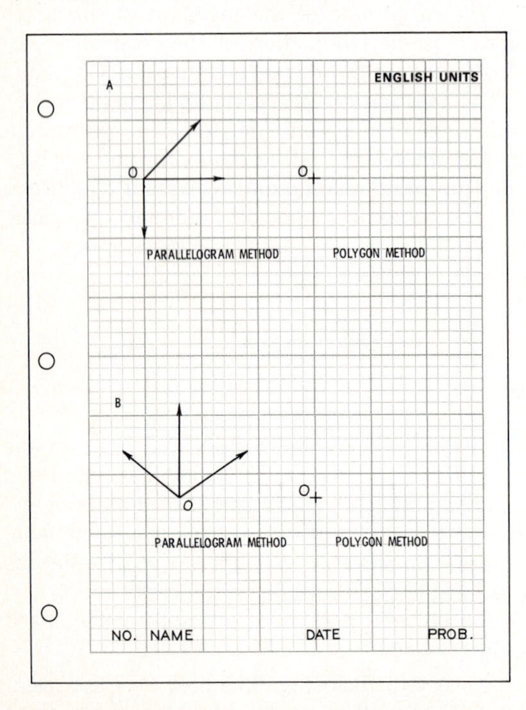

Fig. 8.28 Resultant of concurrent, coplanar vectors.

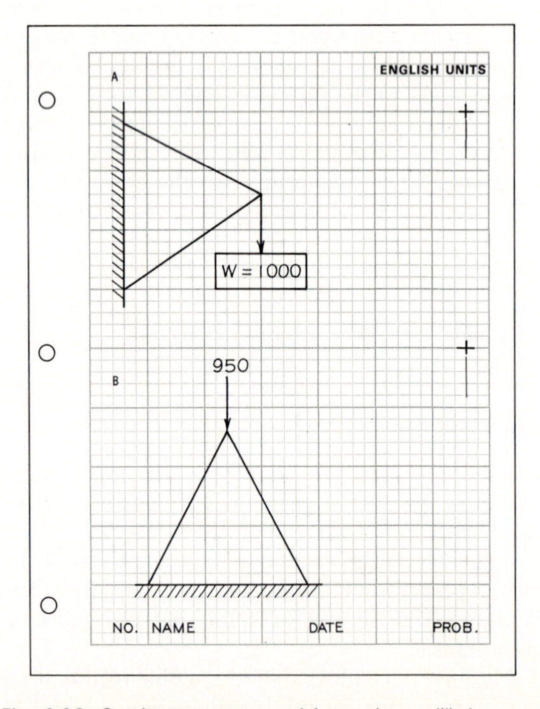

Fig. 8.30 Coplanar, concurrent forces in equilibrium.

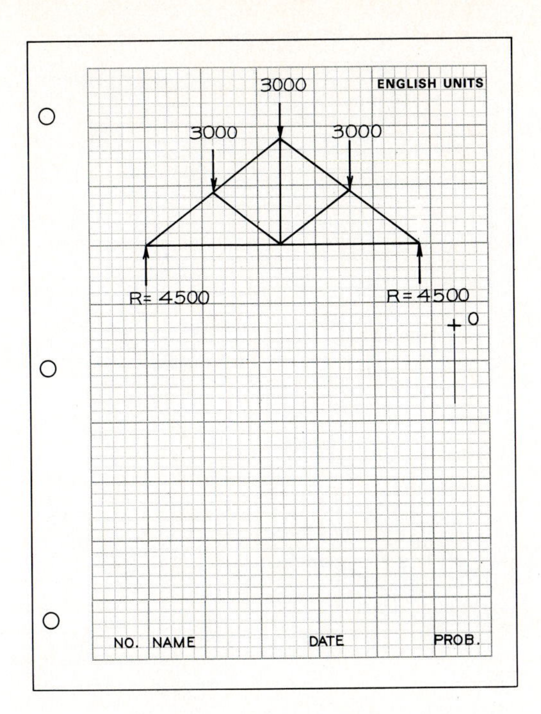

Fig. 8.31 Truss analysis.

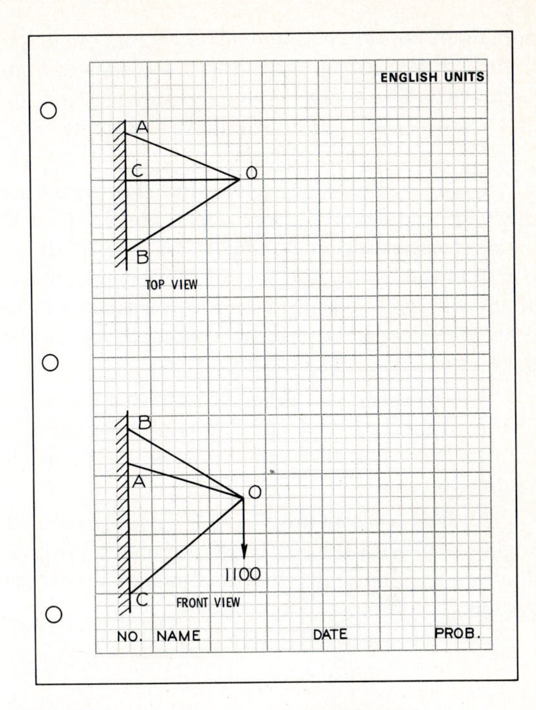

Fig. 8.33 Noncoplanar, concurrent forces in equilibrium.

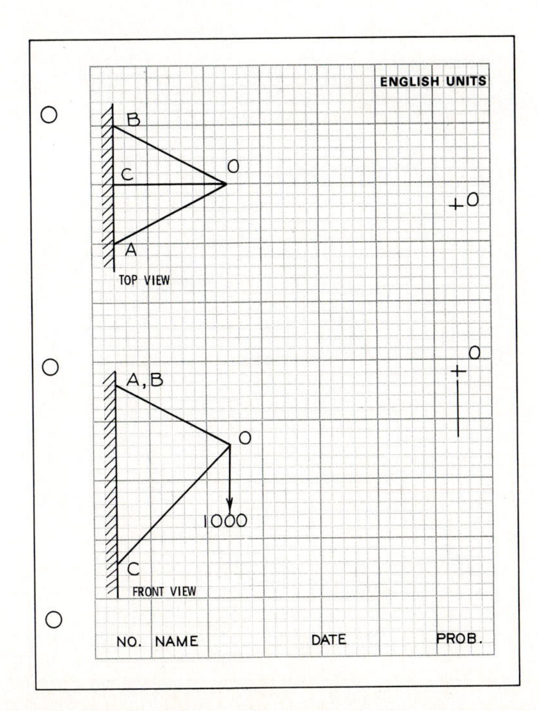

Fig. 8.32 Noncoplanar, concurrent forces in equilibrium.

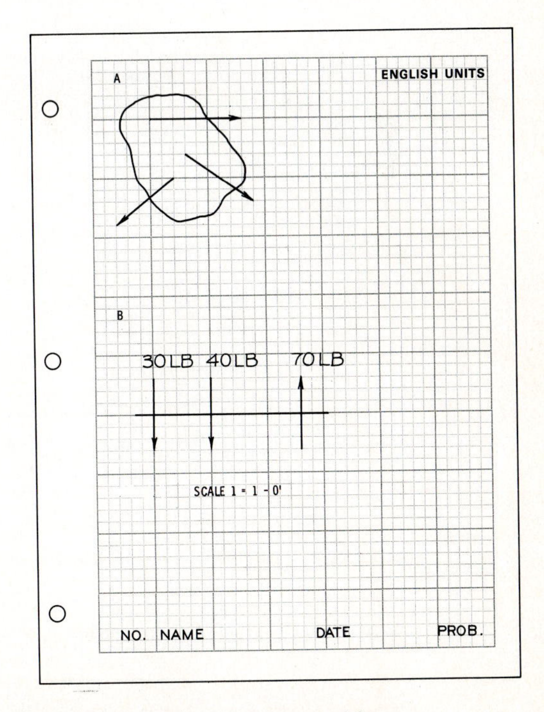

Fig. 8.34 Coplanar, nonconcurrent forces.

nonconcurrent force system. The vectors are given in their true positions and at the true distances from each other. The space diagram is drawn to a scale of 1″ = 1.0′. Draw the vectors to a scale of 1″ = 30 lb. Show all construction.

8. (A) In Fig. 8.35, determine the force that must be applied at *A* to balance the horizontal member supported at *B*. Scale 1″ = 100 lb. (B) In part B of the figure, find the resultants at each end of the horizontal beam. Find the resultant of the downward loads and determine where it would be positioned. Scale: 1″ = 600 lb.

9. Determine the forces in the three members of the tripod in Fig. 8.36. The tripod supports a load of *W* = 250 lb. Find the true lengths of all vectors.

10. The vectors in Fig. 8.37 each make an angle of 60° with the structural member on which they are applied. Find the resultant of this force system.

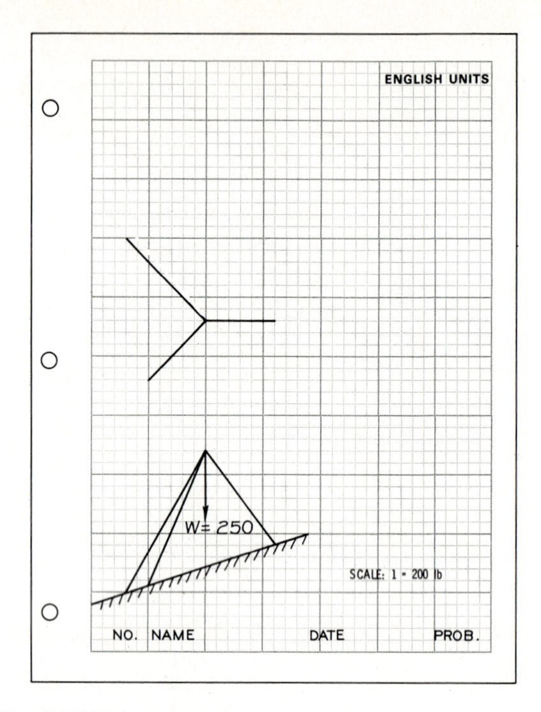

Fig. 8.36 Beam analysis.

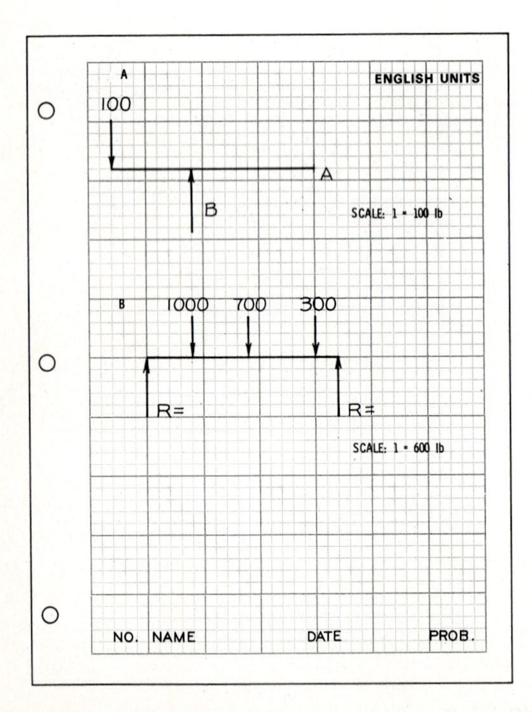

Fig. 8.35 Beam analysis.

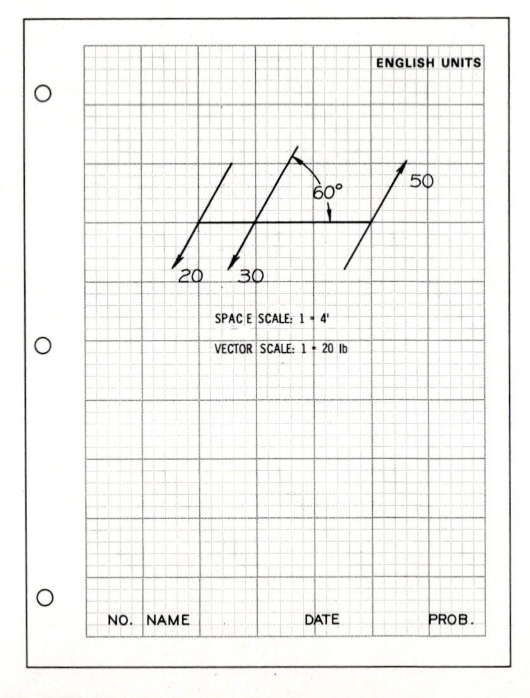

Fig. 8.37 Noncoplanar, concurrent forces in equilibrium.

INTERSECTIONS AND DEVELOPMENTS

9.1 INTRODUCTION

This chapter deals with the methods of finding lines of **intersections** between parts that join together. Usually these parts are made of sheet metal or plywood used for forming concrete to a desired shape.

Once the intersections have been found, flat patterns called **developments** can be found graphically. The patterns can then be laid out on the sheet metal and cut to conform to the desired shape. Consequently, intersection and development problems are closely related. Developments cannot be found until after the intersections are determined.

You can see many examples of intersections and developments in Fig. 9.1 where a refinery is under construction. Although this is a massive project, the principles covered in this chapter must be used in solving problems of this type.

Fig. 9.1 This refinery installation illustrates many examples of the application of principles of intersections and developments.

9.2 INTERSECTIONS OF LINES AND PLANES

The basic steps of finding the intersection between a line and a plane are illustrated in Fig. 9.2. This is a special case since the plane appears as an edge, and the point of intersection can be easily seen in this view, Step 1. It is projected to the

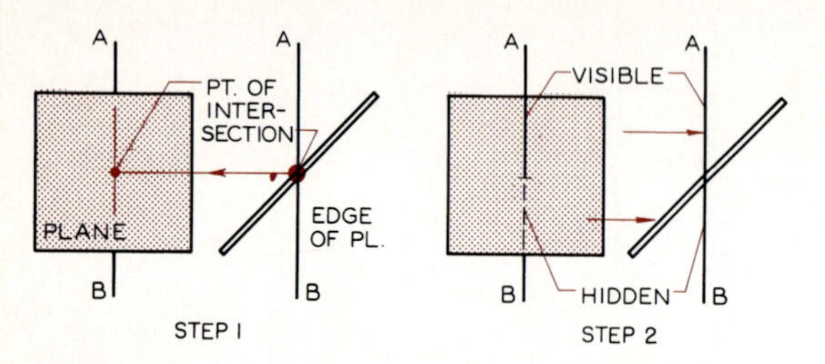

Fig. 9.2 Intersection of a line and a plane

Step 1 The point of intersection can be found in the view where the plane appears as an edge, the side view in this example.

Step 2 Visibility in the front view is determined by looking from the front view to the right side view.

front view, and the visibility of the line is found in Step 2.

This principle is used in Fig. 9.3 to find the line of intersection between two planes. Since *EFGH* appears as an edge in the side view, points of intersection 1 and 2 can be found and projected to the front view and the visibility determined in Step 2. Note that the intersection was found by finding the piercing points of lines *AB* and *DC* and connecting these points.

The intersection of a plane at a corner of a prism results in a line of intersection that bends around the corner (Fig. 9.4). Piercing points 2′ and 1′ are found in Step 1.

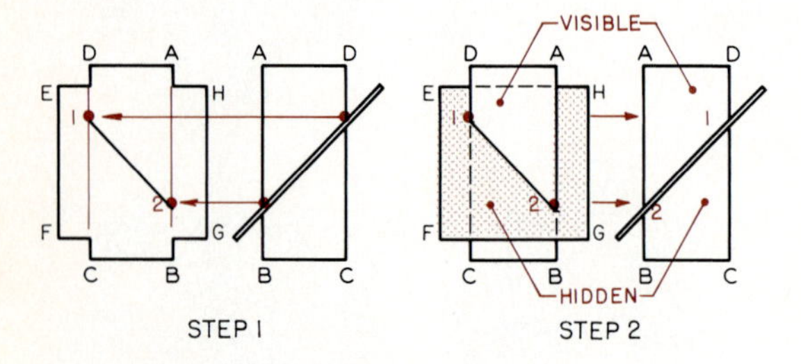

Fig. 9.3 Intersection between planes

Step 1 The points where lines *AB* and *DC* intersect plane *EFGH* are found in the view where the plane appears as an edge. These points are projected to the front view.

Step 2 Line 1–2 is the line of intersection. Visibility is determined by looking from the front view to the right side view.

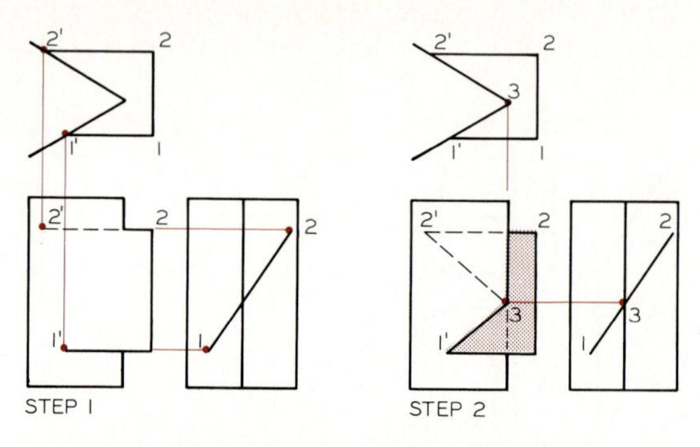

Fig. 9.4 Intersection of a plane at a corner

Step 1 The intersecting plane appears as an edge in the side view. Intersection points 1′ and 2′ are projected from the top and side views to the front view.

Step 2 The line of intersection from 1′ to 2′ must bend around the vertical corner at point 3 in the top and side views. This point is projected to the front view to locate line 1′–2′–3′.

Corner point 3 is seen in the side view of Step 2 where the vertical corner pierces the plane. Point 3 is projected to the corner in the front view. Point 2′ is hidden in the front view since it is on the back side of the assembly.

The intersection of a plane and a prism is found in Fig. 9.5, where the plane appears as an edge. The points of intersection are found for each corner line and are connected; visibility is shown to complete the line of intersection.

An intersection between a plane and a prism is shown in Fig. 9.6 where the vertical corners of the prism are true length in the front view and the plane appears foreshortened in both views. Imaginary cutting planes are passed vertically through the planes of the prism in the top view to find the piercing points of the corners in the front view. The points are connected and the visibility is determined to complete the solution.

The intersection between a foreshortened plane and an oblique prism is found in Fig. 9.7. The plane is found as an edge in a primary auxiliary view. The piercing points of the corners of the prism are located in the auxiliary view and are projected back to the given views.

Points 1, 2, and 3 are projected from the auxiliary view to the given views as examples. Visibility is determined by analysis of crossing lines, as previously covered.

Fig. 9.5 Intersection of a plane and a prism

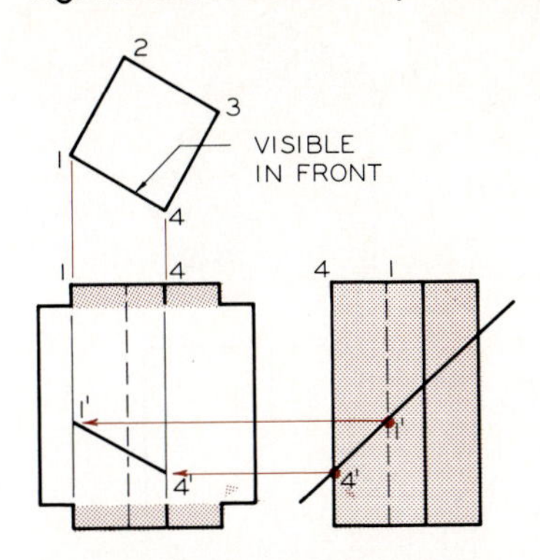

Step 1 Vertical corners 1 and 4 intersect the edge view of the plane in the side view at 1′ and 4′. These points are projected to the front view and are connected with a visible line.

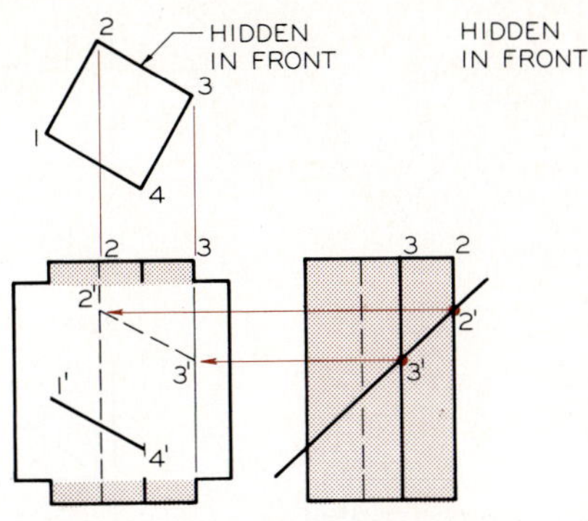

Step 2 Vertical corners 2 and 3 intersect the edge of the inclined plane at 2′ and 3′ in the side view. 2′ and 3′ are connected in the front view with a hidden line. Inspection of the top view tells us that this line is hidden.

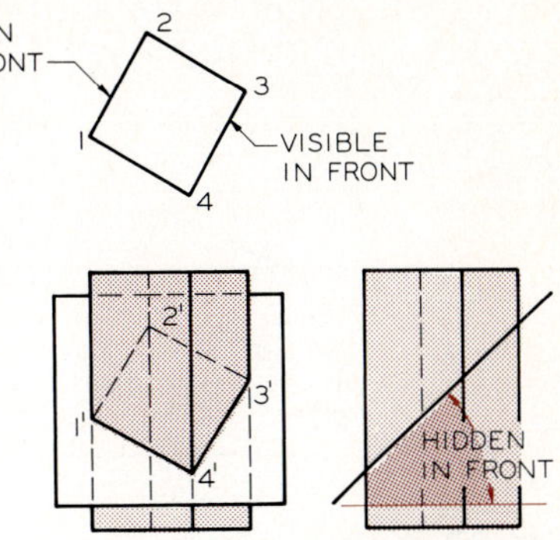

Step 3 Lines 1′–2′ and 3′–4′ are drawn as hidden and visible lines, respectively. Visibility is determined by inspection of the top and side views and by projection to the front view.

Fig. 9.6 Intersection of an oblique plane and a prism

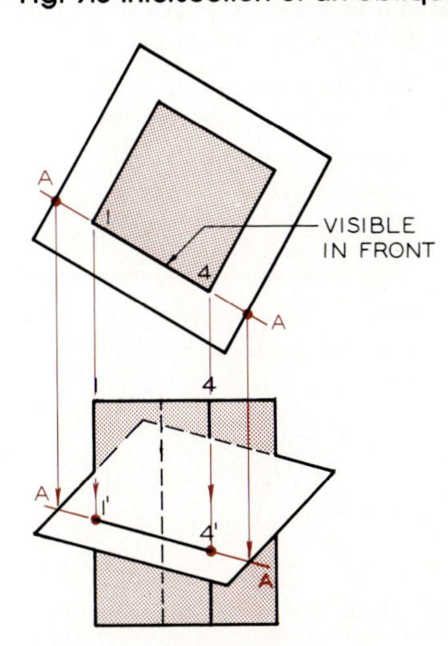

Step 1 Vertical cutting plane *A–A* is passed through the vertical plane, 1–4, in the top view and is projected to the front view. Piercing points 1′ and 4′ are found in this view.

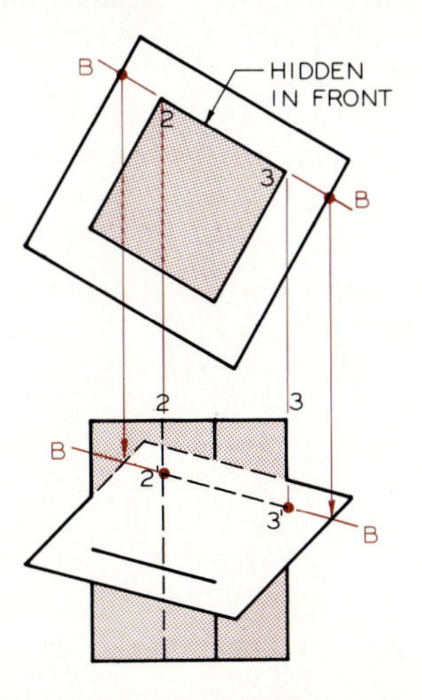

Step 2 Vertical plane *B–B* is passed through the top view of plane 2–3 and is projected to the front view where piercing points 2′ and 3′ are found. Line 2′–3′ is a hidden line.

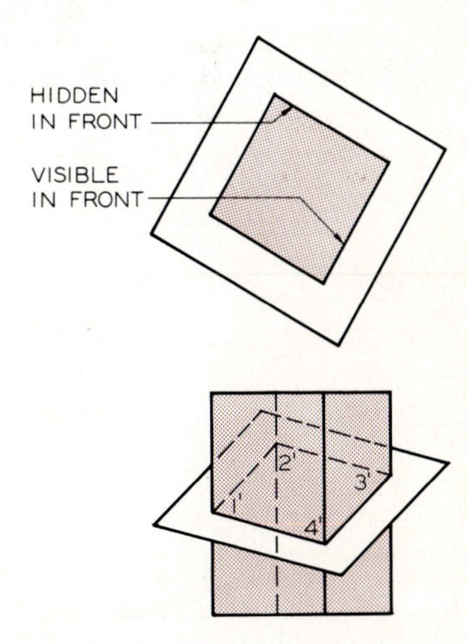

Step 3 The line of intersection is completed by connecting the four points in the front view. Visibility in the front view is found by inspection of the top view.

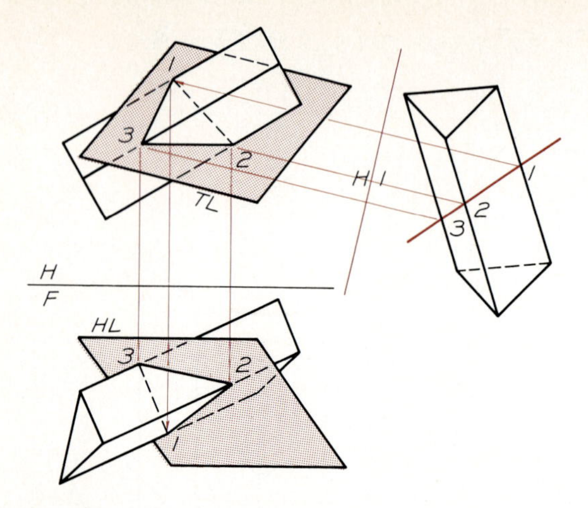

Fig. 9.7 The intersection between a plane and a prism can be found by constructing a view in which the plane appears as an edge.

Fig. 9.8 Three views of intersecting prisms. The points of intersection can be seen where intersecting planes appear as edges.

9.3 INTERSECTIONS BETWEEN PRISMS

The same principles used to find the intersection between a plane and a line are used to find the intersection between two prisms in Fig. 9.8. Piercing points 1, 2, 4, and 6 are found in the front view by projecting from the side and top views. Points 3 and 5 are located in the side view where the lines of intersection 2–4 and 4–6 bend around the vertical corner of the other prism. These points are connected in sequence and visibility is determined.

In Fig. 9.9, an inclined prism intersects a vertical prism. In Step 1, the end view of the inclined prism is found by an auxiliary view. In the auxiliary view, you can see where plane 1–2 bends around corner *AB* at point *X* in Step 2.

Fig. 9.9 Intersection between prisms

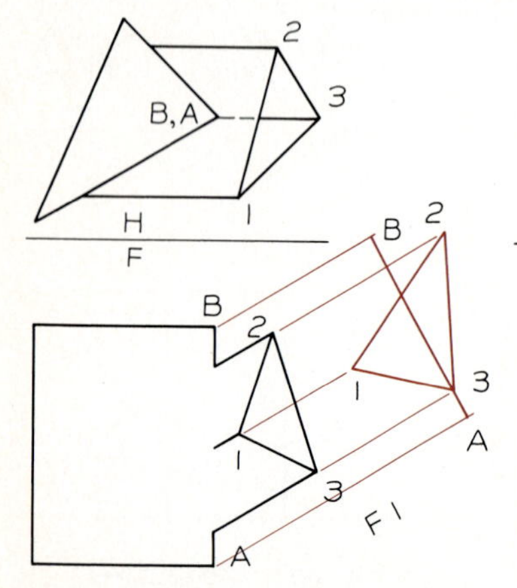

Step 1 Construct the end view of the inclined prism by projecting an auxiliary view from the front view. Show only line *AB* of the vertical prism in the auxiliary view.

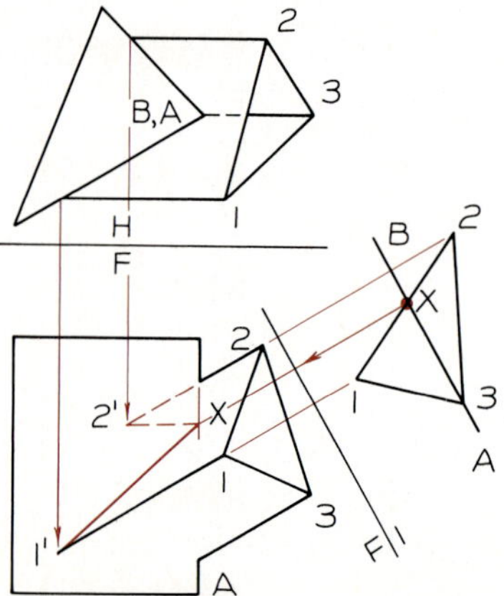

Step 2 Locate piercing points 1′ and 2′ in the top and front views. Intersection line 1′–2′ will not be a straight line, but it will bend around corner *AB* at point *X*, which is projected from the auxiliary view.

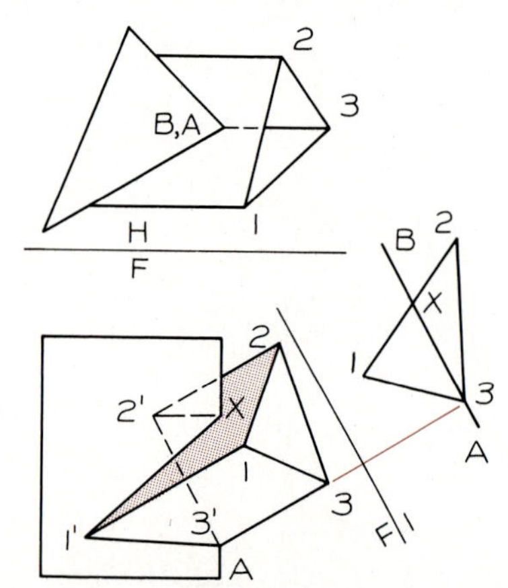

Step 3 Intersection lines from 2′ and 1′ to 3′ do not bend around the corner. Therefore, these are drawn as straight lines. Line 1′–3′ is visible and line 2′–3′ is invisible.

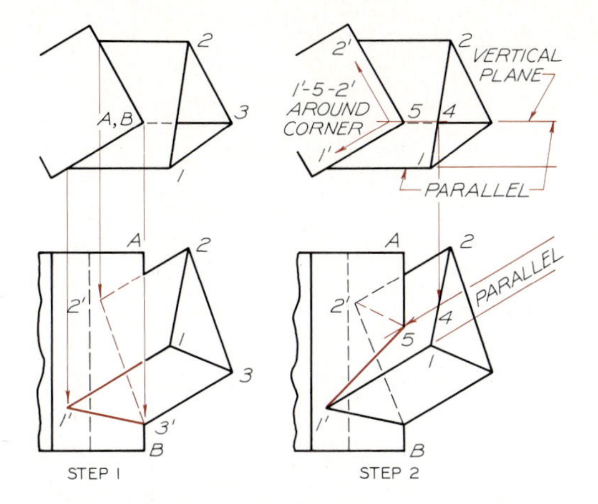

Fig. 9.10 Cutting planes can be passed through one prism parallel to its edges to locate point 5 where the plane wraps around corner 3–4.

Points of intersection 2′ and 3′ are projected from the top to the front view. The line of intersection 1′–X–2′ can be drawn to complete this portion of the line of intersection. The remaining lines, 1′–3′ and 2′–3′, are connected to complete the solution (Step 3).

An alternative method of solving a problem of this type is shown in Fig. 9.10 where vertical cutting planes are used to locate point 5 where the intersection bends around corner *AB*. The plane is passed through the corner and parallel to the sides of the inclined prism in the top view. It is projected to the front view where its projection intersects line *AB* at point 5, where the intersection bends around the corner. The other piercing points are found in the top view and are projected to the front view.

The conduit connector shown in Fig. 9.11 is an example of the application of intersecting planes and prisms.

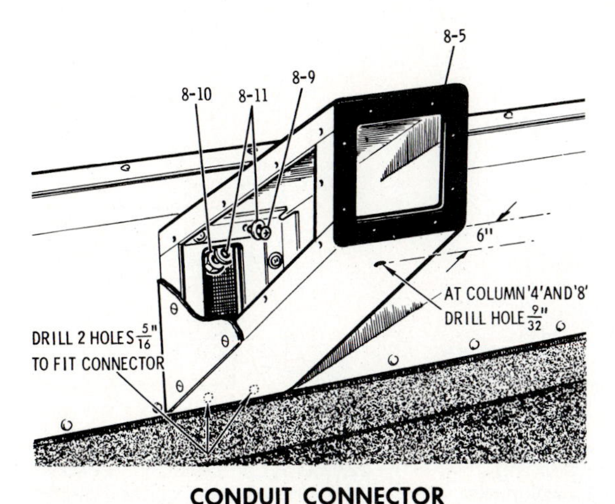

Fig. 9.11 This conduit connector was designed through the use of the principles of intersection of a plane and a prism. (Courtesy of the Federal Aviation Administration.)

9.4 INTERSECTION OF A PLANE AND CYLINDER

The intersections of the components of this gas transmission system in Fig. 9.12 offer numerous examples of problems that were solved using the principles of intersections.

The intersection between a plane and a cylinder is found in Fig. 9.13 where the plane appears as an edge in one of the given views. Cutting planes are passed vertically through the top view

Fig. 9.12 This complex of pipes and vessels contains many applications of intersections. (Courtesy of Trunkline Gas Company.)

Fig. 9.13 Intersection between a cylinder and a plane

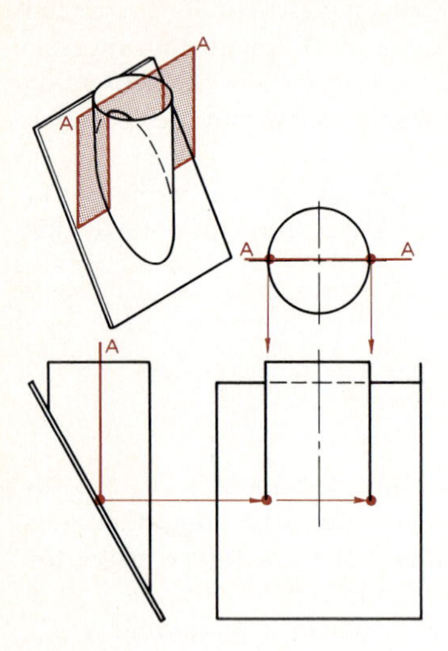

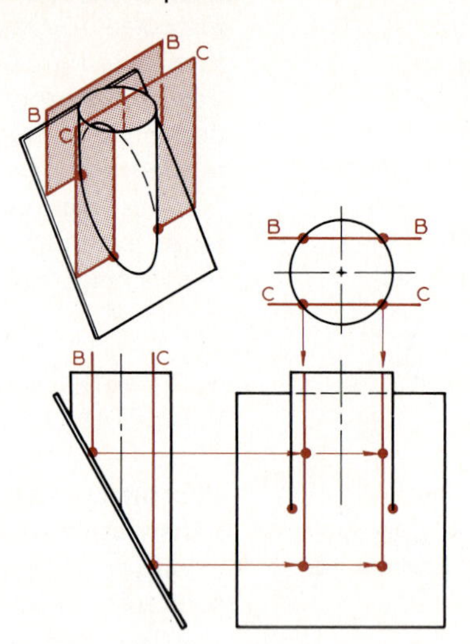

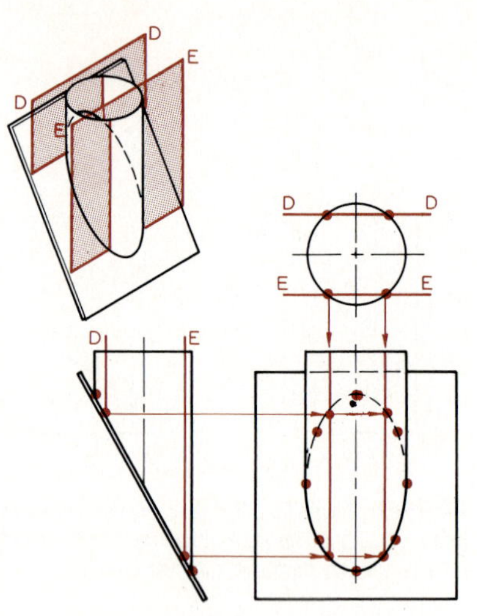

Step 1 A vertical cutting plane, *A–A,* is passed through the cylinder parallel to its axis to find two points of intersection.

Step 2 Two more cutting planes, *B–B* and *C–C,* are used to find four additional points in the top and left-side views; these points are projected to the front view.

Step 3 Additional cutting planes are used to find more points. These points are connected to give an elliptical line of intersection.

Fig. 9.14 Intersection of a cylinder and an oblique plane

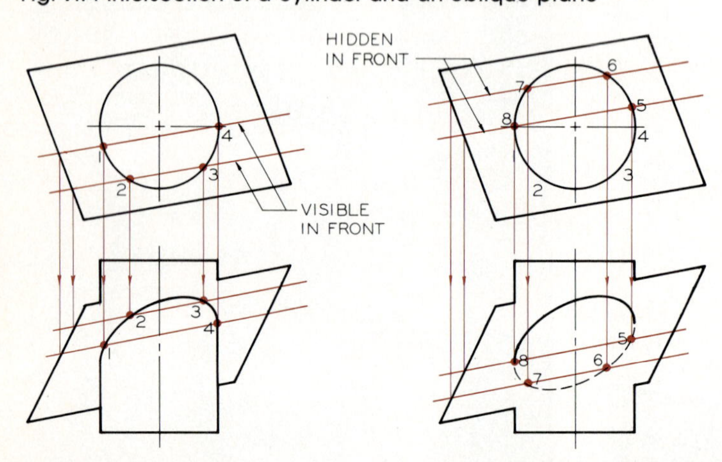

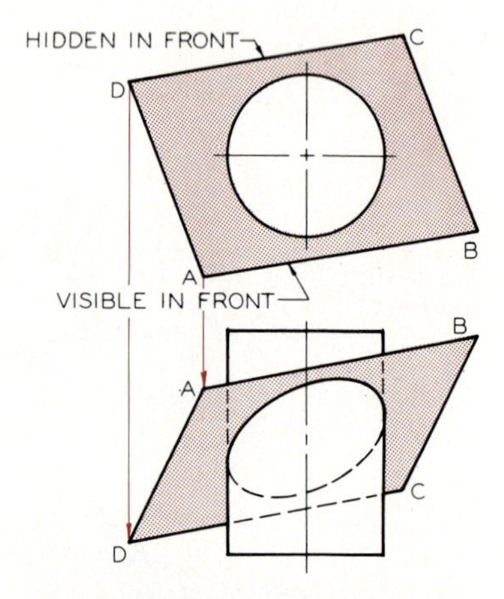

Step 1 Vertical cutting planes are passed through the cylinder in the top view to establish elements on its surface and lines on the oblique plane. Piercing points 1, 2, 3, and 4 are projected to the front view to their respective lines and are connected with a visible line.

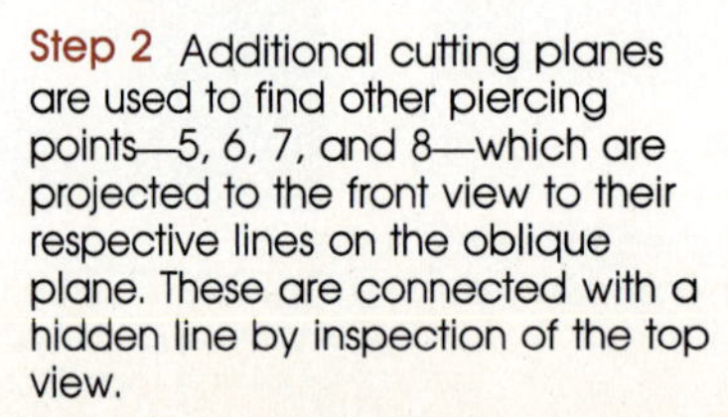

Step 2 Additional cutting planes are used to find other piercing points—5, 6, 7, and 8—which are projected to the front view to their respective lines on the oblique plane. These are connected with a hidden line by inspection of the top view.

Step 3 Visibility of the plane and cylinder is completed in the front view. Line *AB* is found to be visible by inspection of the top view, and line *CD* is found to be hidden.

of the cylinder to establish elements on the cylinder and their piercing points. The piercing points are projected to each view to find the line of intersection, which is an ellipse.

A more general problem is solved in Fig. 9.14 where the cylinder is vertical, but the plane is oblique. Vertical cutting planes are passed through the cylinder and the plane in the top view to find piercing points of the cylinder's elements on the plane. These points are projected to the front view to complete the line of intersection, an ellipse. The more cutting planes that are used, the more accurate will be the plotted line of intersection.

The general case of the intersection between a plane and cylinder is solved in Fig. 9.15 where both are oblique in the given views. The edge view of the plane is found in an auxiliary view. Cutting planes are passed through the cylinder parallel to the cylinder's axis in the auxiliary view to find the piercing points.

The piercing points of the elements are connected to give elliptical lines of intersection in the given views. Visibility is determined by analysis.

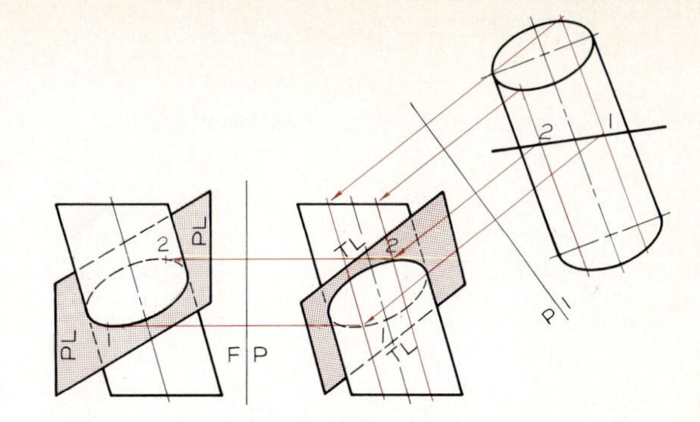

Fig. 9.15 The intersection between an oblique cylinder and an oblique plane can be found by constructing a view that shows the plane as an edge.

9.5 INTERSECTIONS BETWEEN CYLINDERS AND PRISMS

A series of vertical cutting planes are used in Fig. 9.16 to establish lines that lie on the surfaces of the cylinder and prism. A primary auxiliary view is drawn to show the end view of the inclined prism. The vertical cutting planes are shown in this view also, spaced the same distance apart as in the top view (Step 1).

Fig. 9.16 Intersection between a cylinder and a prism

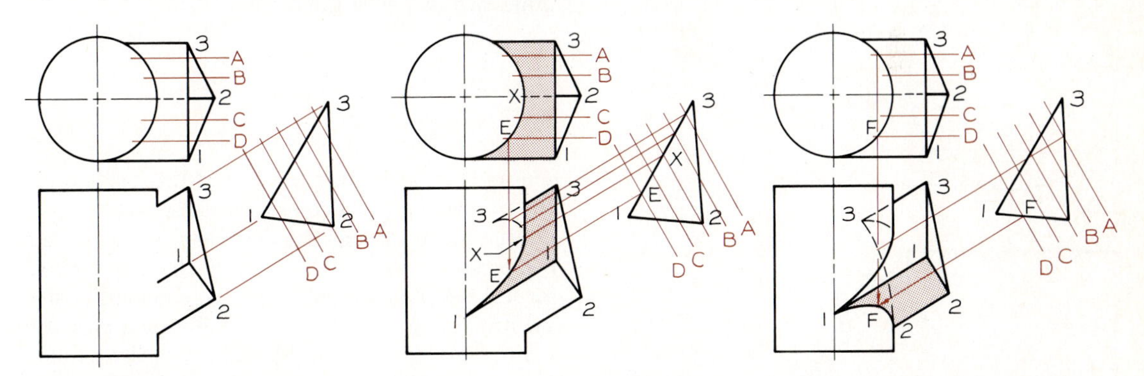

Step 1 Project an auxiliary view of the triangular prism from the front view to show three of its surfaces as edges. Pass frontal cutting planes through the top view of the cylinder and project them to the auxiliary view. The spacing between the planes is equal in both views.

Step 2 Locate points along the line of intersection of plane 1–3 in the top view and project them to the front view. *Example:* Point *E* on cutting plane *D* is found in the top and primary auxiliary views and projected to the front view where the projectors intersect. Point *X* on the centerline is the point where visibility changes from visible to hidden in the front view.

Step 3 Determine the remaining points of intersection by using the same cutting planes. Point *F* is shown in the top and primary auxiliary views and is projected to the front view on line of intersection 1–2. Connect the points and determine visibility. Judgment should be used in spacing the cutting planes so that they will produce an accurate plot of the line of intersection.

Fig. 9.17 Intersection between two cylinders

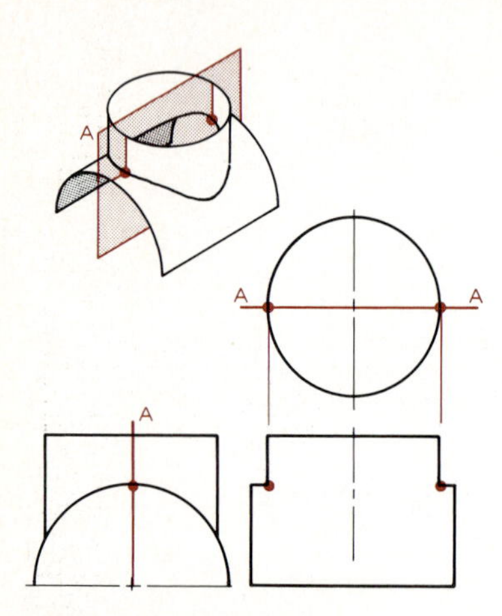

Step 1 A cutting plane, *A–A*, is passed through the cylinders parallel to the axes of both. Two points of intersection are found.

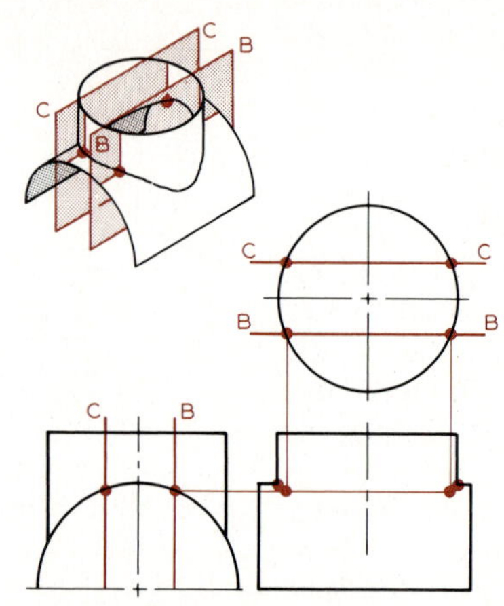

Step 2 Cutting planes *C–C* and *B–B* are used to find four additional points of intersection.

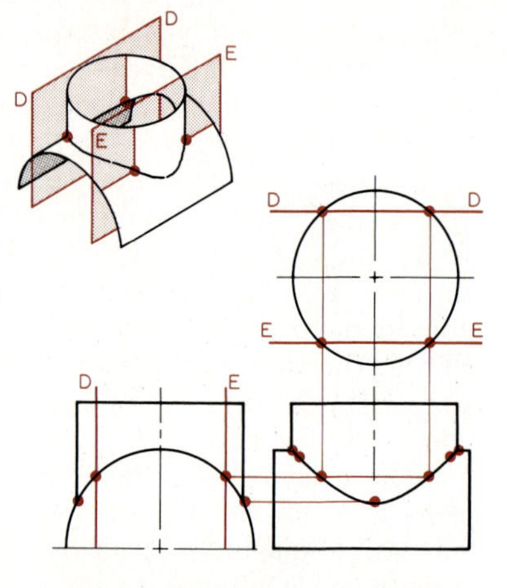

Step 3 Cutting planes *D–D* and *E–E* locate four more points. Points found in this manner are connected to give the line of intersection.

Fig. 9.18 The intersection between these cylinders is found by finding the end view of the inclined cylinder in an auxiliary view. Vertical cutting planes are used to find the piercing points of the elements of the cylinder and the line of intersection.

The line of intersection from 1 to 3 is projected from the auxiliary view to the front view, Step 2, where the intersection is an elliptical curve. The change of visibility of this line is found at point X in the top and auxiliary views, and is projected to the front view. In Step 3, the process is continued to find the lines of intersection of the other two planes of the prism.

9.6 INTERSECTIONS BETWEEN TWO CYLINDERS

The method of finding the line of intersection between two perpendicular cylinders is to pass a cutting plane through the cylinders parallel to the centerlines of each (Fig. 9.17). Each cutting plane locates the piercing points of two elements of one cylinder on an element of the other cylinder. The points are connected and visibility is determined to complete the solution.

The intersection between nonperpendicular cylinders is found in Fig. 9.18 by a series of vertical cutting planes. Each cutting plane is passed through the cylinders parallel to the centerline of each. Points 1 and 2 are labeled on cutting plane *D* as examples of points on the line of intersection. Other points are found in the same manner. The auxiliary view is an optional view that is not required for the solution of this problem, but it assists you in visualizing the problem. Points 1 and 2 are shown on cutting plane D in the auxiliary view, where they can be projected to the front view as a check on the solution found when projecting from the top view.

9.7 INTERSECTIONS BETWEEN PLANES AND CONES

To find points of intersection on a cone, cutting planes can be used that are (1) perpendicular to the cone's axis, or (2) parallel to the cone's axis. Horizontal cutting planes are shown in Fig. 9.19A where they are labeled as H_1 and H_2. The horizontal planes cut circular sections that appear true size in the top view.

The cutting planes in Fig. 9.19B are passed radially through the top view to establish elements on the surface of the cone that are projected to the front view. Points 1 and 2 are found on these elements in both views by projection. These two types of cutting planes are used to solve intersections involving cones.

A series of radial cutting planes is used to find elements on the cone in Fig. 9.20. These elements cross the edge view of the plane in the front view to locate piercing points that are projected to the top view of the same elements. The points are connected to form the line of intersection.

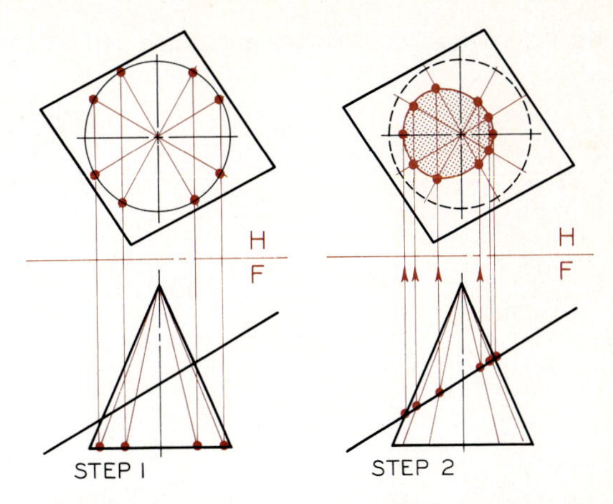

Fig. 9.20 Intersection of a plane and a cone

Step 1 Divide the base into even divisions in the top view and connect these points with the apex to establish elements on the cone. Project these to the front view.

Step 2 The piercing point of each element on the edge view of the plane is projected to the top view to the same elements, where they are connected to form the line of intersection. Visibility is shown to complete the drawing.

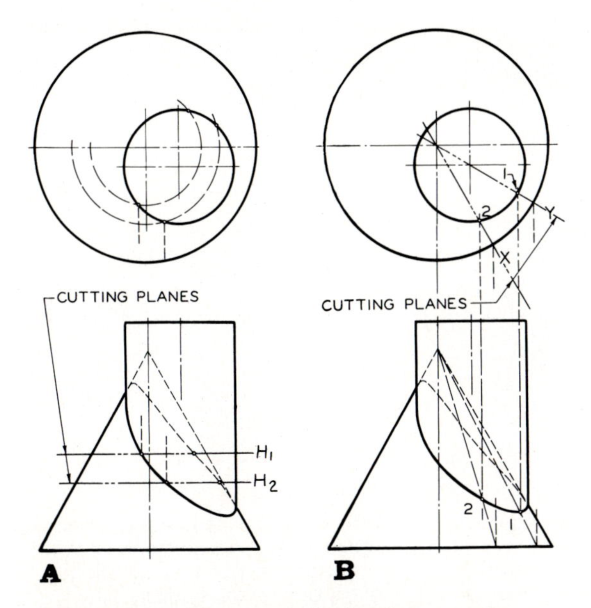

Fig. 9.19 Intersections on conical surfaces can be found by cutting planes that pass through the cone parallel to its base (part A). A second method at B shows radial cutting planes that pass through the cone's centerline and perpendicular to its base.

A cone and an oblique plane intersect in Fig. 9.21, and the line of intersection is found by using a series of horizontal cutting planes. The sections cut by these imaginary planes will be true circles in the top view. Also, the cutting planes locate lines on the oblique plane that intersect the same circular sections cut by each respective cutting plane. The points of intersection are found in the top view and are projected to the front view.

The horizontal cutting-plane method could have been used to solve the example in Fig. 9.20 as an alternative method.

9.8 INTERSECTIONS BETWEEN CONES AND PRISMS

A primary auxiliary view is used to find the end view of the inclined prism that intersects the cone in Fig. 9.22, Step 1. Cutting planes that radiate from the apex of the cone in the top view are

Fig. 9.21 Intersection of an oblique plane and a cone

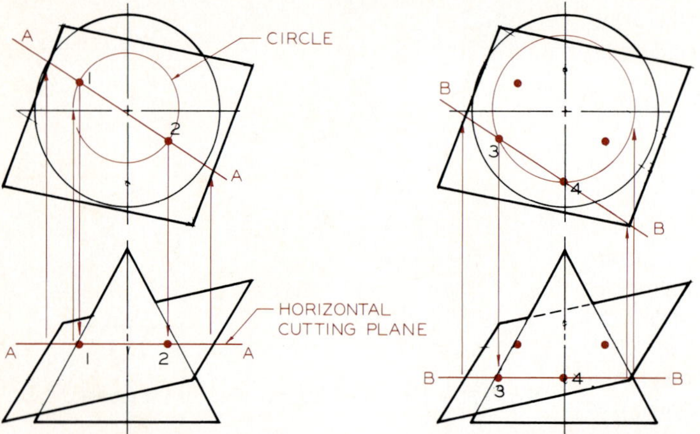

Step 1 A horizontal cutting plane is passed through the front view to establish a circular section on the cone and a line on the oblique plane in the top view. The piercing point of this line must lie on the circular section. Piercing points 1 and 2 are projected to the front view.

Step 2 Horizontal cutting plane *B–B* is passed through the front view in the same manner to locate piercing points 3 and 4 in the top view. These points are projected to the horizontal plane in the front view from the top view.

Step 3 Additional horizontal planes are used to find sufficient points to complete the line of intersection. Determination of the visibility completes the solution.

Fig. 9.22 Intersection between a cone and a prism

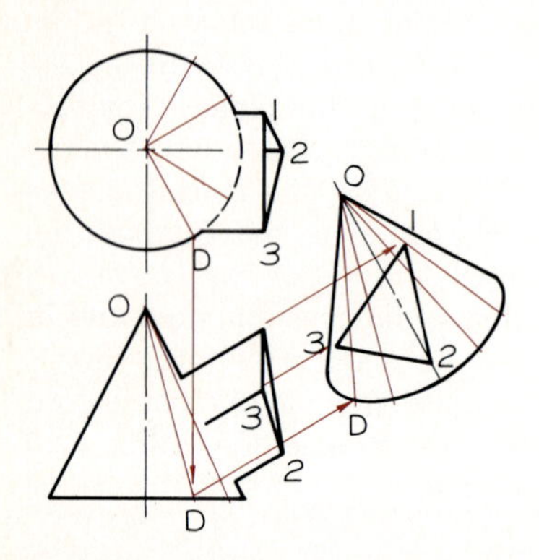

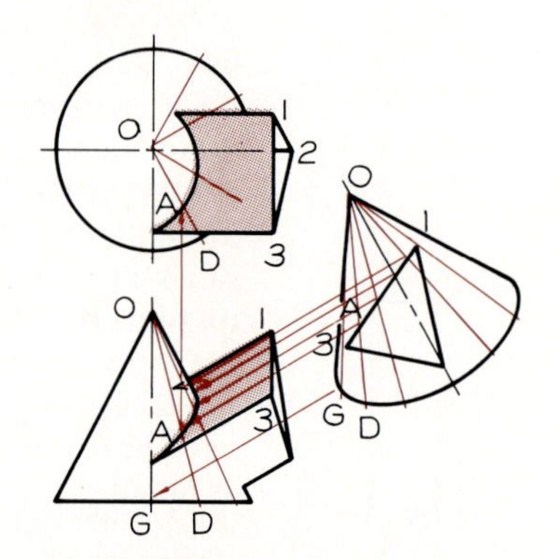

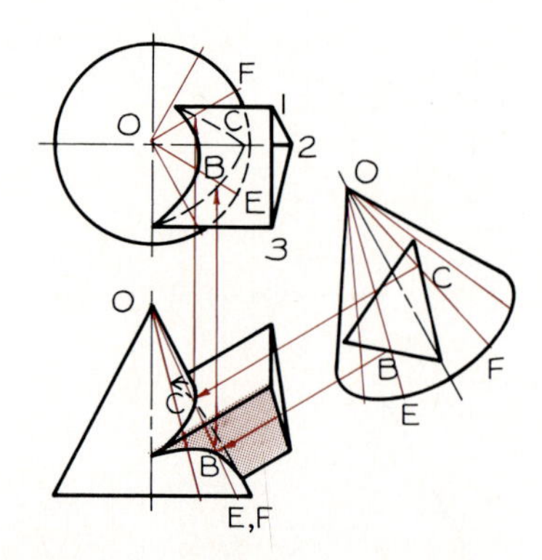

Step 1 Construct a primary auxiliary view to obtain the edge views of the lateral surfaces of the prism. In the auxiliary view, pass cutting planes through the cone which radiate from the apex to establish elements on the cone. Project the elements to the front and auxiliary views.

Step 2 Locate the piercing points of the cone's elements with the edge view of plane 1–3 in the auxiliary view and project them to the front and top views. *Example:* point A lies on element *OD* in the auxiliary view, so it is projected to the front and top views of element *OD*. Locate other points in this manner.

Step 3 Locate the piercing points where the conical elements intersect the edge views of the other planes of the prism in the auxiliary view. *Example:* Point *B* is found on *OE* in the primary auxiliary view and is projected to the front and top views of *OE*. Show visibility of the lines of intersection in each view.

drawn in the auxiliary view to locate elements on the cone's surface that intersect the prism. These elements are drawn in the front view by projection.

Wherever the edge view of plane 1–3 intersects an element in the auxiliary view, the piercing points are projected to the same element in the front and top views, Step 2. An extra cutting plane is passed through point 3 in the auxiliary view to locate an element that is projected to the front and top views. Piercing point 3 is projected to this element in sequence from the auxiliary view to the top view.

This same procedure is used to find the piercing points of the other two planes of the prism in Step 3. All projections of points of intersection originate in the auxiliary view, where the planes of the prism appear as edges.

A similar problem is solved in Fig. 9.23 where a cylinder intersects a cone. The circular view of the cylinder is found in a primary auxiliary view. Points 2 and 2′ are found in the auxiliary view on element *O–X* where a radial cutting plane is passed through apex *O*. Element *O–X* is found in the top and front views by projecting from the auxiliary view to locate points 2 and 2′.

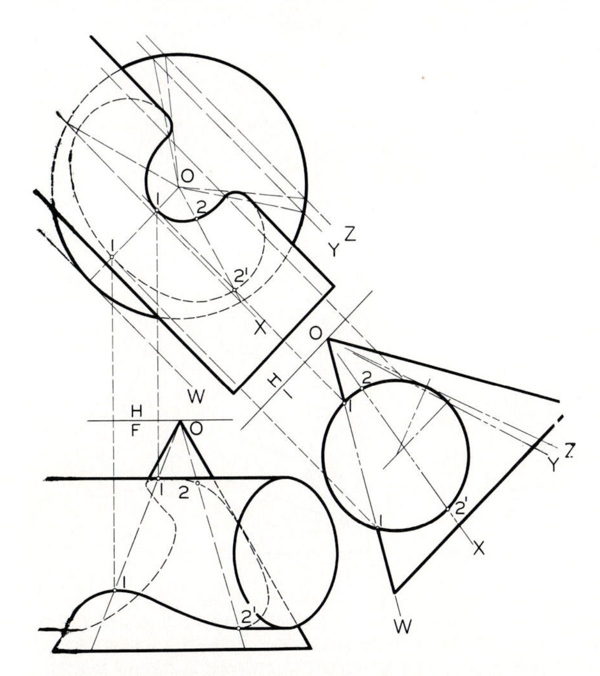

Fig. 9.23 The intersection between these two cylinders is found by projecting an auxiliary view from the top view of the cone to find the circular view of the cylinder. Radial cutting planes are passed through the cone and the cylinder in the auxiliary view to locate piercing points of the cylinder's elements.

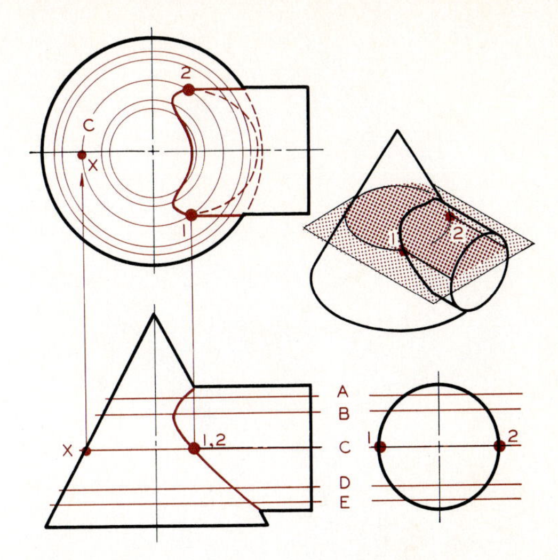

Fig. 9.24 Horizontal cutting planes are used to find the intersection between the cone and the cylinder. The cutting planes form circles in the top view.

In Fig. 9.24, horizontal cutting planes are passed through the cone and the intersecting perpendicular cylinder to locate the line of intersection. A series of circular sections are found in the top view. Points 1 and 2 are found on cutting plane *C* in the top view as examples, and are projected to the front view. Other points are found in this same manner.

This method is feasible only when the centerline of the cylinder is perpendicular to the axis of the cone, so that circular sections can be found in the top view, rather than elliptical sections that would be difficult to draw.

The distributor housing in Fig. 9.25 is an example of an intersection between cylinders and a cone.

9.9 INTERSECTIONS BETWEEN PYRAMIDS AND PRISMS

The intersection between an inclined prism and a pyramid is solved in Fig. 9.26. The end view of the inclined prism is found in the primary auxiliary view and the pyramid is shown in this view also (Step 1). Radial lines *OB* and *OA* are passed through corners 1 and 3 in the auxiliary view

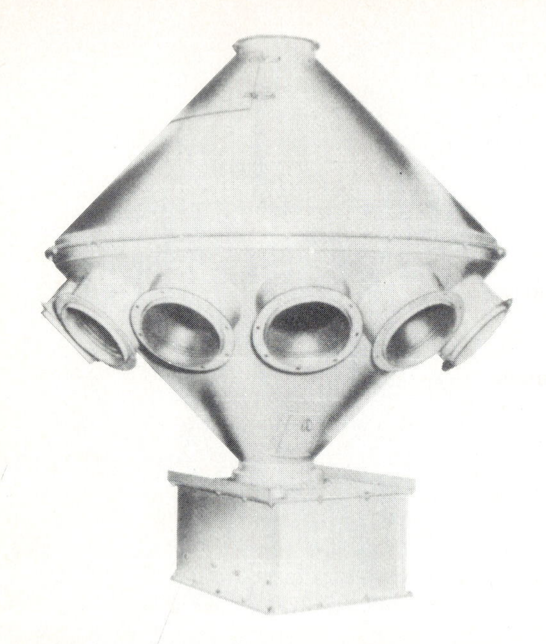

Fig. 9.25 This electrically operated distributor illustrates intersections between a cone and a series of cylinders. (Courtesy of GATX.)

(Step 2). The radial lines are projected from the auxiliary view to the front and top views. Intersecting points 1 and 3 are located on *OB* and *OA* in each of these views by projection. Point 2 is the point where line 1–3 bends around corner *OC*. Lines of intersection 1–4 and 4–3 are found in Step 3; the visibility is determined; and the solution is completed.

A prism that is parallel to the base of a pyramid is shown in Fig. 9.27. Its lines of intersection are found by using a series of horizontal cutting planes that pass through the pyramid parallel to its base to form triangular sections in the top view.

The same cutting planes are passed through the corner lines of the prism in the front and auxiliary views. Each corner edge is extended in the top view to intersect the triangular section formed by the cutting plane passing through it. Point 1 is given as an example.

Corner point *X* is found by passing cutting plane *B* through it in the auxiliary view where it crosses the corner line. This is where the line of intersection of this plane bends around the corner. The radial cutting plane method could have been used as an alternative method to solve this problem.

Fig. 9.26 Intersection between a prism and a pyramid

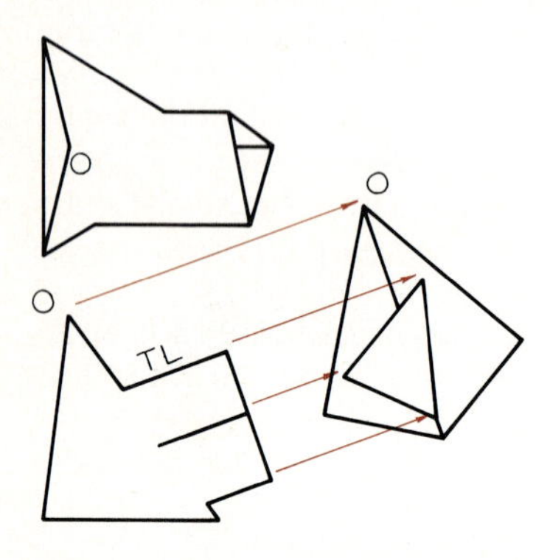

Step 1 Find the edge view of the surfaces of the prism by projecting an auxiliary view from the front view. Project the pyramid into this view also. Only the visible surfaces need be shown in this view.

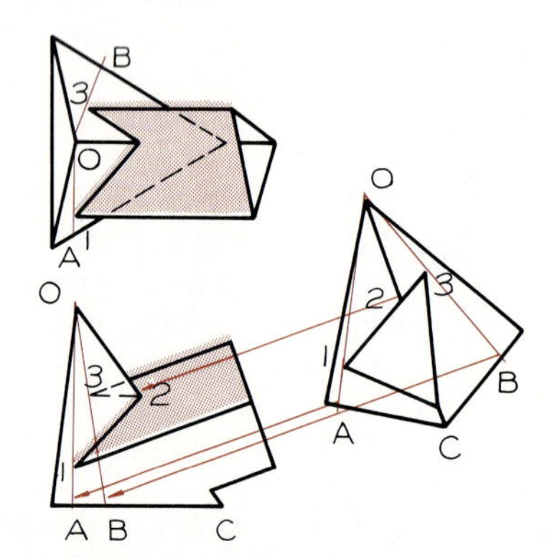

Step 2 Pass planes *A* and *B* through apex *O* and points 1 and 3 in the auxiliary view. Project the intersections of the planes *OA* and *OB* to the front and top views. Project points 1 and 3 to *OA* and *OB* in the principal views. Point 2 lies on line *OC*. Connect points 1, 2, and 3 to give one line of intersection.

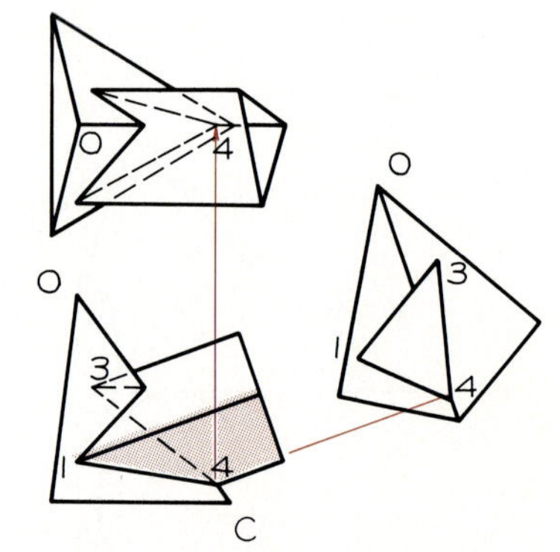

Step 3 Point 4 lies on line *OC* in the auxiliary view. Project this point to the principal views. Connect points 1, 4, and 3 to complete the intersection. Visibility is indicated. Note that these geometric shapes are assumed to be hollow as though constructed of sheet metal.

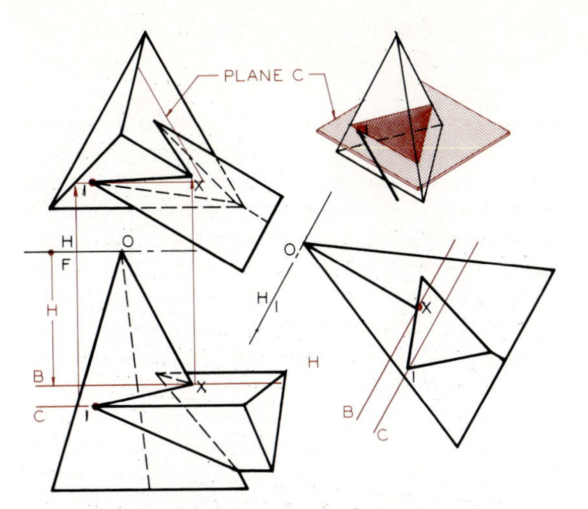

Fig. 9.27 The intersection between this pyramid and prism is found by finding the end view of the prism in an auxiliary view. Horizontal cutting planes are passed through the corner lines of the prism to find the piercing points and the line of intersection.

9.10 INTERSECTIONS BETWEEN SPHERES AND PLANES

The line of intersection between a sphere and a plane is found in Fig. 9.28 where the plane appears as an edge in the front view. Horizontal cutting planes are passed through the front view to form circular sections in the top view. Two points are located on each cutting plane in the top view by projecting from the front view where the cutting plane crosses the edge view of the intersecting plane. The points are connected and the visibility is shown in the top view.

The elliptical intersection could have been drawn with an ellipse template that was selected by measuring the angle between the edge view of the plane in the front view and the projectors from the top view. The major diameter of the ellipse would be equal to the true diameter of the sphere, since the plane passes through the center of the sphere.

Fig. 9.28 Intersection of a sphere and a plane

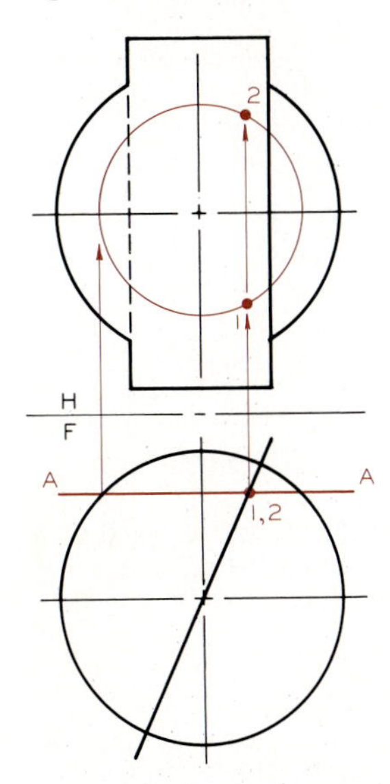

Step 1 Horizontal cutting plane *A–A* is passed through the front view of the sphere to establish a circular section in the top view. Piercing points 1 and 2 are projected from the front view to the top view, where they lie on the circular section.

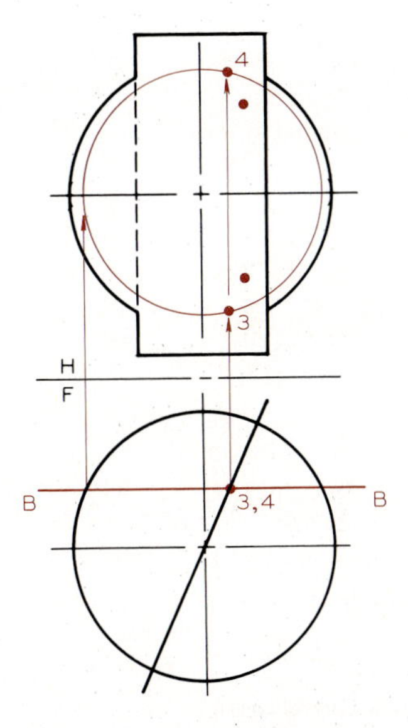

Step 2 Horizontal cutting plane *B–B* is used to locate piercing points 3 and 4 in the top view by projecting to the circular section cut by the plane in the top view. Additional horizontal planes are used to find sufficient points in this manner.

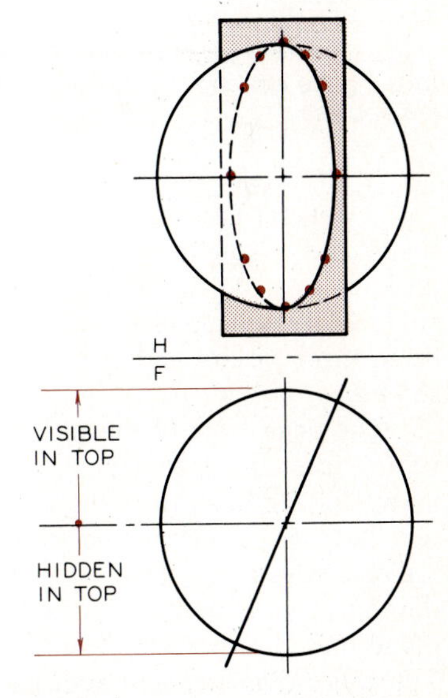

Step 3 Visibility of the top view is found by inspection of the front view. The upper portion of the sphere will be visible in the top view and the lower portion will be hidden.

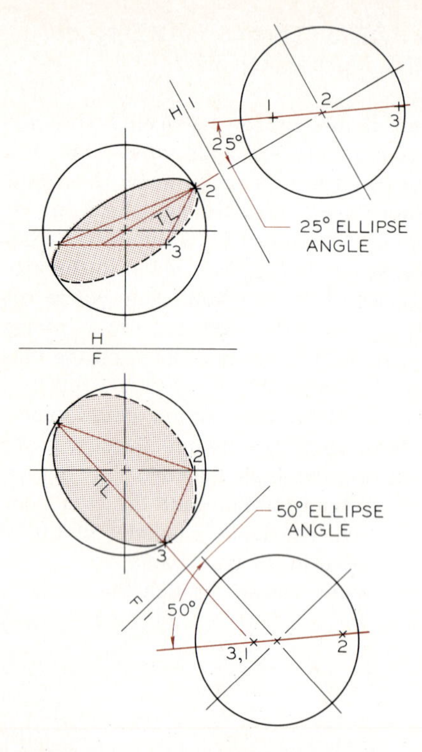

Fig. 9.29 Three points are given on the surface of the sphere through which a circle passes. This plane is found as an edge by projecting from the top and front views. Ellipse angles of 25° and 50° are found for drawing the top and front views of the elliptical intersections.

A general case of the intersection between a plane and a sphere is given in Fig. 9.29 where three points, 1, 2, and 3, are located on the sphere's surface. A circle is to be drawn through these three points and lie on the surface of the sphere.

The edge view of the plane is found by projecting from the top view to a primary auxiliary view. The circle on the sphere through 1, 2, and 3 will have an elliptical line of intersection in the top and front views. The ellipse template angle for the top view is the angle between the edge of the plane and the projector from the top view, 25°. The major diameter is drawn parallel to the true-length lines on plane 1–2–3 in the top view.

The ellipse for the front view of the intersection is found in the same manner by finding the edge view of the plane in an auxiliary view projected from the front view. The ellipse template angle is found to be 50°.

9.11 INTERSECTIONS BETWEEN SPHERES AND PRISMS

The intersection between a sphere and a prism is found in Fig. 9.30 by drawing a series of vertical cutting planes in the top and side views. The planes form circular sections in the front view.

The intersections of the edges with the cutting planes in the side view are projected to their respective circles in the front view. Points 1 and 2 are located on cutting plane *A* in the side view and on the circular path of *A* in the front view. Point *X* in the side view locates the point where the visibility changes in the front view.

Point *Y* in the side view is the point where the visibility of the intersection changes in the top view. Both of these points lie on the centerlines of the sphere on the side view.

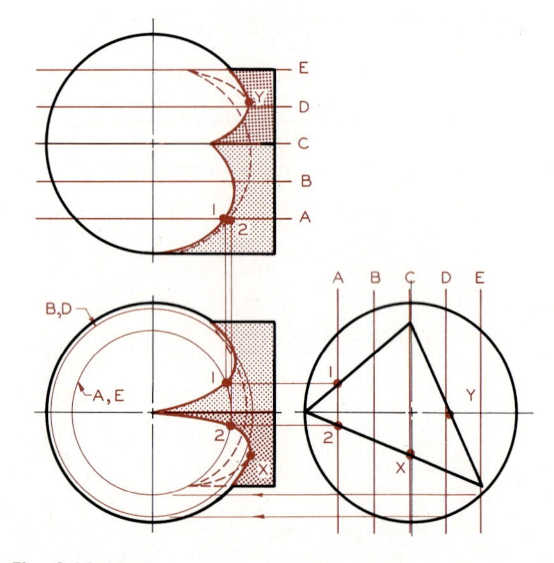

Fig. 9.30 Vertical cutting planes are used to find the intersection between the prism and sphere.

9.12 MISCELLANEOUS INTERSECTIONS

A series of intersections of cutting planes passed through different figures is shown in Fig. 9.31. Radial lines are used to find a hyperbolic section on a cone at A. Horizontal cutting planes are used to locate a section in the top view of a torus (a donut-shaped object) at B.

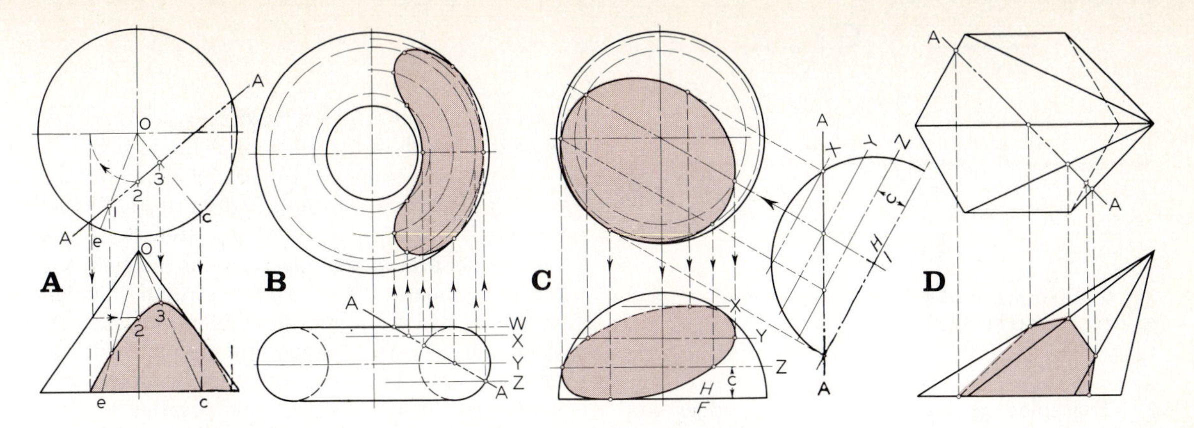

Fig. 9.31 Examples of cutting-plane intersections through four types of figures are shown here. (A) Radial lines are used to find a hyperbolic section through a cone. (B) An inclined plane is shown intersecting a torus. (C) The section cut by a plane through a hemisphere is shown. (D) A section cut by a vertical plane through a pyramid.

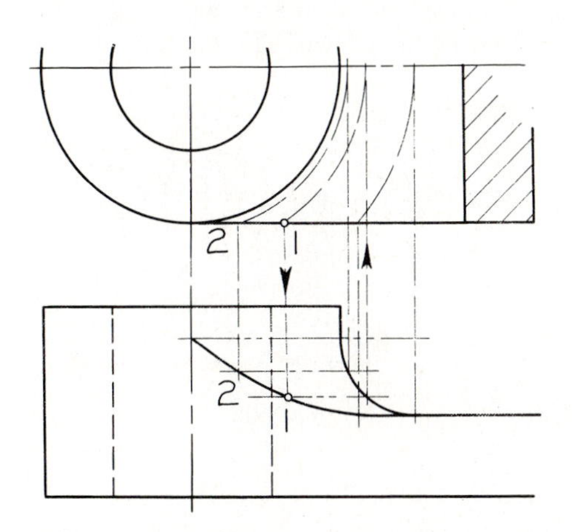

Fig. 9.32 Horizontal cutting planes are used to find the runout (intersection) where surfaces intersect the cylindrical ends of these parts.

Horizontal cutting planes are drawn to locate an elliptical section through a hemisphere at C with a supplementary auxiliary view. The section of a cutting plane through an oblique pyramid is shown at D.

The construction of runouts formed by fillets that intersect cylinders is shown in Fig. 9.32. Horizontal cutting planes are passed through the fillets in the front view and are projected to the top view to locate arcs formed by the cutting planes.

Points along the runout in the front view are found by projecting from the top view. Points 1 and 2 are shown as examples.

9.13 PRINCIPLES OF DEVELOPMENTS

The processing plant shown in Fig. 9.33 illustrates numerous examples of sheet metal shapes that were designed using the principles of developments. In other words, their patterns were laid out on flat stock and then formed to the proper shape by bending and seaming the joints.

Fig. 9.33 Most of the surfaces shown in this refinery were made from flat stock that was fabricated to form these irregular shapes. These flat patterns are called developments.

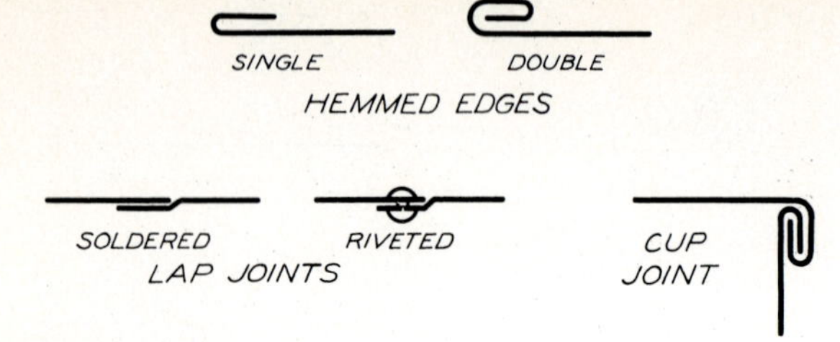

Fig. 9.34 Examples of the types of seams that are used to join developments.

Examples of standard hemmed edges and joints are shown in Fig. 9.34. The application of the sheet metal design will determine the best method of connecting the seams.

The developments of the surfaces of three typical shapes into a flat pattern are shown in Fig. 9.35. The sides of a box are imagined to be unfolded into a common plane. The cylinder is rolled out for a distance that is equal to its circumference. The pattern of a right cone is developed using the length of the element as the radius.

Patterns of shapes with parallel elements such as the prisms and cylinders shown in Fig. 9.36(a) and (b) are begun by constructing stretch-out lines that are parallel to the edge view of the right section of the parts. The distance around the right section is laid off along the stretch-out line. The prism and cylinder at (c) and (d) are inclined; consequently the right sections must be drawn perpendicular to their sides, not parallel to their bases. The distances around the right sections are laid out along these stretch-out lines.

An inside pattern (development) is more desirable than an outside pattern because most bending machines are designed to fold metal so that markings are folded inward, and because markings and scribings will be hidden when the pattern is assembled in final form. The method of denoting

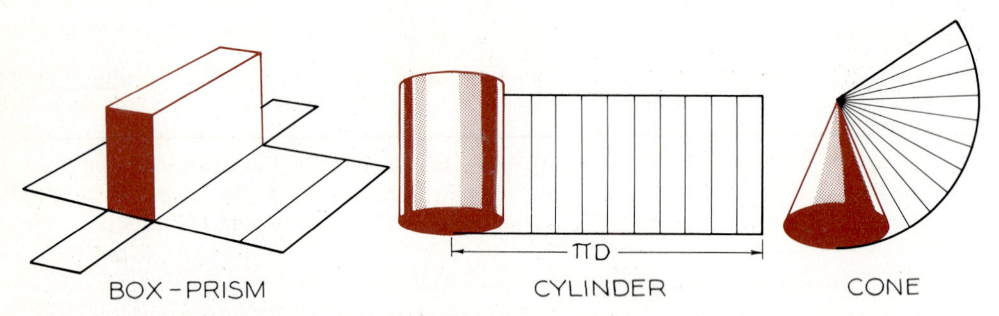

Fig. 9.35 Three standard types of developments: the box, cylinder, and cone.

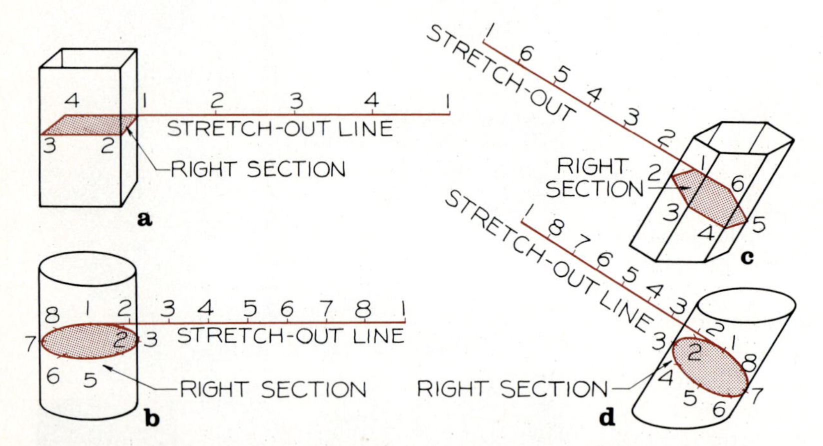

Fig. 9.36 The developments of right prisms and cylinders are found by rolling out the right sections along a stretch-out line (parts A and B). When these figures are oblique, the right sections are found to be perpendicular to the sides of the prism and cylinder. The development is laid out along the stretch-out line that is parallel to the edge view of the right section (parts C and D).

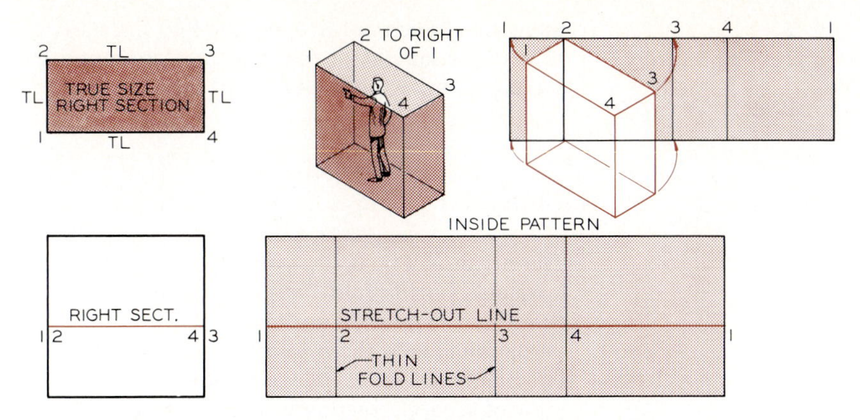

Fig. 9.37 The development of a rectangular prism to give an inside pattern. The stretch-out line is parallel to the edge view of the right section.

a pattern is by a series of lettered or numbered points about its layout. All lines on a development must be true length.

It is economical if seam lines (lines where the pattern is joined) are selected as the shortest lines. This results in the least expense of riveting or welding the pattern together to form the final shape.

9.14 DEVELOPMENT OF PRISMS

A flat pattern for a prism is developed in Fig. 9.37. Since the edges of the prism are vertical in the front view, its right section is perpendicular to these sides. The top view shows the right section true size. The stretch-out line is drawn parallel to the edge of the right section, beginning with point 1.

If an inside pattern is drawn and it is to be laid out to the right, point 2 will be to the right of point 1. This is determined by looking from the inside of the top view where 2 is seen to the right of 1.

Lines 2–3, 3–4, and 4–1 are transferred from the right section in the top view with your dividers to the stretch-out line to locate the fold lines on the pattern. The length of each fold line is found by projecting its true length from the front view. The ends of the fold lines are connected to form the limits of the developed surface. Fold lines are drawn as thin lines and the outside lines of a development are drawn as visible object lines.

Fig. 9.38 Numerous examples of developments fabricated from sheet metal are shown in this plant. (Courtesy of Heat Engineering.)

The complex installation in Fig. 9.38 is composed of many developments ranging from simple prisms to more complicated shapes.

The development of the prism in Fig. 9.39 is similar to the example in Fig. 9.37 except that one end is beveled rather than square. The stretch-out line is drawn parallel to the edge view of the right section in the front view. The true-length distances around the right section are laid off

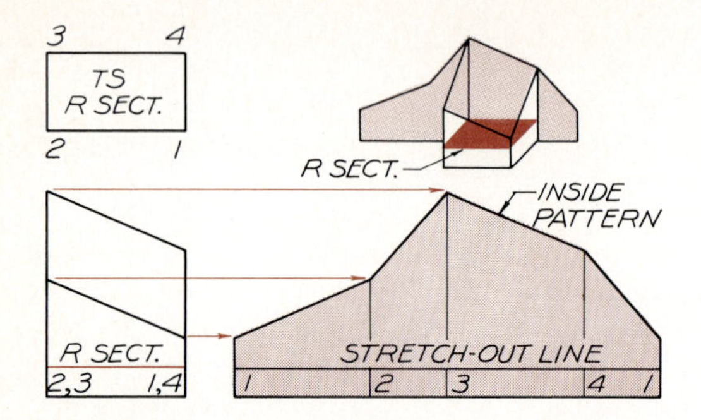

Fig. 9.39 The development of a rectangular prism with a beveled end to give an inside pattern. The stretch-out line is parallel to the right section.

along the stretch-out line and the fold lines are located. The lengths of the fold lines are found by projecting from the front view of these lines.

9.15 DEVELOPMENT OF OBLIQUE PRISMS

The prism in Fig. 9.40 is inclined to the horizontal plane, but its fold lines are true length in the front view. The right section is drawn as an edge perpendicular to these fold lines and the stretch-out line is drawn as shown in Step 1. A true-size view of the right section is constructed in the auxiliary view.

In Step 2, the distances between the fold lines are transferred from the true-size right section to the stretch-out line. The lengths of the fold lines are found by projecting from the front view.

In Step 3, the ends of the prism are found and are attached to the pattern so that they can be folded into position. All lines that are laid out in a flat pattern must be true length.

Fig. 9.40 Development of an oblique prism

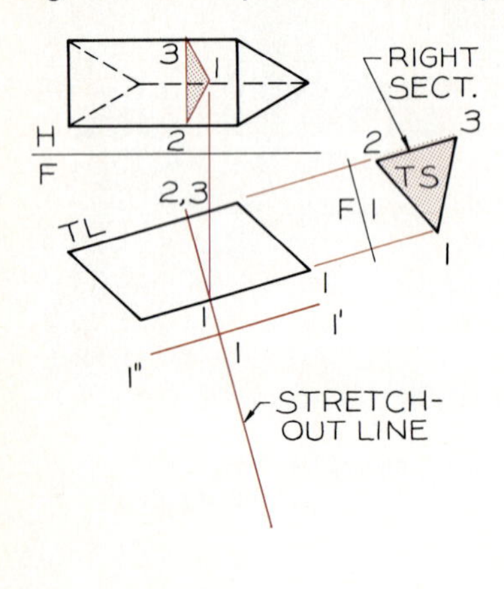

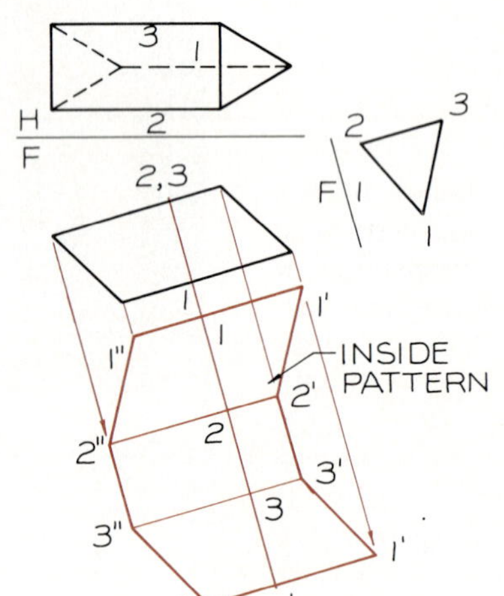

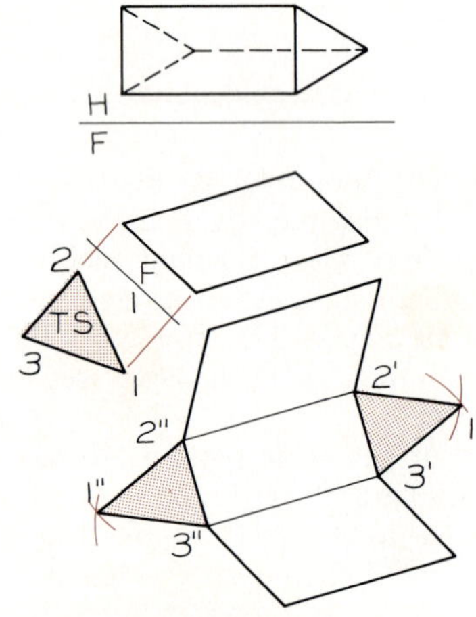

Step 1 The edge view of the right section will appear perpendicular to the true-length axis of the prism in the front view. Determine the true-size view of the right section by constructing an auxiliary view. Draw the stretch-out line parallel to the edge view of the right section. Project bend line 1′–1″ as the first line of the development.

Step 2 Since the pattern is developed toward the right, beginning with line 1′–1″, the next point is found to be line 2′–2″ by referring to the auxiliary view. Transfer true-length lines 1–2, 2–3, and 3–1 from the right section to the stretch-out line to locate the elements. Determine the lengths of the bend lines by projection.

Step 3 Find the true-size views of the end pieces by projecting auxiliary views from the front view. Connect these surfaces to the development of the lateral sides to form the completed pattern. Fold lines are drawn with thin lines, while outside lines are drawn as regular object lines.

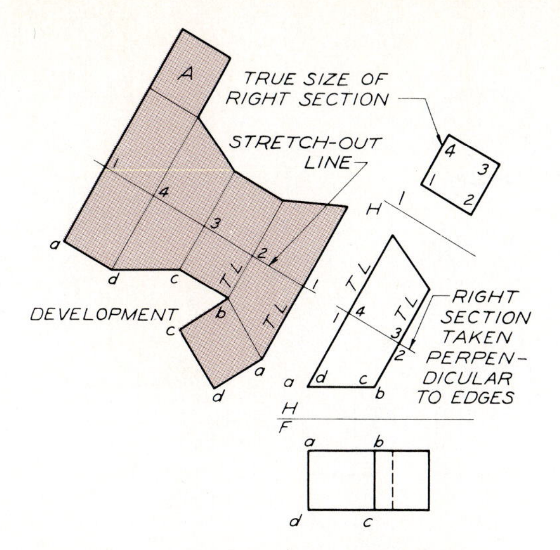

Fig. 9.41 The development of this oblique chute is found by finding the right section true-size in the auxiliary view. The stretch-out line is drawn parallel to the right section.

A similar example is solved in Fig. 9.41. The fold lines are true length in the top view; this enables you to draw the edge view of the right section perpendicular to the fold lines in the top view. The stretch-out line is drawn parallel to the edge view of the section, and the true size of the right section is found in an auxiliary view projected from the top view. The distances about the right section are transferred to the stretch-out line to locate the fold lines. The lengths of the fold lines are found by projecting from the top view. The end portions of the pattern are attached to the pattern to complete the construction.

A prism that does not project true length in either view can be developed as illustrated in Fig. 9.42. The fold lines are found true length in an auxiliary view projected from the front view. The right section will appear as an edge perpendicular to the fold lines in the auxiliary view. The true size of the right section is found in a secondary auxiliary view.

The stretch-out line is drawn parallel to the edge view on the right section. The fold lines are located on the stretch-out line by measuring around the right section in the secondary auxiliary view. The lengths of the fold lines are then projected to the development from the primary auxiliary view.

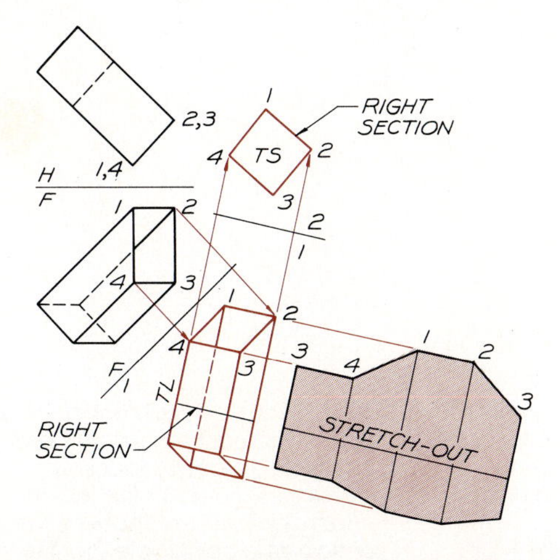

Fig. 9.42 The development of an oblique cylinder is found by finding an auxiliary view in which the fold lines are true length, and a secondary auxiliary view in which the right section appears true size. The stretch-out line is drawn parallel to the right section.

9.16 DEVELOPMENT OF CYLINDERS

The development of a cylinder is found in Fig. 9.43. Since the elements of the cylinder are true length in the front view, the right section will appear as an edge in this view, and true size in the top view. The stretch-out line is drawn parallel to the edge view of the right section, and point 1 is chosen as the beginning point since it is the shortest element. To draw an inside pattern, assume that you are standing on the inside looking at point 1 and you will see that point 2 is to the right of 1. Therefore, the pattern is laid out with point

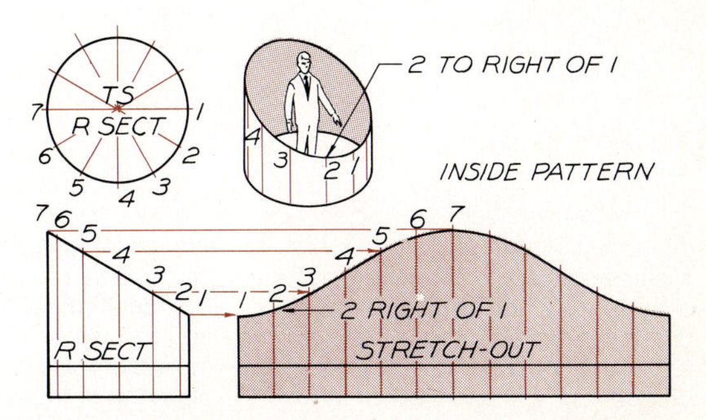

Fig. 9.43 The development of a right cylinder's inside pattern. The stretch-out line is parallel to the right section. Point 2 is to the right of point 1 for an inside pattern.

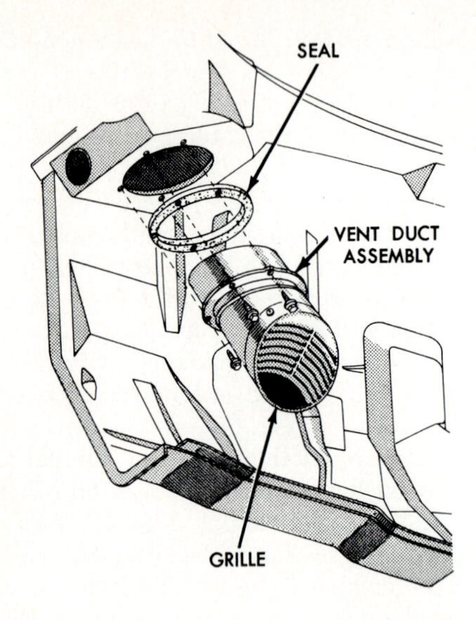

Fig. 9.44 This ventilator air duct was designed through the use of development principles. (Courtesy of Ford Motor Company.)

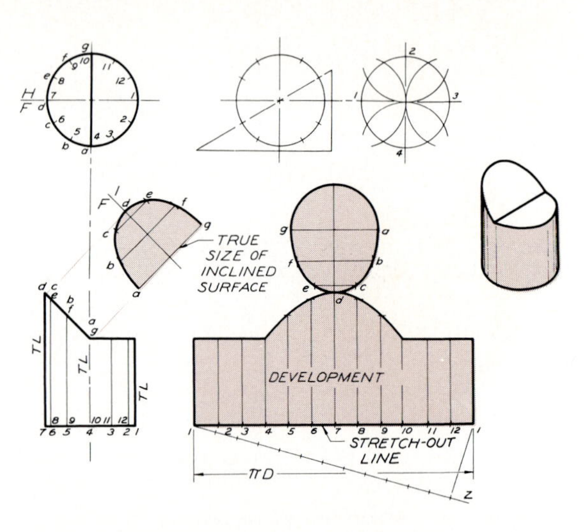

Fig. 9.45 The development of a right cylinder can be found by mathematically locating elements along the stretch-out line, which is equal in length to the circumference of the right section.

Fig. 9.46 Development of an oblique cylinder

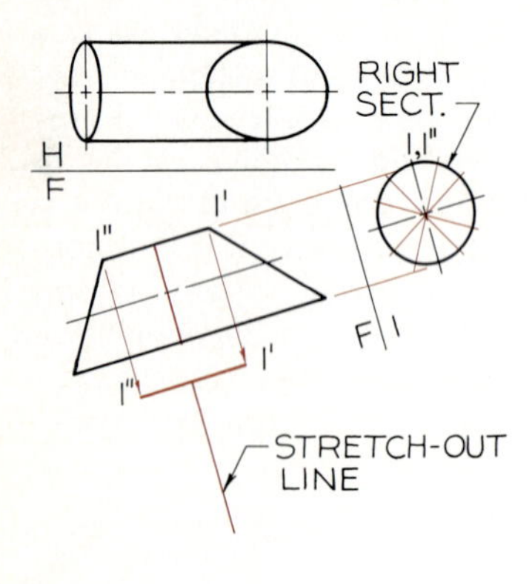

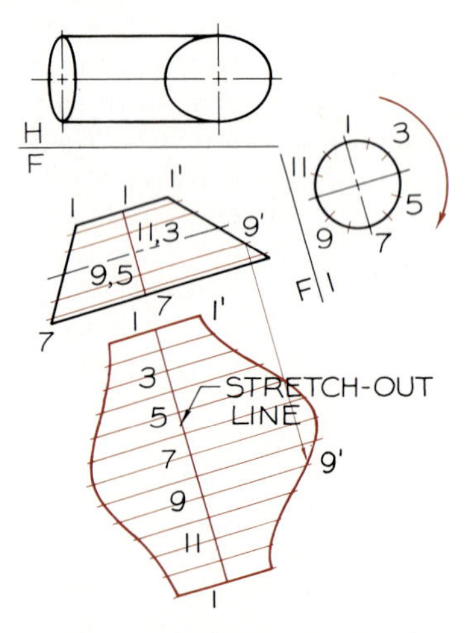

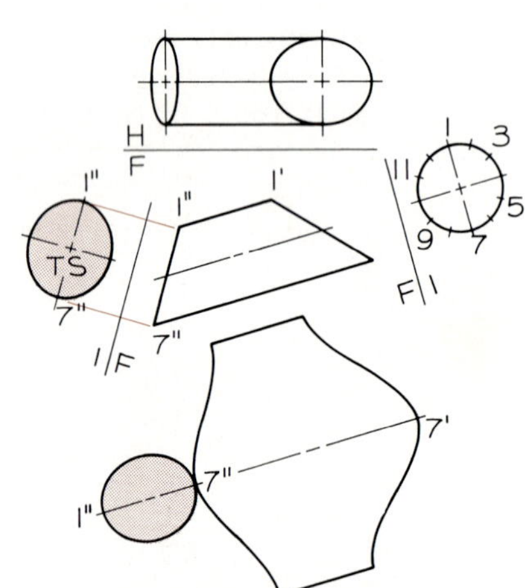

Step 1 The right section appears as an edge in the front view, in which it is perpendicular to the true-length axis. Construct an auxiliary view to determine the true size of the right section. Draw a stretch-out line parallel to the edge view of the right section. Locate element 1′–1″.

Step 2 Divide the true-size right section into equal divisions to locate point views of elements on the cylinder. Project these elements to the front view. Transfer measurements between the elements in the auxiliary view to the stretch-out line to locate the elements. Determine the lengths of the elements by projection to complete the development.

Step 3 The development of the end pieces will require auxiliary views that project these surfaces as ellipses, as shown for the left end. Attach this true-size ellipse to the pattern at a point. Note that the line of departure for the pattern was made along line 1″–1′, the shortest element, for economy.

2 to the right of point 1. This establishes the sequence of locating the elements.

The spacing between the elements in the top view can be conveniently done by drawing radial lines at 15° or 30° intervals. Using this technique, the elements will be equally spaced, making it convenient to lay them out along the stretch-out line. The lengths of the elements are found by projecting from the front view to complete the pattern.

An application of a developed cylinder with a beveled end is the air-conditioning duct from an automobile shown in Fig. 9.44. The development of a similar cylinder is shown in Fig. 9.45. The base of the front view is the edge view of the right section, which appears true size in the top view. Elements are located around the circumference of the right section in the top view. Two alternative methods are shown to illustrate how the elements are located at 30° intervals. One employs the 30°–60° triangle, and the other uses a compass with the radius equal to the radius of the right section.

The stretch-out line is drawn parallel to the right section in the front view. The total length is found mathematically by the formula $C = \pi D$. The stretch-out line is divided into the same number of divisions as there are elements, 12 in this case. This provides a high degree of accuracy in finding the circumference.

The pattern for the end of the cylinder is found by combining the partial top view and the auxiliary view. This is connected to the overall pattern to complete the solution.

9.17 DEVELOPMENT OF OBLIQUE CYLINDERS

The pattern for an oblique cylinder (Fig. 9.46) is found in the same manner as the previous examples, but with the addition of one preceding step. The right section must be found true-size in an auxiliary view. A series of equally spaced elements is located around the right section in the auxiliary view and is projected back to the true-length view, Step 1. The stretch-out line is drawn parallel to the edge of the right section in the front view.

In Step 2, the spacing between the elements is laid out along the stretch-out line, and the elements are drawn through these points perpendicular to the stretch-out line. The lengths of the elements are found by projecting from the front view.

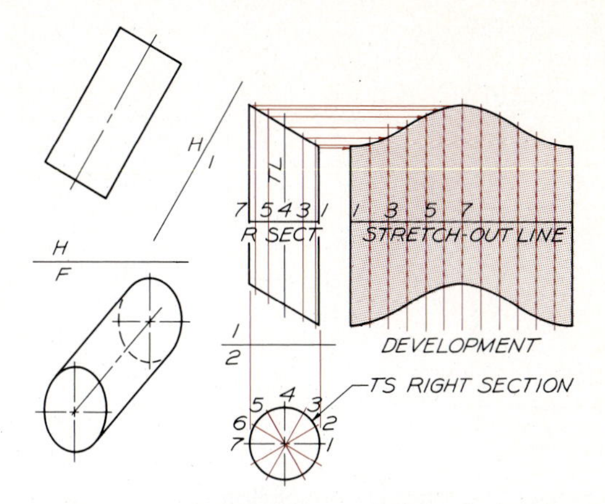

Fig. 9.47 The development of an oblique cylinder is found by constructing an auxiliary view in which the elements appear true length. The right section is found true size in a secondary auxiliary view.

The ends of the cylinder are found in Step 3 to complete the pattern. Only one end pattern is shown as an example.

The oblique cylinder in Fig. 9.47 is a more general case where the elements are not true length in the given views. A primary auxiliary view is used to find a view where the elements are true length, and a secondary auxiliary view is drawn to find the true-size view of the right section. The stretch-out line is drawn parallel to the edge view of the right section in the primary auxiliary view and the elements are located along this line by transferring their distances apart from the true-size right section.

The elements are drawn perpendicular to the stretch-out line. The length of each element is found by projecting from the primary auxiliary view. The endpoints are connected with a smooth curve to complete the pattern.

9.18 DEVELOPMENT OF PYRAMIDS

All lines used to draw a pattern must be true length. Pyramids have only a few lines that are true length in the given views; for this reason, the sloping corner lines must be found true-length at the outset.

The corner lines of a pyramid can be found by revolution, as shown in Fig. 9.48. Line 0–5 is revolved in the frontal position of 0–5′ in the top

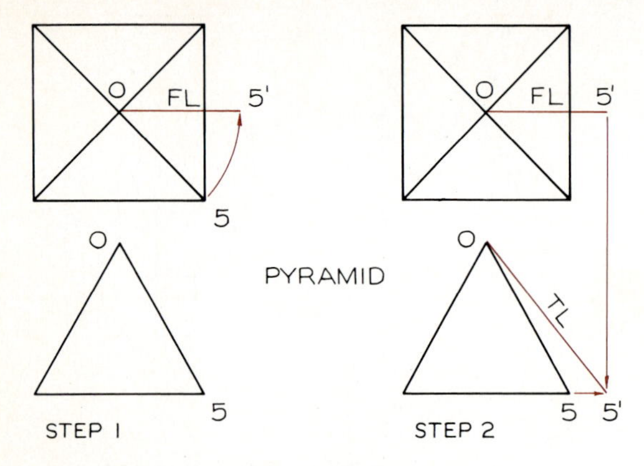

Fig. 9.48 True length by revolution

Step 1 Corner 0–5 of a pyramid is found true length by revolving it into the frontal plane in the top view, 0–5′.

Step 2 Point 5′ is projected to the front view where 0–5′ is true length.

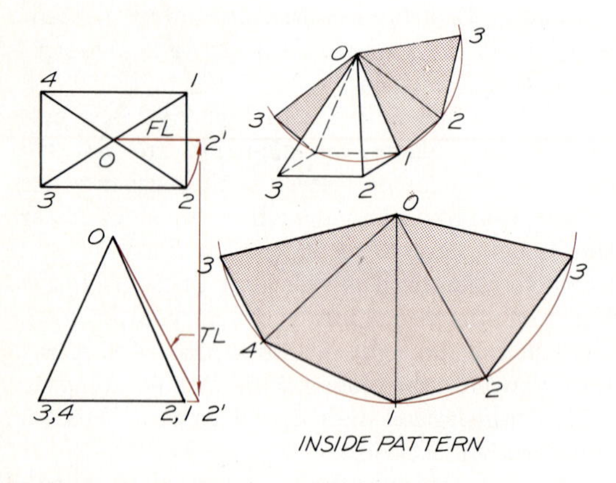

Fig. 9.49 Development of a right pyramid.

view, Step 1. Since 0–5′ is a frontal line, it will be true-length in the front view, Step 2.

The development of a pyramid is given in Fig. 9.49. Lines 0–1 and 0–2 are revolved into the frontal plane in the top view to find their true lengths in the front view. All bend lines are equal in length since this is a right pyramid. Line 0–1 is used as a radius to construct the base circle for drawing the development. Distance 1–2 is transferred from the base in the top view to the development, where it forms a chord on the base circle. Lines 2–3, 3–4, and 4–1 are found in the same manner and in sequence. The bend lines are drawn as thin lines from the base to the apex, point *O*.

A variation of this problem is given in Fig. 9.50, in which the pyramid has been truncated or cut at an angle to its axis. The development of the inside pattern is found in the same manner as the previous example; however, an additional step is required to establish the upper lines of the development. The true-length lines from the apex to points 1′, 2′, 3′, and 4′ are found by revolution. These distances are laid on their respective lines of the pattern to locate the upper limits of the pattern.

The mounting pads in Fig. 9.51 are sections of pyramids that intersect an engine body.

The development of an oblique pyramid is shown in Fig. 9.52. In Step 1, the corner lines are

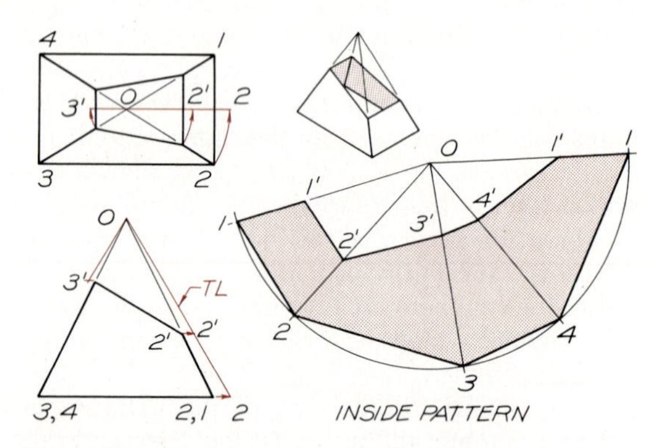

Fig. 9.50 The development of an inside pattern of a truncated pyramid. The corner lines are found true length by revolution.

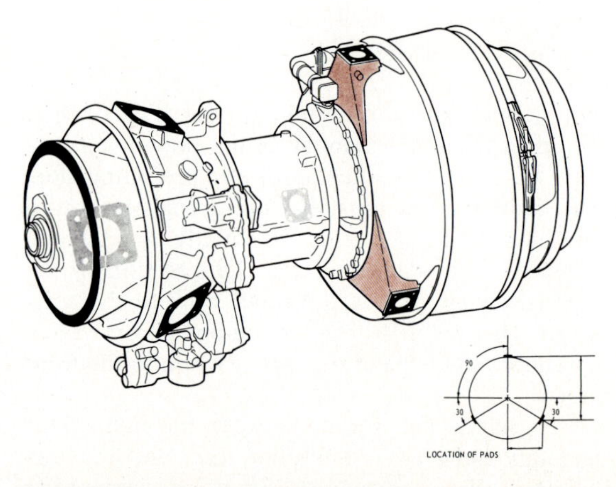

Fig. 9.51 Examples of pyramid shapes in the design of mounting pads for an engine. (Courtesy of Avco Lycoming.)

Fig. 9.52 Development of an oblique pyramid

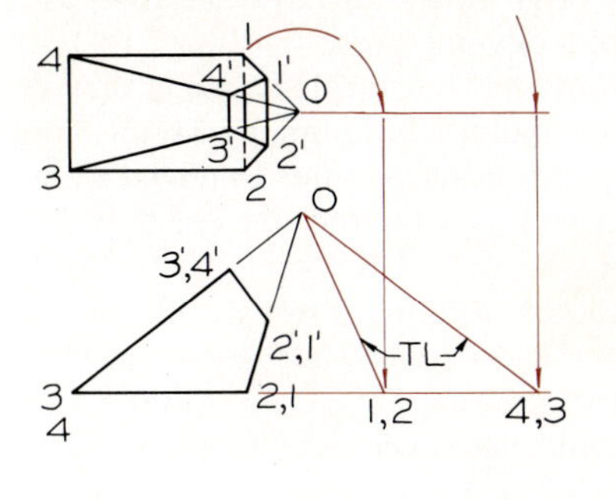

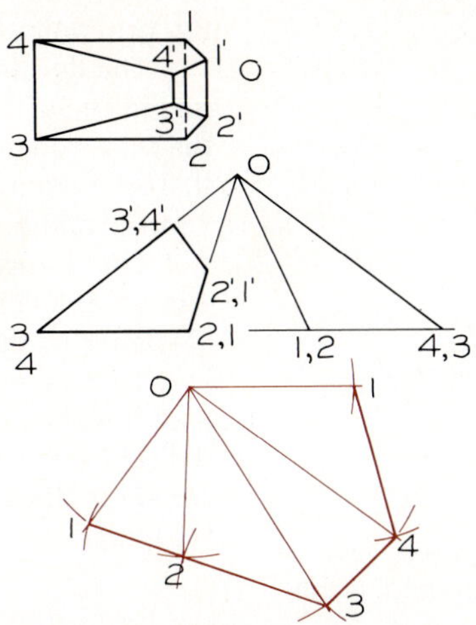

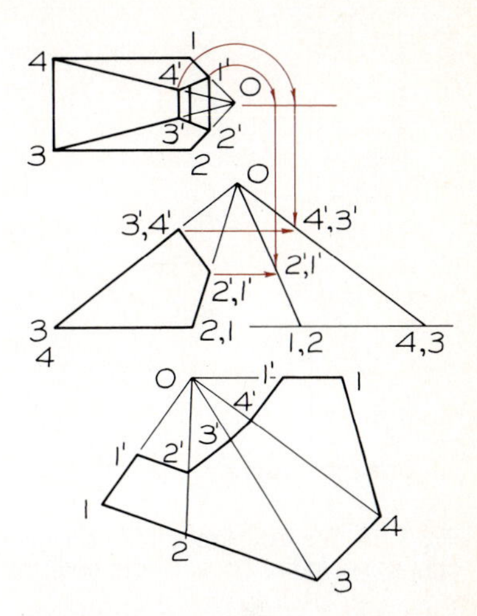

Step 1 Revolve each of the bend lines in the top view until they are parallel to the frontal plane. Project to the front view where the true-length views of the revolved lines can be found. Let point 0 remain stationary but project points 1, 2, 3, and 4 horizontally in the front view to the projectors from the top view.

Step 2 The base lines appear true length in the top view. Using these true-length lines from the top view and the revolved lines in the front view, draw the development triangles. All triangles have one side and point 0 in common. This gives a development of the surface, excluding the truncated section.

Step 3 The true lengths of the lines from point 0 to 1′, 2′, 3′, and 4′ are found by revolving these lines. These distances are laid off from point 0 along their respective lines to establish points along the upper edge of the pattern. The points are then sequentially connected by straight lines.

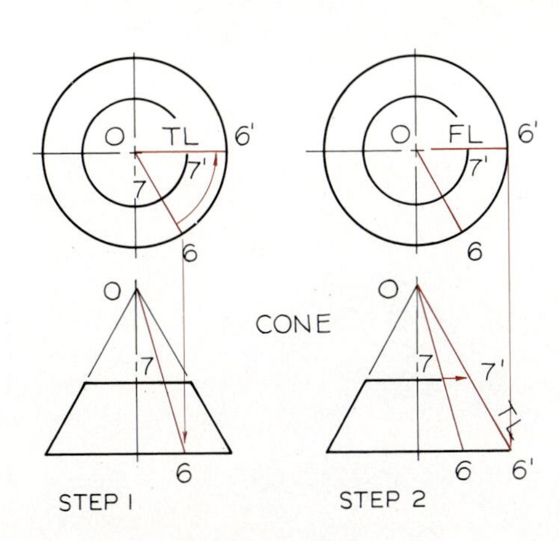

Fig 9.53 True length by revolution

Step 1 An element of a cone, 0–6, is revolved into a frontal plane in the top view.

Step 2 Point 6′ is projected to the front view where it is the outside element of the cone and is true length. Line 0–7′ is true length and is found by projecting to the outside element in the front view.

found true length by revolution. Using these true-length lines and those that are given in the principal views, the pattern is drawn by triangulation using a compass, Step 2. In Step 3, the upper limits of the pattern are found by measuring along the fold lines from point *O*.

9.19 DEVELOPMENT OF CONES

All elements of a right cone are equal in length as illustrated in Fig. 9.53 where *O*–6 is found true length by revolution. When revolved to *O*–6′ position, it is a frontal line and is therefore true length in the front view where it is an outside element of the cone. Point 7′ is found by projecting horizontally to element *O*–6′.

The right cone in Fig. 9.54 is developed by dividing the base into equally spaced elements in the top view and by projecting these to the base in the front view. These elements radiate to the apex at *O*. The outside elements in the front view, *O*–10 and *O*–4, are true length.

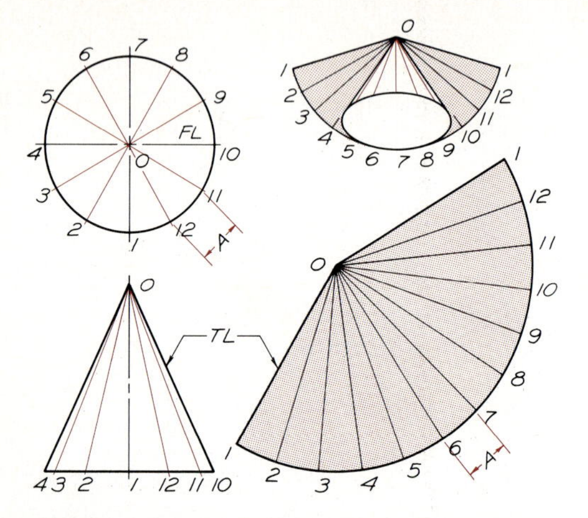

Fig. 9.54 The development of an inside pattern of a right cone. In this case, all elements are equal in length.

Using element O–10 as a radius, draw the base arc of the development. The elements are located along the base arc equal to the chordal distances between the points around the base in the top view. You can see that this is an inside pattern by inspecting the top view where point 2 is to the right of point 1 when viewed from the inside. Always label the points of a pattern in order to designate whether it is an inside or outside pattern.

The sheet metal conical vessel in Fig. 9.55 is an example of a large vessel that was designed using principles of a development.

The development of a truncated cone is shown in Fig. 9.56. The pattern is found by laying out the total cone ignoring the portions removed from it. This is found by following the steps of the previous example.

The hyperbolic sections through the front view of the cone can be found on their respective elements in the top and front views. Lines O–2′ and O–3′ are projected horizontally to the true length element O–1 in the front view, where they will appear true length. These distances are measured off along their respective elements in the development to establish a smooth curve.

9.20 DEVELOPMENT OF OBLIQUE CONES

The development of an oblique cone is found in Fig. 9.57. The elements of this cone are of varying lengths, but the pattern will be symmetrical about an axis.

Elements are located in the top view by dividing the base into equal arcs and drawing the elements to point O, the apex. These elements are

Fig. 9.55 An example of a conical shape that was formed from steel panels by applying principles of developments.

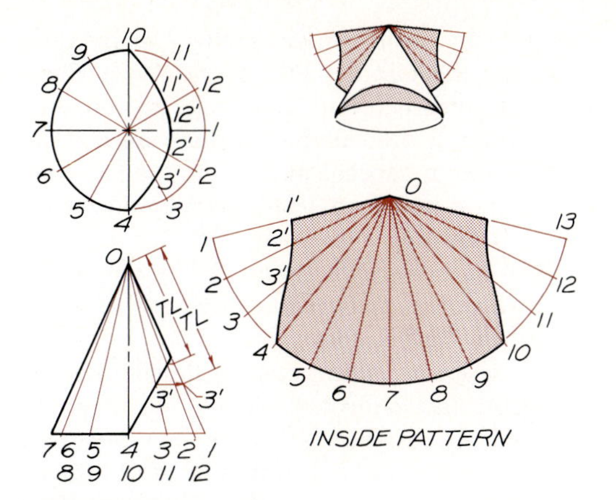

Fig. 9.56 The development of a conical surface with a side opening.

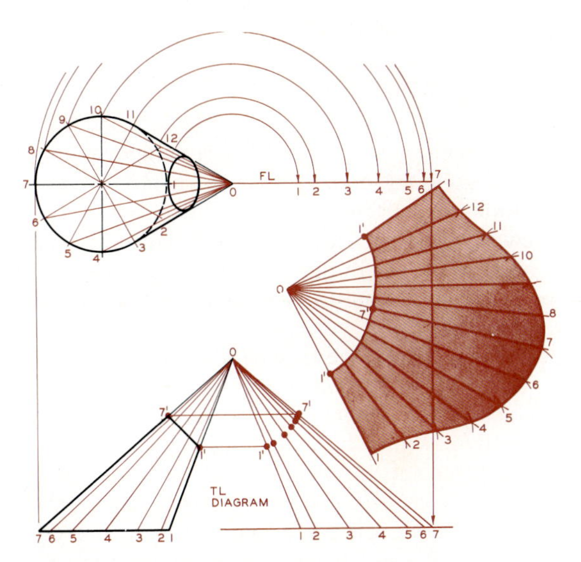

Fig. 9.57 The development of an oblique cone. All elements must be found true length by a true-length diagram before the pattern can be constructed.

projected to the frontal view. Each element is found true length by revolving it into a frontal plane in the top view. When projected to the front view, the revolved lines are found true length in a true-length diagram.

The development is begun by constructing a series of triangles using the true-length elements and the chordal distances found on the base in the top view. Line O–1 is chosen for the line of separation since it is the shortest element. The base of the pattern is drawn as a smooth curve.

The distances from point O to the upper cut through the cone are found by projecting to their respective elements in the true-length diagram from the front view. Line O–7′ is shown as an example. These shorter elements are located on their respective elements in the pattern to give the upper limits of the pattern. This will be an irregular curve rather than an arc because this is not a right cone.

9.21 DEVELOPMENT OF WARPED SURFACES

The geometric shape in Fig. 9.58 is an approximate cone with a warped surface; it is similar to the oblique cone in Fig. 9.57. The development of this surface will be an approximation, since a warped surface cannot be laid out on a flat surface.

The surface is divided into a series of triangles in the top and front views by dividing the upper and lower views into equal sectors. The true lengths of all lines are found in the true-length diagrams drawn on both sides of the front and by projecting horizontally from the ends of the lines. If necessary, review Fig. 9.57 to see how the true-length diagrams were found.

The chordal distances between the points on the base appear true-length in the top view since the base is horizontal. However, an auxiliary view

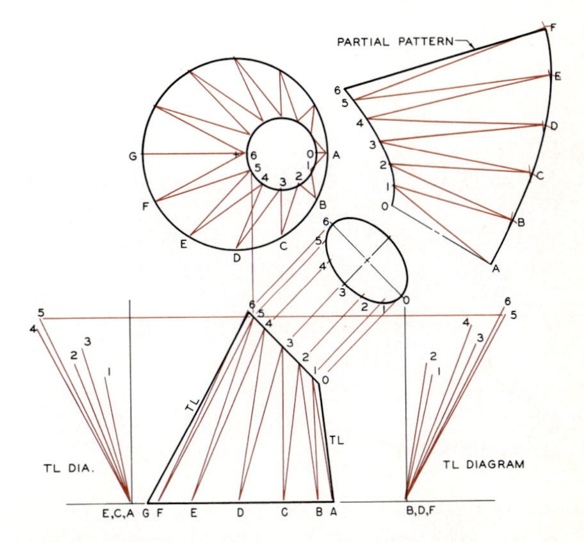

Fig. 9.58 The development of a partial pattern of a warped surface.

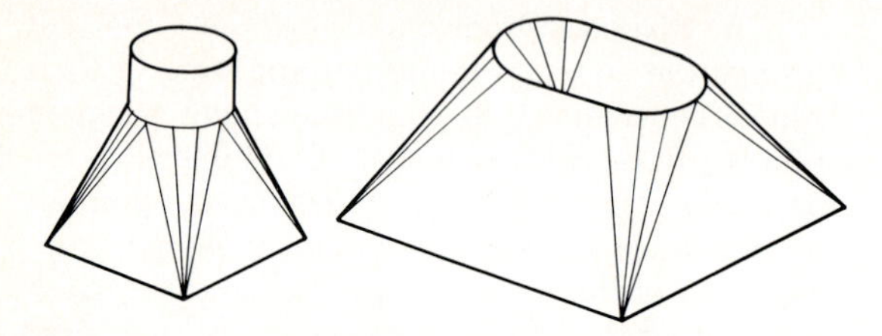

Fig. 9.59 Examples of transition pieces that join together parts that have different cross section.

is necessary to find the true distance about the upper surface of the object.

The developed surface is found by triangulation using true-length lines from (1) the true-length diagram, (2) the horizontal base in the top view, and (3) the primary auxiliary view. Each point should be carefully labeled as it is laid out to avoid confusion.

9.22 DEVELOPMENT OF TRANSITION PIECES

A transition piece is a form that transforms the section at one end to a different shape at the other (Fig. 9.59). Huge transition pieces can be seen in the industrial installation in Fig. 9.60.

The development of a transition piece is shown in Fig. 9.61. In Step 1, radial elements are drawn from each corner to the equally spaced points on the circular end of the piece. Each of these elements is found true length by revolution.

Fig. 9.60 Transition-piece developments are used to join a circular shape with a rectangular section. (Courtesy of Western Precipitation Group, Joy Manufacturing Company.)

In Step 2, the true-length lines are used with the true-length chordal distance in the top view to lay out a series of adjacent triangles to form the pattern beginning with element *A*–2.

The triangles *A*–1–2 and *G*–3–4 are added at each end of the pattern to complete the development of the half pattern.

A similar transition piece is developed in Fig. 9.62 using the same techniques of construction. Elements are established at the corners of the given views and are found true length in the true-length diagrams by revolution. By triangulation, using the true-length lines, the full pattern is drawn to complete the development.

9.23 DEVELOPMENT OF SPHERES–ZONE METHOD

In Fig. 9.63, a development of a sphere is found by the zone method. A series of parallels, called latitudes in cartography, are drawn in the front view. Each is spaced an equal distance, D, apart along the surface in the front view. Distance D can be found mathematically to improve the accuracy of this step.

Cones are passed through the sphere's surface so that they pass through two parallels at the outer surface of the sphere. The largest cone with element R_1 is found by extending it through where the equator and the next parallel intersect on the sphere's surface in the front view until R_1 intersects the extended centerline of the sphere. Elements R_2, R_3, and R_4 are found by repeating this process.

The development is begun by laying out the largest zone, using R_1 as the radius, on the arc that represents the base of an imaginary cone. The breadth of the zone is found by laying off distance D from the front view to the development and drawing the upper portion of the zone with a radius equal to R_1–D, using the same center. No regard is given to finding the arc lengths at this point.

The next zone is drawn using the radius R_2 with its center located on a line through the center of arc R_1. The center of R_2 is positioned along this line such that the arc to be drawn will be tangent to the preceding arc, which was drawn with ra-

Fig. 9.61 Development of a transition piece

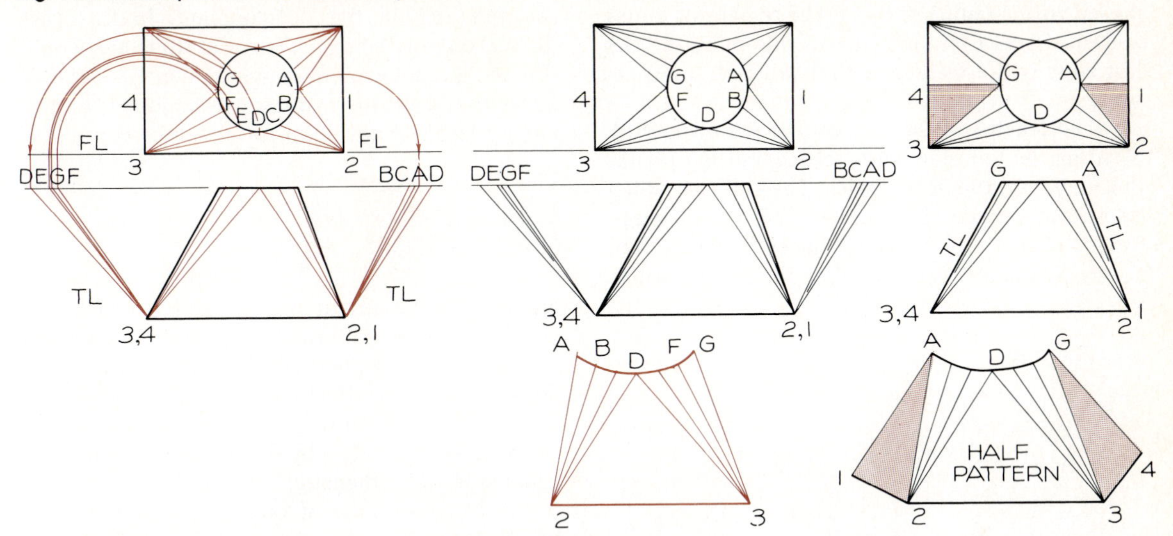

Step 1 Divide the circular edge of the surface into equal parts in the top view. Connect these points with bend lines to the corner points, 2 and 3. Find the true length of these lines by revolving them into a frontal plane and projecting them to the front view. These lines represent elements on the surface of a cone.

Step 2 Using the TL lines found in the TL diagram and the lines on the circular edge in the top view, draw a series of triangles, which are joined together at common sides, to form the development. *Example:* Arcs *2D* and *2C* are drawn from point 2. Point C is found by drawing arc *DC* from point *D* to find point *C*. Line *DC* is TL in the top view.

Step 3 Construct the remaining planes, *A*–1–2 and *G*–3–4, by triangulation to complete the inside half-pattern of the transition piece. Draw the fold lines as thin lines at the places where the surface is to be bent slightly. The line of departure for the pattern is chosen along *A*–1, the shortest possible line, for economy.

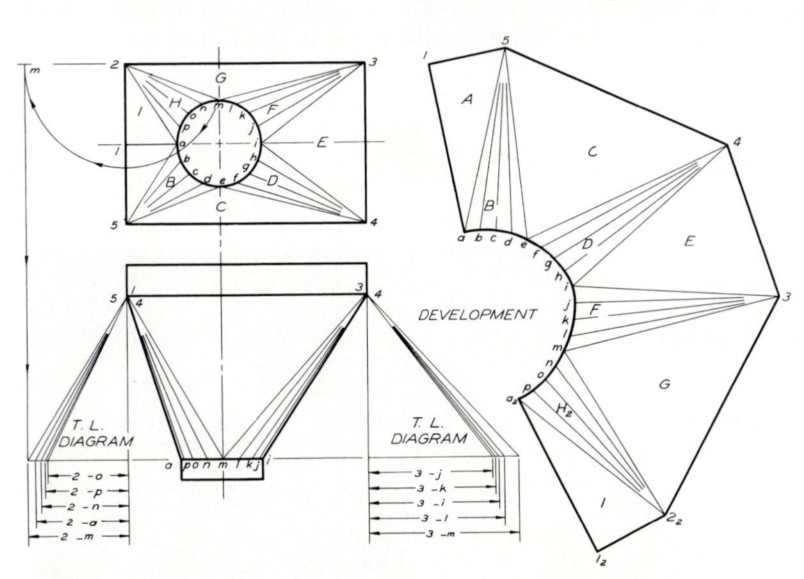

Fig. 9.62 The development of this transition piece is found by finding the conical elements at the corner's true-length in a true-length diagram. Line 2—*m* is found true-length as an example. The pattern is constructed by a series of triangulations.

dius R_1–D. The upper arc of this second zone is drawn with a radius of R_2–D. The remaining zones are constructed successively in this manner. The last zone will appear as a circle with R_4 as its radius.

The lengths of the arcs can be established by dividing the top view with vertical cutting planes that radiate through the poles. These lines, which lie on the surface of the sphere, are called longitudes in cartography. Arc distances S_1, S_2, S_3, and S_4 are found on each parallel in the top view.

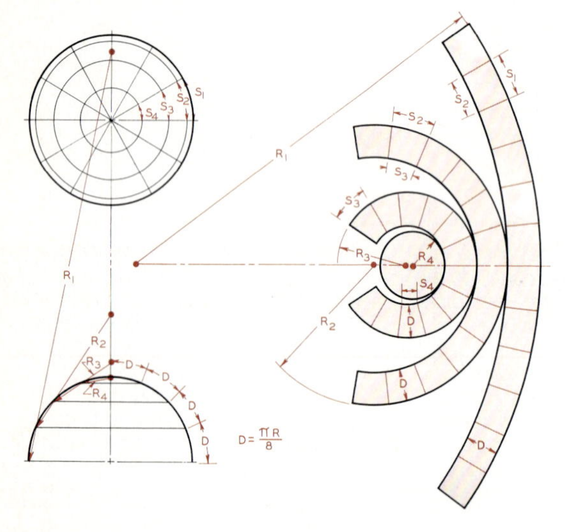

Fig. 9.63 The zone method of finding the inside development of a spherical pattern.

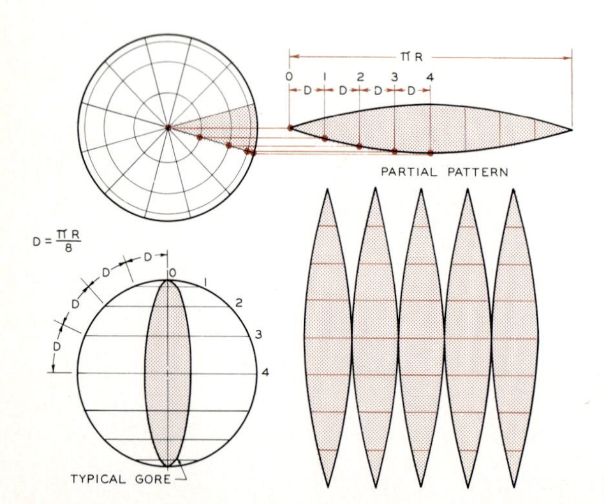

Fig. 9.64 The gore method of developing an inside pattern of a sphere.

These distances are measured off on the constructed arcs in the development. In this case there are 12 divisions, but smaller divisions would provide a more accurate measurement. A series of zones found in this manner can be joined to give an approximate development of a sphere.

9.24 DEVELOPMENT OF SPHERES—GORE METHOD

Figure 9.64 illustrates an alternative method of developing a flat pattern for a sphere. This method uses a series of spherical elements called gores. Equally spaced vertical cutting planes are passed through the poles in the top view. Parallels are located in the front view by dividing the surface into equal zones of dimension *D*.

A true-size view of one of the gores is projected from the top view. Dimensions can be checked mathematically for all points. A series of these gores is laid out in sequence to complete the pattern.

9.25 DEVELOPMENT OF ELBOWS

An elbow is a cylindrical shape that turns a 90° angle. An elbow that is also a transition piece is shown in Fig. 9.65.

The method of constructing the pattern for an elbow is illustrated in Fig. 9.66. In Step 1, the 90° arc is drawn and the cylinder is drawn to size about its centerline. Divide this arc into one divi-

Fig. 9.65 The radial bend of the cylinder was developed using the technique covered in Fig. 9.66.

Fig. 9.66 Construction of an elbow

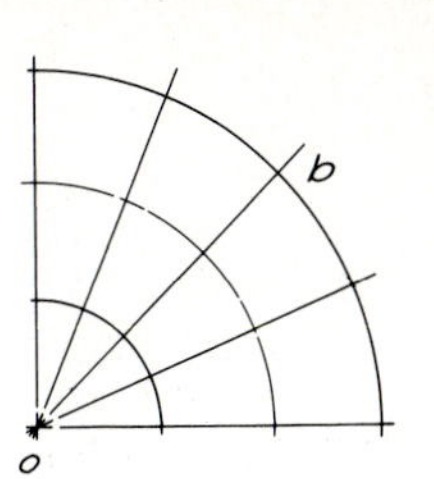

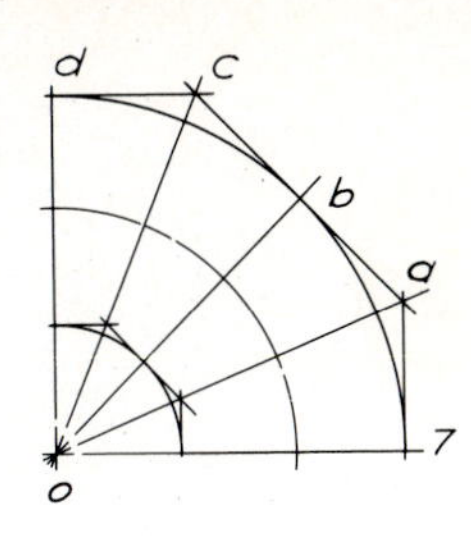

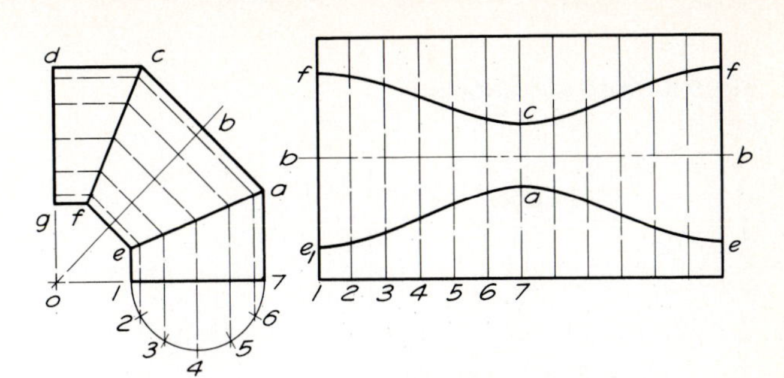

Step 1 Divide the arc into one division less than the number of pieces desired in the elbow (two in this case) with *Ob*. Bisect the two arcs.

Step 2 Draw a tangent line through point *b* to intersect tangents *dc* and *a*-7.

Step 3 Draw a semicircle using 1–7 as the diameter. This is half of the right section that will be used to space the elements along the stretch-out line.

Step 4 The three patterns can be cut from a rectangular piece of material whose height is equal to the sum of *a*–7, *e*–*f* and *cd*. The curves for each cut are found by transferring the lengths of the elements to parallel element lines in the pattern.

sion less than the number of pieces that will be used in the elbow. Since three pieces are to be used in this example, the arc is divided into two with line *O*–*b*. The two arcs are then bisected.

In Step 2, perpendiculars are drawn tangent to the arc at the division lines. In Step 3, a half-circular view of the cylinder is drawn for locating equally spaced elements that are projected back to the given view parallel to the elements of the three pieces of the elbow. The elements will be true-length in this view.

In Step 4, the flat pattern is developed, with the patterns of the two short pieces of the elbow located at the top and bottom of the rectangular piece. The middle segment is located between these two smaller patterns so that there will be no wasted material, and only two cuts will be necessary.

9.26 DEVELOPMENT OF STRAPS

Figure 9.67 illustrates the steps of finding the development of a strap that has been bent to serve as a support bracket, between point *A* on a vertical surface and point *B* on a horizontal surface. In Step 1, the strap is drawn in the side view where the strap appears as an edge using the specified radius to show the bend. The bend is divided into equal arcs and is developed into a flat piece in the vertical plane through point *A*. Point *B′* is located in this view.

In Step 2, the location of the hole at *B′* is found in the front view by projection. The true-size development of the strap is drawn in this view and it is projected back to the side view to complete that view of the strap.

In Step 3, the projected view of the strap in its bent position in the front view can be found by using projectors from the side view and the true pattern of the strap.

9.27 INTERSECTIONS AND DEVELOPMENTS IN COMBINATION

Intersections between parts must be found before developments of each can be completed. An example of this type is shown in Fig. 9.68.

The intersection between the two prisms is found in the given top and front views. The pattern of the vertical prism is found to the left of the front view and the intersections are shown to indicate cuts that must be made in the pattern.

The pattern of the inclined prism cannot be found without the construction of an auxiliary view in which the fold lines appear true length. The true-size right section is found in a secondary auxiliary view. The fold lines of the pattern are laid off along a stretch-out line by transferring the distances around the right section to it.

The resulting two patterns can be cut out and folded to form shapes that will intersect as shown in the top and front views.

Fig. 9.67 Strap development

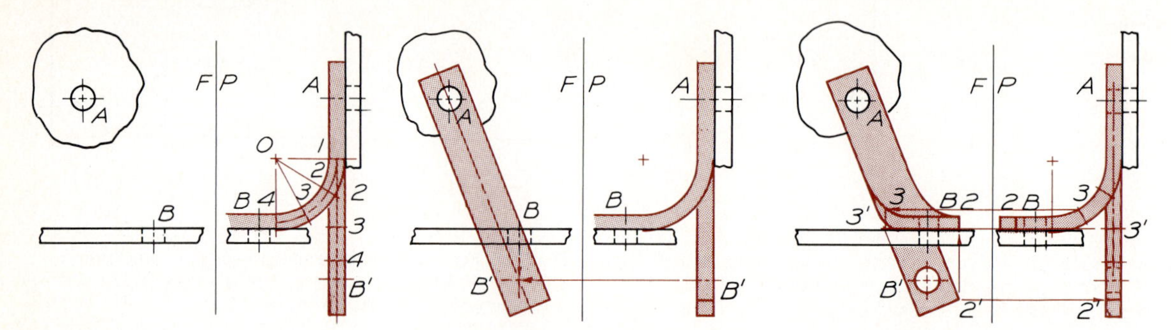

Step 1 Construct the edge view of the strap in the side view using the specified radius of bend. Locate 1, 2, 3, and 4 on the neutral axis at the bend. Revolve this portion of the strap into the vertical plane and measure the distances along this view of the neutral axis. Check the arc distances by mathematics. The hole is located at *B'* in this view.

Step 2 Construct the front view of *B'* by revolving point *B* parallel to the profile plane until it intersects the projector from *B'* in the side view. Draw the centerline of the true-size strap from *A* to *B'* in the front view. Add the outline of the strap around this centerline and the holes at each end, allowing enough material to provide sufficient strength.

Step 3 Determine the projection of the strap in the front view by projecting points from the given views. Points 3 and 2 are shown in the views to illustrate the system of projection used. The ends of the strap are drawn in each view to form true projections.

Fig. 9.68 Two prisms intersect in this example. Their intersections are found and the developments of each are found by using the principles covered in this chapter. The development of the inclined prism is found by using primary and secondary auxiliary views.

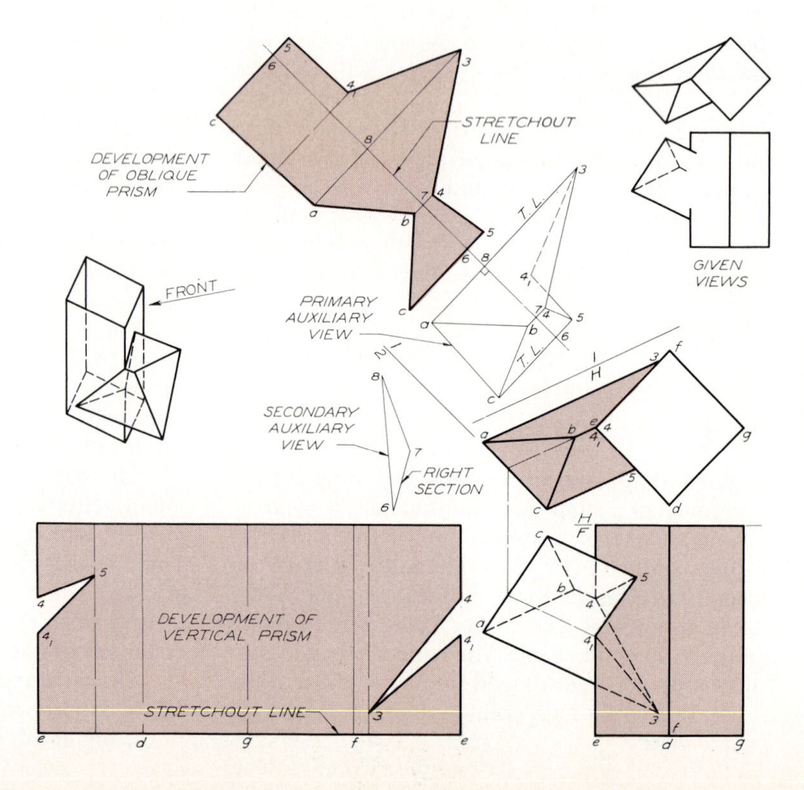

PROBLEMS

These problems are designed to fit on Size A and Size B sheets. Some problems can be grouped two or more on a single sheet. For example, two problems might be solved on a Size A sheet and four on a Size B sheet. The proper layout of a problem should be considered as an important part of its solution.

Intersections

1–5. Lay out two problems on Size A sheets. Use the given scales to transfer the dimensions from the views to the drawing. In Problems 1–3, the labeled points represent points on cutting planes that pass through the objects. Find their lines of intersection and show visibility.

6–7. For Problem 6, lay out the top and front views twice on a Size A sheet, one above the other on the sheet. Using the auxiliary sections assigned, A, B, C, D, or E, find the intersection in the front and top views. Lay out the views of Problem 7 in the same manner and use right sections A through E as assigned to find the intersections.

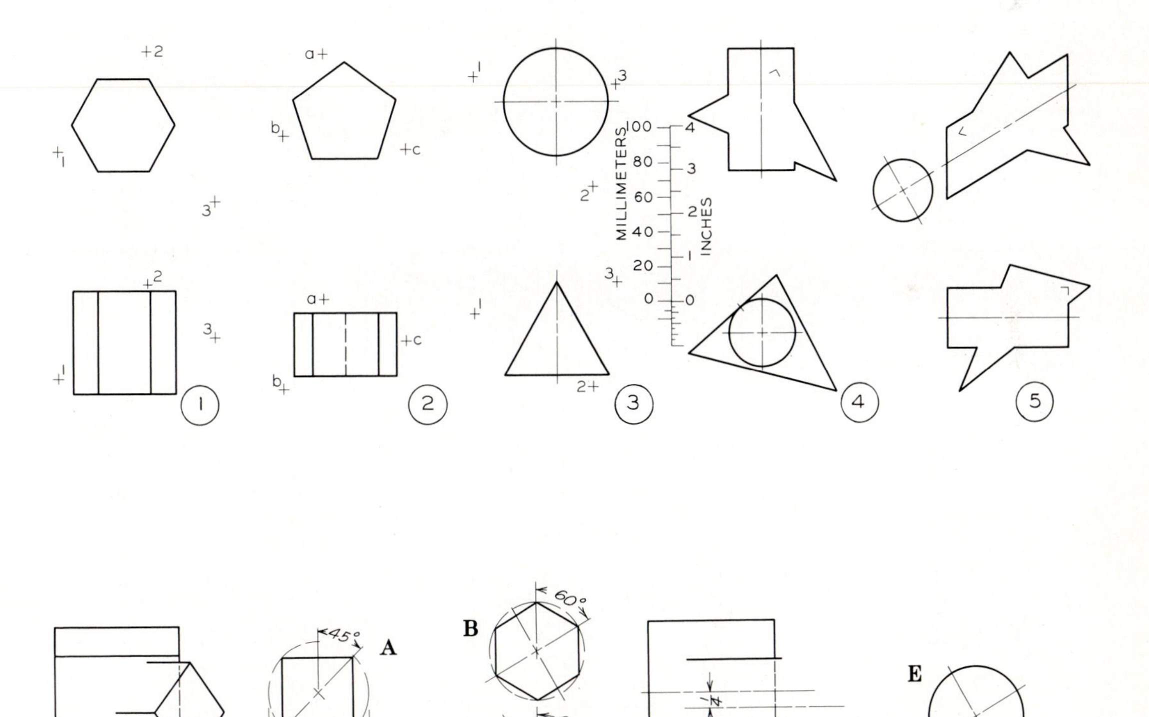

8–20. Lay out one problem per Size A sheet. Find the intersections and show the visibility for each problem.

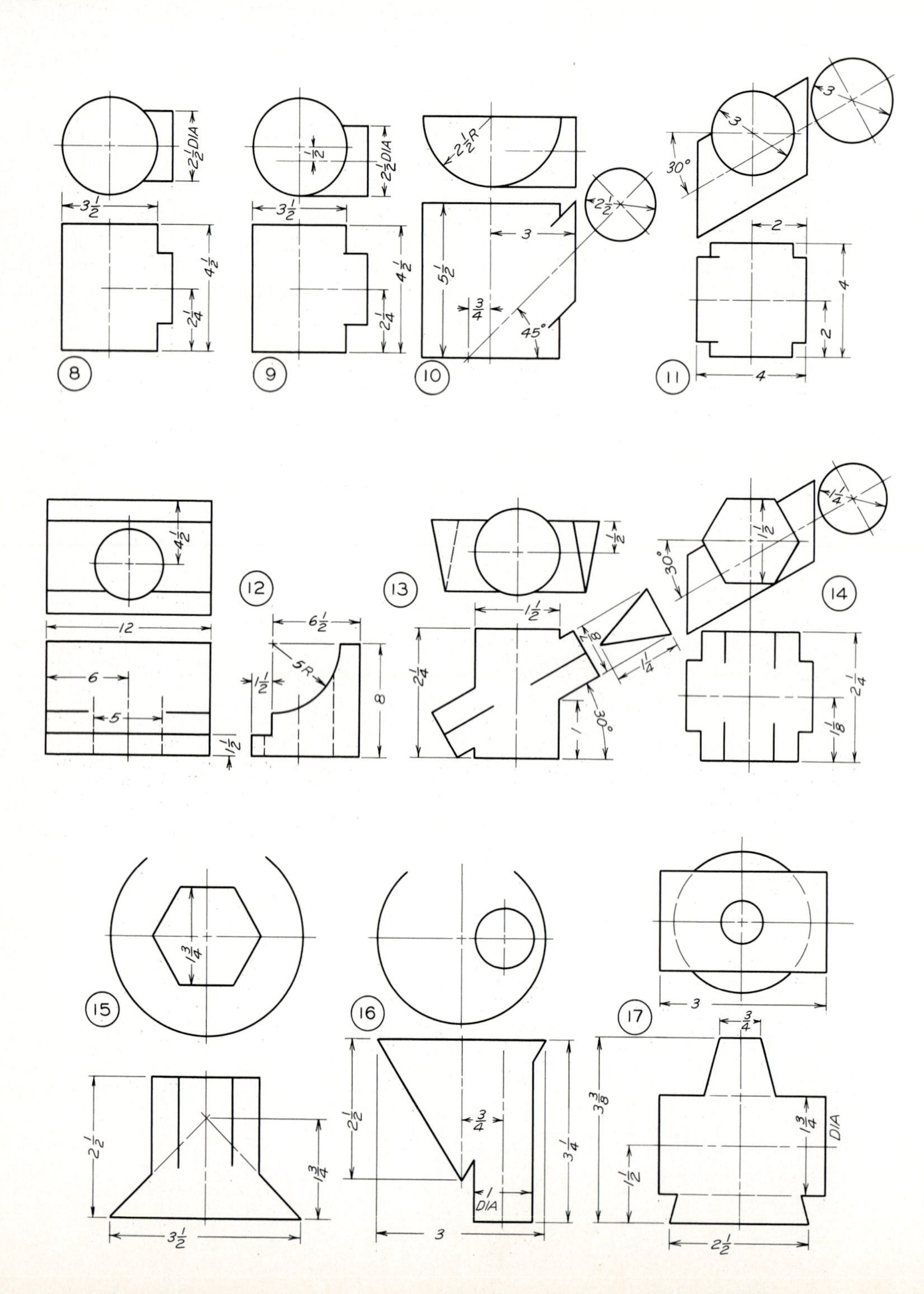

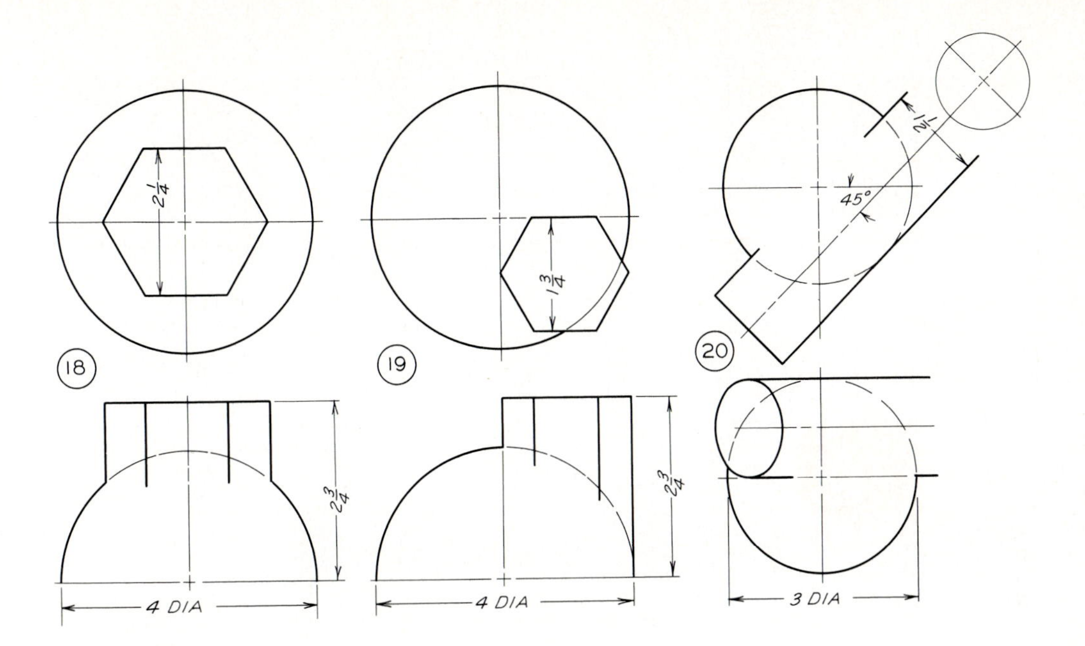

Developments

21–47. Lay out two problems per Size A sheet after dividing the sheet in half by a division line. Find the flat inside patterns of each and label the fold lines in all views.

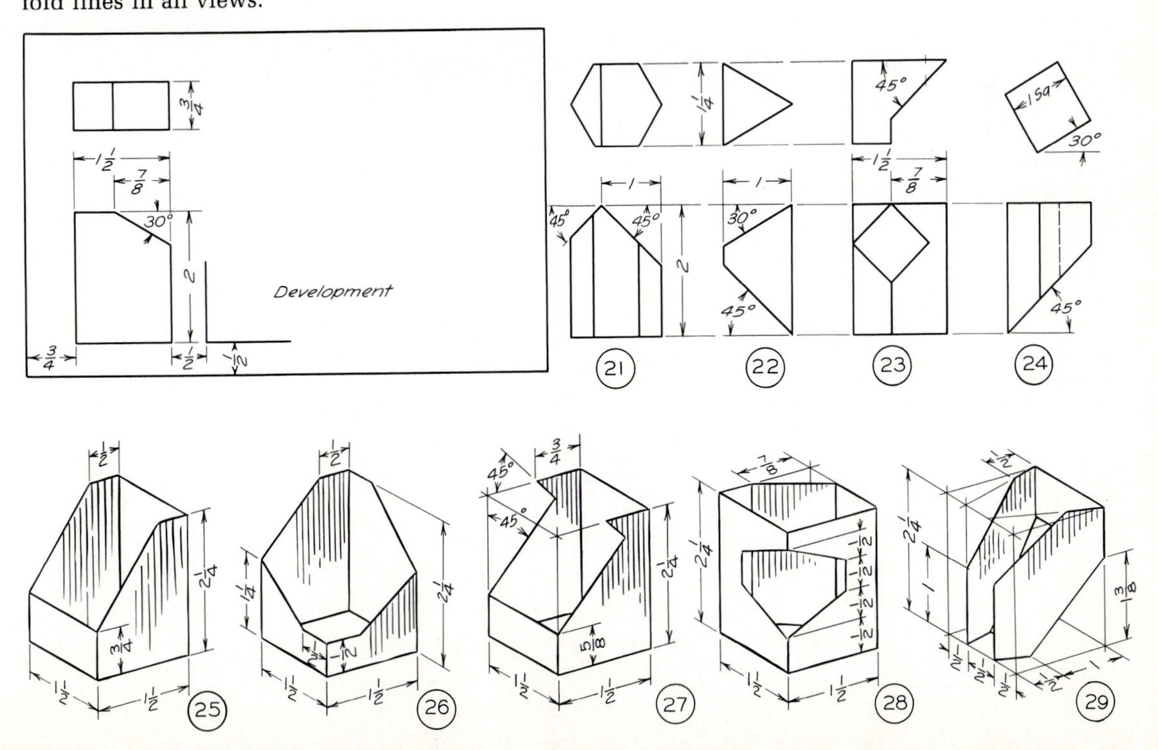

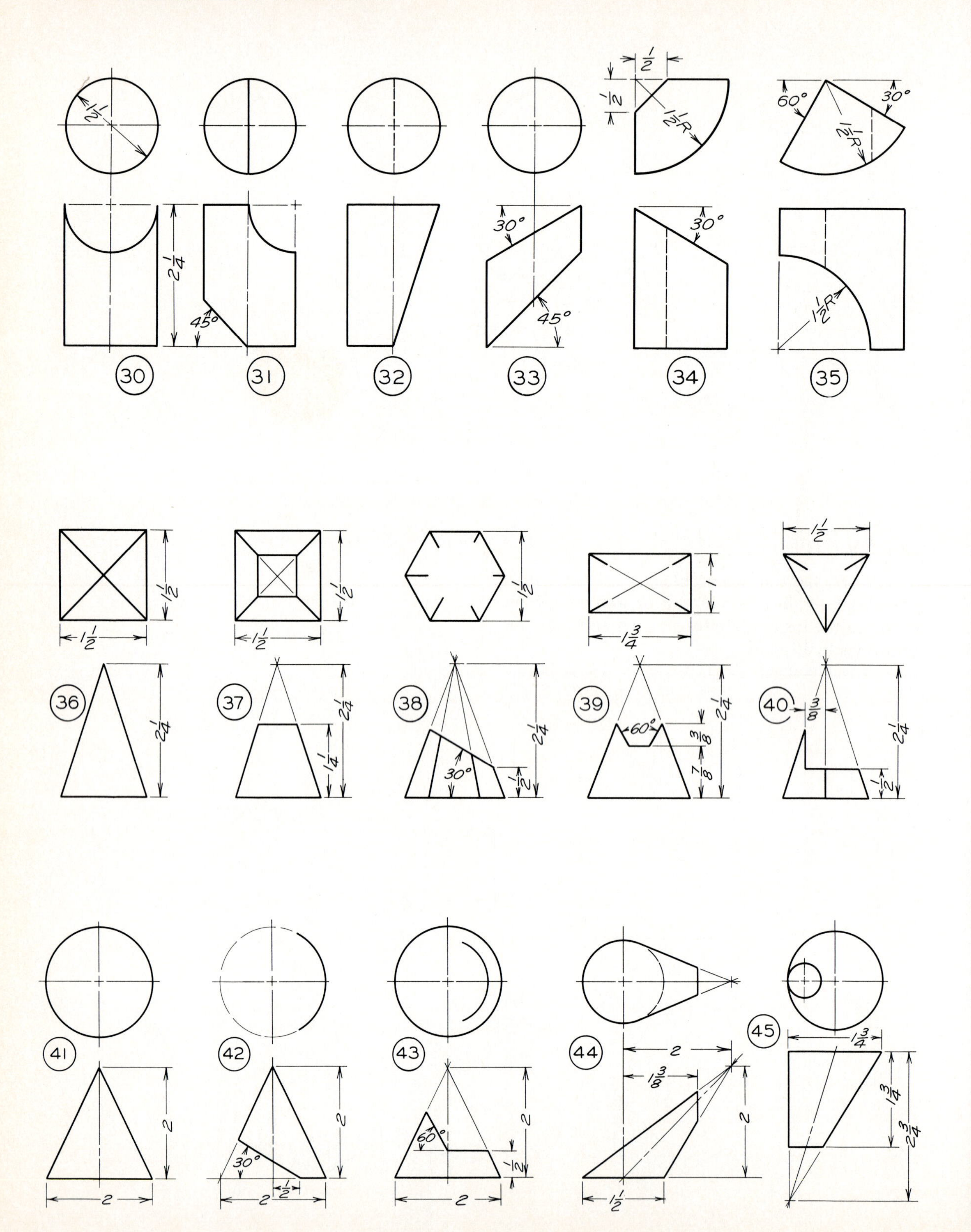
30
31
32
33
34
35
36
37
38
39
40
41
42
43
44
45

48–51. Lay out two problems per Size A sheet. Find the inside developments of each and remove the ends of the object that have been cut away by the cutting planes. Label the points in all views.

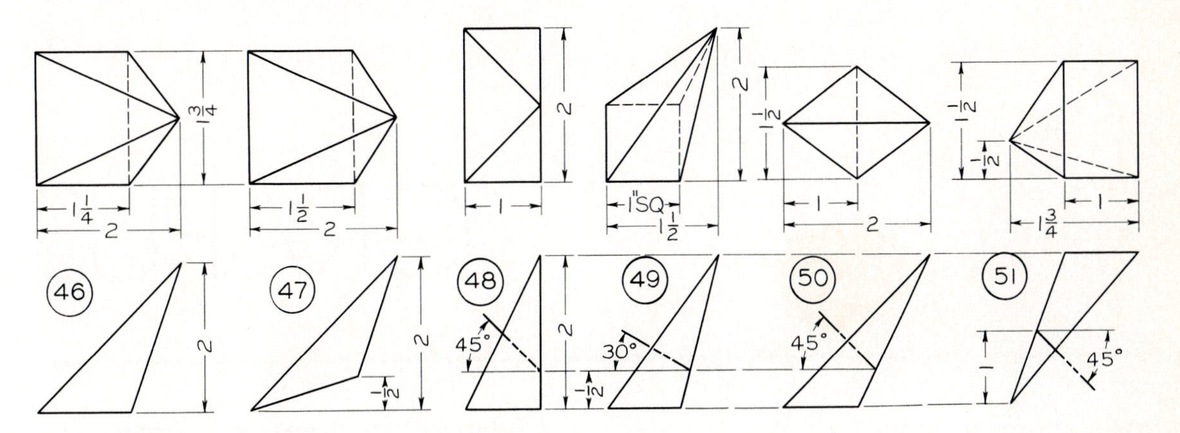

Combination Problems

52–63. Using Problems 6–17, lay them out with one per size B sheet. Find the intersections and inside developments of each. Label the points in all views.

10 GRAPHS

10.1 INTRODUCTION

Data and information expressed as numbers and words are difficult to analyze or evaluate unless they are transcribed into graphical form. Drawings of data shown graphically are called **graphs.** They are sometimes called **charts,** an acceptable term, but one that is more appropriate when referring to maps, which are specialized forms of graphs.

Graphs are helpful in the communication of data to others (Fig. 10.1); consequently, this is a popular means of briefing other people on trends that would otherwise be difficult to communicate. The trends of a plotted curve on a graph can be compared to the expressions on a person's face, which is a graph of sorts that reveals a person's feelings (Fig. 10.2). For example, a flat curve shows no change while an upwardly inclined curve indicates a positive increase. A downward curve, on the other hand, represents a downward trend and a negative result.

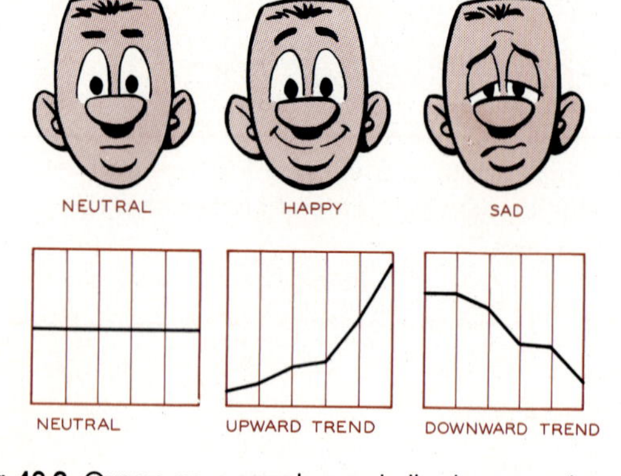

Fig. 10.2 Curves on a graph are similar to expressions on a face.

Fig. 10.1 Graphs are helpful in presenting technical data to one's associates.

This chapter will deal with the more commonly used graphs. The basic types are:

1. Pie graphs,
2. Bar graphs,
3. Linear coordinate graphs,
4. Logarithmic coordinate graphs,
5. Semilogarithmic coordinate graphs, and
6. Schematics and diagrams.

10.2 SIZE PROPORTIONS OF GRAPHS

Graphs may be used to illustrate technical reports that are reproduced in quantity, and they may be used for projection by slide or overhead projectors. In all cases, the proportion of the graph must be determined so that it will match the proportion of the space or the format of the visual aid.

If a graph is to be photographed by a 35 mm camera (Fig. 10.3), the graph must conform to the standard size of the 35 mm film that is used. This proportion is approximately 2 × 3, as shown in Fig. 10.4.

The proportions of the area in which the graph is to be drawn can be enlarged or reduced by using the diagonal-line method, as illustrated in Fig. 10.4. The horizontal dimension of the slide opening is extended to the right and the left edge is extended upward. Any point on either of these extended lines is projected to the diagonal, and then to the other extended line, to give an area of an equal proportion.

Fig. 10.3 When graphs are drawn to be photographed, they must be laid out at a proportion that will match the proportion of the film in the camera.

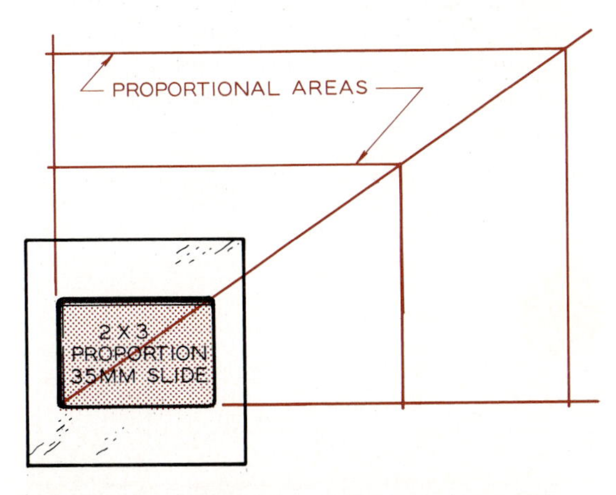

Fig. 10.4 This diagonal-line method can be used for constructing areas whose sides are proportional to those of a 35 mm slide.

10.3 PIE GRAPHS

Pie graphs compare the relationship of parts to a whole when there are only a few parts. Figure 10.5 shows the distribution of skilled workers employed in industry; this graph gives a good visual comparison of these groups.

The method of drawing a pie graph is shown in Fig. 10.6. Note that the given data does not give as good an impression of the comparisons as does the pie graph, even though the data is quite simple.

To facilitate lettering within narrow spaces, the thin sectors should be placed as nearly horizontal as possible. This provides more room for the label and the percentage. The actual percent-

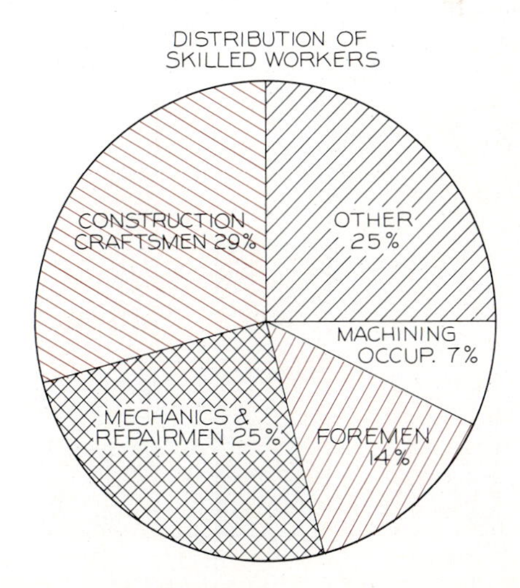

Fig. 10.5 A pie graph shows the relationship of the parts to a whole. It is effective only when there are a few parts.

Fig. 10.6 Drawing pie graphs

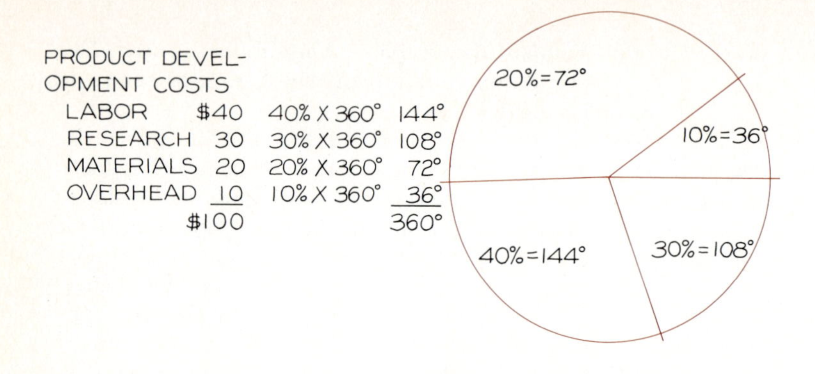

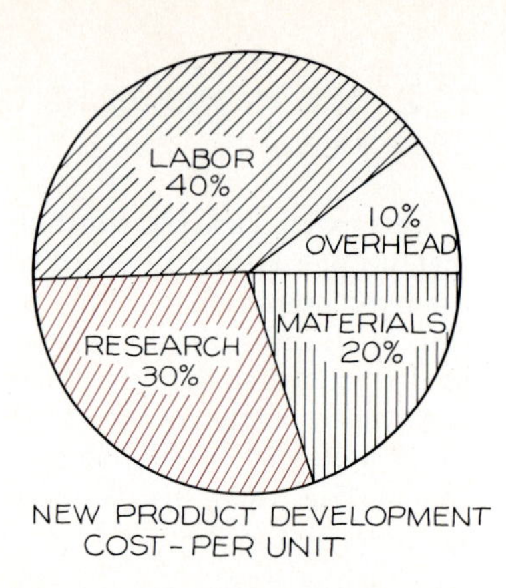

Step 1 The total sum of the parts is found, and the percentage of each is found. Each percentage is multiplied times 360° to find the angle of each sector of the pie graph.

Step 2 The circle is drawn to represent the pie graph. Each sector is drawn using the degrees found in Step 1. The smaller sectors should be placed as nearly horizontal as possible.

Step 3 The sectors are labeled with their proper names and percentages. In some cases it might be desirable to include the exact numbers in each sector as well.

age should be given in all cases, and it may be desirable to give the actual numbers or values as well in each sector.

10.4 BAR GRAPHS

Bar graphs are effective to compare values, especially since they are well understood by the general public. For example, the production of timber for various uses is compared in Fig. 10.7. In this example, the bars not only show the overall production (the total lengths of the bars), but the portions of the total devoted to the three uses of the timber.

A bar graph can be composed of a single bar (Fig. 10.8). The total length of the bar is 100% and the bar is divided into lengths that are proportional to the percentages represented by each of the three parts of the bar.

The method of constructing a bar graph is given in Fig. 10.9. The title of the graph is placed inside the graph where space is available in this case. The title could have been placed under or over the graph.

For bar graphs, the data should be sorted by arranging the bars in ascending or descending order, since it is desirable to know how the data represented by the bars rank from category to category

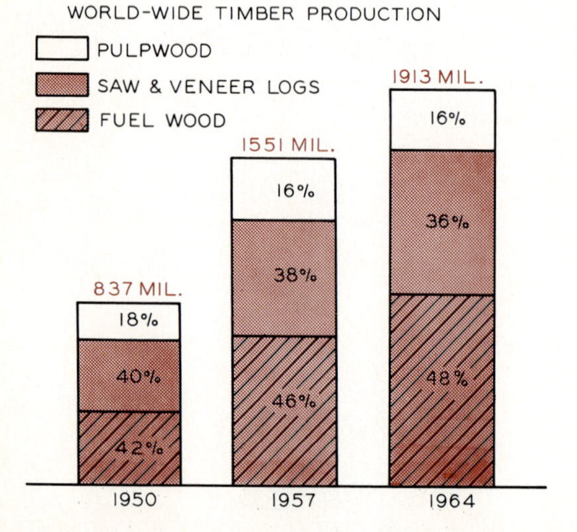

Fig. 10.7 In this example, each bar represents 100% of the total amount, and each bar represents different totals.

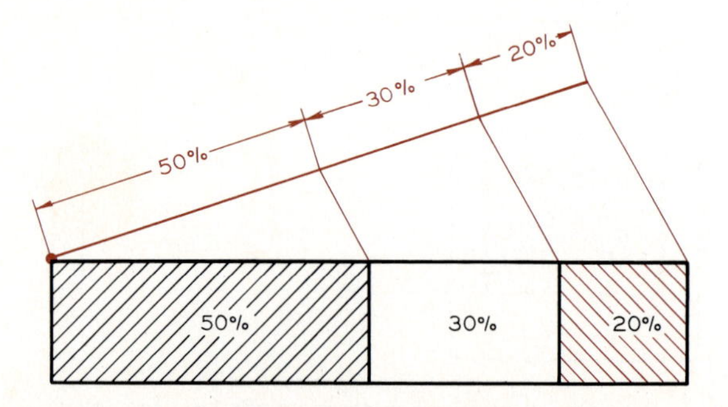

Fig. 10.8 The method of constructing a single bar where the sum of all of the parts will be 100%.

Fig. 10.9 Construction of a bar graph

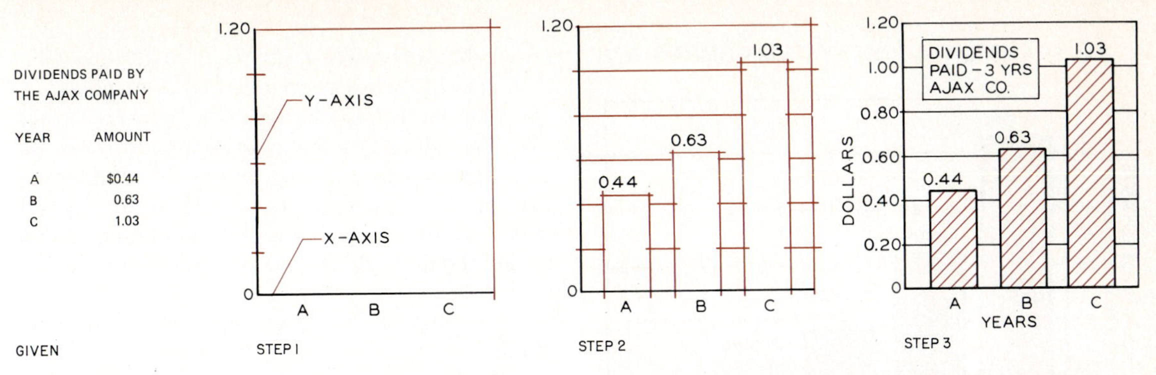

Given These data are to be plotted as a bar graph.

Step 1 Lay off the vertical and horizontal axes so that the data will fit on the grid. Make the bars begin at zero.

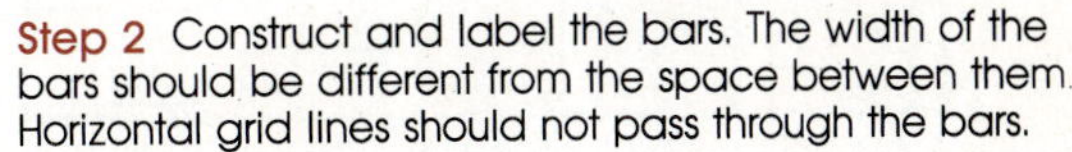

Step 2 Construct and label the bars. The width of the bars should be different from the space between them. Horizontal grid lines should not pass through the bars.

Step 3 Strengthen lines, title the graph, label the axes, and crosshatch the bars.

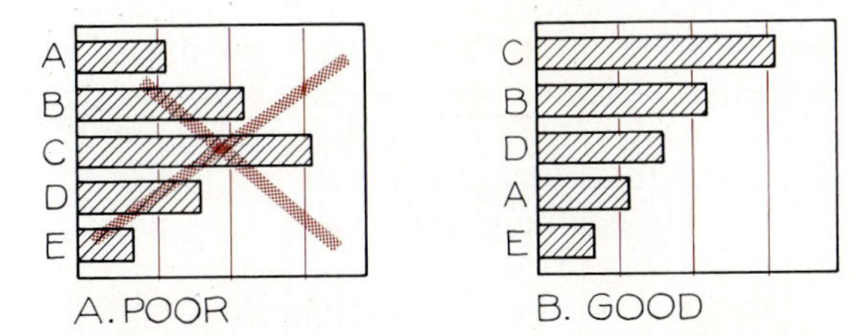

Fig. 10.10 The bars at A are arranged alphabetically. The resulting graph is not as easy to evaluate as the one at B, where the bars have been sorted and arranged in descending order.

(Fig. 10.10). An arbitrary arrangement of bars, alphabetically or numerically, results in a graph that is more difficult to evaluate than the descending arrangement at B.

If the data are sequential and involve time, such as sales per month, it would be less effective to rank the data in ascending order because it is more important to see variations in the data as related to periods of time. The determination of the method of arranging the bars depends upon the purpose of the graph and will be left to the judgment of the person constructing the graph.

Bars in a bar graph may be horizontal, as shown in Fig. 10.11, or vertical as shown in Fig. 10.12. In both of these cases, the bars are arranged in order of length of the bars for ease of comparison and ranking of the data. It is desirable for the bars of a graph to begin at zero to show a true comparison in the data.

10.5 LINEAR COORDINATE GRAPHS

A typical coordinate graph is given in Fig. 10.13 with the accompanying notes that explain its important features. The divisions along an axis of the graph should be equal; in other words, the scale is linear. The other axis is also divided into equal units, and therefore, it is a linear scale also.

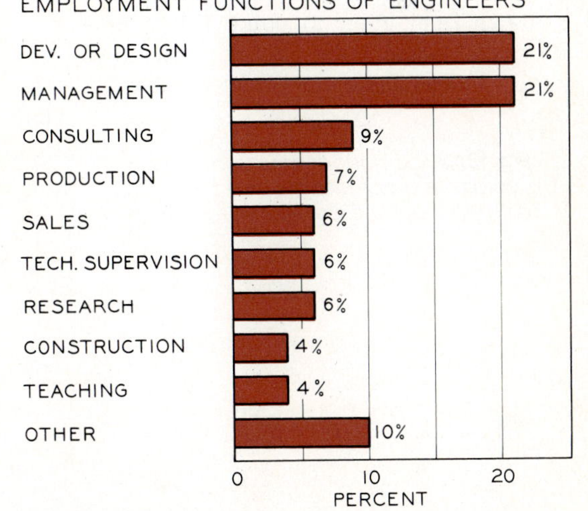

Fig. 10.11 A horizontal bar graph that is arranged in descending order to show where engineers are employed.

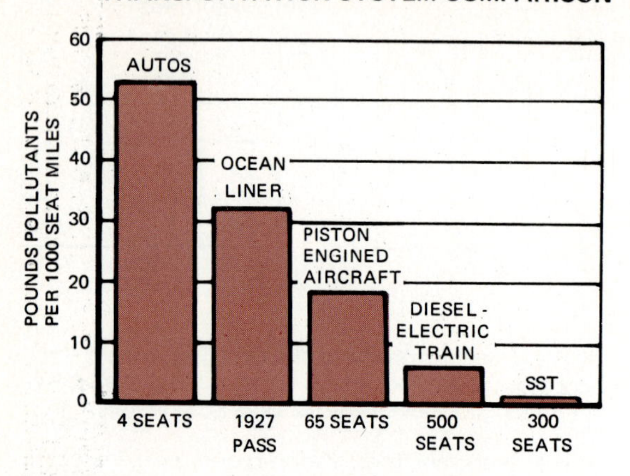

Fig. 10.12 This bar graph has the bars arranged in descending order to compare several sources of pollution. (Courtesy of Boeing Company.)

Points are plotted on the grid by using two measurements, called coordinates, made along each axis. The plotted points are indicated by using easy-to-draw symbols, such as circles, that can be drawn with a template.

The horizontal line of the graph is called the abscissa or x-axis, and the vertical scale at the left is called the ordinate or y-axis. In mathematics, data is plotted in terms of x- and y-coordinates on a graph.

Once the points have been plotted, the curve is drawn from point to point. (The line that is drawn to represent the plotted points is called a curve whether it is a smooth or broken line.) The curve should not close up the plotted points, but they should be left as open circles or symbols.

The curve must be drawn as a heavy prominent line, since it is the most important part of the graph and shows the data. In Fig. 10.13, there are two curves; therefore, it is helpful if they are drawn as different types of lines to distinguish between them. Each is labeled with a note and a leader.

The title of the graph is placed inside the graph in a box to explain the graph. It is a good rule to give enough information on a graph to make it understandable without additional text.

Units are given along the x- and y-axes with labels that designate the units that the graph is comparing.

Broken-Line Graphs

The steps of drawing a linear coordinate graph are shown in Fig. 10.14. In this case, the points are connected with a broken-line curve since the data points are ten years apart on the x-axis. Thus it is impossible to assume that the change in the data is a smooth, continual progression from point to point.

For the best appearance the plotted points should not be crossed by the curve or the grid lines of the graph (Fig. 10.15). Each circle or symbol used to plot points should be about 1/8″ (3 mm) in diameter.

Different symbols can be used to plot points, along with distinctively different lines to represent the curves. Several approved symbols and lines are shown in Fig. 10.16.

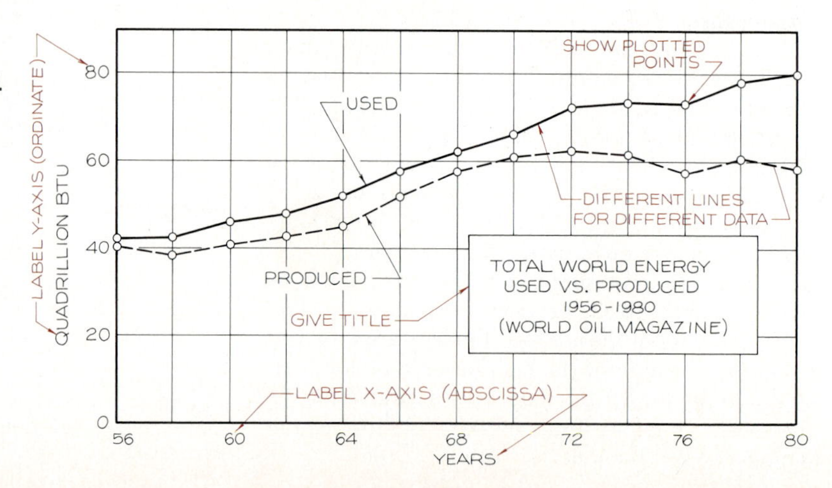

Fig. 10.13 The basic linear coordinate graph with the important features identified.

Fig. 10.14 Construction of a broken-line graph

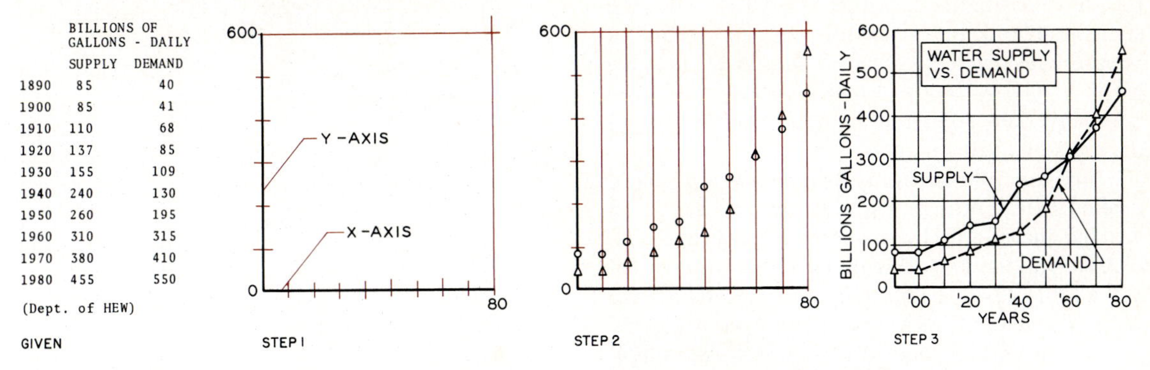

	BILLIONS OF GALLONS - DAILY SUPPLY	DEMAND
1890	85	40
1900	85	41
1910	110	68
1920	137	85
1930	155	109
1940	240	130
1950	260	195
1960	310	315
1970	380	410
1980	455	550

(Dept. of HEW)

Given A record of water supply and water demand since 1890 plotted as a line graph.

Step 1 The vertical and horizontal axes are laid off to provide space for the largest values.

Step 2 The points are plotted directly over the respective years. Different symbols are used for each curve.

Step 3 The data points are connected with straight lines, the axes are labeled, the graph is titled, and the lines are strengthened.

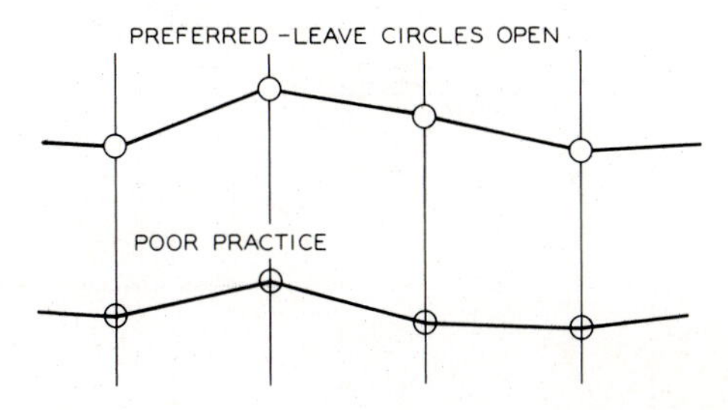

Fig. 10.15 The curve of a graph should be drawn from point to point, but it should not close up the symbols used to locate the plotted points.

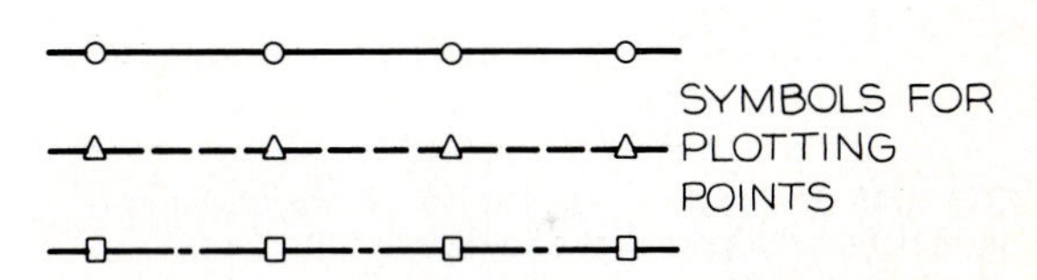

Fig. 10.16 Any of these symbols or lines can be used effectively to represent different curves on a single graph. The symbols are about 1/8" (3 mm) in diameter.

Fig. 10.17 Title placement on a graph

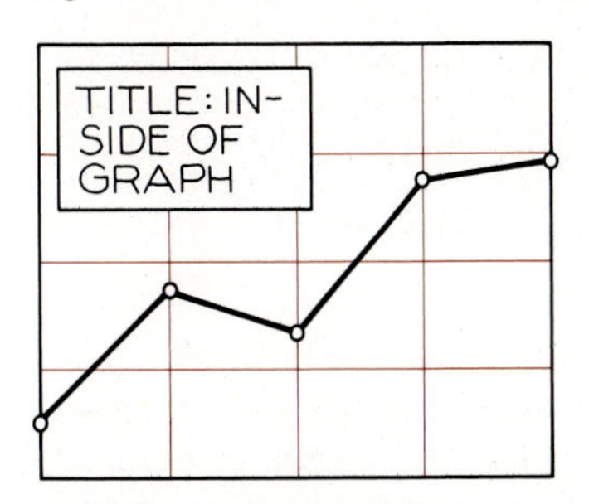

A. The title of a graph can be lettered inside a box placed within the area of the graph. The perimeter lines of the box should not coincide with grid lines within the graph.

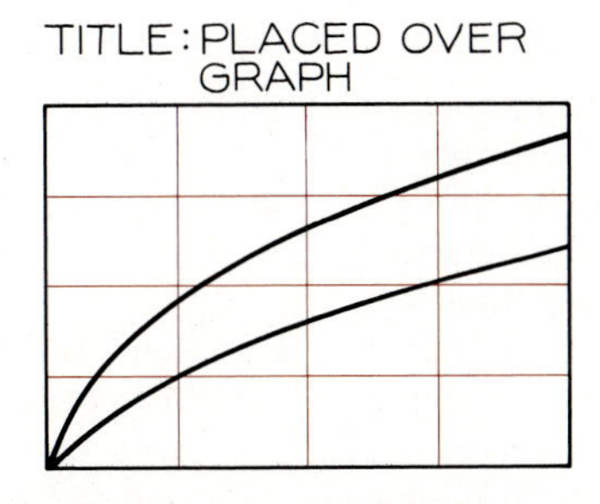

B. The title can be placed above the graph. The title should be drawn in 1/8" letters, or slightly larger.

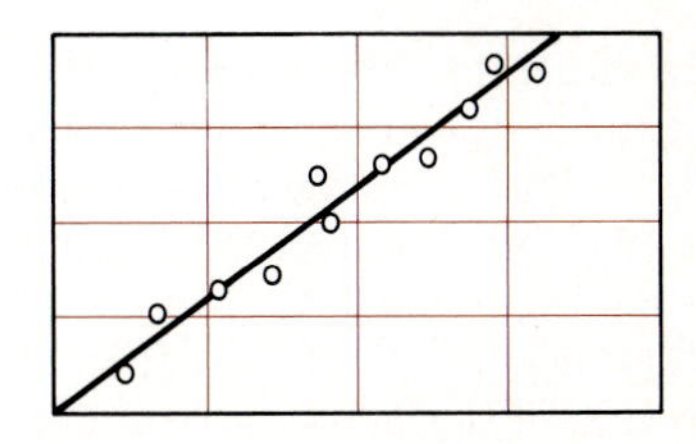

TITLE: PLACED UNDER GRAPH

C. The title can be placed under a graph. It is good practice to be consistent when a series of graphs is used in the same report.

The title of a graph can be placed in any of the three positions shown in Fig. 10.17. The title should never be one as meaningless as "graph" or "coordinate graph." Instead, it should explain the graph by giving the important information such as company, date, source of the data, and the general comparisons being shown.

The example in Fig. 10.18A shows a properly labeled axis. You can see that the axis at part B has too many grid lines and too many units labeled along the axis; this clutters the graph without adding to its value. On the other hand, the units selected at C make it difficult to easily interpolate between the labeled values. For example, it is more difficult to locate a value such as 22 by eye on this scale than on the one at A.

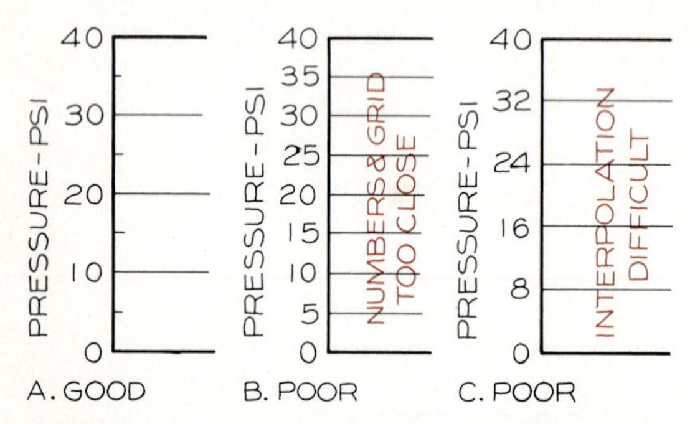

Fig. 10.18 The scale at A is the best. It has about the right number of grid lines and divisions, and the numbers are given in well-spaced, easy-to-interpolate form. The numbers at B are too close and there are too many grid lines. The units at C are given in units that make interpolation difficult by eye.

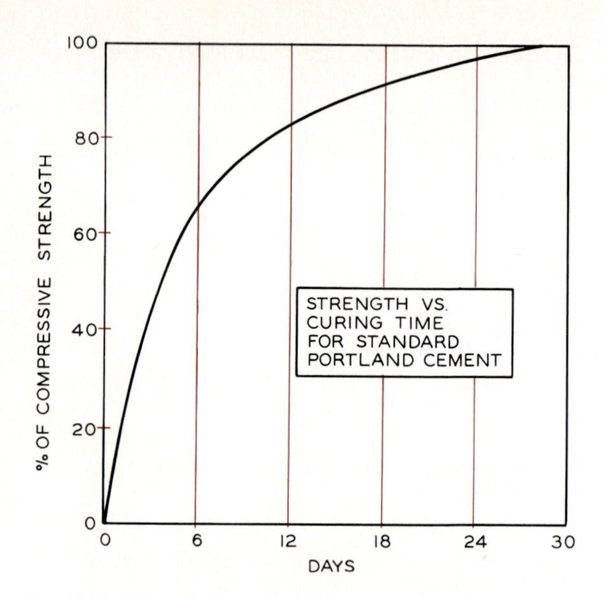

Fig. 10.19 When the process that is graphed involves gradual, continuous changes in relationships, the curve should be drawn as a smooth line.

Smooth-Line Graphs

In order to determine whether the points on a graph will be connected with a broken-line curve as previously shown, or a smooth-line curve as shown in Fig. 10.19, you must have an understanding of the data. You instinctively understand that the strength of cement and its curing time will result in a smooth, continuous relationship that should be connected by a smooth curve. Even if the data points do not plot to lie on the curve, you know that the deviation of the points from this curve is due to errors of measurement or the methods used in collecting the data.

Similarly, the strength of clay tile, as related to its absorption characteristics, is an example of data that yields a smooth curve (Fig. 10.20). Note that the plotted data does not lie on the curve. Since you know that the relationship is a continuous one that should be connected with a smooth curve, the *best curve* is drawn to interpret the data to give an average representation of the points.

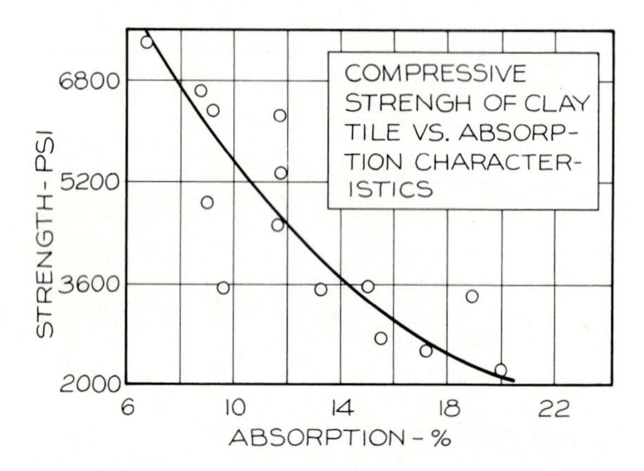

Fig. 10.20 If it is known that a relationship plotted on a graph should yield a smooth gradual curve, a smooth-line "best" curve is drawn to represent the average of the plotted points. You must use your judgment and knowledge of the data in cases of this type.

For the same reasons, you know that there is a smooth-line curve relationship between miles

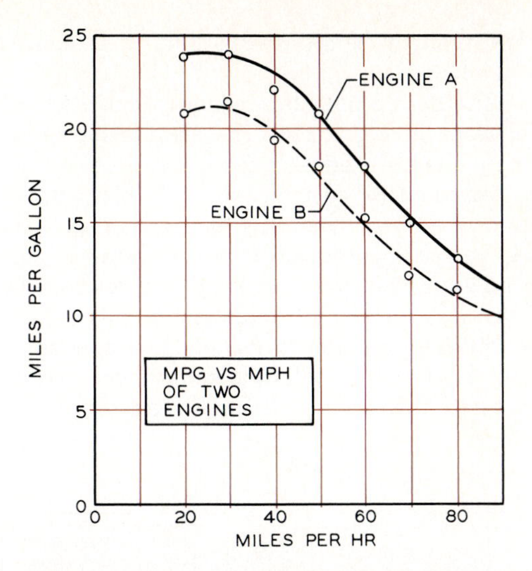

Fig. 10.21 These are "best" curves that approximate the data without necessarily passing through each data point. Inspection of the data tells you that this curve should be a smooth-line curve rather than a broken-line curve.

per gallon and the speed at which a car is driven. Two engines are compared in Fig. 10.21 with two smooth-line curves. The effect of speed on several automotive characteristics is compared in Fig. 10.22.

When a smooth-line curve is used to connect data points, there is the implication that you can *interpolate* between the plotted points to estimate other values. Points connected by a broken-line curve imply that you cannot interpolate between the plotted points.

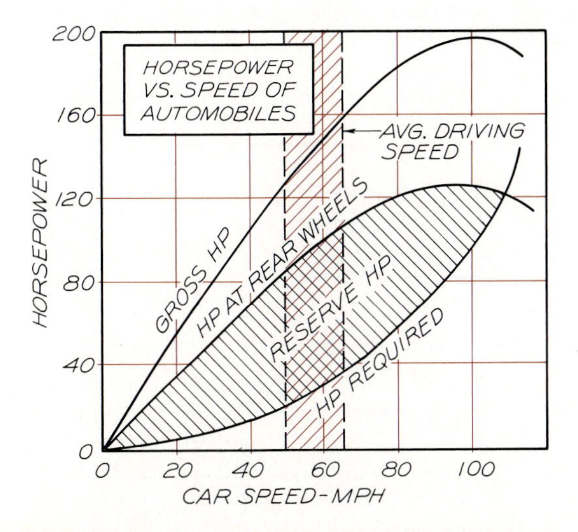

Fig. 10.22 A linear coordinate graph is used here to analyze data affecting the design of an automobile's power system.

Straight-Line Graphs

Some graphs have neither broken-line curves nor smooth-line curves, but straight-line curves as shown in the example in Fig. 10.23. You can determine a third value from the two given values using this graph. For example, if you are driving 70 miles per hour and it takes 5 seconds to react to apply your brakes, you will have traveled 550 feet in this interval of time.

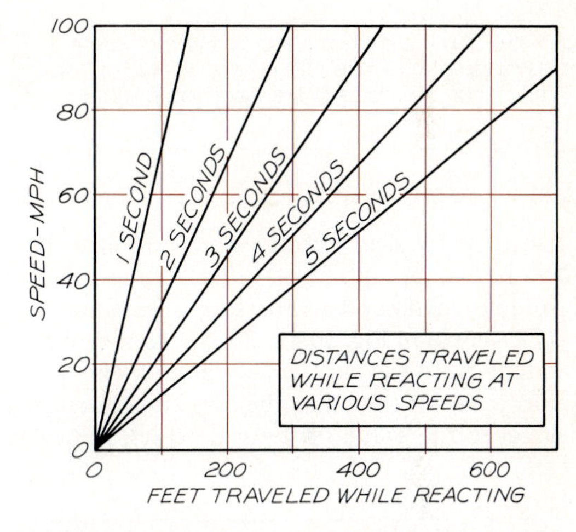

Fig. 10.23 A graph can be used to determine a third value when two variables are known. Taking this information from a graph is easier than computing each answer separately.

Two-Scale Coordinate Graphs

Graphs can be drawn with different scales in combination, such as the one shown in Fig. 10.24. The vertical scale at the left is in units of pounds, and the one at the right is expressed in units of degrees of temperature. Both curves are drawn using their respective *y*-axes, and each curve is labeled.

Care must be taken to avoid confusing the reader of a graph of this type. However, these graphs are effective when comparing related variables such as the drag force and air temperature of a type of automobile tire, as shown in this example.

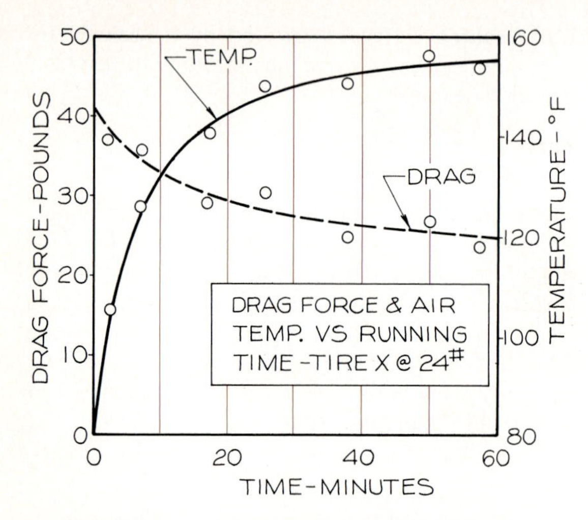

Fig. 10.24 This is a composite graph with different scales along each *y*-axis. The curves are labeled so reference can be made to the applicable scale.

Optimization Graphs

Graphs can be used effectively to optimize data. For example, the optimization of the depreciation of an automobile and its increase in maintenance costs is shown in Fig. 10.25. These two sets of data are plotted and the curves cross at an x-axis value of four years. At this time, the cost of maintenance is equal to the value of the car, which indicates that this might be a desirable time to exchange it for a new one.

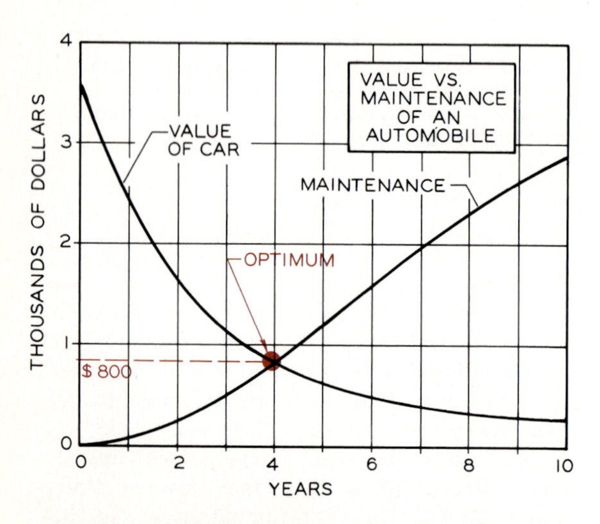

Fig. 10.25 This graph shows the optimum time to sell a car, based on the intersection of two curves that represent the depreciation of the car's value and its increasing maintenance costs.

Another optimization graph is constructed in two steps in Fig. 10.26. The manufacturing cost per unit is reduced as more units are made, but the warehousing cost increases. A third curve is found in Step 2, by adding the two curves. Value A is shown to illustrate how this value is transferred with your dividers to add it to the manufacturing cost curve. The "total" curve tells you that the optimum number to manufacture at a time is about 11,000 units. When more or fewer are manufactured, the expense per unit is greater.

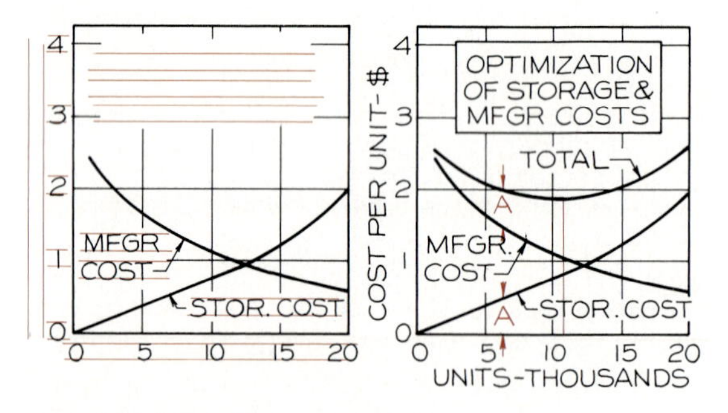

Fig. 10.26 Optimization graphs

Step 1 Lay out the graph and plot the given curves.

Step 2 Add the two curves to find a third curve. Distance A is shown transferred to locate a point on the third curve. The lowest point of the "total" curve is the optimum point of 11,000 units.

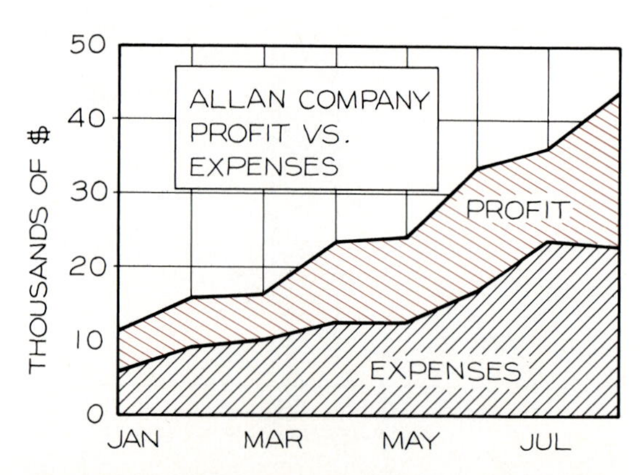

Fig. 10.27 This graph is a combination of a coordinate graph and an area graph. The upper curve represents the total of two values plotted, one above the other.

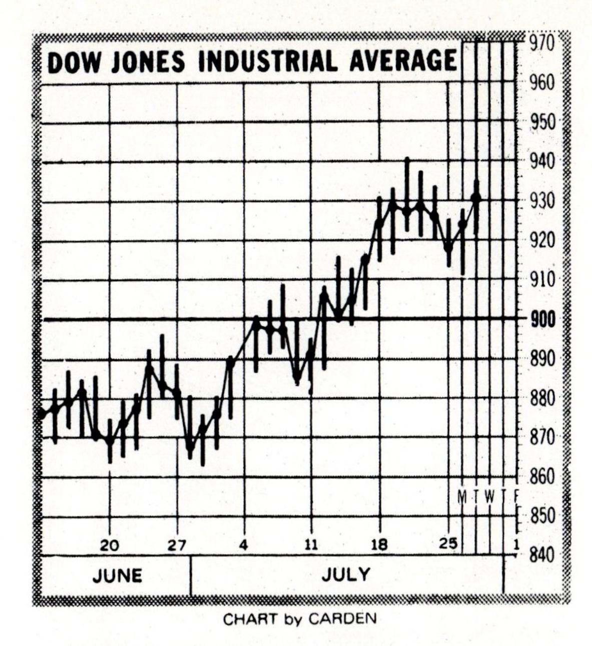

Fig. 10.28 This graph is a combination of a coordinate graph and a bar graph. The bars represent the ranges of selling during a day, and the broken-line curve connects the points at which the market closed each day.

Composite Graphs

The graph in Fig. 10.27 is a composite between an area graph and a coordinate graph. The lower curve is plotted first. The upper curve is found by adding the values to the lower curve so that the two areas represent the data. The upper curve is equal to the sum of the two y-values.

The graph in Fig. 10.28 is a combination of a coordinate graph and a bar graph that is used to show the Dow-Jones Industrial stock average. The bars represent the daily ranges in the index. The broken-line curve connects the points where the market closed for each day.

Break-Even Graphs

Coordinate graphs are helpful in evaluating marketing and manufacturing costs that are used to determine the selling cost for a product. The break-even graph in Fig. 10.29 is drawn to reveal that 10,000 units must be sold at $3.50 each to cover the manufacturing and development costs. Sales in excess of 10,000 result in profit.

A second type of break-even graph (Fig. 10.30) uses the cost of manufacturing per unit versus the number of units produced. In this example, the development costs must be incorporated into the unit costs. The manufacturer can determine how many units must be sold to break even at a given

Fig. 10.29 Break-even graph

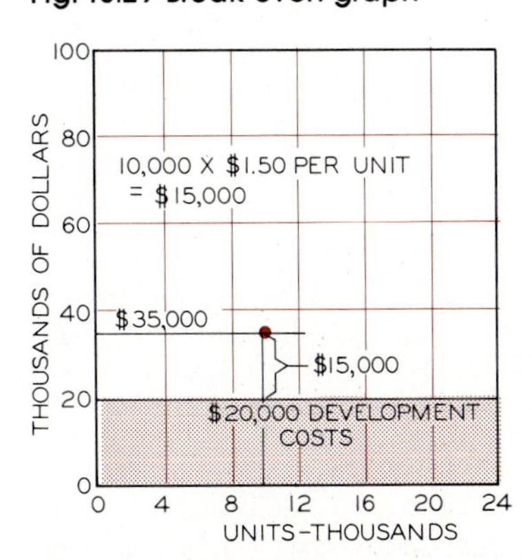

Step 1 The graph is drawn to show the cost ($20,000 in this case) of developing the product. It is determined that each unit would cost $1.50 to manufacture. This is a total investment of $35,000 for 10,000 units.

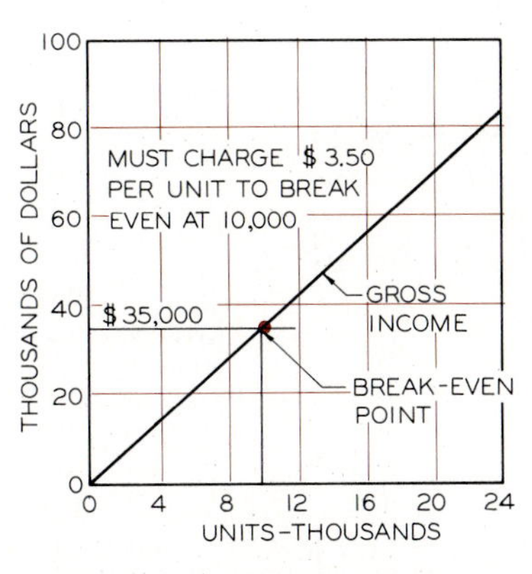

Step 2 In order for the manufacturer to break even at 10,000, the units must be sold for $3.50 each. Draw a line from zero through the break-even point for $35,000.

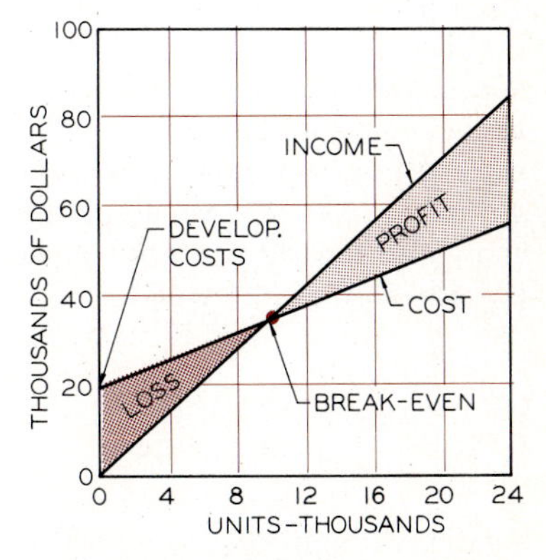

Step 3 The loss is $20,000 at zero units and becomes progressively less until the break-even point is reached. The profit is the difference between the cost and income to the right of the break-even point.

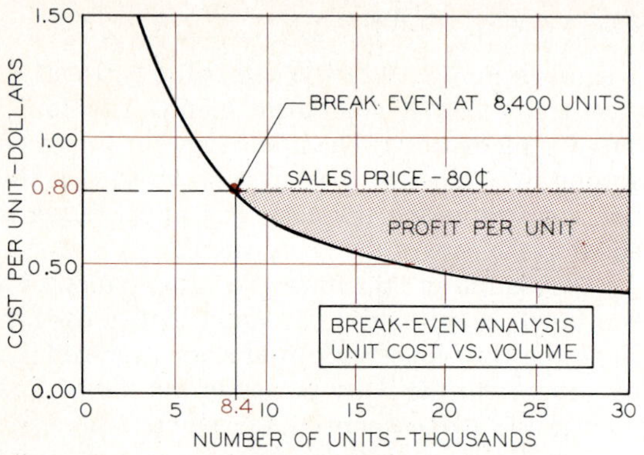

1.50
COST PER UNIT - DOLLARS
BREAK EVEN AT 8,400 UNITS
1.00
0.80
SALES PRICE - 80¢
PROFIT PER UNIT
0.50
BREAK-EVEN ANALYSIS
UNIT COST VS. VOLUME
0.00
0 5 8.4 10 15 20 25 30
NUMBER OF UNITS - THOUSANDS

Fig. 10.30 The break-even point can be found on a graph that shows the relationship between the cost per unit, which includes the development cost, and the number of units produced. The sales price is a fixed price. The break-even point is reached when 8400 units have been sold at 80¢ each.

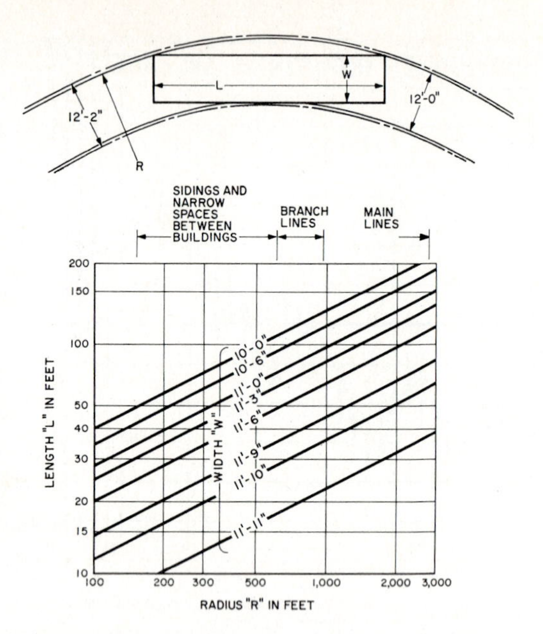

Fig. 10.31 This logarithmic graph shows the maximum load projection of 12 feet in relation to the length of a railroad car and the radius of the curve. (Courtesy of *Plant Engineering.*)

price, or the price per unit if a given number is selected. In this example, a sales price of 80¢ requires that 8400 units be sold to break even.

10.6 LOGARITHMIC COORDINATE GRAPHS

Both scales of a logarithmic grid are calibrated into divisions that are equal to the logarithms of the units represented. Commercially printed logarithmic grid paper is available in many variations that can be used for graphing data.

The graph in Fig. 10.31 has a logarithmic grid and shows the geometry of standard railroad cars as they relate to the tracks so that there will not be more than a maximum projection width of 12 feet around curves. Extremely large values can be shown on logarithmic grids since the lengths are considerably compressed.

10.7 SEMILOGARITHMIC COORDINATE GRAPHS

Semilogarithmic graphs are referred to as ratio graphs because they give graphical representations of ratios. One scale, usually the vertical scale, is logarithmic, and the other is linear (divided into equal divisions). Two curves that are parallel on a semilogarithmic graph have equal percentage increases.

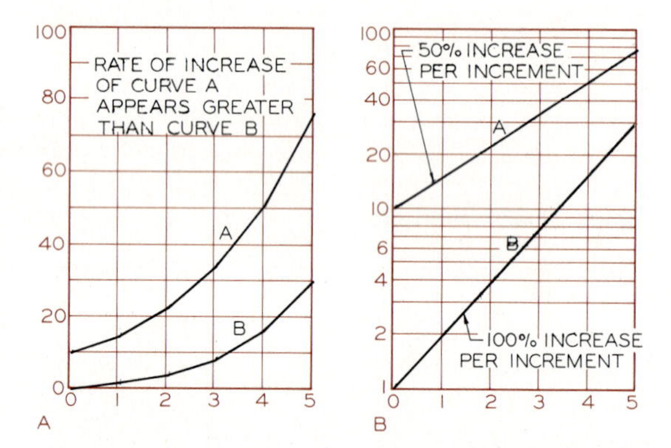

Fig. 10.32 When plotted on a standard grid, curve *A* appears to be increasing at a greater rate than curve *B*. However, the true rate of increase can be seen when the same data are plotted on a semilogarithmic graph in part B.

In Fig. 10.32, the same data is shown plotted on a linear grid and on a semilogarithmic grid. The semilogarithmic graph reveals that the percent of change from 0 to 5 is greater for Curve B than for Curve A, since Curve B is steeper. This comparison was not apparent in the plot on the linear grid.

The relationship between the linear scale and the logarithmic scale is shown in Fig. 10.33. Equal divisions along the linear scale have unequal ratios, and equal divisions along the log scale have equal ratios.

Log scales can be drawn to have one or many cycles. Each cycle increases by a factor of 10. For example, the scale in Fig. 10.34A is a three-cycle scale, and the one at B is a two-cycle scale. When these must be drawn to a special length, commercially printed log scales can be used to graphically transfer the calibrations to the scale being drawn (Fig. 10.34C).

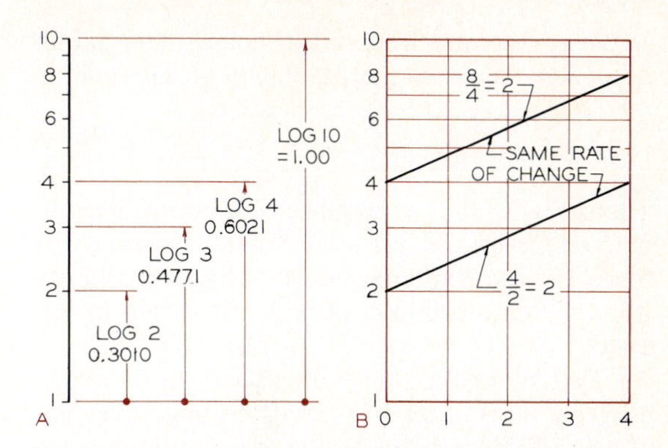

Fig. 10.35 A number's logarithm is used to locate its position on a log scale (A). This makes it possible to see the true rate of change at any location on a semilogarithmic graph (B).

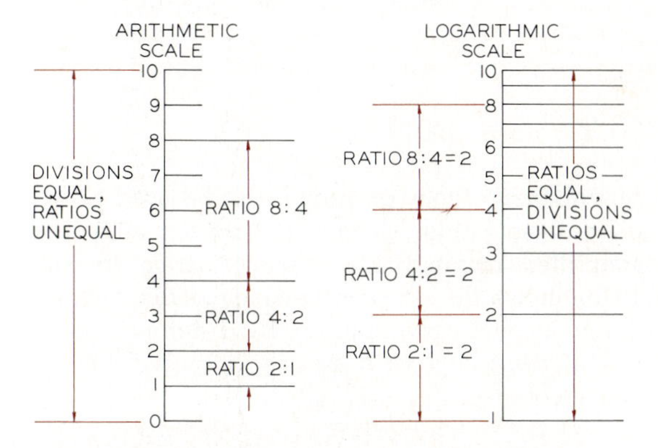

Fig. 10.33 The spacings on an arithmetic scale are equal, with unequal ratios between points. The spacings on logarithmic scales are unequal, but equal spaces represent equal ratios.

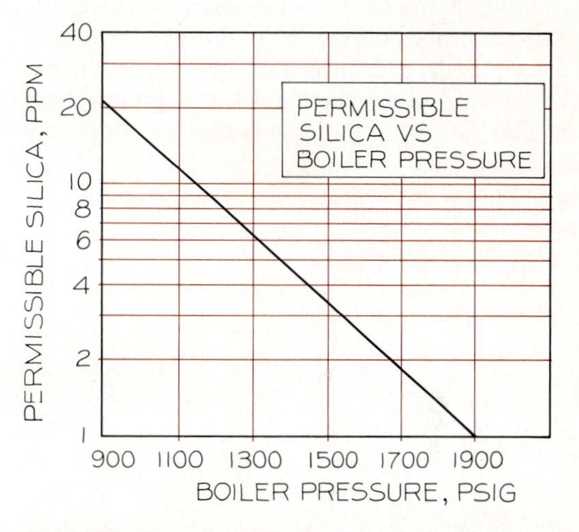

Fig. 10.36 A semilogarithmic graph is used to compare the permissible silica (parts per million) in relation to the boiler pressure.

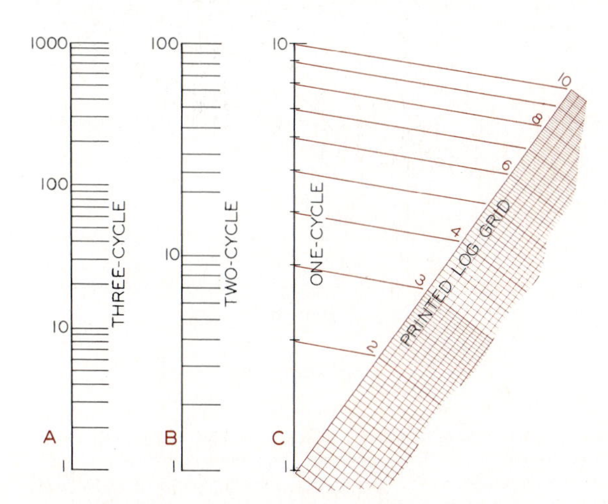

Fig. 10.34 Logarithmic paper can be purchased or drawn using several cycles. Three-, two-, and one-cycle scales are shown here. Calibrations can be drawn on a scale of any length by projecting from a printed scale as shown in part C.

In Fig. 10.35, the calibrations along the log scale are separated by the difference in their logarithms. The logarithms are laid off using a convenient scale that is calibrated in decimal divisions.

It can be seen in Fig. 10.35B that parallel straight-line curves yield equal ratios of increase. Figure 10.36 is an example of a semilogarithmic graph used to present industrial data.

Semilog graphs have several disadvantages. The most critical one is that they are misunderstood by many people who do not recognize them as being different from linear coordinate graphs. Also, zero values cannot be shown on log scales.

Percentage Graphs

The percent that one number is of another, or the percent increase of one number that is greater than the other, can be determined by using a semilogarithmic graph. Examples of both are shown in Fig. 10.37.

Data plotted in Step 1 are used to find the percent that 30 is of 60, two points on the curve. The vertical distance between the two points is equal to the difference of their logarithms. This distance is subtracted from the log of 100 at the right of the graph. This gives a value of 50% as a direct reading.

In Step 2, the percent of increase between two points is transferred from the grid to the lower end of the log scale and measured upward. It is measured upward since the increase is greater than zero. The percent of increase is measured directly from scale. These methods can be used to find percent increases or decreases of any sets of points on the grid.

10.8 POLAR GRAPHS

Polar graphs are drawn with a series of concentric circles with the origin at the center. Lines are drawn from the center toward the perimeter of the graph, where the data can be plotted through 360° by measuring values from the origin. The illumination of a lamp is shown in Fig. 10.38, where the maximum lighting of the lamp is 550 lumens at 35° from the vertical, for example.

This type of graph is used to plot the areas of illumination of all types of lighting fixtures. Polar graph paper is available commercially for drawing graphs of this type.

10.9 SCHEMATICS

Miscellanous types of graphs can be used to explain organizations, events, and relationships in a simplified manner. The block diagram in Fig. 10.39 shows the progressive steps of the completion of a construction project. Each step is blocked in and connected with arrows to explain the sequence of events.

Fig. 10.37 Percentage graphs

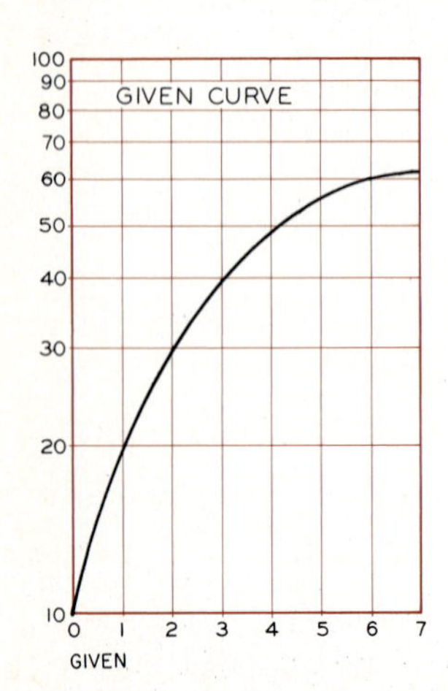

Given The data are plotted on a semilogarithmic graph to enable you to determine percentages and ratios in much the same manner that you use a slide rule.

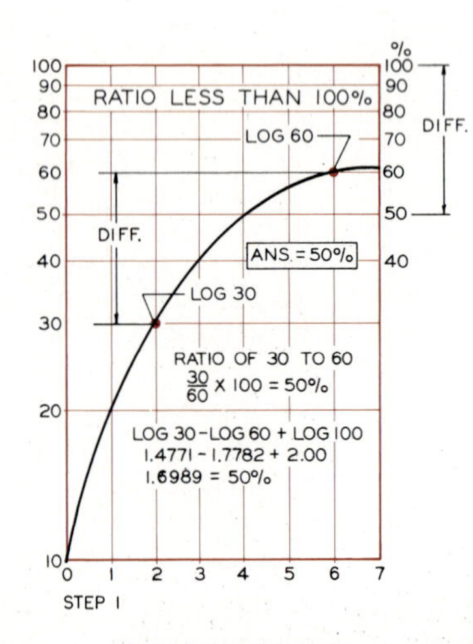

Step 1 In finding the percent that a smaller number is of a larger number, you know that the percent will be less than 100%. The log of 30 is subtracted from the log of 60 with dividers and is transferred to the percent scale at the right, where 30 is found to be 50% of 60.

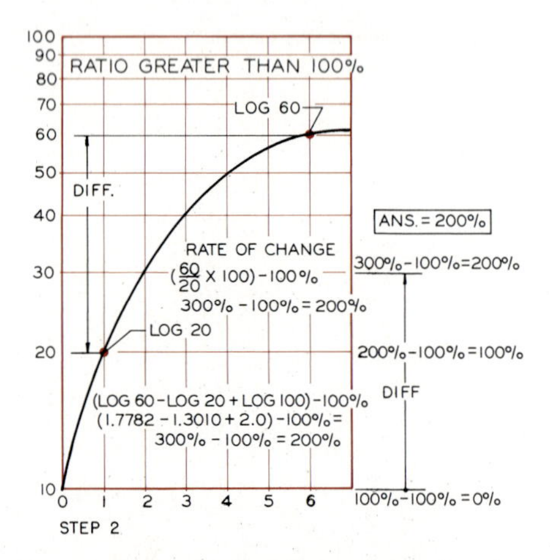

Step 2 To find the percent of increase, a smaller number is divided into a larger number to give a value greater than 100%. The difference between the logs of 60 and 20 is found with dividers, and is measured upward from 100% at the right, to find that the percent of increase is 200%.

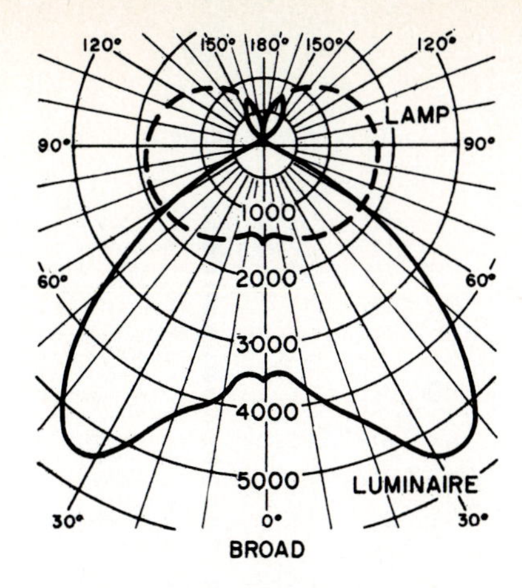

Fig. 10.38 A polar graph is used to show the illumination characteristics of luminaires.

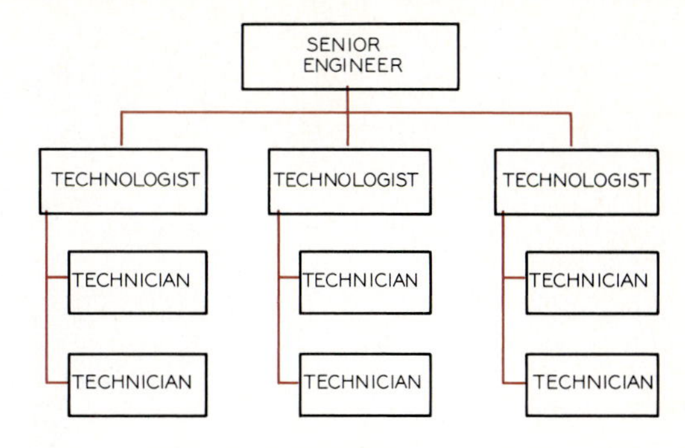

Fig. 10.40 This schematic shows the organization of a design team in a block diagram.

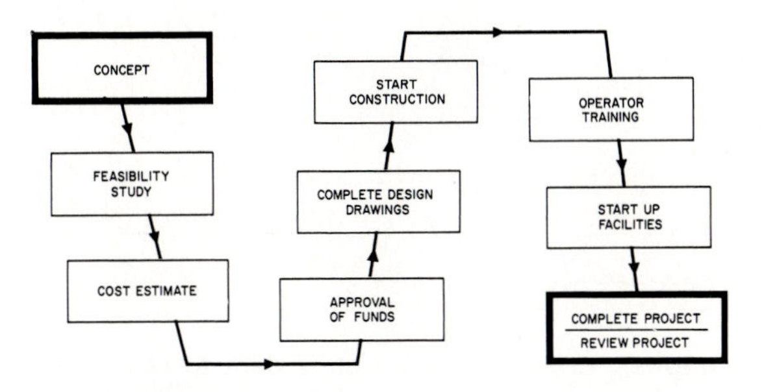

Fig. 10.39 This schematic shows a block diagram of the steps required to complete a project. (Courtesy of *Plant Engineering.*)

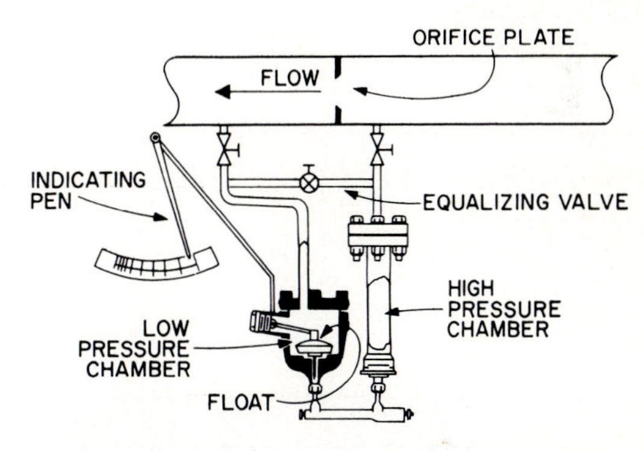

Fig. 10.41 A schematic showing the components of a gauge that measures the flow in a pipeline. (Courtesy of *Plant Engineering.*)

The organization of a company or a group of people can be depicted in an organizational chart of the type shown in Fig. 10.40. The offices represented by the blocks at the lower part of the graph are responsible to the blocks above them as they are connected by lines of authority. All blocks are finally connected with lines that converge at the top to the principal office in charge of all those below. The lines connecting the blocks also suggest the routes for communication from one to another in an upward or downward direction.

The schematic in Fig. 10.41 is not a graph, nor is it a true view of the apparatus. Instead, it is a schematic that effectively shows how the parts and their functions relate to each other.

Geographical graphs are used to combine maps and other relationships such as weather (Fig. 10.42). Different symbols are used to represent the annual rainfall that various areas of the nation receive. The graph in Fig. 10.43 shows the locations of reclamation dams by using circular symbols.

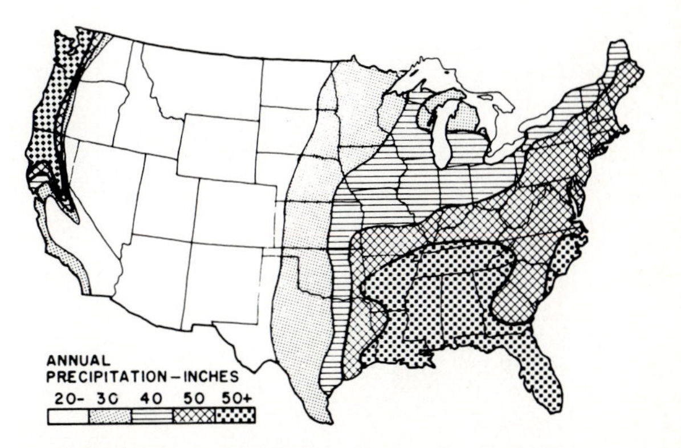

Fig. 10.42 A map chart that shows the weather characteristics of various geographical areas. (Courtesy of the *Structural Clay Products Institute.*)

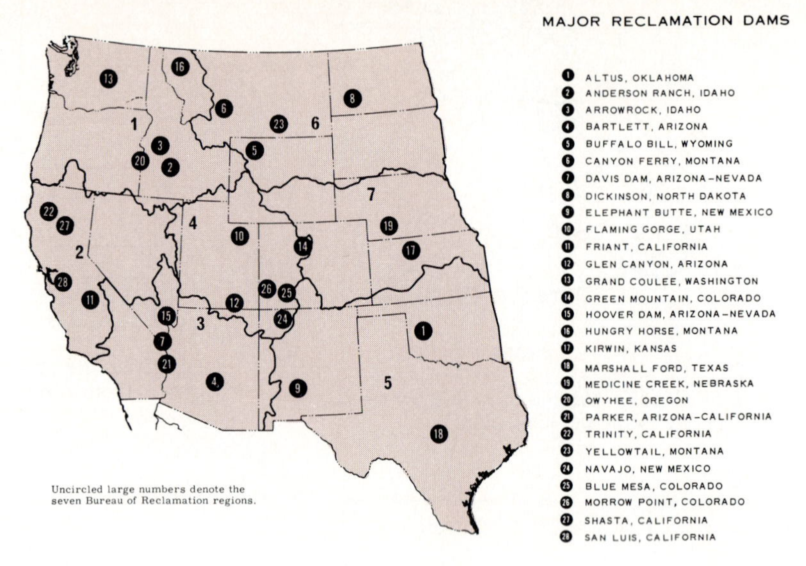

Fig. 10.43 A chart that locates the various reclamation dams in the western portion of the United States. (Couresy of the *Bureau of Reclamation.)*

10.10 HOW TO LIE WITH GRAPHS

Graphs can be used to distort data to the extent that the graph is actually lying. You should become familiar with these techniques so you will be able to properly interpret the data shown by a graph.

The three bar graphs in Fig. 10.44 present the same data, but a different impression is given by each. The upper graph gives the impression that Fuel B is almost five times better than Fuel A. This is because the bars do not begin at zero, thereby distorting a true comparison.

The center graph begins at zero, which gives a true comparison between the two bars. Fuel B is shown to be about 35% better than Fuel A.

The lower graph de-emphasizes the difference between the two fuels since the x-axis is much longer than is necessary. The implication of this horizontal scale is that a car might be expected to get over 90 miles per gallon. The bars appear insignificant on this graph, and even though they are drawn accurately, the difference in their lengths appears much less than in the other two graphs.

The width of bars and the colors used can give a misleading impression in a bar graph. Beware of graphs in which the bars run out of the bounds of the graph. These seldom give a true graphical picture.

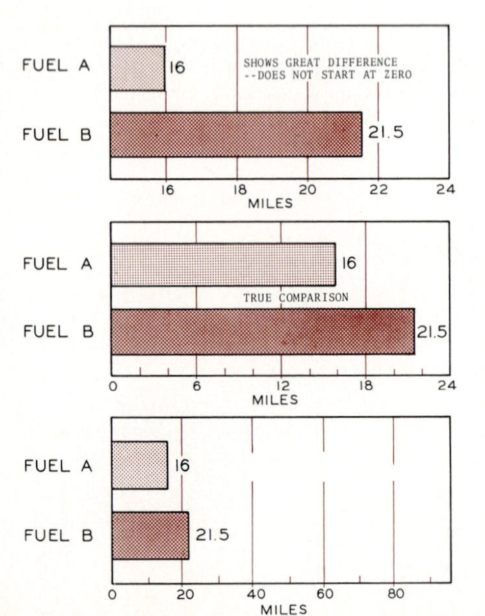

Fig. 10.44 All three graphs show the same data, but each gives a different impression of the data.

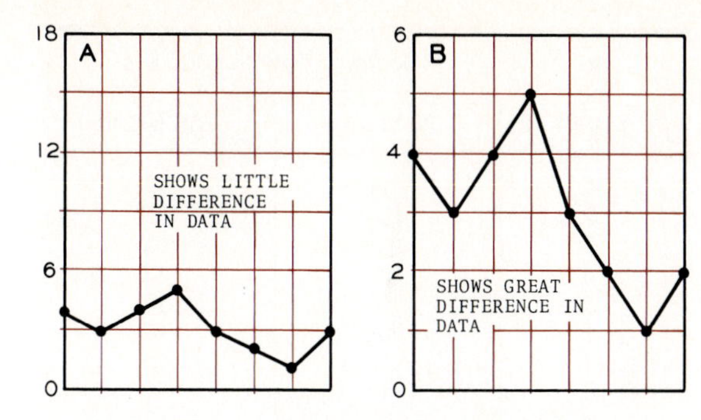

Fig. 10.45 Due to a variation in the scales, the change in the curve at B appears greater than the curve at A, although the data is the same in each.

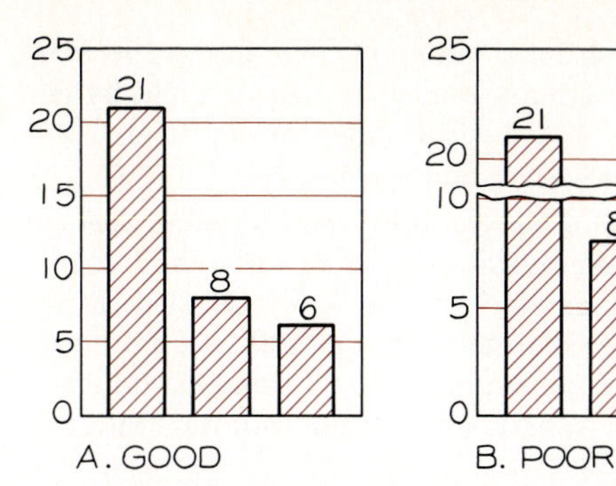

Fig. 10.47 This bar graph distorts the data at B because a portion of the graph has been removed, negating the pictorial value of the graph.

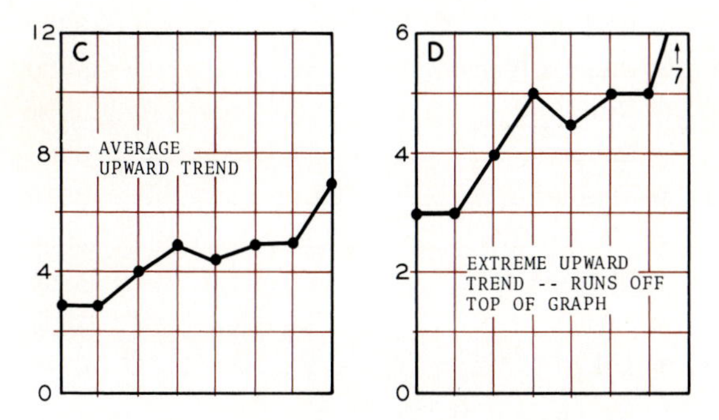

Fig. 10.46 The upward trend of the data at D appears greater than the curve at C. Both graphs show the same data.

Identical data are plotted on two graphs with different *y*-axes in Fig. 10.45. Graph A shows only a little variation in the data. Graph B gives a more dramatic effect because the expanded vertical scale emphasizes the difference in the data.

The two graphs in Fig. 10.46 give a different appearance even though the same data are plotted on both. The data in Graph D appear to have a more significant rate of increase because of the selection of the vertical scale. The curve of Graph D is drawn to run off the top of the graph, giving the impression that the increase is too great to be contained on a graph.

Another method of misrepresenting data is the removal of a portion of the graph (Fig. 10.47). This distorts the comparison between the bars and negates the value of the graph.

PROBLEMS

The problems below are to be drawn on Size A sheets (8½″ × 11″) in pencil or ink as specified. Follow the techniques that were covered in this chapter, and the examples that are given as the problems are being solved.

Pie Graphs

1. Draw a pie graph that compares the employment of male youth between the ages of 16 and 21: Operators—25%; Craftsmen—9%; Professionals, technicians, and managers—6%; Clerical and sales—17%; Service—11%; Farm workers—13%; and Laborers—19%.

2. Draw a pie graph that shows the relationship between the following members of the technological team: Engineers—985,000; Technicians—932,000; and Scientists—410,000.

3. Construct a pie graph of the following percentages of the employment status of graduates of two-year technician programs one year after graduation; Employed—63%; Continuing full-time study—23%; Considering job offers—6%; Military—6%; and Other—2%.

4. Construct a pie graph that shows the relationship between the types of degrees held by engineers in aeronautical engineering: Bachelor—65%; Master—29%; and Ph.D.—6%.

5. Draw a pie graph for the following average annual expenditures of a state on public roads. The approximate figures are: Construction—$13,600,000; Maintenance—$7,100,000; Purchase and upkeep of equipment—$2,900,000; Bonds—$5,600,000; Engineering and administration—$1,600,000.

6. Draw a pie that shows the data given in Problem 10.

7. Draw a pie that shows the data given in Problem 11.

Bar Graphs

8. Draw a bar graph that depicts the unemployment rate of high-school graduates and dropouts in various age categories. The age groups and the percent of unemployment of each group are given in the following table:

Ages	Percent of labor force	
	Graduates	Dropouts
16–17	18	22
18–19	12.5	17.5
20–21	8	13
22–24	5	9

9. Draw a single bar that represents 100% of a die casting alloy. The proportional parts of the alloy are: Tin—16%; Lead—24%; Zinc—38.8%; Aluminum—16.4%; and Copper—4.8%.

10. Draw a bar graph that compares the number of skilled workers employed in various occupations. Arrange the graph for ease of interpretation and comparison of occupations. Use the following data: Carpenters—82,000; All-round machinists—310,000; Plumbers—350,000; Bricklayers—200,000; Appliance servicers—185,000; Automotive mechanics—760,000; Electricians—380,000; and Painters—400,000.

11. Draw a bar graph that represents the flow of a river in cubic feet per second (cfs) as shown in the following table. Show bars that represent the data for ten days in the first month only. Omit the second month.

Day of month	Rate of flow in 1000 cfs	
	1st Month	2nd Month
1	19	19
2	130	70
3	228	79
4	129	33
5	65	19
6	32	14
7	17	15
8	13	11
9	22	19
10	32	27

12. Draw a bar graph that shows the airline distances in statute miles from New York to the cities listed in the table below. Arrange in ascending or descending order.

- Berlin....................3965
- Buenos Aires....................5300
- Honolulu....................4960
- London....................3465
- Manila....................8510
- Mexico City....................2090
- Moscow....................4665
- Paris....................3634
- Tokyo....................6740

13. Draw a bar graph that compares the corrosion resistance of the materials listed in the table below:

	Loss in weight %	
	In atmosphere	In sea water
Common Steel	100	100
10% Nickel Steel	70	80
25% Nickel Steel	20	55

14. Draw a bar graph using the data in Problem 1.

15. Draw a bar graph using the data in Problem 2.

16. Draw a bar graph using the data in Problem 3.

17. Construct a rectangular grid graph to show the accident experience of Company A. Plot the numbers of disabling accidents per million

Table 10.1

	1890	1900	1910	1920	1930	1940	1950	1960	1970	1980
Supply	80	90	110	135	155	240	270	315	380	450
Demand	35	35	60	80	110	125	200	320	410	550

person-hours of work on the y-axis. Years will be plotted on the x-axis. Data: 1970—1.21; 1971—0.97; 1972—0.86; 1973—0.63; 1974—0.76; 1975—0.99; 1976—0.95; 1977—0.55; 1978—0.76; 1979—0.68; 1980—0.55; 1981—0.73; 1982—0.52; 1983—0.46.

Linear Coordinate Graphs

18. Using the data given in Table 10.1, draw a linear coordinate graph that compares the supply and demand of water in the United States from 1890 to 1980. Supply and demand are given in billions of gallons of water per day.

19. Present the data in Table 10.2 in a linear coordinate graph to decide which lamps should be selected to provide economical lighting for an industrial plant. The table gives the candlepower directly under the lamps (0°) and at various angles from the vertical when the lamps are mounted at a height of 25 feet.

20. Construct a linear coordinate graph that shows the relationship in energy costs (mills per kilowatt-hour) and the percent capacity of two types of power plants. Plot energy costs along the y-axis, and the capacity factor along the x-axis. The plotted curve will compare the costs of a nuclear plant with a gas- or oil-fired plant. Data for a gas-fired plant: 17 mills, 10%; 12 mills, 20%; 8 mills, 40%; 7 mills, 60%; 6 mills, 80%; 5.8 mills, 100%. Nuclear plant data: 24 mills, 10%; 14 mills, 20%; 7 mills, 40%; 5 mills, 60%; 4.2 mills, 80%; 3.7 mills, 100%.

21. Plot the data from Problem 17 as a linear coordinate graph.

22. Construct a linear coordinate graph to show the relationship between the transverse resilience in inch-pounds (y-axis) and the single-blow impact in foot-pounds (x-axis) of gray iron. Data: 21 fp, 375 ip; 22 fp, 350 ip; 23 fp, 380 ip; 30 fp, 400 ip; 32 fp 420 ip; 33 fp, 410 ip; 38 fp, 510 ip; 45 fp, 615 ip; 50 fp, 585 ip; 60 fp, 785 ip; 70 fp, 900 ip; 75 fp, 920 ip.

23. Draw a linear composite coordinate graph to compare the two sets of data in the following table: capacity vs. diameter, and capacity vs. weight of a brine cooler. The horizontal scale is to be tons of capacity, and the vertical scales are to be outside diameter on the left, and weight (cwt) on the right.

Tons refrigerating capacity	Outside diameter, inches	Weight, cwt
15	22	25
30	28	46
50	34	73
85	42	116
100	46	136
130	50	164
160	58	215
210	60	263

Use 20 × 20 graph paper 8½ × 11. Horizontal scale of 1″ = 40 tons. Vertical scales of 1″ = 10″ of outside DIA, and 1″ = 40 cwt.

24. Draw a linear coordinate graph that shows the voltage characteristics for a generator as given in the following table of values: Abscissa—arma-

Table 10.2

Angle with vertical	0	10	20	30	40	50	60	70	80	90
Candlepower (thous.) 2–400W	37	34	25	12	5.5	2.5	2	0.5	0.5	0.5
Candlepower (thous.) 1–1000W	22	21	19	16	12.3	7	3	2	0.5	0.5

ture current in amperes (I_a); ordinate—terminal voltage in volts (E_t).

I_a	E_t	I_a	E_t	I_a	E_t
0	288	31.1	181.8	41.5	68
5.4	275	35.4	156	40.5	42.5
11.8	257	39.7	108	39.5	26.5
15.6	247	40.5	97	37.8	16
22.2	224.5	40.7	90	13.0	0
26.2	217	41.4	77.5		

25. Draw a linear coordinate graph for the centrifugal pump test data in the table below. The units along the x-axis are to be gallons per minute. There will be four curves to represent the variables given.

Gallons per min.	Discharge pressure	Water HP	Electric HP	Efficiency %
0	19.0	0.00	1.36	0.00
75	17.5	0.72	2.25	32.0
115	15.0	1.00	2.54	39.4
154	10.0	1.00	2.74	36.5
185	5.0	0.74	2.80	26.5
200	3.0	0.63	2.83	2.22

26. Draw a linear coordinate graph that compares two of the values shown in the table below—ultimate strength and elastic limit—with degrees of temperature labeled along the x-axis.

°F	Ultimate strength	Elastic limit	Elongation %	Red of area %	Brinell hardness no.
400	257,500	208,000	10.8	31.3	500
500	247,000	224,500	12.5	39.5	483
600	232,500	214,000	13.3	42.0	453
700	207,500	193,500	15.0	47.5	410
800	180,500	169,000	17.0	52.5	358
900	159,500	146,500	18.5	56.5	313
1000	142,400	128,500	20.3	59.2	285
1100	126,500	114,000	23.0	60.8	255
1200	114,500	96,500	26.3	67.8	230
1300	108,000	85,500	25.8	58.3	235

27. Draw a linear coordinate graph that compares two of the values shown in the table in Problem 26—percent of elongation and percent of reduction of area of the cross section—with the degrees of temperature along the x-axis.

Break-Even Graphs

28. Construct a break-even graph that shows the earnings for a new product that has a development cost of $12,000. It will cost 50¢ each to manufacture, and you wish to break even at 8000. What would be the profit at a volume of 20,000 and at 25,000?

29. Same as Problem 28 except that the development costs are $80,000, the manufacturing cost is $2.30 each, and the desired break-even point is 10,000. What would be the profit at a volume of 20,000 and at 30,000? What sales price would be required to break even at 10,000 units?

30. A manufacturer has incorporated the manufacturing and development costs into a cost-per-unit estimate. He wishes to sell the product at $1.50 each. Construct a graph of the following data. On the y-axis plot cost per unit in dollars; on the x-axis, number of units in thousands. Data: 1000, $2.55; 2000, $2.01; 3000, $1.55; 4000, $1.20; 5000, $0.98; 6000, $0.81; 7000, $0.80; 8000, $0.75; 9000, $0.73; 10,000, $0.70. How many must be sold to break even? What will be the total profit when 9000 are sold?

31. The cost per unit to produce a product by a manufacturing plant is given below. Construct a break-even graph with the cost per unit plotted on the y-axis and and the number of units on the x-axis. Data: 1000, $5.90; 2000, $4.50; 3000, $3.80; 4000, $3.20; 5000, $2.85; 6000, $2.55; 7000, $2.30; 8000, $2.17; 9000, $2.00; 10,000, $1.95.

Logarithmic Graphs

32. Using the data given in Table 10.3, construct a logarithmic graph where the vibration amplitude (A) is plotted as the ordinate and vibration frequency (F) as the abscissa. The data for Curve 1 represent the maximum limits of machinery in good condition with no danger from vibration. The data for Curve 2 are the lower limits of machinery that is being vibrated excessively to the danger point. The vertical scale should be three cycles and the horizontal scale two cycles.

Table 10.3

F	100	200	500	1000	2000	5000	10,000
$A(1)$	0.0028	0.002	0.0015	0.001	0.0006	0.0003	0.00013
$A(2)$	0.06	0.05	0.04	0.03	0.018	0.005	0.001

33. Plot the data below on a two-cycle log graph to show the current in amperes (y-axis) versus the voltage in volts (x-axis) of precision temperature-sensing resistors. Data: 1 volt, 1.9 amps; 2 volts, 4 amps; 4 volts, 8 amps; 8 volts, 17 amps; 10 volts, 20 amps; 20 volts, 30 amps; 40 volts, 36 amps; 80 volts, 31 amps; 100 volts, 30 amps.

34. Plot the data from Problem 18 as a logarithmic graph.

35. Plot the data from Problem 24 as a logarithmic graph.

Semilogarithmic Graphs

36. Construct a semilogarithmic graph with the y-axis a two-cycle log scale from 1 to 100 and the x-axis a linear scale from 1 to 7. Plot the data below to show the survivability of a shelter at varying distances from a one-megaton air burst. The data consists of overpressure in psi along the y-axis, and distance from ground zero in miles along the x-axis. The data points represent an 80% chance of survival of the shelter. Data: 1 mile, 55 psi; 2 miles, 11 psi; 3 miles, 4.5 psi; 4 miles, 2.5 psi; 5 miles, 2.0 psi; 6 miles, 1.3 psi.

37. The growth of two divisions of a company, Division A and Division B, is given in the data below. Plot the data on a rectilinear graph and on a semilog graph. The semilog graph should have a one-cycle log scale on the y-axis for sales in thousands of dollars, and a linear scale on the x-axis showing years for a six-year period. Data in dollars: 1 yr, A = $11,700 and B = $44,000; 2 yr, A = $19,500 and B = $50,000; 3 yr, A = $25,000 and B = $55,000; 4 yr, A = $32,000 and B = $64,000; 5 yr, A = $42,000 and B = $66,000; 6 yr, A = $48,000 and B = $75,000. Which division has the better growth rate?

38. Draw a semilog chart showing probable engineering progress. Use the following indices: 40,000 B.C. = 21; 30,000 B.C. = 21.5; 20,000 B.C. = 22; 16,000 B.C. = 23; 10,000 B.C. = 27; 6000 B.C. = 34; 4000 B.C. = 39; 2000 B.C. = 49; 500 B.C. = 60; A.D. 1900 = 100. Horizontal scale 1″ = 10,000 years. Height of cycle = about 5″. Two-cycle printed paper may be used if available.

39. Plot the data from Problem 24 as a semilogarithmic graph.

40. Plot the data from Problem 26 as a semilogarithmic graph.

Percentage Graphs

41. Plot the data given in Problem 18 on a semilog graph. What is the percent of increase in the demand for water from 1890 to 1920? What percent of the demand is the supply for the following years: 1900, 1930, and 1970?

42. Using the graph plotted in Problem 37, determine the percent of increase of Division A and Division B from year 1 to year 4. Also, what percent of sales of Division A are the sales of Division B at the end of year 2? At the end of year 6?

43. Plot two values from Problem 26—water horsepower and electric horsepower—on semilog paper compared with gallons per minute along the x-axis. What is the percent that water horsepower is of the electric horsepower when 75 gallons per minute are being pumped? What is the percent increase of the electric horsepower from 0 to 185 gallons per minute?

Organizational Charts

44. Draw an organization chart for a city government organized as follows: The electorate elects school board, city council, and municipal court officers. The city council is responsible for the civil service commission, city manager, and city planning board. The city manager's duties cover finance, public safety, public works, health and welfare, and law.

45. Draw an organization chart for a manufacturing plant. The sales manager, chief engineer, treasurer, and general manager are responsible to the president. Each of these officers has a department force. The general manager has three department heads: master mechanic, plant superintendent, and purchasing agent. The plant superintendent has charge of the shop foremen, under whom are the working forces, and also has direct charge of the shipping, tool and die, inspection, order, and stores and supplies departments.

Polar Graphs

46. Construct a polar graph of the data given in Problem 19.

47. Construct a polar graph of the following illumination, in lumens at various angles, emitted from a luminaire. The zero-degrees position is vertically under the overhead lamp. Data: 0°, 12,000; 10°, 15,000; 20°, 10,000; 30°, 8000; 40°, 4200; 50°, 2500; 60°, 1000; 70°, 0. The illumination is symmetrical about the vertical.

11

NOMOGRAPHY

11.1 NOMOGRAPHY

An additional aid in analyzing data is a graphical computer called a **nomogram** or **nomograph.** Basically, a nomogram or "number chart" is any graphical arrangement of calibrated scales and lines that may be used to permit calculations, usually those of a repetitive nature.

The term "nomogram" is frequently used to denote a specific type of scale arrangement called an alignment graph. Typical examples of alignment graphs are shown in Fig. 11.1. Many other types are also used that have curved scales or other scale arrangements for more complex problems. The discussion of nomograms in this chapter will be limited to the simpler conversion, parallel-scale, and N-type graphs and their variations.

Using an Alignment Chart

An alignment graph is usually constructed to solve for one or more unknowns in a formula or empirical relationship between two or more quantities. For example, it can be used to convert degrees Celsius to degrees Fahrenheit, to find the size of a structural member to sustain a certain load, and so on. An alignment graph is read by placing a straightedge, or by drawing a line called

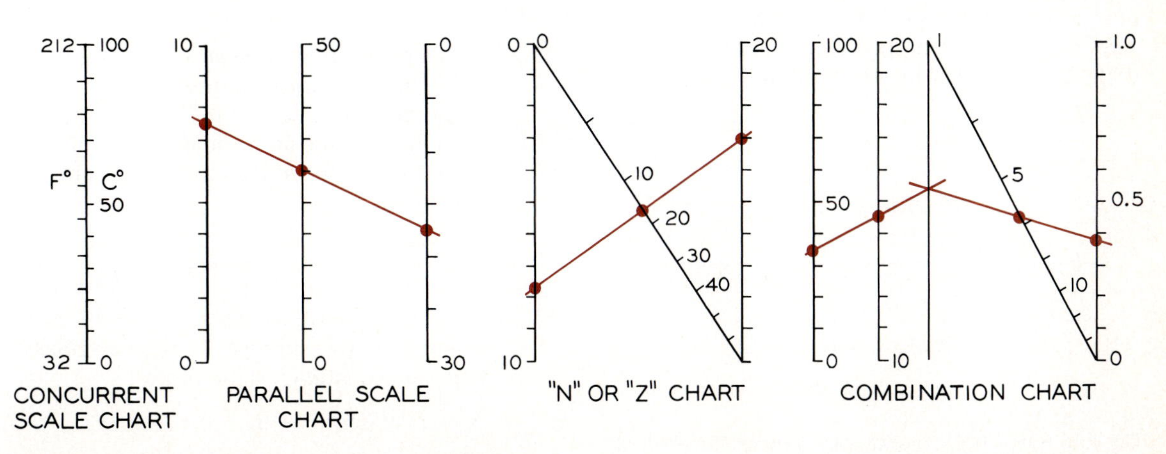

Fig. 11.1 Typical examples of types of alignment charts.

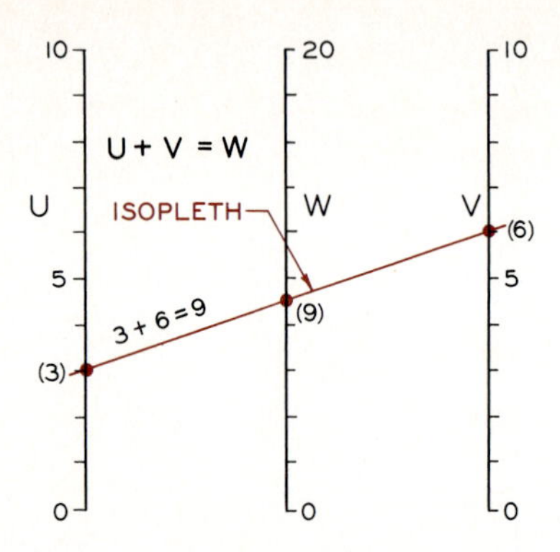

Fig. 11.2 Use of an isopleth to solve graphically for unknowns in the given equation.

an **isopleth,** across the scales of the chart and reading corresponding values from the scale on this line. The example in Fig. 11.2 shows readings for the formula $U + V = W$.

11.2 ALIGNMENT-GRAPH SCALES

To construct any alignment graph, you must first determine the graduations of the scales. Alignment-graph scales are called **functional scales.** A functional scale is one that is graduated according to values of some *function* of a variable, but *calibrated* with values of the variable. A functional scale for $F(U) = U^2$ is illustrated in Fig. 11.3. It can be seen in this example that if a value of $U = 2$ was substituted into the equation, the position of U on the functional scale would be 4 units from zero, or $2^2 = 4$. This procedure can be repeated with all values of U by substitution.

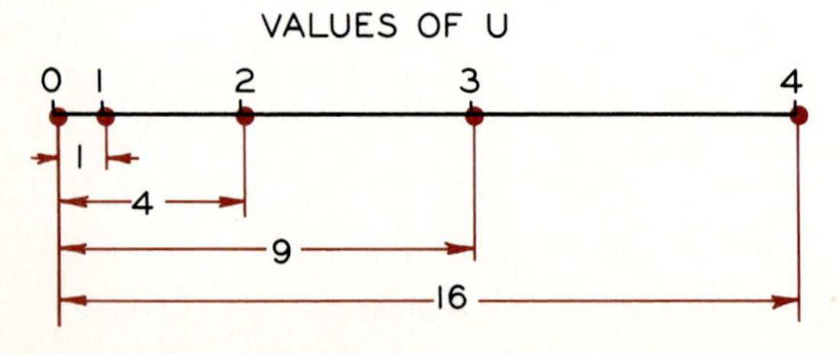

Fig. 11.3 Functional scale for units of measurement that are proportional to $F(U) = U^2$.

The Scale Modulus

Since the graduations on a functional scale are spaced in proportion to values of the function, a proportionality, or scaling factor, is needed. This constant of proportionality is called the **scale modulus** and it is given by the equation

$$m = \frac{L}{F(U_2) - F(U_1)} \tag{1}$$

where

m = scale modulus, in inches per functional unit,
L = desired length of the scale, in inches,
$F(U_2)$ = function value at the end of the scale,
$F(U_1)$ = function value at the start of the scale.

For example, suppose that we are to construct a functional scale for $F(U) = \sin U$, with $0° \leqslant U \leqslant 45°$ and a scale 6″ in length. Thus $L = 6''$, $F(U_2) = \sin 45° = 0.707$, $F(U_1) = \sin 0° = 0$. Therefore, Eq. (1) can be written in the following form by substitution:

$$m = \frac{6}{0.707 - 0} = 8.49 \text{ inches per (sine) unit.}$$

The Scale Equation

Graduation and calibration of a functional scale are made possible by a **scale equation.** The general form of this equation may be written as a variation of Eq. (1) in the following form:

$$X = m[F(U) - F(U_1)] \tag{2}$$

where

X = distance from the measuring point of the scale to any graduation point,
m = scale modulus,
$F(U)$ = functional value at the graduation point,
$F(U_1)$ = functional value at the measuring point of the scale.

For example, a functional scale is constructed for the previous equation, $F(U) = \sin U$ $(0° \leqslant U \leqslant 45°)$. It has been determined that $m = 8.49$, $F(U) = \sin U$, and $F(U_1) = \sin 0° = 0$. Thus by substitution the scale equation, (2), becomes

$$X = 8.49 (\sin U - 0) = 8.49 \sin U.$$

Using the eqaution, we can substitute values of U

Table 11.1

U	0°	5°	10°	15°	20°	25°	30°	35°	40°	45°
X	0	0.74	1.47	2.19	2.90	3.58	4.24	4.86	5.45	6.00

and construct a table of positions. In this case, the scale is calibrated at 5° intervals, as reflected in Table 11.1.

The values of X from the table give the positions, in inches, for the corresponding graduations, measured from the start of the scale (U = 0°); see Fig. 11.4. It should be noted that the measuring point does *not* need to be at one end of the scale, but it is usually the most convenient point, especially if the functional value is zero at that point.

A graphical method of locating the functional values along a scale can be found as shown in Fig. 11.5 by the proportional-line method. The sine functions are measured off along a line at 5° intervals, with the end of the line passing through the 0° end of the scale. The functions are transferred from the inclined line with parallel lines back to the scale where the functions are represented and labeled.

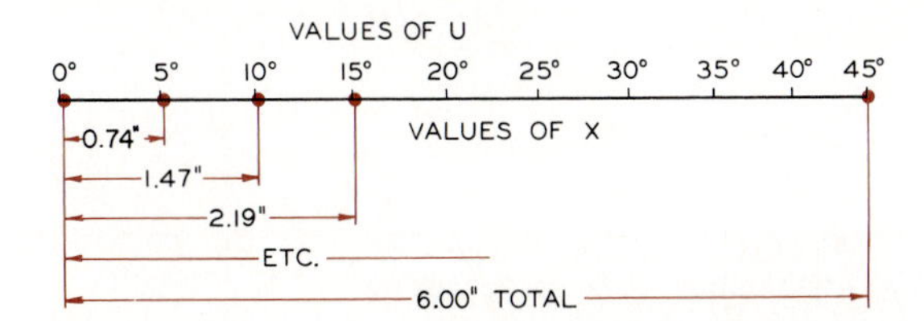

Fig. 11.4 Construction of a functional scale using values from Table 11.1, which were derived from the scale equation.

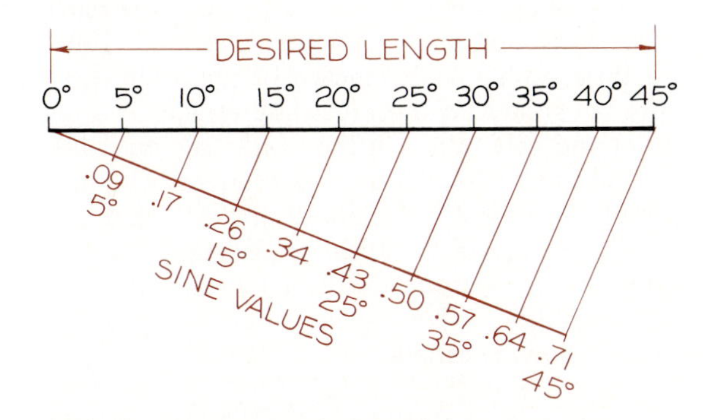

Fig. 11.5 A functional scale that shows the sine of the angles from 0° to 45° can be drawn graphically by the proportional-line method. The scale is drawn to a desired length and the sine values of angles at 5° intervals are laid off along a construction line that passes through the 0° end of the scale. These values are projected back to the scale.

11.3 CONCURRENT SCALES

Concurrent scales are useful in the rapid conversion of one value into terms of a second system of measurement. Formulas of the type $F_1 = F_2$, which relate two variables, can be adapted to the concurrent-scale format. Typical examples might be the Fahrenheit–Celsius temperature relation.

$$°\mathrm{F} = \frac{9}{5}°\mathrm{C} + 32$$

or the area of a circle,

$$A = \pi r^2.$$

Design of a concurrent-scale chart involves the construction of a functional scale for each side of the mathematical formula in such a manner that the *position* and *lengths* of each scale coincide. For example, to design a conversion chart 5″ long that will give the areas of circles whose radii range from 1 to 10, we first write $F_1(A) = A$, $F_2(r) = \pi r^2$, and $r_1 = 1$, $r_2 = 10$. The scale modulus for r is

$$\begin{aligned} m_r &= \frac{L}{F_2(r_2) - F_2(r_1)} \\ &= \frac{5}{\pi(10)^2 - \pi(1)^2} = 0.0161. \end{aligned}$$

Thus the scale equation for r becomes

$$\begin{aligned} X_r &= m_r[F_2(r) - F_2(r_1)] \\ &= 0.0161\,[\pi r^2 - \pi(1)^2] \\ &= 0.0161\,\pi(r^2 - 1) \\ &= 0.0505(r^2 - 1). \end{aligned}$$

A table of values for X_r and r may now be completed as shown in Table 11.2. The r-scale can be

Table 11.2

r	1	2	3	4	5	6	7	8	9	10
X_r	0	0.15	0.40	0.76	1.21	1.77	2.42	3.18	4.04	5.00

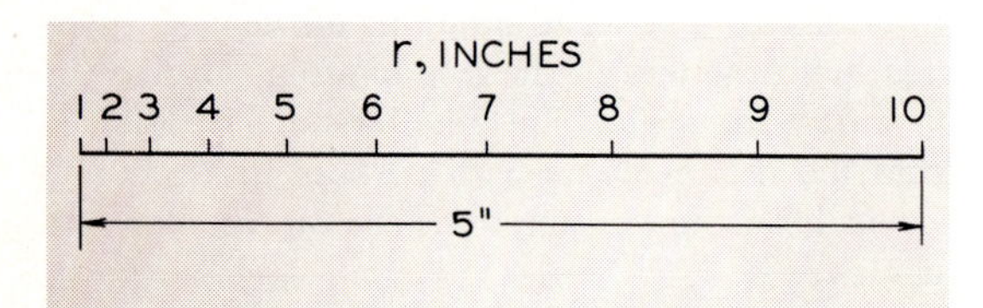

Fig. 11.6 Calibration of one scale of a concurrent scale graph using values from Table 11.2.

drawn from this table, as shown in Fig. 11.6. From the original formula, $A = \pi r^2$, the limits of A are found to be $A_1 = \pi = 3.14$ and $A_2 = 100\pi = 314$. The scale modulus for concurrent scales is always the same for equal-length scales; therefore $m_A = m_r = 0.0161$, and the scale equation for A becomes

$$\begin{aligned} X_A &= m_A[F_1(A) - F_1(A_1)] \\ &= 0.0161(A - 3.14). \end{aligned}$$

The corresponding table of values is then computed for selected values of A, as shown in Table 11.3.

The A-scale is now superimposed on the r-scale; its calibrations have been placed on the other side of the line to facilitate reading (Fig. 11.7). It may be desired to expand or contract one of the scales, in which case an alternative arrangement may be used, as shown in Fig. 11.8. The two scales are drawn parallel at any convenient distance, and calibrated in *opposite* directions. A different scale modulus and corresponding scale equation must be calculated for each scale if they are *not* the same length.

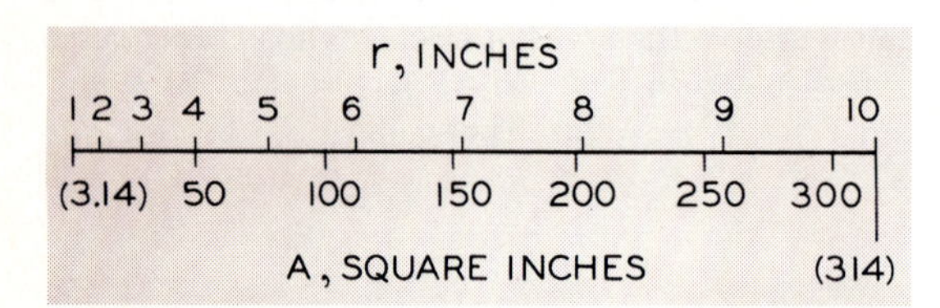

Fig. 11.7 The completed concurrent scale graph for the formula $A = \pi r^2$. Values for the A-scale are taken from Table 11.3.

A graphical method can be used to construct concurrent scales as shown in Fig. 11.9 by using the proportional-line method. Since there are 101.6 mm in 4 inches, the units of millimeters can be located on the upper side of the inch scale by projecting to the scale with a series of parallel projectors.

11.4 CONSTRUCTION OF ALIGNMENT GRAPHS WITH THREE VARIABLES

For a formula of three functions (of one variable each), the general approach is to select the lengths and positions of *two* scales according to the range of variables and size of the graph desired. These are then calibrated by means of the scale equations, as shown in the preceding article. The position and calibration of the third scale will then depend upon these initial constructions. Although definite mathematical relationships exist that may be used to locate the third scale, graphical constructions are simpler and usually less subject to error. Examples of the various forms are presented in the following articles.

Table 11.3

A	(3.14)	50	100	150	200	250	300	(314)
X_1	0	0.76	1.56	2.36	3.16	3.96	4.76	5.00

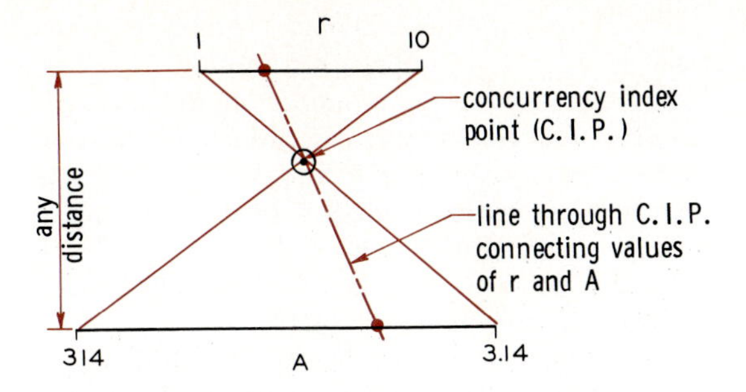

Fig. 11.8 Concurrent scale graph with unequal scales.

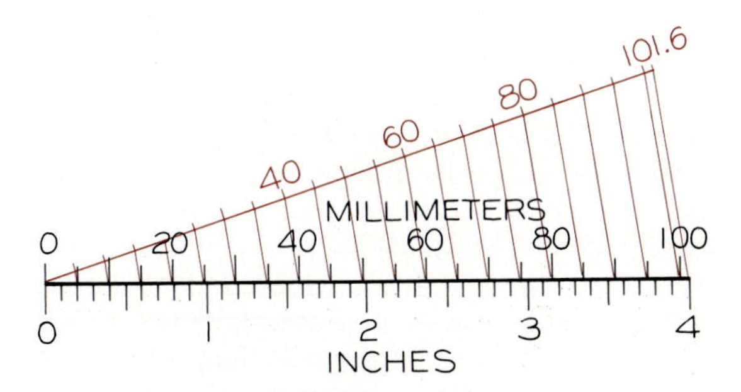

Fig. 11.9 The proportional-line method can be used to construct an alignment graph that converts inches to millimeters. This requires that the units at each end of the scales be known. For example, there are 101.6 millimeters in 4 inches.

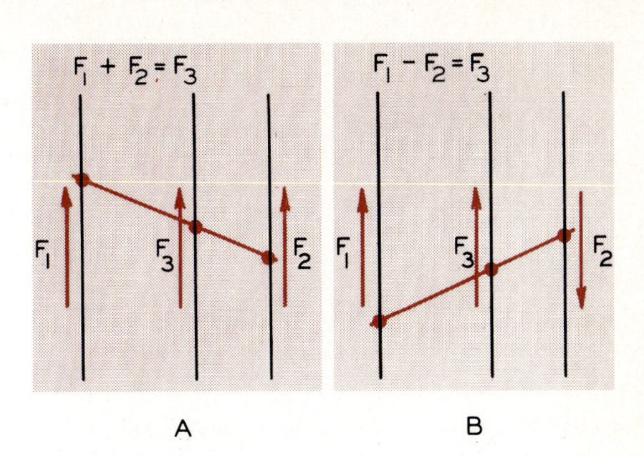

Fig. 11.10 Two common forms of parallel-scale alignment nomographs.

11.5 PARALLEL-SCALE NOMOGRAPHS

Many engineering relationships involve three variables that can be computed graphically on a repetitive basis. Any formula of the type $F_1 + F_2 = F_3$ may be represented as a parallel-scale alignment chart, as shown in Fig. 11.10A. Note that all scales increase (functionally) in the same direction and that the function of the middle scale represents the *sum* of the other two. Reversing the direction of any scale changes the sign of its func-

Fig. 11.11 Parallel-scale nomogram (linear)

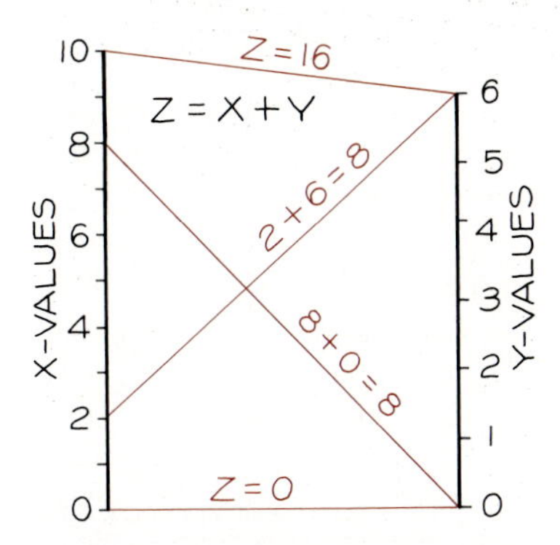

Step 1 Two parallel scales are drawn at any length and calibrated. The location of the parallel *Z*-scale is found by selecting two sets of values that will give the same value of *Z*, 8 in this case. The ends of the *Z*-scale will be 0 and 16, the sum of the end values of *X* and *Y*.

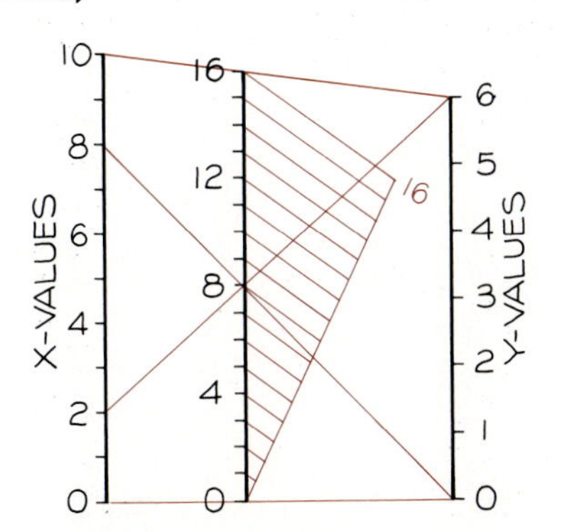

Step 2 The *Z*-scale is drawn through the point located in Step 1 parallel to the other scales. The scale is calibrated from 0 to 16 by using the proportional-line method. Note that the two sets of *X*- and *Y*-values cross at 8, the sum of each set.

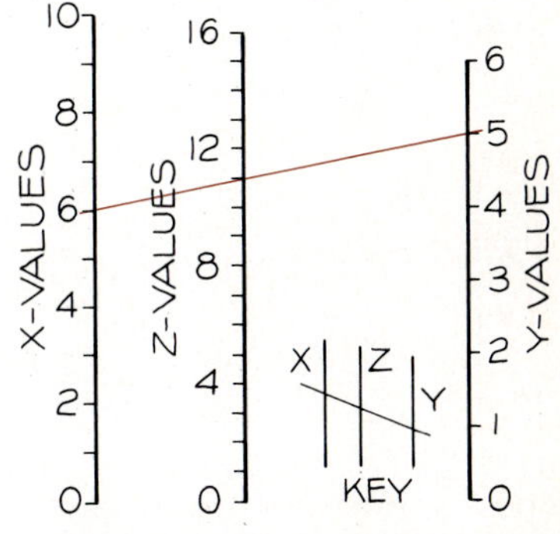

Step 3 The *Z*-scale is labeled and calibrated with easy-to-read units. A key is drawn to show how the nomograph is used. If the *Y*-scale were calibrated with 0 at the upper end instead of the bottom, a different *Z*-scale could be computed and the nomograph could be used for $Z = X - Y$.

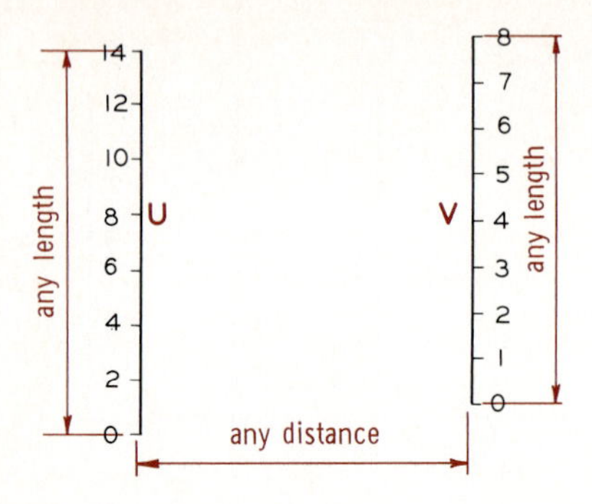

Fig. 11.12 Calibration of the outer scales for the formula $U + 2V = 3W$, where $0 \leq U \leq 14$ and $0 \leq V \leq 8$.

tion in the formula, as for $F_1 - F_2 = F_3$ in Fig. 11.10B.

To illustrate this type of alignment graph we will use the formula $Z = X + Y$ as illustrated in Fig. 11.11. The outer scales for X and Y are drawn and calibrated. They can be drawn to any length and positioned any distance apart as shown in Fig. 11.12. Two sets of data that yield a Z of 8 in Step 1 are used to locate the parallel Z-scale. In Step 2, the Z-scale is drawn and divided into 16 equal units. The finished nomograph in Step 3 can be used to add various values of X and Y to find their sums along the Z-scale.

A more complex alignment graph is illustrated in Fig. 11.13 where the formula $U + 2V = 3W$ is expressed in the form of a nomograph.

First it is necessary to determine and calibrate the two outer scales for U and V; we can make them any convenient length and any convenient distance apart, as shown in Fig. 11.12. These scales are used as the basis for the step-by-step construction shown in Fig. 11.13.

The limits of calibration for the middle scale are found by connecting the endpoints of the outer scales and substituting these values into the formula. Here, W is found to be 0 and 10 at the extreme ends (Step 1). Two pairs of corresponding values of U and V are selected that will give the same value of W. For example, values of $U = 0$ and $V = 7.5$ give a value of 5 for W. We also find that $W = 5$ when $U = 14$ and $V = 0.5$. This should be verified by substitution before continuing with construction. We connect these corresponding pairs of values with isopleths to locate their intersection and the position of the W-scale.

Since the W-scale is linear ($3W$ is a linear function), it may be subdivided into uniform intervals (Step 2). For a nonlinear scale, the scale modulus (and the scale equation) may be found in Step 2 by substituting its length and its two end values into Eq. (1) of Section 11.2. The nomograph

Fig. 11.13 Parallel-scale nomograph (linear)

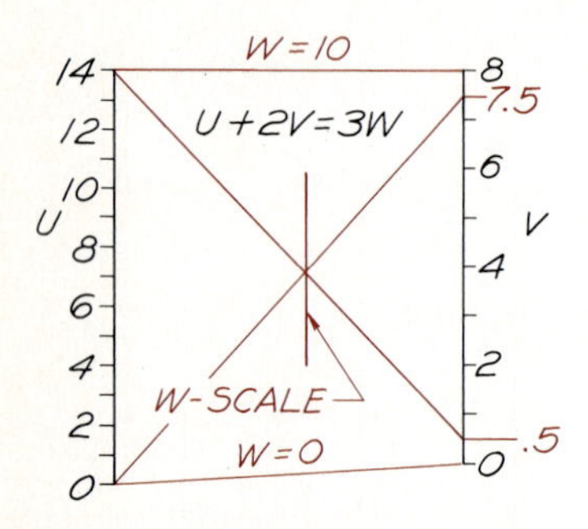

Step 1 Substitute the end values of the U- and V-scales into the formula to find the end values of the W-scale: $W = 10$ and $W = 0$. Select two sets of U and V that will give the same value of W. *Example:* When $U = 0$ and $V = 7.5$, W will equal 5, and when $U = 14$ and $V = 0.5$, W will equal 5. Connect these sets of values; the intersection of their lines locates the position of the W-scale.

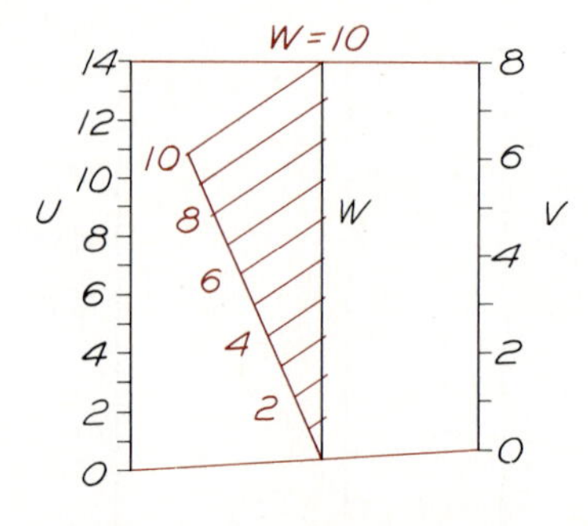

Step 2 Draw the W-scale parallel to the outer scales; its length is controlled by the previously established lines of $W = 10$ and $W = 0$. Since this scale is 10 linear divisions long, divide it graphically into ten units as shown. This will be a linear scale constructed as shown in Fig. 11.11.

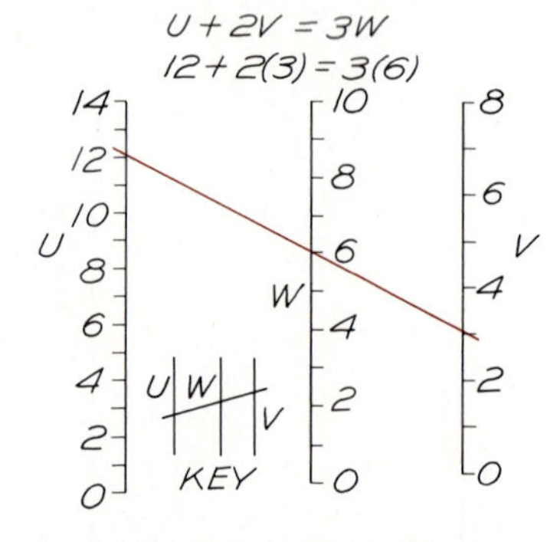

Step 3 The nomogram can be used as illustrated by selecting any two known variables and connecting them with an isopleth to determine the third unknown. A key is always included to illustrate how the nomogram is to be used. An example of $U = 12$ and $V = 3$ is shown to verify the accuracy of the graph.

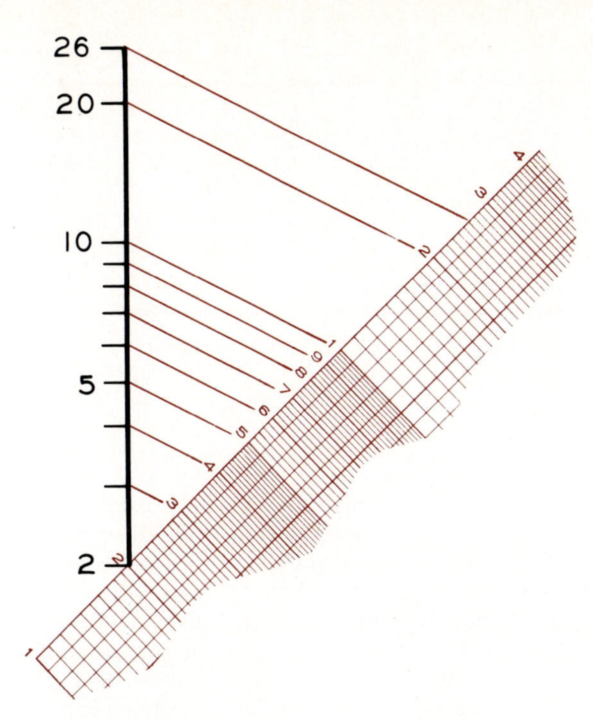

Fig. 11.14 Graphical calibration of a scale using logarithmic paper.

can be used to determine an infinite number of problem solutions when sets of two variables are known, as illustrated in Step 3.

Parallel-Scale Graph with Logarithmic Scales

Problems involving formulas of the type $F_1 \times F_2 = F_3$ can be solved in a manner very similar to the example given in Fig. 11.1 when logarithmic scales are used.

The first step in drawing a nomograph with logarithmic scales is learning how to transfer logarithmic functions to the scale. The graphical method is shown in Fig. 11.14, where units along the scale are found by projecting from a printed logarithmic scale with parallel lines. The mathematical method can also be used to locate these logarithmic spacings.

The formula $Z = XY$ is converted into a nomograph in Fig. 11.15. In Step 1, the X and Y scales are drawn within the desired limits from 1 to 10 on each. Sets of values of X and Y that yield the same value of Z, 10 in this case, are used to locate the Z-axis. The limits of the Z-axis are 1 and 100.

In Step 2, the Z-axis is drawn and calibrated as a two-cycle log scale. In Step 3, a key is given to explain how an isopleth is used to add the log-

Fig. 11.15 Parallel-scale nomogram (logarithmic)

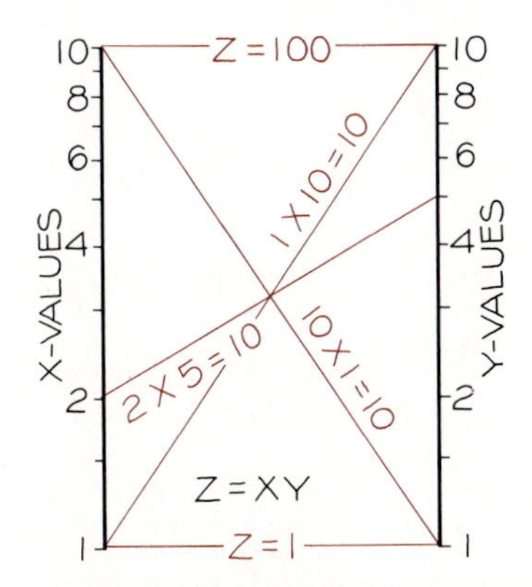

Step 1 To find scales that will solve the equation, $Z = XY$, parallel log scales are drawn. Sets of X and Y points that yield the same value of Z, 10 in this example, are drawn. Their intersection locates the Z-scale. The end values of the scale are 1 and 100.

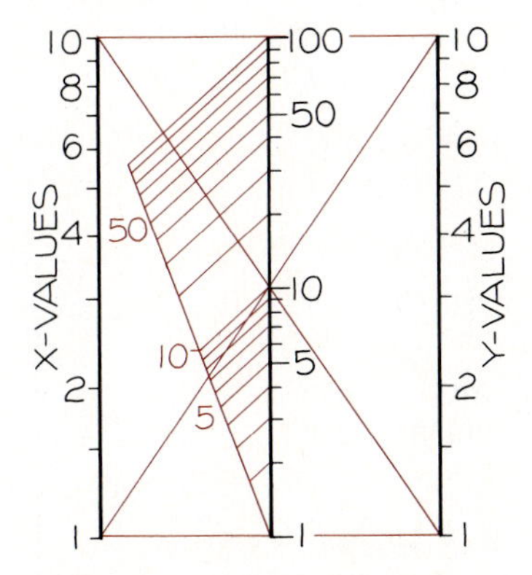

Step 2 The Z-axis is graphically calibrated as a two-cycle logarithmic scale from 1 to 100. This scale is parallel to the X- and Y-scales.

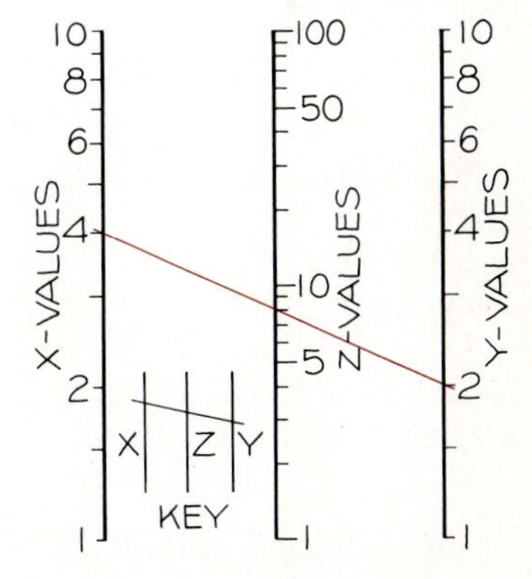

Step 3 A key is drawn to explain how the nomograph is used. An example isopleth is drawn to show that $4 \times 2 = 8$. By reversing the numerical order of the Y-value scale and computing a different Z-scale, the nomogram could be used for the equation $Z = X/Y$.

Table 11.4

S	0.1	0.2	0.3	0.4	0.5	0.6	0.7	0.8	0.9	1.0
X_s	0	1.80	2.88	3.61	4.19	4.67	5.07	5.42	5.72	6.00

Table 11.5

T	1	2	4	6	8	10	20	40	60	80	100
X_T	0	0.91	1.80	2.33	2.71	3.00	3.91	4.81	5.33	5.77	6.00

arithms of X and Y to give the log of Z. When logarithms are added the result is multiplication. Had the Y-axis been calibrated in the opposite direction with 1 at the upper end and 10 at the lower end, a new Z-axis could have been calibrated and the nomograph used for the formula, $Z = Y/X$, since it would be subtracting logarithms.

A more advanced example of this type of problem is the formula $R = S\sqrt{T}$, for $0.1 \leq S \leq 1.0$ and $1 \leq T \leq 100$. Assume the scales to be 6″ long. These scales need not be equal except for convenience. This formula may be converted into the required form by taking common logarithms of both sides, which gives

$$\log R = \log S + \tfrac{1}{2} \log T.$$

Thus we have

$$F_1(S) + F_2(T) = F_3(R), \qquad (1)$$

where $F_1(S) = \log S$, $F_2(T) = \frac{1}{2} \log T$, and $F_3(R) = \log R$. The scale modulus for $F_1(S)$ is, from Eq. (1),

$$m_s = \frac{6}{\log 1.0 - \log 0.1} = \frac{6}{0 - (-1)} = 6 \qquad (2)$$

Choosing the scale measuring point from $S = 0.1$, we find from Eq. (2) that the scale equation for $F_1(S)$ is

$$X_s = 6(\log S - \log 0.1) = 6(\log S + 1) \qquad (3)$$

Similarly, the scale modulus for $F_2(T)$ is

$$m_T = \frac{6}{\frac{1}{2}\log 100 - \frac{1}{2}\log 1} = \frac{6}{\frac{1}{2}(2) - \frac{1}{2}(0)} = 6 \qquad (4)$$

Thus, the scale equation, measuring from $T = 1$, is

$$X_T = 6(\tfrac{1}{2}\log T - \tfrac{1}{2}\log 1) = 3 \log T \qquad (5)$$

The corresponding tables for the two scale equations may be computed as shown in Tables 11.4 and 11.5. We shall position the two scales 5″ apart, as shown in Fig. 11.16. The logarithmic scales are graduated using the values in Tables 11.4 and 11.5. The step-by-step procedure for constructing the remainder of the nomogram is given in Fig. 11.17 using the two outer scales determined here.

The end values of the middle (R) scale are found from the formula $R = S\sqrt{T}$ to be $R = 1.0\sqrt{100} = 10$ and $R = 0.1\sqrt{1} = 0.1$. Choosing a value of $R = 1.0$, we find pairs of S and T might be $S = 0.1$, $T = 100$, and $S = 1.0$, $T = 1.0$ that yield $R = 1$. We connect these pairs with isopleths in Step 1 and position the middle scale at their intersection. The R-scale is drawn parallel to the outer scales and is calibrated by deriving its

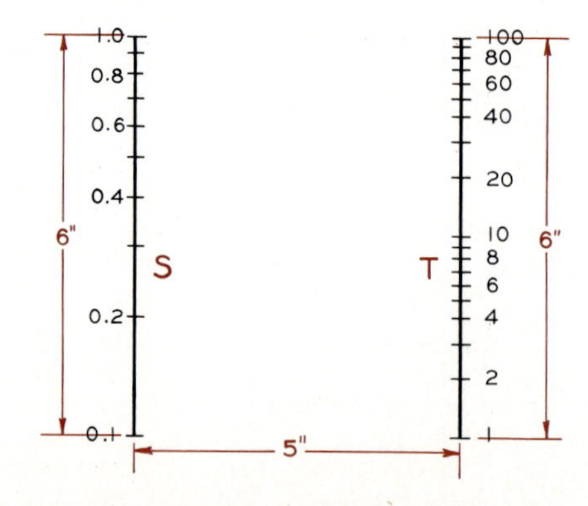

Fig. 11.16 Calibration of the outer scales for the formula $R = S\sqrt{T}$, where $0.1 \leq S \leq 1.0$ and $1 \leq T \leq 100$.

Fig. 11.17 Parallel-scale chart (logarithmic)

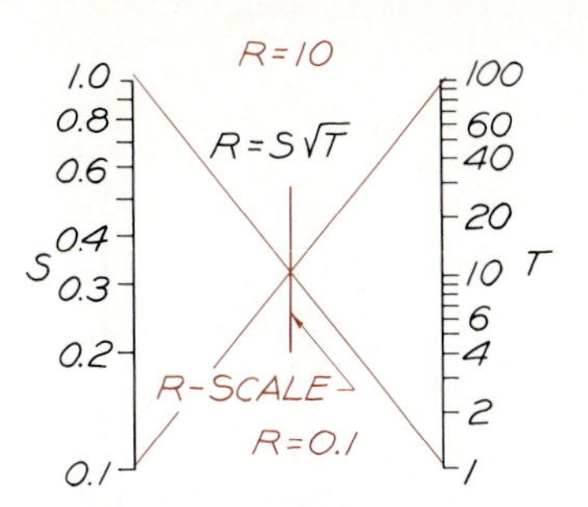

Step 1 Connect the end values of the outer scales to determine the extreme values of the R-scale, $R = 10$ and $R = 0.1$. Select corresponding values of S and T that will give the same value of R. Values of $S = 0.1$, $T = 100$ and $S = 1.0$, $T = 1.0$ give a value of $T = 1.0$. Connect the pairs to locate the position of the R-scale.

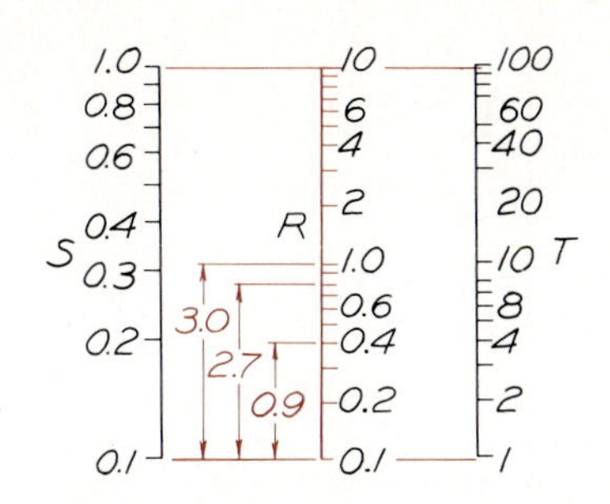

Step 2 Draw the R-scale to extend from 0.1 to 10. Calibrate it by substituting values determined from its scale equation. These values have been computed and tabulated in Table 11.6. The resulting tabulation is a logarithmic, two-cycle scale.

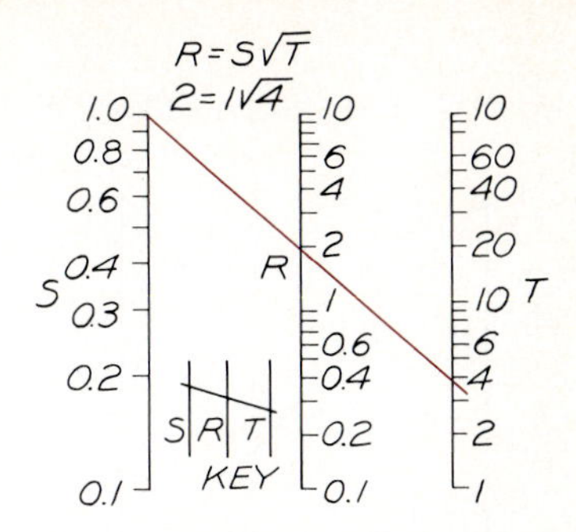

Step 3 Add labels to the finished nomogram and draw a key to indicate how it is to be used. An isopleth has been used to determine R when $S = 1.0$ and $T = 4$. The result of 2 is the same as that obtained mathematically, thus verifying the accuracy of the chart. Other combinations can be solved in this same manner.

scale modulus:

$$m_R = \frac{6}{\log 10 - \log 0.1} = \frac{6}{1 - (-1)} = 3.$$

Thus its scale equation (measuring from $R = 0.1$) is

$$X_R = 3(\log R - \log 0.1) = 3(\log R + 1.0)$$

Table 11.6 is computed to give the values for the scale. These values are applied to the R-scale as shown in Step 2. The finished nomogram can be used as illustrated in Step 3 to compute the unknown variables when two variables are given.

Note that this example illustrates a general method of creating a parallel-scale graph for all formulas of the type $F_1 + F_2 = F_3$ through the use of a table of values computed from the scale equation.

11.6 N- OR Z-GRAPHS

Whenever F_2 and F_3 are linear functions, we can partially avoid using logarithmic scales for formulas of the type

$$F_1 = \frac{F_2}{F_3}.$$

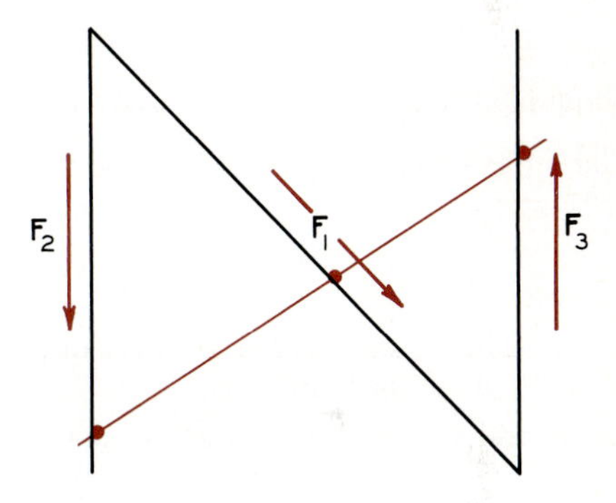

Fig. 11.18 An N-graph for solving an equation of the form $F_1 = F_2/F_3$.

Instead, we use an N-graph, as shown in Fig. 11.18. The outer scales, or "legs" of the N are functional scales and will therefore be linear if F_2 and F_3 are linear, whereas if the same formula were drawn as a parallel-scale graph, all scales would have to be logarithmic.

Table 11.6

R	0.1	0.2	0.4	0.6	0.8	1.0	2.0	4.0	6.0	8.0	10.0
X_R	0	0.91	1.80	2.33	2.71	3.00	3.91	4.81	5.33	5.71	6.00

Fig. 11.19 An N-chart nomograph

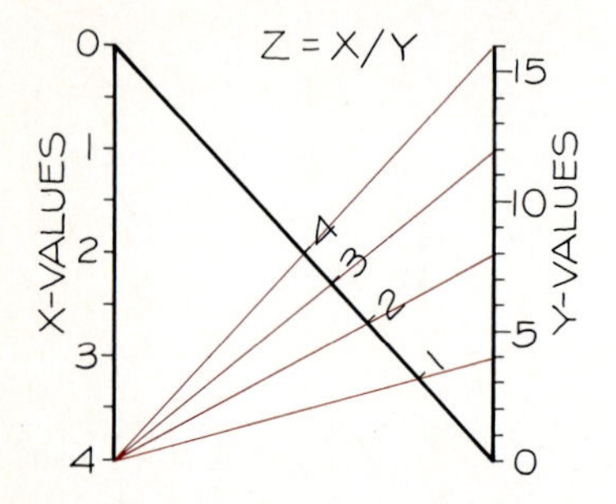

Step 1 An *N*-nomograph for the equation of *Z-Y/X* can be drawn with two parallel scales. The zero ends of each scale are connected with a diagonal scale. Isopleths are drawn to locate units along the diagonal.

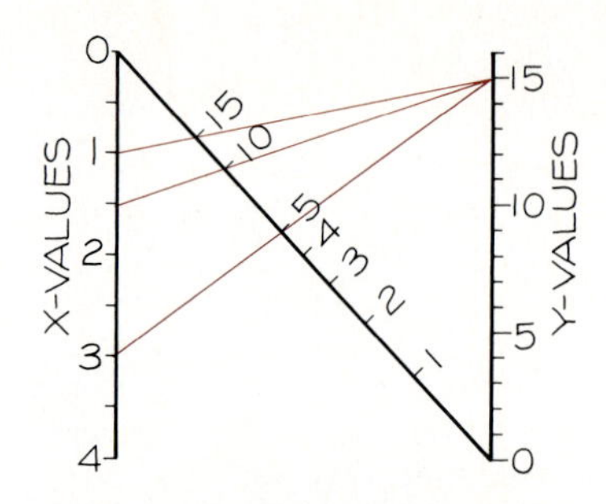

Step 2 Additional isopleths are drawn to locate other units along the diagonal. It is important that the units labeled on the diagonal be whole units that are easy to interpolate between when the nomogram is used.

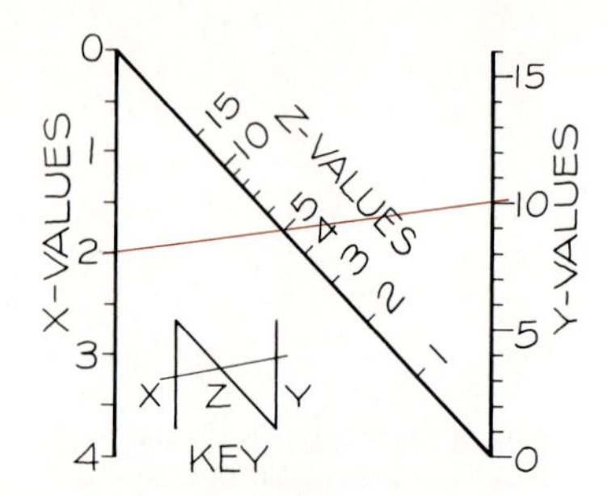

Step 3 The diagonal scale is labeled and a key is drawn to illustrate how the nomogram is used. An example isopleth is drawn to show that 10 ÷ 2 = 5. Note that the accuracy of the N-chart is greater at the 0 end of the diagonal. It approaches infinity at the other end.

Some main features of the N-chart are:

1. The outer scales are parallel functional scales of F_2 and F_3.
2. They increase (functionally) in *opposite* directions.
3. The diagonal scale connects the (functional) *zeros* of the outer scale.
4. In general, the diagonal scale is not a functional scale for the function F_1 and is generally nonlinear.

Construction of an N-graph is simplified by the fact that locating the middle (diagonal) scale is usually less of a problem than it is for a parallel-scale graph. Calibration of the diagonal scale is most easily accomplished by graphical methods.

The steps in constructing a basic N-graph of the equation $Z = Y/X$ are shown in Fig. 11.19. In Step 1, the diagonal is drawn to connect the zero ends of the scales. Whole values are located along the diagonal by using combinations of X- and Y-values. This process of locating values is continued in Step 2. It is important that the units located along the diagonal are whole values that are easy to interpolate between. For example, fractional units such as 1.36 and 2.25 would be complex values that would give a scale that would be difficult to use.

In Step 3, the diagonal is labeled and a key is given to explain how to use the nomograph. A sample isopleth is given that verifies the correctness of the graphical relationship between the scales.

A more advanced N-graph is constructed (see Fig. 11.21) for the equation

$$A = \frac{B + 2}{C + 5}$$

where $0 \leq B \leq 8$ and $0 \leq C \leq 15$. This equation follows the form of

$$F_1 = \frac{F_2}{F_3}$$

where $F_1(A) = A$, $F_2(B) = B + 2$, and $F_3(C) = C$

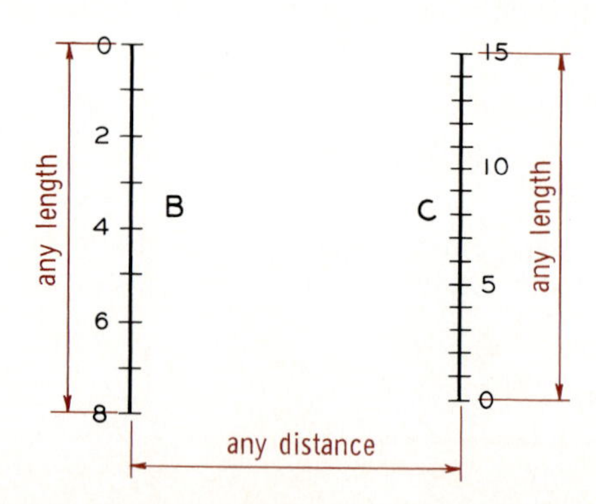

Fig. 11.20 Calibration of the outer scales of an N-graph for the equation $A = (B + 2)/(C + 5)$.

+ 5. Thus the outer scales will be for $B + 2$ and $C + 5$, and the diagonal scale will be for A.

Construction is begun in the same manner as for a parallel-scale graph by selecting the layout of the outer scales (Fig. 11.20). As before, the limits of the diagonal are determined by connecting the endpoints on the outer scales, giving $A = 0.1$ for $B = 0$, $C = 15$ and $A = 2.0$ for $B = 8$, $C = 0$, as shown in Step 1 of Fig. 11.21.

The diagonal scale is located by finding the *function zeros* of the outer scales, i.e., the points where $B + 2 = 0$ or $B = -2$, and $C + 5 = 0$ or $C = -5$. The diagonal scale is drawn by connecting these points as shown in Step 1. Calibration of the diagonal scale is most easily accomplished by substituting into the formula. Select the upper limit of an outer scale, for example, $B = 8$. This gives the formula

$$A = \frac{10}{C + 5}.$$

Solve this equation for the other scale variable:

$$C = \frac{10}{A} - 5.$$

Using this as a "scale equation," make a table of values for the desired values of A and corresponding values of C (up to the limit of C in the chart), as shown in Table 11.7. Connect isopleths from $B = 8$ to the tabulated values of C. Their intersections with the diagonal scale give the required calibrations for approximately half the diagonal scale, as shown in Step 2 of Fig. 11.21.

The remainder of the diagonal scale is calibrated by substituting the end value of the other outer scale ($C = 15$) into the formula, giving

$$A = \frac{B + 2}{20}$$

Solving this for B yields

$$B = 20A - 2.$$

Table 11.7

A	2.0	1.5	1.0	0.9	0.8	0.7	0.6	0.5
C	0	1.67	5.0	6.11	7.50	9.28	11.7	15.0

Fig. 11.21 Construction of an N-graph for the equation $A = (B + 2)/(C + 5)$

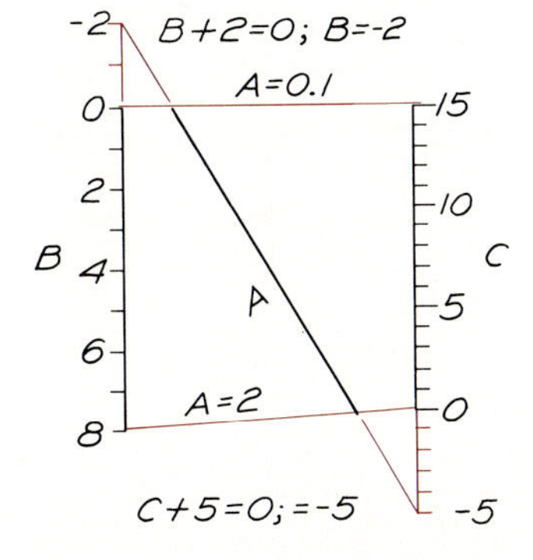

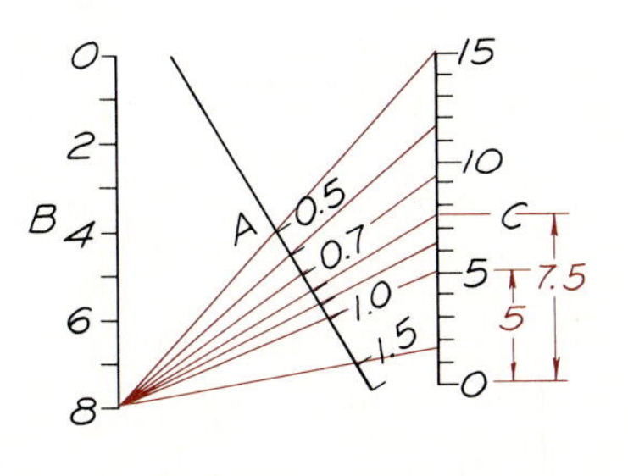

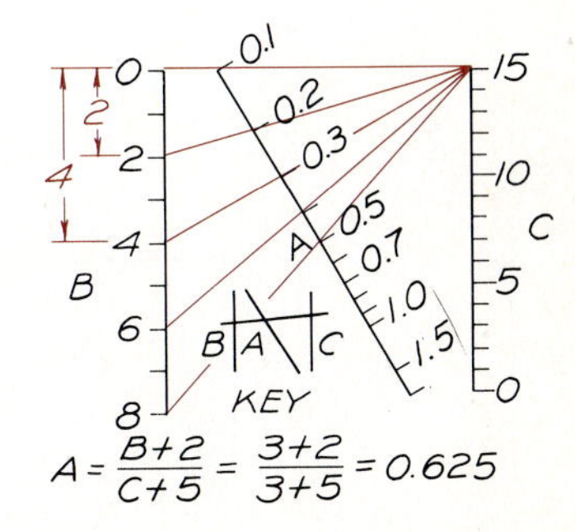

Step 1 Locate the diagonal scale by finding the functional zeros of the outer scales. This is done by setting $B + 2 = 0$ and $C + 5 = 0$, which gives a zero value for A when $B = -2$ and $C = -5$. Connect these points with diagonal scale A.

Step 2 Select the upper limit of one of the outer scales, $B = 8$ in this case, and substitute it into the given equation to find a series of values of C for the desired values of A, as shown in Table 11.7. Draw isopleths from $B = 8$ to the values of C to calibrate the A-scale.

Step 3 Calibrate the remainder of the A-scale in the same manner by substituting $C = 15$ into the equation to determine a series of values on the B-scale for desired values on the A-scale, as listed in Table 11.8. Draw isopleths from $C = 15$ to calibrate the A-scale as shown. Draw a key to indicate how the nomogram is to be used.

Table 11.8

A	0.5	0.4	0.3	0.2	0.1
B	8.0	6.0	4.0	2.0	0

A table for the desired values of A can be constructed as shown in Table 11.8. Isopleths connecting $C = 15$ with the tabulated values of B will locate the remaining calibrations on the A-scale, as shown in Step 3.

11.7 COMBINATION FORMS OF ALIGNMENT GRAPHS

The types of graphs discussed above may be used in combination to handle different types of formulas. For example, formulas of the type $F_1/F_2 = F_3/F_4$ (four variables) may be represented as *two* N-charts by the insertion of a "dummy" function. To do this, let

$$\frac{F_1}{F_2} = S \quad \text{and then} \quad S = \frac{F_3}{F_4}.$$

Each of these may be represented as shown in Step 1 of Fig. 11.22, where one N-graph is inverted and rotated 90°. In this way, the charts may be superimposed as shown in Step 2 if the S-scales are of equal length. The S-scale, being a "dummy" scale, does not need to be calibrated; it is merely a "turning" scale for intermediate values of S that do not actually enter into the formula itself. The chart is read with *two* isopleths that connect the four variable values and cross on the S-scale as shown in Step 3. Nomograms of this form are commonly called *ratio graphs*.

Formulas of the type $F_1 + F_2 = F_3F_4$ are handled similarly. As in the preceding example, a "dummy" function is used: $F_1 + F_2 = S$ and $S = F_3F_4$. In order to apply the superimposition principle, a more equitable arrangement is obtained by rewriting the equations as $F_2 = S - F_1$ and $F_3 = S/F_4$. These two equations then take the form of a parallel-scale nomogram and an N-graph, respectively, as shown Step 1 in Fig. 11.23. Again the S-scales must be identical but need not be calibrated. The nomograms are superimposed in Step 2. The S-scale is used as a "turning" scale for the two isopleths, as shown in Step 3. Many other combinations are possible, limited only by the ingenuity of the nomographer in adapting formulas and scale arrangements to specific needs.

Fig. 11.22 Four-variable graph

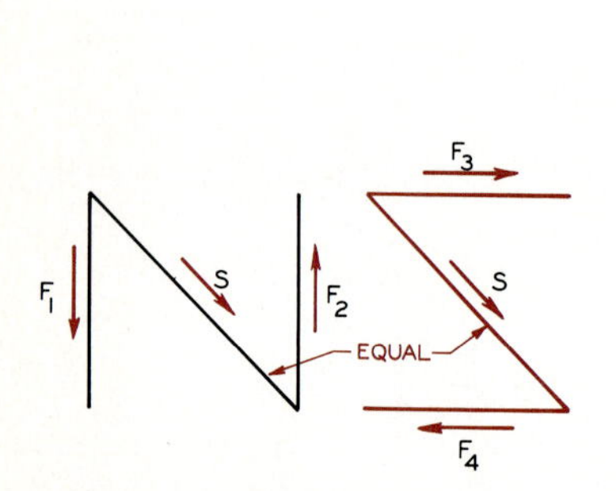

Step 1 A combination chart can be developed to handle four variables in the form $F_1/F_2 = F_3/F_4$ by developing two N-graphs in the forms $F_1/F_2 = S$ and $F_3/F_4 = S$, where S is a dummy scale of equal length in both charts.

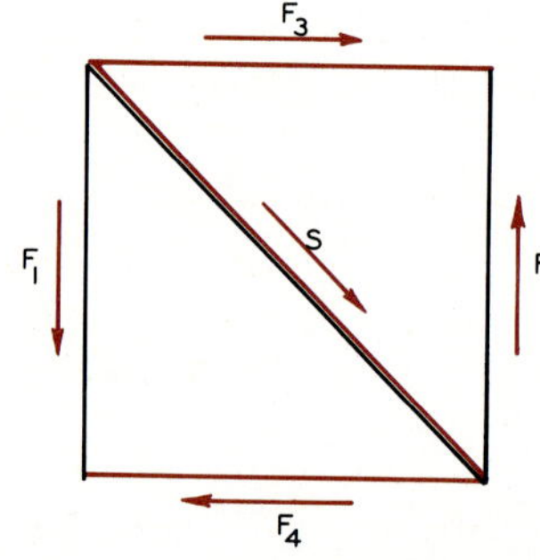

Step 2 If equal-length scales are used in each of the N-graphs and if the S-scales are equal, then the charts can be overlapped so that each is common to the S-scale.

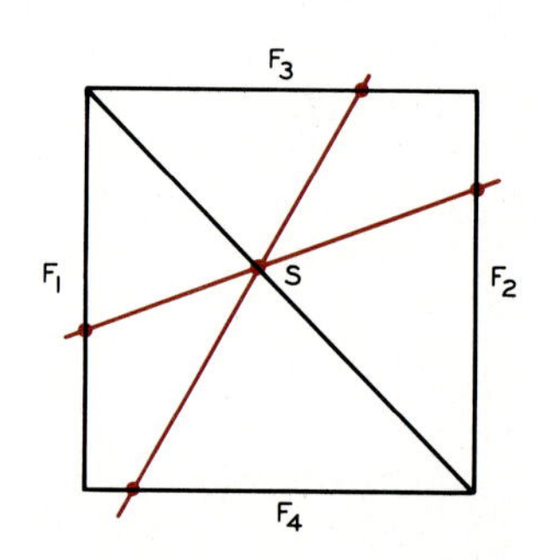

Step 3 Two lines (isopleths) are drawn to cross at a common point on the S-scale. Numerous combinations of the four variables can be read on the surrounding scales. The S-scale need not be calibrated, since no values are read from it.

Fig. 11.23 Combination parallel-scale chart and N-graph

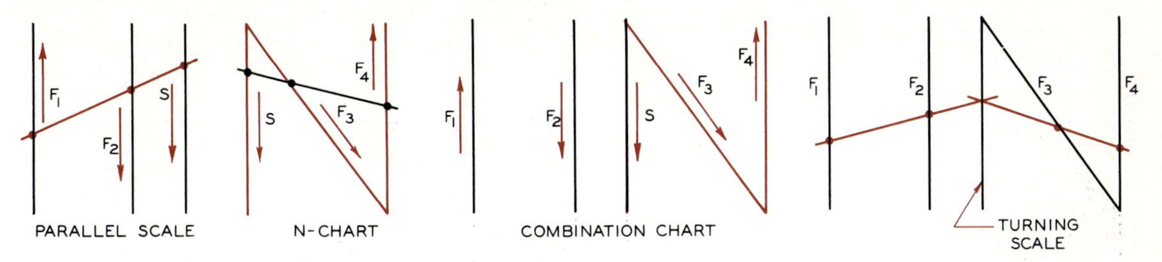

Step 1 Formulas of the type $F_1 + F_2 = F_3F_4$ can be combined into one nomogram by constructing a parallel-scale chart and an N-graph with an equal *S*-scale (a dummy scale).

Step 2 By superimposing the two equal *S*-scales, the two nomograms are combined into a combination chart. The *S*-scale need not be calibrated, since values are not read from it.

Step 3 The addition of the variables can be handled at the left of the chart. The *S*-scale is the turning scale from which the N-graph can be used to find the unknown variables.

PROBLEMS

The following problems are to be solved on Size A sheets (8½″ × 11″) using the principles covered in this chapter. Problems that involve geometric construction and mathematical calculations should show the construction and calculations as part of the solutions. If the mathematical calculations are extensive, these should be included on a separate sheet.

Concurrent Scales

Construct concurrent scales for converting the following relationships of one type of unit to another. The range of units for the scales is given for each.

1. Kilometers and miles: 1.609 km = 1 mile, from 10 to 100 miles.

2. Liters and U.S. gallons: 1 liter = 0.2692 U.S. gallons, from 1 to 10 liters.

3. Knots and miles per hour: 1 knot = 1.15 miles per hour, from 0 to 45 knots.

4. Horsepower and British thermal units: 1 horsepower = 42.4 btu, from 0 to 1200 hp.

5. Centigrade and Fahrenheit: °F = 9⁄5°C + 32, from 32°F to 212°F.

6. Radius and area of a circle: Area = πr^2, from r = 0 to 10.

7. Inches and millimeters: 1 inch = 25.4 millimeters, from 0 to 5 inches.

8. Numbers and their logarithms: (use logarithm tables), numbers from 1 to 10.

Addition and Subtraction Nomographs

Construct parallel-scale nomographs to solve the following addition and subtraction problems.

9. $A = B + C$, where $B = 0$ to 10 and $C = 0$ to 5.

10. $Z = X + Y$, where $X = 0$ to 8 and $Y = 0$ to 12.

11. $Z = Y - X$, where $X = 0$ to 6 and $Y = 0$ to 24.

12. $A = C - B$, where $C = 0$ to 30 and $B = 0$ to 6.

13. $3W = 2V + U$, where $U = 0$ to 12 and $V = 0$ to 9.

14. $W = 3U + V$, where $U = 0$ to 10 and $V = 0$ to 10.

15. Electrical current at a circuit junction:

$$I = I_1 + I_2.$$

I = current entering the junction in amperes
I_1 = current leaving the junction, varying from 2 to 15 amps
I_2 = current leaving junction, varying from 7 to 36 amps.

16. Pressure change in fluid flowing in a pipe:

$$\Delta P = P_2 - P_1.$$

ΔP = pressure change between two points in pounds per square inch,
P_1 = pressure upstream, varying from 3 psi to 12 psi,
P_2 = pressure downstream, varying from 10 psi to 15 psi.

Multiplication and Division: Parallel Scales

Construct parallel-scale nomographs with logarithmic scales that will perform the following multiplication and division operations:

17. Area of a rectangle: Area = Height × Width, where $H = 1$ to 10 and $W = 1$ to 12.

18. Area of a triangle: A = Base × Height/2, where $B = 1$ to 10 and $H = 1$ to 5.

19. Electrical potential between terminals of a conductor: $E = IR$.

E = electrical potential in volts,
I = current, varying from 1 to 10 amperes,
R = resistance, varying from 5 to 30 ohms.

20. Pythagorean theorem: $C^2 = A^2 + B^2$.

C = hypotenuse of a right triangle in centimeters,
A = one leg of the right triangle, varying from 5 to 50 cm,
B = second leg of the right triangle, varying from 20 to 80 cm.

21. Allowable pressure on a shaft bearing: $P = ZN/100$.

P = pressure in pounds per square inch,
Z = viscosity of lubricant from 15 to 50 cp-(centipoises),
N = angular velocity of shaft from 10 to 1000 rpm.

22. Miles per gallon and automobile travels: mgp = miles/gallon. Miles vary from 1 to 500 and gallons from 1 to 24.

23. Cost per mile (cpm) of an automobile: cpm = cost/miles. Miles vary from 1 to 500 and cost varies from $1 to $28.

24. $R = S\sqrt{T}$ where S varies from 1 to 10 and T from 1 to 10.

25. Angular velocity of a rotating body: $W = V/R$.

W = angular velocity, in radians per second,
V = Peripheral velocity, varying from 1 to 100 meters per second,
R = radius, varying from 0.1 to 1 meter.

N-Graphs

Construct N-graphs that will solve the following equations.

26. Stress = P/A: where P varies from 0 to 1000 psi and A varies from 0 to 15 square inches.

27. Volume of a cylinder: $V = \pi r^2 h$

V = volume in cubic inches,
r = radius, varying from 5 to 10 feet,
h = height, varying from 2 to 20 feet.

28. Same as Problem 17.

29. Same as Problem 18.

30. Same as Problem 19.

31. Same as Problem 20.

32. Same as Problem 21.

33. Same as Problem 22.

34. Same as Problem 23.

35. Same as Problem 24.

36. Same as Problem 25.

Combination Nomographs

37. Construct a combination nomograph to express the law of sines: $a/\text{sine } A = b/\text{sine } B$. Assume that a and b vary from 0 to 10, and that A and B vary from 0° to 90°.

38. Construct a combination nomograph to determine the velocity of sound in a solid, using the formula

$$C = \sqrt{\frac{E + 4\mu/3}{p}}$$

where E varies from 10^6 to 10^7 psi, μ varies from 1×10^6 to 2×10^6 psi, and C varies from 1000 to 1500 fps. (*Hint*: rewrite the formula as $C^2p = E + 4/3\ \mu$.)

12

EMPIRICAL EQUATIONS AND CALCULUS

12.1 EMPIRICAL DATA

Data gathered from laboratory experiments and tests of prototypes or from actual field tests are called **empirical data.** Often empirical data can be transformed to equation form by means of one of three types of equations to be covered here.

The analysis of empirical data begins with the plotting of the data on rectangular grids, logarithmic grids, and semilogarithmic grids. Curves are then sketched through each point to determine which of the grids renders a straight-line relationship (Fig. 12.1). When the data plots as a straight line, its equation may be determined. Each curve appears as a straight line in one of the graphs. We use this straight-line curve to write an equation for the data.

12.2 SELECTION OF POINTS ON A CURVE

Two methods of finding the equation of a curve are (1) the selected-points method and (2) the slope-intercept method. These are compared on a linear graph in Fig. 12.2.

Selected-Points Method

Two widely separated points, such as (1, 30) and (4, 60) can be selected on the curve. These points are substituted in the equation below:

$$\frac{Y - 30}{X - 1} = \frac{60 - 30}{4 - 1}$$

The resulting data for the equation is

$$Y = 10X + 20.$$

Slope-Intercept Method

To apply the slope-intercept method, the intercept on the y-axis where $X = 0$ must be known. If the x-axis is logarithmic, then the log of $X = 1$ is 0 and the intercept must be found above the value of $X = 1$.

In the slope-intercept method in Fig. 12.2, the data do not intercept the y-axis; therefore, the curve must be extended to find the intercept $B = 20$. The slope of the curve is found ($\Delta Y/\Delta X$) and substituted into the slope-intercept form to give the equation as $Y = 10X + 20$.

Other methods of converting data to equations are used, but the two methods illustrated here make the best use of the graphical process and are the most direct methods of introducing these concepts.

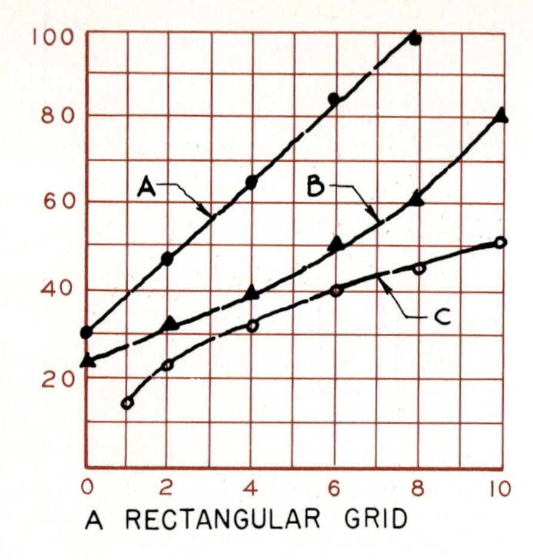

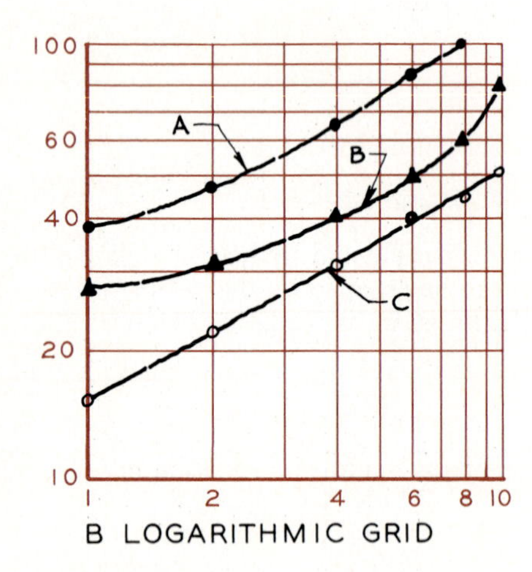

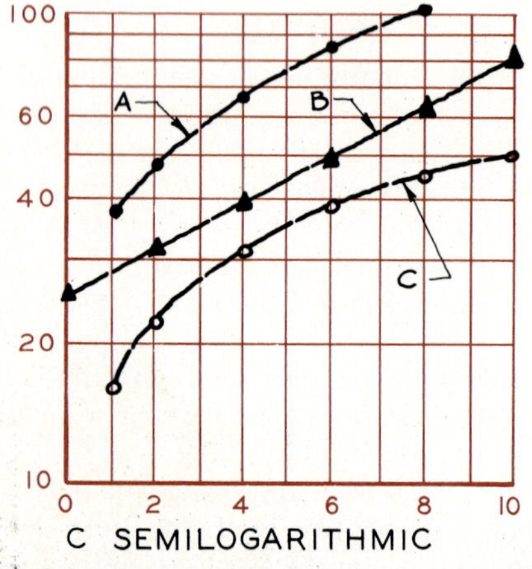

Fig. 12.1 Empirical data are plotted on each of these types of grids to determine which will render a straight-line plot. If the data can be plotted as a straight line on one of these grids, their equation can be found.

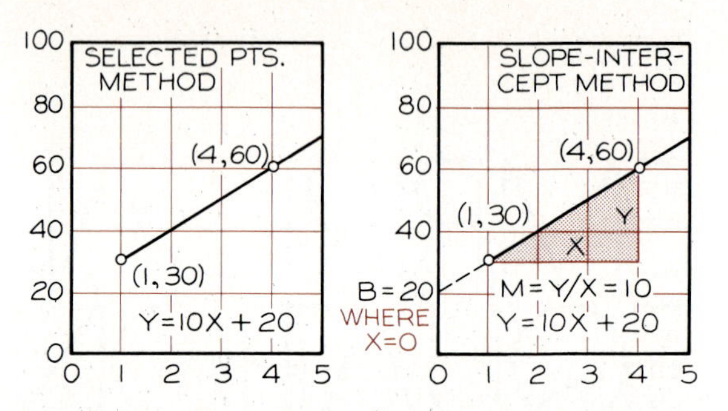

Fig. 12.2 The equation of a straight line on a grid can be determined by selecting any two points on the line. The slope-intercept method requires that the intercept be found where $X = 0$ on a semilog grid. This requires the extension of the curve to the Y-axis.

12.3 THE LINEAR EQUATION: $Y = MX + B$

The curve fitting the experimental data plotted in Fig. 12.3 is a straight line; therefore, we may assume that these data are linear, meaning that each measurement along the y-axis is directly proportional to x-axis units. We may use the slope-intercept form or the selected-points method to find the equation for the data.

In the slope-intercept method, two known points are selected along the curve. The vertical and horizontal differences between the coordinates of each of these points are determined to establish the right triangle shown in part A of the

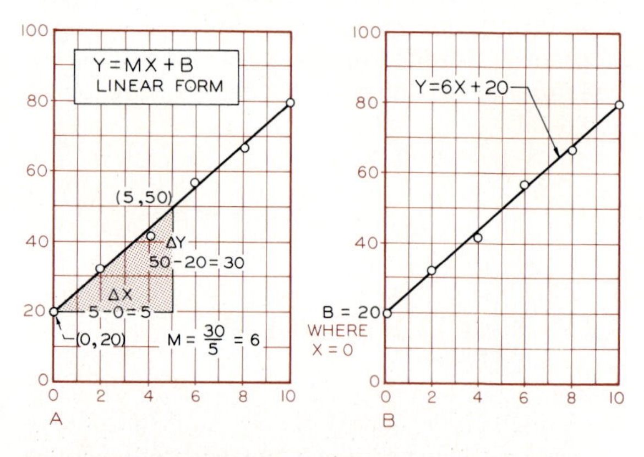

Fig. 12.3 (A) A straight line on an arithmetic grid will have an equation in the form $Y = MX + B$. The slope, M, is found to be 6. (B) The intercept, B, is found to be 20. The equation is written as $Y = 6X + 20$.

figure. In the slope-intercept equation, $Y = MX + B$, M is the tangent of the angle between the curve and the horizontal, B is the intercept of the curve with the y-axis where $X = 0$, and X and Y are variables. In this example $M = 30/5 = 6$ and the intercept is 20.

If the curve has sloped downward to the right, the slope would have been negative. By substituting this information into the slope-intercept equation, we obtain $Y = 6X + 20$, from which we can determine values of Y by substituting any value of X into the equation.

The selected-points method could also have been used to arrive at the same equation if the intercept were not known. By selecting two widely separated points such as (2, 32) and (10, 80), one can write the equation in this form:

$$\frac{Y - 32}{X - 2} = \frac{80 - 32}{10 - 2}, \quad \therefore Y = 6X + 20,$$

which results in the same equation as was found by the slope-intercept method ($Y = MX + B$).

12.4 THE POWER EQUATION: $Y = BX^M$

Since the data shown plotted on a rectangular grid in Fig. 12.4 do not form a straight line, they cannot be expressed in the form of a linear equation. However, when the data are plotted on a logarithmic grid, they form a straight line (Step 1). Therefore, we express the data in the form of a power equation in which Y is a function of X raised to a given power, or $Y = BX^M$. The equation of the data is obtained in much the same manner as was the linear equation, using the point where the curve intersects the Y-axis where $X = 0$, and letting M equal the slope of the curve. Two known points are selected on the curve to form the slope triangle. The engineers' scale can be used, when the cycles along the X- and Y-axes are equal, to measure the slope between the coordinates of the two points.

If the horizontal distance of the right triangle is drawn to be 1 or 10 or a multiple of 10, the vertical distance can be read directly. In Step 2, the slope M (tangent of the angle) is found to be 0.54. The intercept B is 7; thus the equation is $Y = 7X^{0.54}$, which can be evaluated for each value of Y by converting this power equation into the logarithmic form of log Y:

$$\log Y = \log B + M \log X.$$
$$\log Y = \log 7 + 0.54 \log X.$$

Note that when the slope-intercept method is used, the intercept is found on the y-axis where $X = 1$. In Fig. 12.5, the y-axis at the left of the graph has an X-value of 0.1; consequently, the intercept is located midway across the graph where

Fig. 12.4 The power equation, $Y = BX^M$

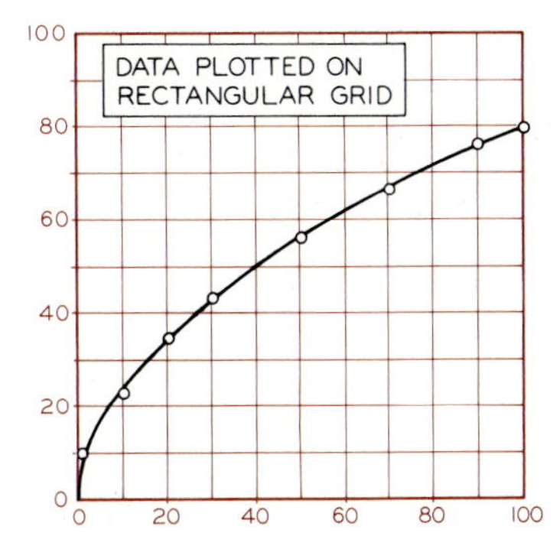

Given The data plotted on the rectangular grid give an approximation of a parabola. Since the data does not form a straight line on the rectangular grid, the equation will not be linear.

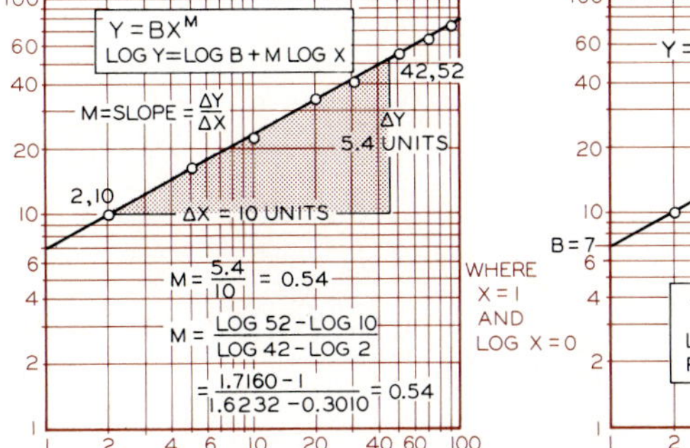

Step 1 The data forms a straight line on a logarithmic grid, making it possible to find its equation. The slope, M, can be found graphically with an engineer's scale, setting dX at 10 units and measuring the slope (dY) using the same scale.

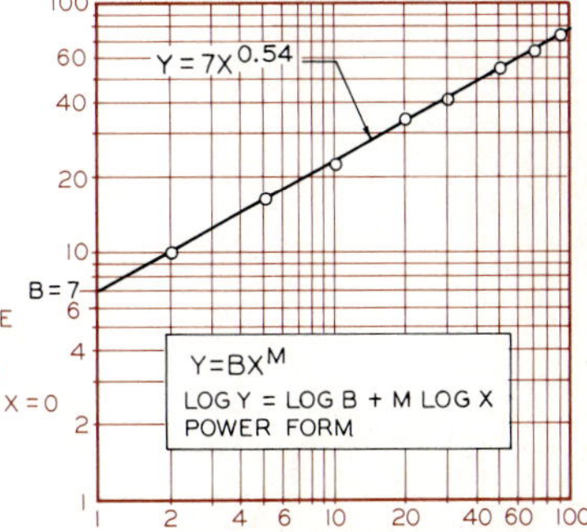

Step 2 The intercept $B = 7$ is found on the y-axis where $X = 1$. The slope and intercept are substituted into the equation, which then becomes $Y = 7X^{0.54}$.

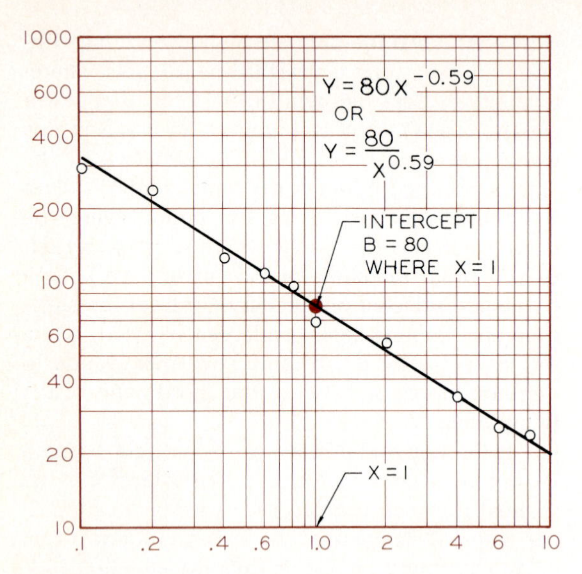

Fig. 12.5 When the slope-intercept equation is used, the intercept can be found only where $X = 1$. Therefore, in this example the intercept is found at the middle of the graph.

$X = 1$. This is analogous to the linear form of the equation, since the log of 1 is 0. The curve slopes downward to the right; thus the slope, M, is negative. The selected-points method can be applied to find the equation of the data as discussed in the previous article.

Base-10 logarithms are used in these examples, but natural logs could be used with e (2.718) as the base.

12.5 THE EXPONENTIAL EQUATION: $Y = BM^X$

The experimental data plotted in Fig. 12.6 form a curve, indicating that they are not linear. When the data are plotted on a semilogarithmic grid (Step 1) they approximate a straight line for which we can write the equation $Y = BM^X$, where B is the Y-intercept of the curve and M is the slope of the curve. The procedure for deriving the equation is shown in Step 1, in which two points are selected along the curve so that a right triangle can be drawn to represent the differences between the coordinates of the points selected. The slope of the curve is found to be

$$\log M = \frac{\log 40 - \log 6}{8 - 3} = 0.1648$$

Fig. 12.6 The exponential equation: $Y = BM^x$

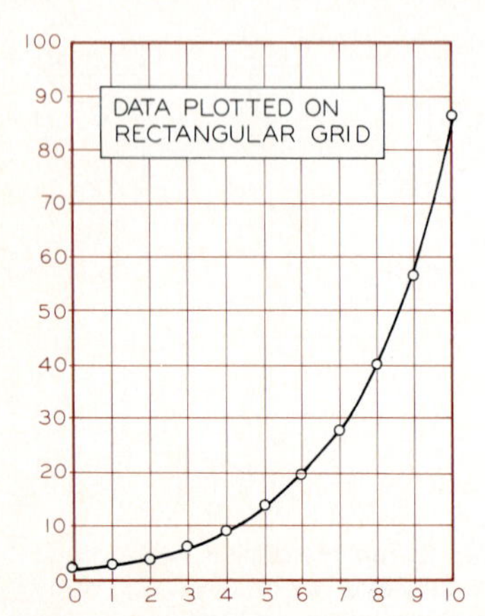

Given These data do not give a straight line on either a rectangular grid or a logarithmic grid. However, when plotted on a semilogarithmic grid, they give a straight line.

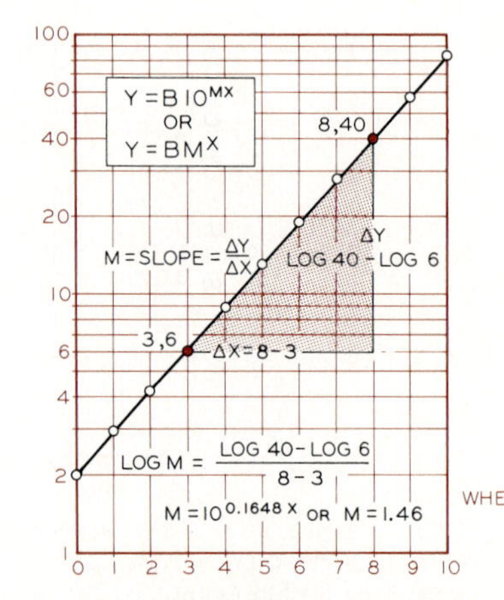

Step 1 The slope must be found by mathematical calculations; it cannot be found graphically. The slope may be written in either of the forms shown here.

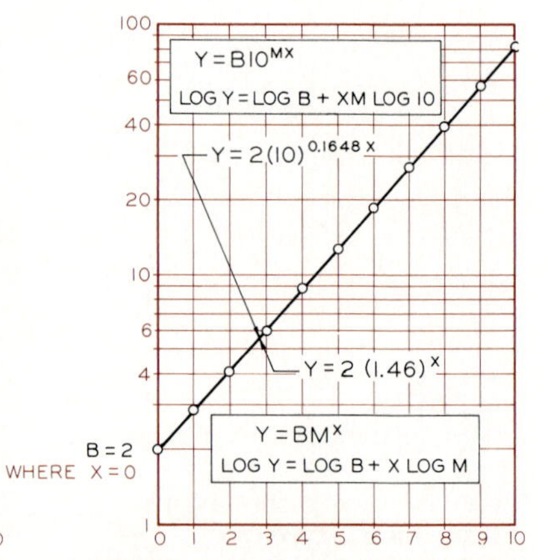

Step 2 The intercept $B = 2$ is found where $X = 0$. The slope, M, and the intercept, B, are substituted into the equation to give $Y = 2(10)^{0.1648X}$ or $Y = 2(1.46)^X$.

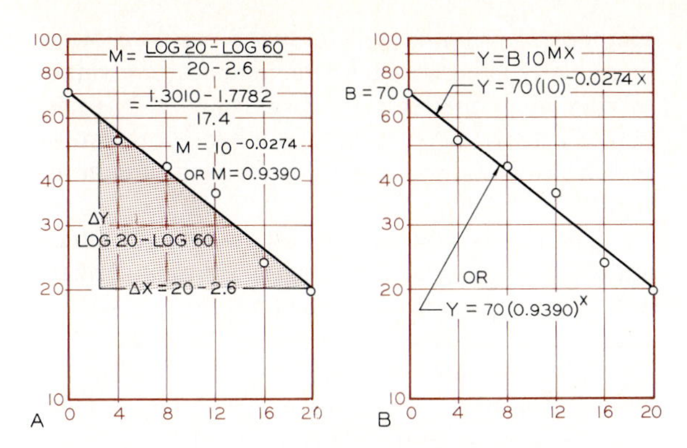

Fig. 12.7 When a curve slopes downward to the right, its slope is negative as calculated in part A. Two forms of a final equation are shown in part B by substitution.

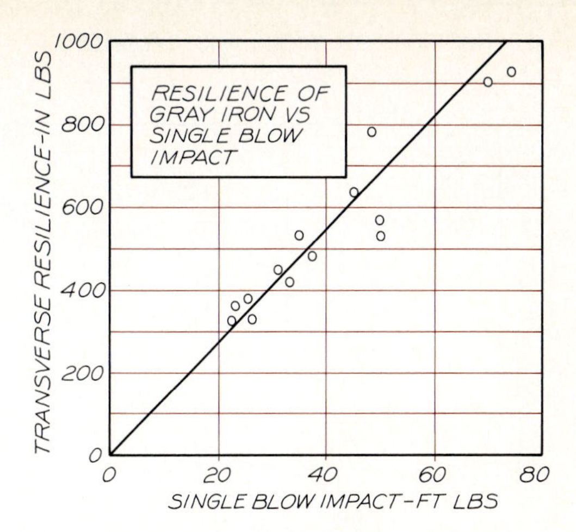

Fig. 12.8 The relationship between the transverse strength of gray iron and impact resistance results in a straight line with an equation of the form $Y = MX + B$.

or

$$M = (10)^{0.1648} = 1.46$$

The value of M can be substituted in the equation in the following manner:

$$Y = BM^X \quad \text{or} \quad Y = 2(1.46)^X,$$
$$Y = B(10)^{MX} \quad \text{or} \quad Y = 2(10)^{0.1648X},$$

where X is a variable that can be substituted into the equation to give an infinite number of values for Y. We can write this equation in its logarithmic form, which enables us to solve it for the unknown value of Y for any given value of X. The equation can be written as

$$\log Y = \log B + X \log M$$

or

$$\log Y = \log 2 + X \log 1.46.$$

The same methods are used to find the slope of a curve with a negative slope. The curve of the data in Fig. 12.7 slopes downward to the right; therefore, the slope is negative. Two points are selected in order to find which is the antilog of -0.0274. The intercept, 70, can be combined with the slope, M, to find the final equations as illustrated in Step 2 of Fig. 12.7.

12.6 APPLICATIONS OF EMPIRICAL GRAPHS

Figure 12.8 is an example of how empirical data can be plotted to compare the transverse strength and impact resistance of gray iron. Note that the data are somewhat scattered, but the best curve is drawn. Since the curve is a straight line on a linear graph, the equation of these data can be found by the equation

$$Y = MX + B.$$

Figure 12.9 is an example of how empirical data can be plotted to compare the specific weight (pounds per horsepower) of generators and hy-

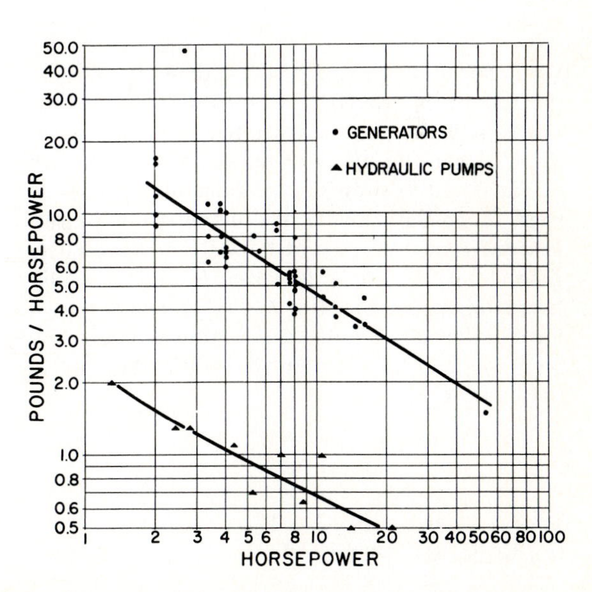

Fig. 12.9 Empirical data plotted on a logarithmic grid, showing the specific weight versus horsepower of electric generators and hydraulic pumps. The curve is the average of joints plotted. (Courtesy of General Motors Corporation, *Engineering Journal.*)

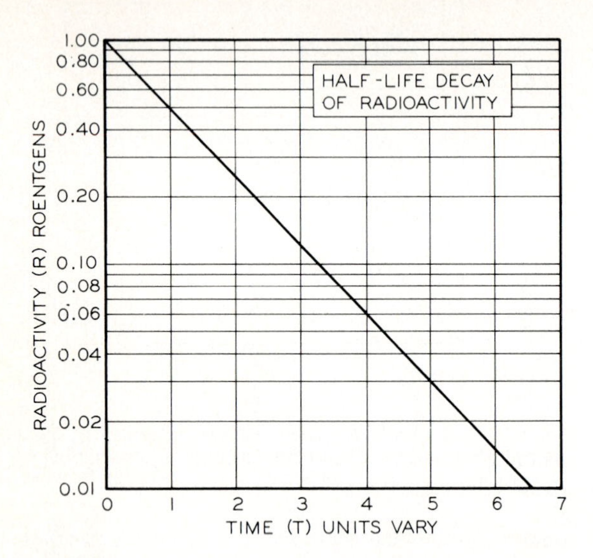

Fig. 12.10 The relative decay of radioactivity is plotted as a straight line on this semilog graph, making it possible for its equation to be found in the form $Y = BM^X$.

draulic pumps versus horsepower. Note that the weight of these units decreases linearly as the horsepower increases. Therefore, these data can be written in the form of the power equation

$$Y = BX^M$$

We obtain the equation of these data by applying the procedures covered in Section 12.3 and thus mathematically analyze these relationships.

The half-life decay of radioactivity is plotted in Fig. 12.10 to show the relationship of decay to time. Since the half-life of different isotopes varies, different units would have to be assigned to time along the X-axis; however, the curve would be a straight line for all isotopes. The exponential form of the equation discussed in Section 12.5 can be applied to find the equation for these data in the form of

$$Y = BM^X.$$

12.7 INTRODUCTION TO GRAPHICAL CALCULUS

Engineers, designers, and technicians must often deal with relationships between variables that must be solved using the principles of calculus. If the equation of the curve is known, traditional methods of calculus will solve the problem. However, many experimental data cannot be converted to standard equations. In these cases, it is desirable to use the graphical method of calculus which provides relatively accurate solutions to irregular problems.

The two basic forms of calculus are (1) differential calculus, and (2) integral calculus. Differential calculus is used to determine the rate of change of one variable with respect to another. For example, the curve plotted in Fig. 12.11A represents the relationship between two variables. Note that the Y-variable increases as the X-variable increases. The rate of change at any instant along the curve is the slope of a line that is tangent to the curve at that particular point. This exact slope is often difficult to determine graphically; consequently, it can be approximated by constructing a chord at a given interval, as shown in Fig. 12.11A. The slope of this chord can be measured by finding the tangent of $\Delta Y/\Delta X$.

This slope can represent miles per hour, weight versus length, or a number of other meaningful rates that are important to the analysis of data.

Integral calculus is the reverse of differential calculus. Integration is the process of finding the area under a given curve, which can be thought of generally as the product of the two variables plotted on the x- and y-axes. The area under a curve is approximated by dividing one of the variables into a number of very small intervals, which become small rectangular areas at a particular zone under the curve, as shown in Fig. 12.11B. The bars are extended so that as much of the square end of

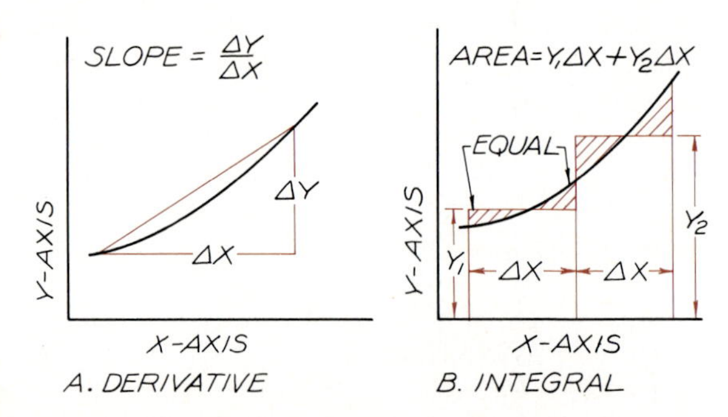

Fig. 12.11 The derivative of a curve is the change at any point that is the slope of the curve, Y/X. The integral of a curve is the cumulative area enclosed by the curve, which is the summation of the products of the areas.

the bar is under the curve as above the curve and the average height of the bar is, therefore, near its midpoint.

12.8 GRAPHICAL DIFFERENTIATION

Graphical differentiation is defined as the determination of the rate of change of two variables with respect to each other at any given point. Figure 12.12 illustrates the preliminary construction of the derivative scale that would be used to plot a continuous derivative curve from the given data.

Step 1 The original data are plotted graphically and the axes are labeled with the proper units of measurement. A chord can be constructed to estimate the maximum slope by inspection. In the given curve, the maximum slope is estimated to be 2.3. A vertical scale is constructed in excess of this to provide for the plotting of slopes that may exceed the estimate. This ordinate scale is drawn to a convenient scale to facilitate measurement.

Step 2 A known slope is plotted on the given data grid. This slope need not be related to the curve in any way. In this case, the slope can be read directly as 1.

Step 3 The pole can be found by drawing a line from the ordinate of 1 (the known slope) on the derivative scale parallel to the slope line. These similar triangles are used to obtain the pole, which will be used in determining the derivative curve.

The steps in completing the graphical differentiation are given in Fig. 12.13. Note that the same horizontal intervals used in the given curve are projected directly beneath on the derivative scale. The maximum slope of the data curve is estimated to be slightly less than 12. A scale is selected that will provide an ordinate that will accommodate the maximum slope. A line is drawn from point 12 on the ordinate axis of the derivative grid that is parallel to the known slope on the given curve grid. The point of intersection of this line and the extension of the x-axis is the pole point.

A series of chords is constructed on the given curve. Lines are constructed parallel to these chords through point P and extended to the y-axis of the derivative grid to locate bars at each interval. Note that the interval between 0 and 1 was divided in half to provide a more accurate plot.

Fig. 12.12 Scales for graphical differentiation

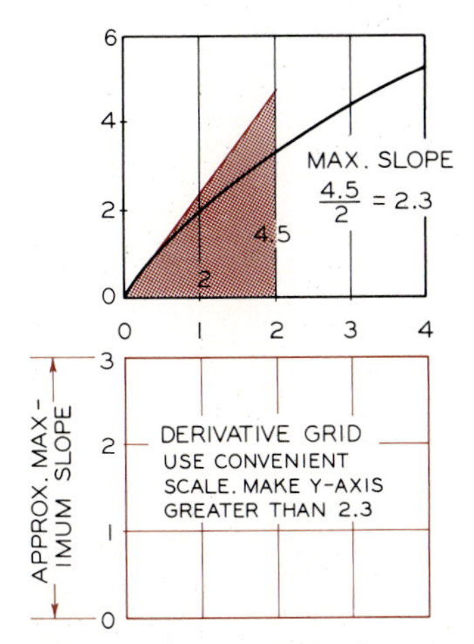

Step 1 The maximum slope of a curve is found by constructing a line tangent to the curve where it is steepest. The maximum slope, 2.3, is found and the derivative grid is laid off with a maximum ordinate of 3.0.

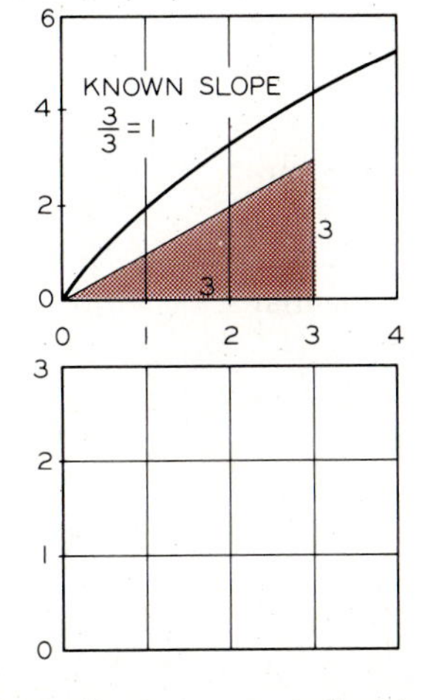

Step 2 A known slope is found on the given grid; this value is 1 in this example. The known slope has no relationship to the curve at this point.

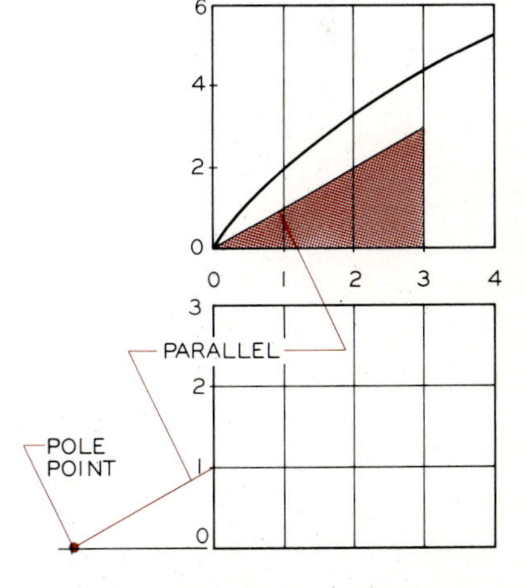

Step 3 Construct a line from 1 on the Y-axis of the derivative grid that is parallel to the slope of the triangle in the given grid. This line locates the pole point where it crosses the extension of the X-axis.

Fig. 12.13 Graphical differentiation

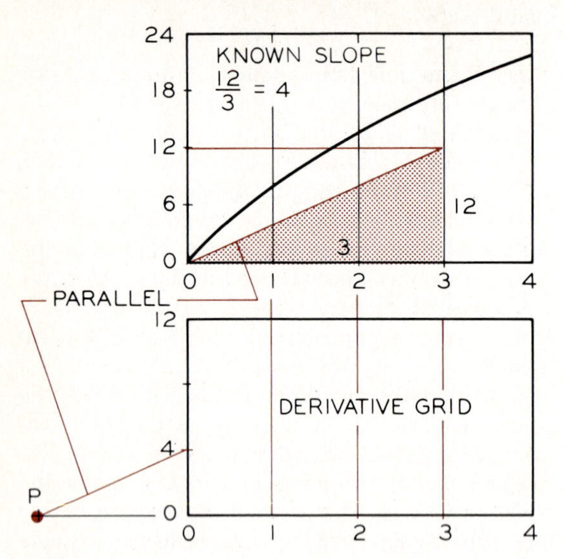

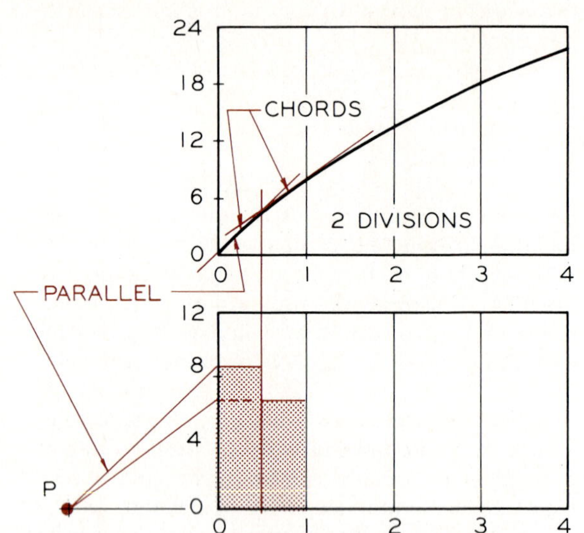

Required Find the derivative curve of the given data.

Step 1 Find the derivative grid and the pole point using the construction illustrated in Fig. 12.12.

Step 2 Construct a series of chords between selected intervals on the given curve and draw lines parallel to these chords through point P on the derivative grid. These lines locate the heights of bars in their respective intervals. The first interval is divided in two since the curve is changing sharply in this interval.

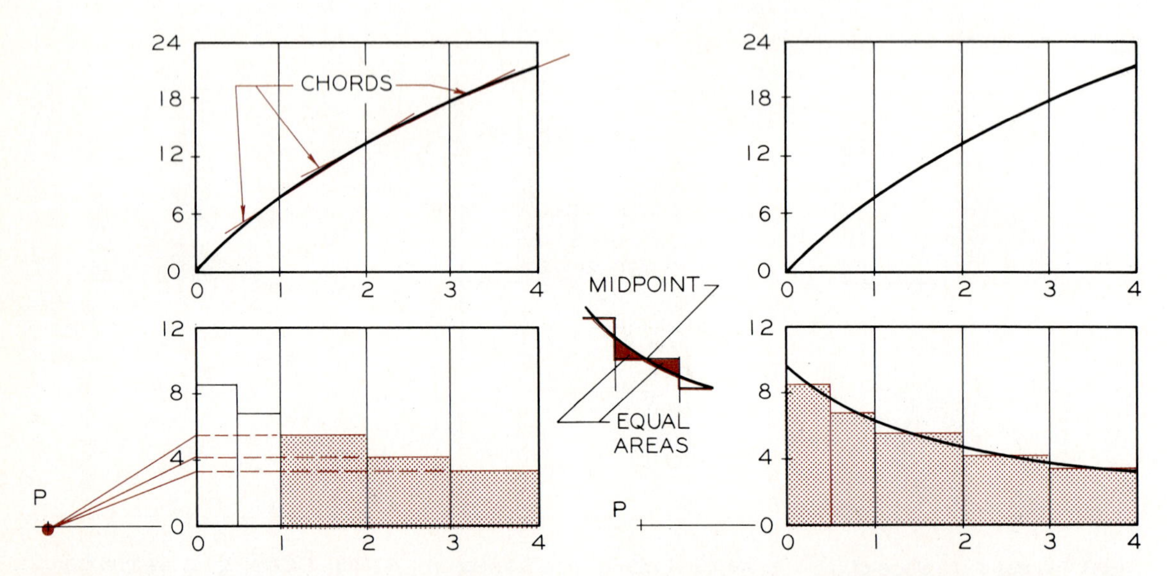

Step 3 Additional chords are drawn in the last two intervals. Lines parallel to these chords are drawn from the pole point to the Y-axis to find the heights of additional bars in their respective intervals.

Step 4 The vertical bars represent the different slopes of the given curve at different intervals. The derivative curve is drawn through the midpoints of the bars so that the areas under and above the bars are approximately equal.

The curve is the sharpest in this interval. A smooth curve is constructed through the top of these bars in such a manner that the area above the horizontal top of the bar is the same as that below it. This curve represents the derivative of the given data. The rate of change, $\Delta Y/\Delta X$, can be found at any interval of the variable X by reading directly from the graph at the value of X in question.

12.9 APPLICATIONS OF GRAPHICAL DIFFERENTIATION

The mechanical handling shuttle shown in Fig. 12.14 is used to convert rotational motion into controlled linear motion. A scale drawing of the linkage components is given so that graphical analysis can be applied to determine the motion resulting from this system.

The linkage is drawn to show the end positions of point P, which will be used as the zero point for plotting the travel versus the degrees of revolution. Since rotation is constant at one revolution per three seconds, the degrees of revolution can be converted to time, as shown in the data curve given at the top of Fig. 12.15. The drive crank, R_1, is revolved at 30° intervals, and the distance that point P travels from its end position is plotted on the graph, as shown in the given data. This gives the distance-versus-time relationship.

We determine the ordinate scale of the derivative grid by estimating the maximum slope of the given data curve, which is found to be a little less than 100 in./sec. A slope of 40 is drawn on the given data curve; this will be used in determining the location of pole P in the derivative grid. From point 40 on the derivative ordinate scale, we draw a line parallel to the known slope, which is found

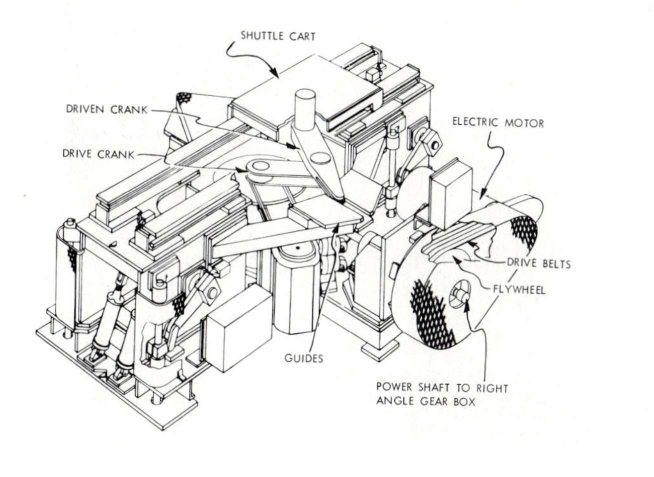

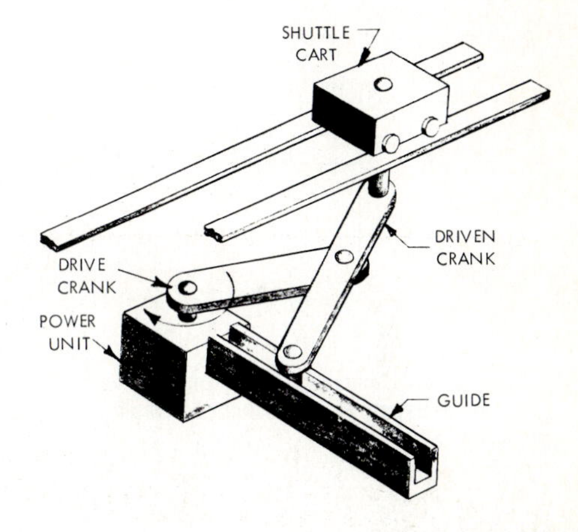

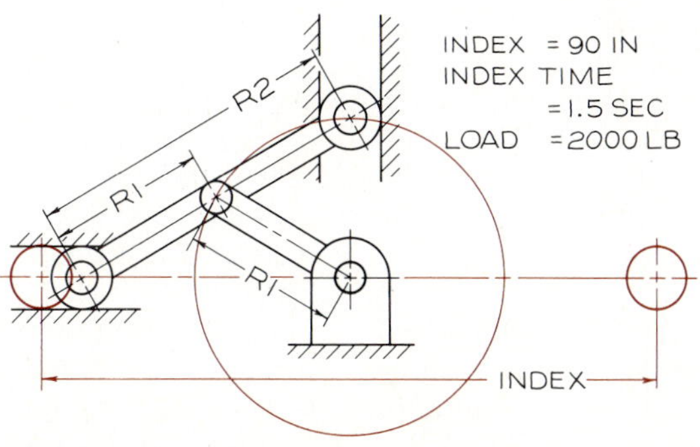

Fig. 12.14 A pictorial and scale drawing of an electrically powered mechanical handling shuttle used to move automobile parts on an assembly line. (Courtesy of General Motors Corporation.)

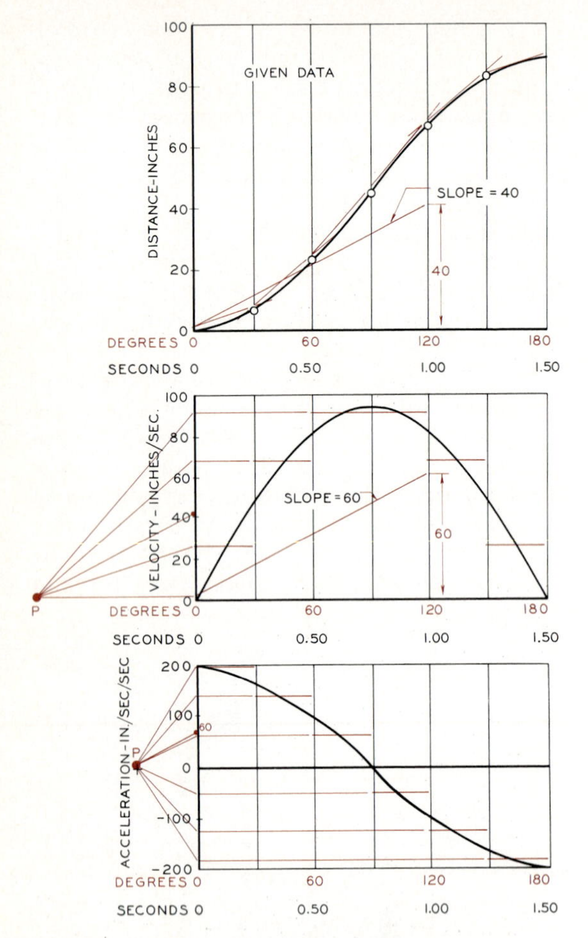

Fig. 12.15 Graphical determination of velocity and acceleration of the mechanical handling shuttle by differential calculus.

on the given grid. Point P is the point where this line intersects the extension of the x-axis.

A series of chords are drawn on the given curve to approximate the slope at various points. Lines are constructed through point P of the derivative scale parallel to the chord lines of the given curve and extended to the ordinate scale. The points thus obtained are then projected across to their respective intervals to form vertical bars. A smooth curve is drawn through the top of each of the bars to give an average of the bars. This curve can be used to find the velocity of the shuttle in inches per second at any time interval.

The construction of the second derivative curve, the acceleration, is very similar to that of the first derivative. By inspecting the first derivative, we estimate the maximum slope to be 200 in./sec/sec. An easily measured scale is established for the ordinate scale of the second derivative curve. Point P is found in the same manner as for the first derivative.

Chords are drawn at intervals on the first derivative curve. Lines are drawn parallel to these chords from point P in the second derivative curve to the y-axis, where they are projected horizontally to their respective intervals to form a series of bars. A smooth curve is drawn through the tops of the bars to give a close approximation of the average areas of the bars. Note that a minus scale is given for the acceleration curve to indicate deceleration.

The maximum acceleration is found to be at the extreme endpoints and the minimum acceleration is at 90°, where the velocity is the maximum. It can be seen from the velocity and acceleration plots that the parts being handled by the shuttle are accelerated at a rapid rate until the maximum velocity is attained at 90°, at which time deceleration begins and continues until the parts come to rest.

12.10 GRAPHICAL INTEGRATION

Integration is the process of determining the area (product of two variables) under a given curve. For example, if the y-axis were pounds and the x-axis were feet, the integral curve would give the product of the variables, foot-pounds, at any interval of feet along the x-axis. Figure 12.16 depicts the method of constructing scales for graphical integration.

Step 1 It is customary to locate the integral curve above the given data curve, since the integral will be an equation raised to a higher power. A line is drawn through the given data curve to approximate the total area under the curve. The approximate area is 4 times 5, or 20, square units of area. The ordinate scale is drawn on the integral graph in excess of 20 units to provide a margin for any overage. The horizontal scale intervals are projected from the given graph to the integral grid.

Step 2 The ordinate at any point on the integral scale will have the same numerical value as the area under the curve as measured from the origin

Fig. 12.16 Scales for graphical integration

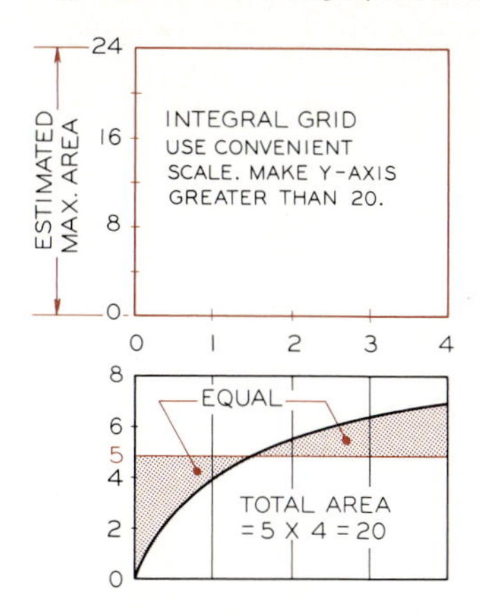

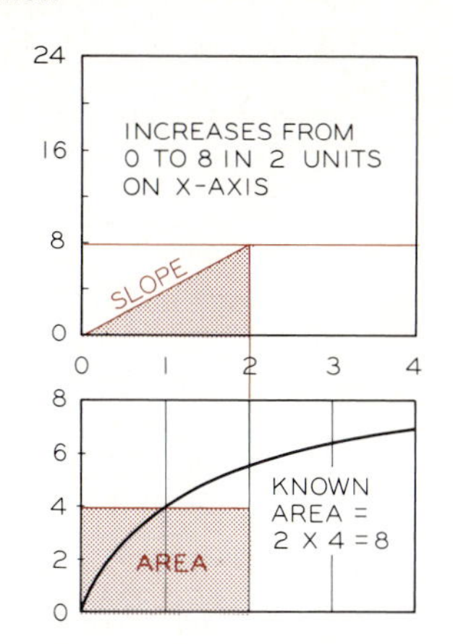

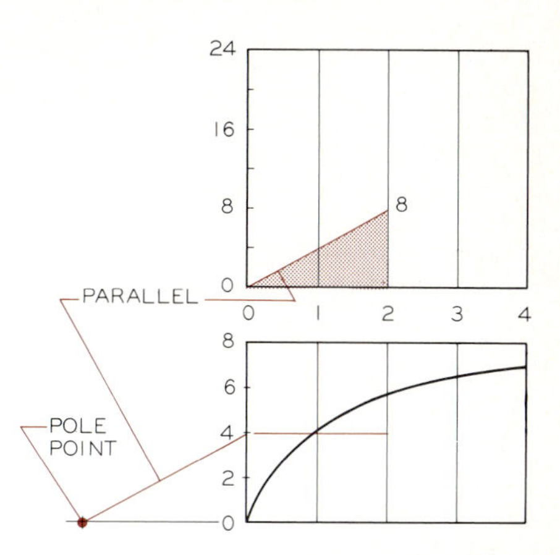

Step 1 To determine the maximum value on the *Y*-axis of the integral grid, a line is drawn to approximate the area under the curve. This is found to be 20, and the *Y*-axis is constructed with 24 as the maximum value.

Step 2 A known area, 8 in this case, is found in the given grid. A slope line from 0 to 8 is constructed in the integral grid directly above the known area, which establishes the integral for this model.

Step 3 A line is drawn from 4 on the *Y*-axis of the given grid parallel to the slope line in the integral grid. This line from 4 crosses the extension of the *X*-axis to locate the pole point.

to that point on the given data grid. The ordinate at point 2 on the x-axis directly above the rectangle must be equal to its area of 8. A slope is drawn from the origin to the ordinate of 8.

Step 3 Point *P* is found by drawing a line from point 4 on the given grid parallel to the slope established in the integral grid. This line intersects the extension of the x-axis at point *P*. This pole point will be used to find the integral curve.

The technique illustrated in Fig. 12.17 can be applied to most integration problems. The equation of the given curve is $Y = 2X^2$, which can also be integrated mathematically as a check.

From the given grid, the total area under the curve can be estimated to be less than 40 units. This value becomes the maximum height of the *y*-axis on the integral curve. A convenient scale is selected, units are assigned to the ordinate, and the pole point, *P*, is found.

A series of vertical bars is constructed to approximate the areas under the curve at these intervals. The narrower the bars, the more accurate will be the resulting calculations. Notice that the interval between 1 and 2 was divided in half to provide a more accurate plot. The top lines of the bars are extended horizontally to the *y*-axis, where the points are then connected by lines to point *P*. Lines are drawn parallel to *AP*, *BP*, *CP*, *DP*, and *EP* in the integral grid to correspond to the respective intervals in the given grid. The intersection points of the chords are connected by a smooth curve—the integral curve. This curve gives the cumulative product of the *X*- and *Y*-variables at any value along the x-axis. For example, the area under the curve at $X = 3$ can be read directly as 18.

Mathematical integration gives the following result for the area under the curve from 0 to 3:

$$\text{Area } A = \int_0^3 Y\,dX, \quad \text{where } Y = 2X^2;$$

$$A = \int_0^3 2X^2 dX = \tfrac{2}{3}X^3 \Big|_0^3 = 18.$$

12.11 APPLICATIONS OF GRAPHICAL INTEGRATION

Integration is commonly used in the study of the strength of materials to determine shear, moments, and deflections of beams. An example problem of this type is shown in Fig. 12.18 in which a truck exerts a total force of 36,000 lb on a beam that is used to span a portion of a bridge. The first step is to determine the resultants supporting each end of the beam.

Fig. 12.17 Graphical integration

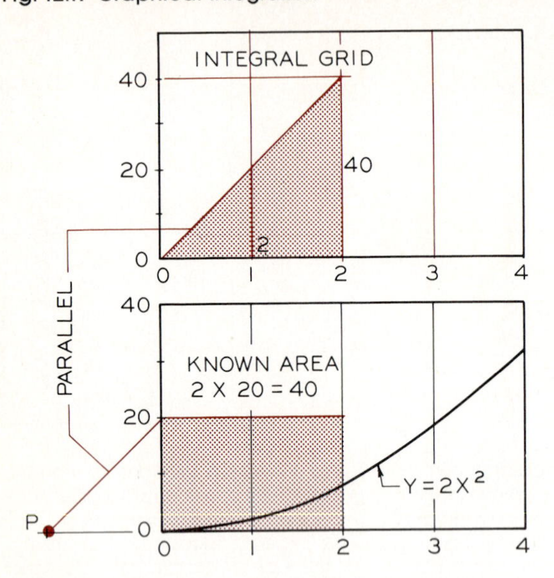

Required Plot the integral curve of the given data.

Step 1 Find the pole point, *P*, using the technique illustrated in Fig. 12.16.

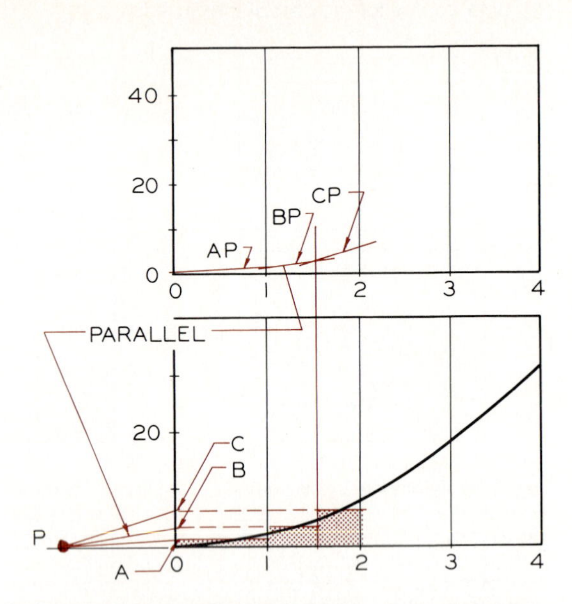

Step 2 Construct bars at intervals to approximate the areas under the curve. The interval from 1 to 2 was divided in half to improve the accuracy of the approximation. The heights of the bars are projected to the *Y*-axis and lines are drawn to the pole point. Sloping lines *AP*, *BP*, and *CP* are drawn in their respective intervals and parallel to the lines drawn to the pole, *P*.

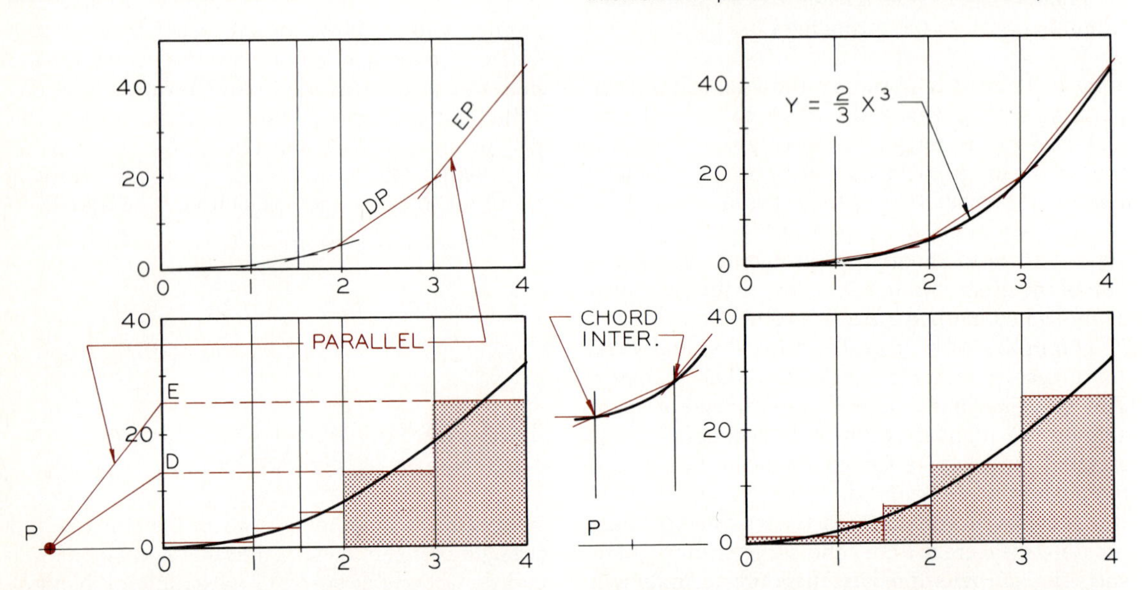

Step 3 Additional bars are drawn from 2 to 4 on the *X*-axis. The heights of the bars are projected to the *Y*-axis and rays are drawn to the pole point, *P*. Lines *DP* and *EP* are drawn in their respective intervals and parallel to their rays in the integral grid.

Step 4 The straight lines connected in the integral grid represent chords of the integral curve. Construct the integral curve to pass through the points where the chords intersect. Any ordinate value on the integral curve represents the cumulative area under the given curve from zero to that point on the *X*-axis.

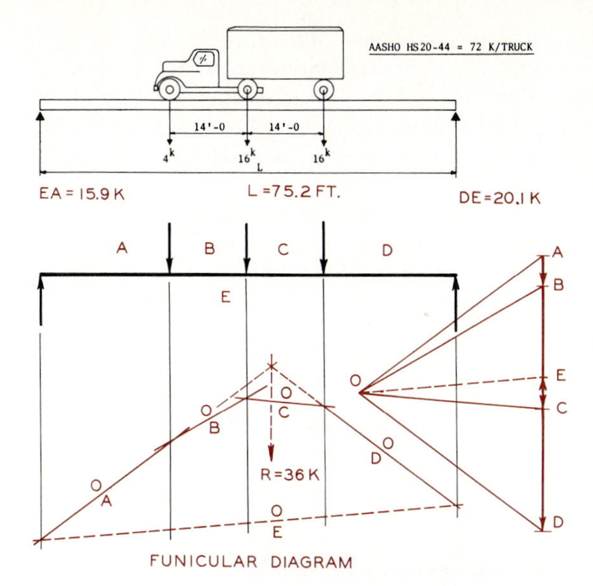

Fig. 12.18 Determination of the forces on a beam of a bridge and its total resultant.

A scale drawing of the beam is made with the loads concentrated at their respective positions. A force diagram is drawn using Bow's notational system for laying out the vectors in sequence. Pole point *O* is located, and rays are drawn from the ends of each vector to point *O*. The lines of force in the load diagram at the top of the figure are extended to the funicular diagram.

Then lines are drawn parallel to the rays between the corresponding lines of force. For example, ray *OA* is drawn in the *A*-interval in the funicular diagram. The closing ray of the funicular diagram, *OE,* is transferred to the vector diagram by drawing a parallel through point *O* to locate point *E*. Vector *DE* is the right-end resultant of 20.1 kips (one kip equals 1000 lb) and *EA* is the left-hand resultant of 15.9 kips. The origin of the resultant force of 36 kips is found by extending *OA* and *OD* in the funicular diagram to their point of intersection.

From the load diagram shown in Fig. 12.19 we can, by integration, find the shear diagram, which indicates the points in the beam where failure is most critical. Since the applied loads are concentrated rather than uniformly applied, the shear diagram will be composed of straight-line segments. In the shear diagram, the left-end resultant of 15.9 kips is drawn to scale from the axis. The first load of 4 kips, acting in a downward direction, is subtracted from this value directly over its point of application, which is projected from the load diagram. The second load of 16 kips also exerts a downward force and so is subtracted from the 11.9 kips (15.9 − 4). The third load of 16 kips is also subtracted, and the right-end resultant will bring the shear diagram back to the x-axis. It can be seen that the beam must withstand maximum shear at each support and minimum shear at the center.

The moment diagram is used to evaluate the bending characteristics of the applied loads in foot-pounds at any interval along the beam. The ordinate of any *X*-value in the moment diagram must represent the cumulative foot-pounds in the shear diagram as measured from either end of the beam.

Pole point *P* is located in the shear diagram by applying the method described in Fig. 12.16. A rectangular area of 200 ft-kips is found in the shear diagram. We estimate the total area to be less than 600 ft-kips; so we select a convenient scale that will allow an ordinate scale of 600 units for

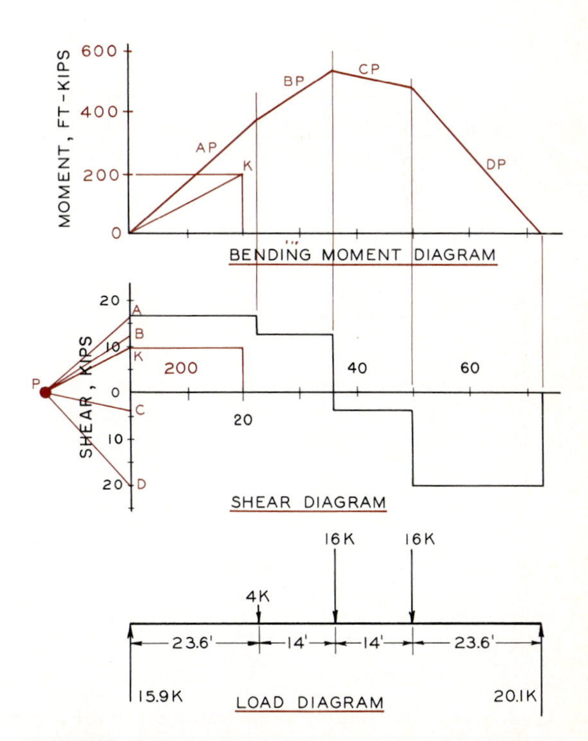

Fig. 12.19 Determination of shear and bending moment by graphical integration.

the moment diagram. We draw a known area of 200 (10 × 20) on the shear diagram. A diagonal line in the moment diagram is drawn that slopes upward from 0 to 200, where $X = 20$. The diagonal, *OK*, is transferred to the shear diagram, where it is drawn from the ordinate of the given rectangle to point P on the extension of the x-axis. Rays *AP*, *BP*, *CP*, and *DP* are found in the shear diagram by projecting horizontally from the various values of shear.

In the moment diagram, these rays are then drawn in their respective intervals to form a straight-line curve that represents the cumulative area of the shear diagram, which is in units of ft-kips. Maximum bending will occur at the center of the beam, where the shear is zero. The bending is scaled to be about 560 ft-kips. The beam selected for this span must be capable of withstanding a shear of 20.1 kips and a bending moment of 560 ft-kips.

PROBLEMS

The following problems are to be solved on Size A sheets (8½″ × 11″). Solutions involving mathematical calculations should show these calculations on separate sheets if space is not available on the sheet where the graphical solution is drawn. Legible lettering practices and principles of good layout should be followed when constructing these solutions.

Empirical Equations—Logarithmic

1. Find the equation of the data shown in the following table. The empirical data compare input voltage (*V*) with input current in amperes (*I*) to a heat pump.

y-axis	V	0.8	1.3	1.75	1.85
x-axis	I	20	30	40	45

2. Find the equation of the data in the following table. The empirical data give the relationship between peak allowable current in amperes (*I*) with the overload operating time in cycles at 60 cycles per second (*C*).

y-axis	I	2000	1840	1640	1480	1300	1200	1000
x-axis	C	1	2	5	10	20	50	100

3. Find the equation of the data in the following table. The empirical data for a low-voltage circuit breaker used on a welding machine give the maximum loading during weld in amperes (rms) for the percent of duty (pdc).

y-axis	rms	7500	5200	4400	3400	2300	1700
x-axis	pdc	3	6	9	15	30	60

4. Construct a three-cycle times three-cycle logarithmic graph to find the equation of a machine's vibration during operation. Plot vibration displacement in mills along the y-axis and vibration frequency in cycles per minute (cpm) along the x-axis. Data: 100 cpm, 0.80 mills; 400 cpm, 0.22 mills; 1000 cpm, 0.09 mills; 10,000 cpm, 0.009 mills; 50,000 cpm, 0.0017 mills.

5. Find the equation of the data in the following table that compares the velocities of air moving over a plane surface in feet per second (*v*) at different heights in inches (*y*) above the surface. Plot y-values on the y-axis.

y	v
0.1	18.8
0.2	21.0
0.3	22.6
0.4	24.1
0.6	26.0
0.8	27.3
1.2	29.2
1.6	30.6
2.4	32.4
3.2	33.7

6. Find the equation of the data in the following table that shows the distance traveled in feet

(s) at various times in seconds (t) of a test vehicle. Plot s on the y-axis and t on the x-axis.

t	s
1	15.8
2	63.3
3	146.0
4	264.0
5	420.0
6	580.0

Empirical Equations—Linear

7. Construct a linear graph to determine the equation for the yearly cost of a compressor in relationship to the compressor's size in horsepower. The yearly cost should be plotted on the y-axis and the compressor's size in horsepower on the x-axis. Data: 0 hp, $0; 50 hp, $2100; 100 hp, $4500; 150 hp, $6700; 200 hp, $9000; 250 hp, $11,400. What is the equation of these data?

8. Construct a linear graph to determine the equation for the cost of soil investigation by boring to determine the proper foundation design for varying sizes of buildings. Plot the cost of borings in dollars along the y-axis and the building area in sq ft along the x-axis. Data: 0 sq ft, $0; 25,000 sq ft, $35,000; 50,000 sq ft, $70,000; 750,000 sq ft, $100,000; 1,000,000 sq ft, $130,000.

9. Find the equation of the empirical data plotted in Fig. 12.8.

10. Plot the data in the table below on a linear graph and determine its equation. The empirical data show the deflection in centimeters of a spring (d) when it is loaded with different weights in kilograms (W). Plot W along the x-axis and d along the y-axis.

W	d
0	0.45
1	1.10
2	1.45
3	2.03
4	2.38
5	3.09

11. Plot the data in the table below on a linear graph and determine its equation. The empirical data show the temperatures that are read from a Fahrenheit thermometer at B and a centigrade thermometer at A. Plot the A-values along the x-axis and the B-values along the y-axis.

°A	°B
−6.8	20.0
6.0	43.0
16.0	60.8
32.2	90.0
52.0	125.8
76.0	169.0

Empirical Equations—Semilogarithmic

12. Construct a semilog graph of the following data to determine their equation. The y-axis should be a two-cycle log scale and the x-axis a 10-unit linear scale. Plot the voltage (E) along the y-axis and time (T) in sixteenths of a second along the x-axis to represent resistor voltage during capacitor charging. Data: 0, 10 volts; 2, 6 volts; 4, 3.6 volts; 6, 2.2 volts; 8, 1.4 volts; 10, 0.8 volts.

13. Find the equation of the data that is plotted in Fig. 12.10.

14. Construct a semilog graph of the following data to determine their equation. The y-axis should be a three-cycle log scale and the x-axis a linear scale from 0 to 250. These data give a comparison of the reduction factor, R (y-axis), with the mass thickness per square foot (x-axis) of a nuclear protection barrier. Data: 0, 1.0R; 100, 0.9R; 150, 0.028R; 200, 0.009R; 300, 0.0011R.

15. An engineering firm is considering its expansion by reviewing its past sales that are shown in the table below. Their years of operation are represented by x and N is their annual profit in tens of thousands. Plot x along the x-axis and N along the y-axis and determine the equation of their progress.

x	N
1	0.05
2	0.08
3	0.12
4	0.20
5	0.32
6	0.51
7	0.80
8	1.30
9	2.05
10	3.25

Empirical Equations—General Types

16–21. Plot the experimental data shown in Table 12.1 on the grid where the data will appear as straight-line curves. Determine the equations of the data.

Calculus—Differentiation

22. Plot the equation $Y = X^3/6$ as a rectangular graph. Graphically differentiate the curve to determine the first and second derivatives.

23. Plot the following data on a graph and find the derivative curve of the data on a graph placed below the first: $Y = 2X^2$.

24. Plot the following equation on a graph and find the derivative curve of the data on a graph placed below the first: $4Y = 8 - X^2$.

25. Plot the following data on a graph and find the derivative curve of the data on a graph placed below the first: $3Y = X^2 + 16$.

26. Plot the following data on a graph and find the derivative curve of the data on a graph placed below the first: $X = 3Y^2 - 5$.

Calculus—Integration

27. Plot the following equation on a graph and find the integral curve of the data on a graph placed above the first: $Y = X^2$.

28. Plot the following equation on a graph and find the integral curve of the data on a graph placed above the first: $Y = 9 - X^2$.

29. Plot the following equation on a graph and find the integral curve of the data on a graph placed above the first: $Y = X$.

30. Using graphical calculus, analyze a vertical strip 12″ wide on the inside face of the dam in Fig. 12.20. The force on this strip will be 52.0 lb/in. at the bottom of the dam. The first graph will be pounds per inch (ordinate) versus height in

Table 12.1

A	x	0	40	80	120	160	200	240	280			
	y	4.0	7.0	9.8	12.5	15.3	17.2	21.0	24.0			
B	x	1	2	5	10	20	50	100	200	500	1000	
	y	1.5	2.4	3.3	6.0	9.2	15.0	23.0	24.0	60	85	
C	x	1	5	10	50	100	500	1000				
	y	3	10	19	70	110	400	700				
D	x	2	4	6	8	12	14					
	y	6.5	14.0	32.0	75.0	320	710					
E	x	0	2	4	6	8	10	12	14			
	y	20	34	53	96	115	270	430	730			
F	x	0	1	2	3	4	5	6	7	8	9	10
	y	1.8	2.1	2.2	2.5	2.7	3.0	3.4	3.7	4.1	4.5	5.0

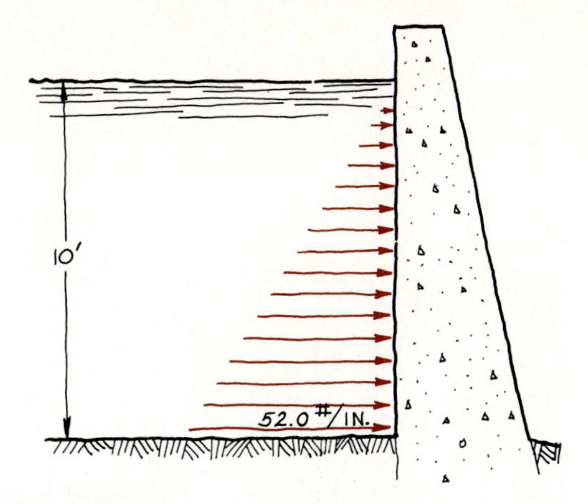

Fig. 12.20 Pressure on a 12"-wide section of a dam. (Problem 30.)

inches (abscissa). The second graph will be the integral of the first to give shear in pounds (ordinate) versus height in inches (abscissa). The third will be the integral of the second graph to give the moment in inch-pounds (ordinate) versus height in inches (abscissa). Convert these scales to give feet instead of inches.

31. A plot plan shows that a tract of land is bounded by a lake front (Fig. 12.21). By graphical integration, determine a graph that will represent the cumulative area of the land from point A to E. What is the total area? What is the area of each lot?

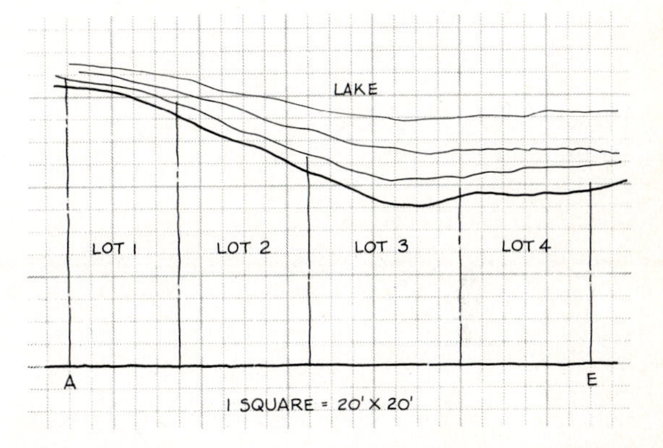

Fig. 12.21 Plot plan of a tract bounded by a lake front. (Problem 31.)

13

COMPUTER GRAPHICS

WILLIAM A. ZAGGLE
Texas A&M University

13.1 INTRODUCTION

Computers, in conjunction with graphical plotters, have already been adopted in many drafting departments. This trend is expected to increase as the price of computer equipment declines and the state of the art advances.

The first application of computer graphics where a drawing was made came about when a pen was used to replace the cutting tool in an automated milling machine. This was done to determine economically the path of the machine tool prior to cutting metal. This system soon evolved into a method of computer graphics whereby a drawing could be made by a plotter driven by a computer program.

Computer-aided manufacturing (CAM) involves computerized production machinery. Initially, this was confined to metal-cutting machines such as lathes, drills, and milling machines. Modern CAM systems make use of and control complicated robots used to assemble delicate printed circuit boards as well as heavy steel auto parts. These robots can have as many as ten arms that may weld, drill holes and tighten bolts in a sequence of steps.

Computer-aided design (CAD) is the computer-aided process of solving design problems in all areas of engineering. Computer-aided design equipment enables the designer to analyze and design a part in an accurate and rapid manner. The specifications of the design can also be stored, and then recalled for further modification and evaluation at a later date. The graphic display of the CAD system aids designers in viewing and studying their designs as they are being developed.

Computer-aided design drafting (CADD) is used to produce the final working drawings once the design has been finalized. The designer can display the drawings on a screen, called a cathode-ray tube (CRT), before the final copies of the drawings are drawn on a sheet by the plotter. Some industries use programs similar to those written for making the drawings, to drive the automated machinery in the shop and make the finished part, thereby eliminating the need for drawings made on paper.

This chapter will touch only briefly on the many applications in which the various phases of computer graphics are involved. A general introduction to programming for CAD applications will be presented to familiarize you with some of the fundamental uses of computer graphics.

13.2 APPLICATIONS OF COMPUTER GRAPHICS

A natural application of computer graphics is the design and drawing of printed circuits of the type shown in Fig. 13.1. Printed circuits are drawn up to five times full size, and are then reduced photographically. A computer-driven plotter will

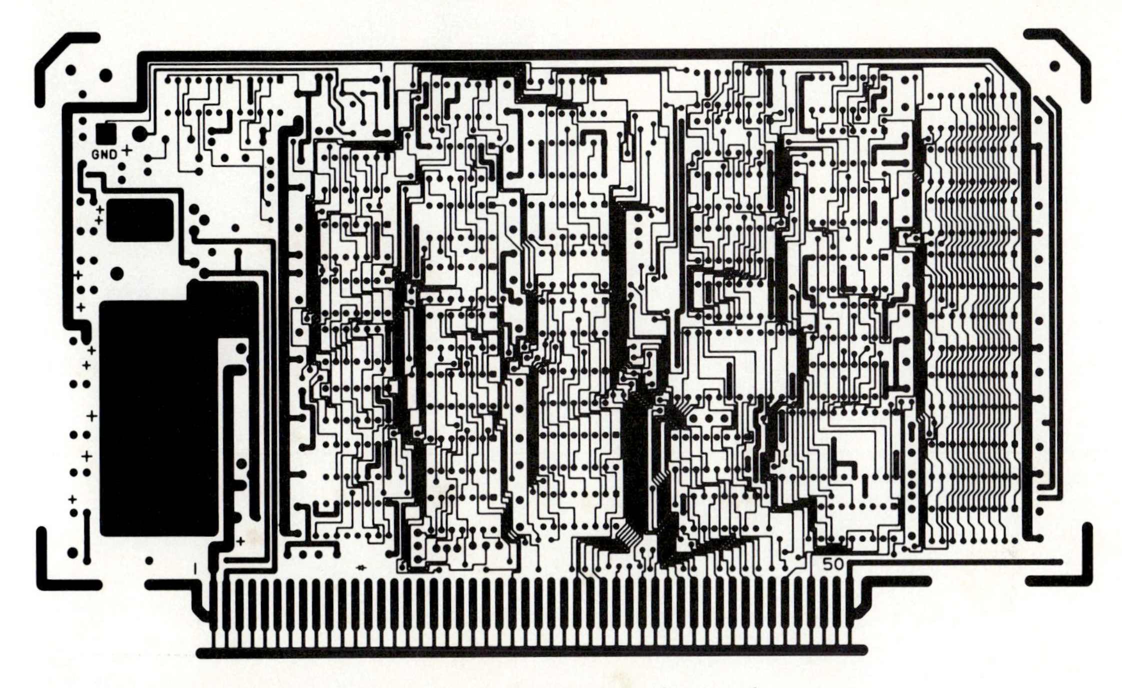

Fig. 13.1 An often-used application of computer graphics is the drawing of printed circuits for electronic systems. They can be drawn very accurately, photographically reduced, and then fabricated. (Courtesy of Computel Engineering, Inc.)

Fig. 13.2 The designer is using computer graphics to plot a piping system in color on the screen of a video display. (Courtesy of Applicon.)

draw this circuit within accuracy of approximately 0.001 in.

Piping systems can be designed and represented in both orthographic and pictorial views with all of the components shown. Once the system has been completed, the design and its specifications can be stored for rapid recall when the system needs to be reexamined. An example piping system is shown in Fig. 13.2.

Computer graphics is also used advantageously as an aid in finite element analysis, where a series of elements is used to represent an irregular three-dimensional shape. In Fig. 13.3, the designer is digitizing (assigning numerical coordinates to) the elements of a gun mount by working from a multiview drawing at his right. The mount is then displayed by the computer system as an isometric on the screen of the CRT.

Clothing patterns for a wide selection of graduated sizes (Fig. 13.4) can be drawn and cut by an application of computer graphics. The automatic cutter follows the computer-generated path to cut the most economical patterns from a section of material.

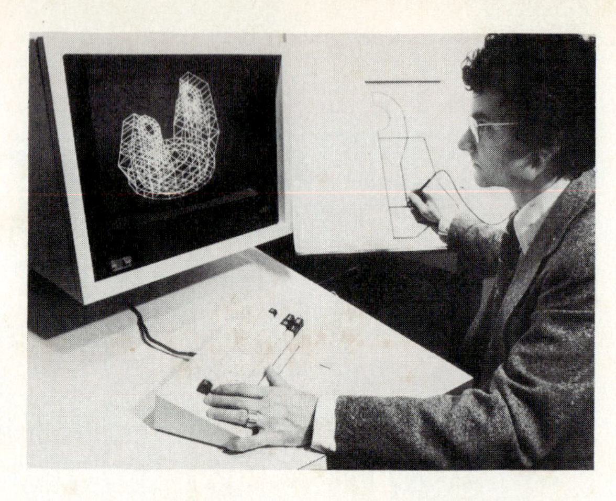

Fig. 13.3 This engineer has constructed a 3-D finite element model of a gun mount on the Applicon display by digitizing a multiview engineering drawing. (Courtesy of Applicon.)

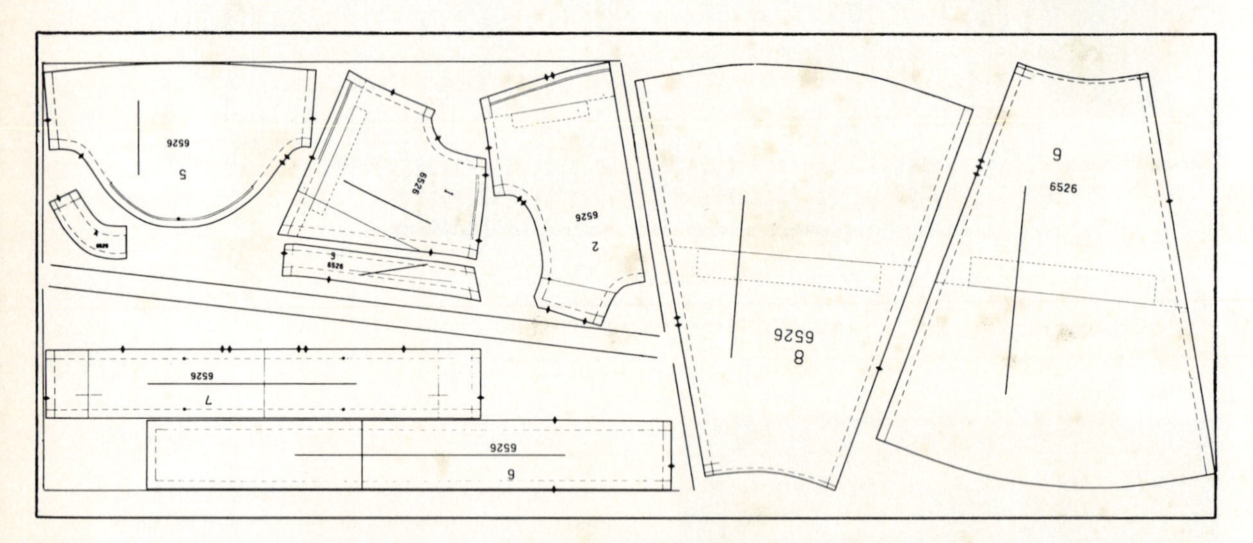

Fig. 13.4 This clothing pattern was designed, plotted and cut out by utilizing computer graphics on a Versatec plotter. Programs have been written that will plot this same pattern for graduated sizes of clothing. (Courtesy of Versatec.)

Fig. 13.5 This CalComp Graphic 7 system is a typical microcomputer system that is comprised of the keyboard, CRT, and computer. (Courtesy of California Computer Products, Inc.)

Pictorials drawn by the computer are very beneficial to the technical as well as the nontechnical person in observing a three-dimensional design. Most 3-D computer graphics systems permit the observer to view a design from any angle as it is revolved on the screen.

13.3 HARDWARE SYSTEMS

A basic computer graphics system consists of a *computer, terminal, plotter,* and *printer.* Additional devices such as *digitizers* and *light pens* may be used for direct input of graphics information (Fig. 13.5).

Computer

Computers are the devices that receive the input of the programmer, execute the programs, and then produce some form of useful output. The largest computers, called *mainframes,* are big, fast, powerful, and expensive. The smallest computers, called *microcomputers,* have recently come into widespread use for personal and small business applications. Microcomputers are much cheaper than mainframes, and they are excellent for graphical applications where massive data storage and high speed is not essential, yet low price and small size is important.

The *minicomputer* is a medium-sized computer that lies between the mainframe and the microcomputer in performance and cost.

Terminal

The terminal is the device by which the user communicates with the computer. It usually consists of a keyboard with some type of output device such as a typewriter or television screen (Fig. 13.6). Most graphics terminals use one of three types of television screens or CRTs (cathode-ray tubes). One type is **raster scanned,** which means that the picture display is being refreshed or scanned from left to right and top to bottom at a rate of about 60 times per second.

A second type is the **storage tube** where the image is drawn on the screen much like a drawing on an erasable blackboard.

The third and most powerful type of screen is **vector refreshed,** which means that each line in the picture is being continuously redrawn by the computer. The vector refresh type requires more computer power than is available from the smaller minicomputers and the microcomputers.

Fig. 13.6 An engineer is shown working at a computer graphics system's terminal where he has access to a keyboard and a display screen. (Courtesy of Computervision.)

Plotters

The plotter is the machine that is directed by the computer and the program to make a drawing. The two basic types of plotters are **flatbed** and **drum.** The flatbed plotter is a large flat bed to which the drawing paper is attached. A pen is moved about over the paper in a raised or lowered position to complete the drawing (Fig. 13.7). This type of plotter is well suited to plotting on a variety of types of surfaces in addition to paper. The drum plotter uses a special type of paper that is held on a spool and rolled over a rotating drum (Fig. 13.8). The drum rotates in two directions as the pen that is suspended above the surface of the paper moves left or right along the drum.

Printers

The printer can be thought of as a typewriter that is operated by the program and computer. Many varieties are available in a wide range of prices. The speed and type quality offered by the printer are what usually determine the purchase price. One of the most important applications of the printer in a computer system is the printing of a copy of a computer program that can be used for

Fig. 13.7 A close-up of the flat-bed plotter that uses four pens of different colors to plot the output from the computer. (Courtesy of Computervision.)

review and modification by the programmer in "hard copy" form.

Digitizer

The digitizer is a device that inputs the coordinates of points of a drawing into the computer by tracing the drawing located on a digitizer board. A pen-like stylus, connected to the computer, is often used for this purpose. Some systems using the refreshed or scanned types of CRT terminals use a *light pen* that can change or establish points on the screen. These points can be either graphical data or commands used by the computer. The light pen is often considered to be a digitizing device.

Fig. 13.8 A drum plotter used for plotting the output from the computer. (Courtesy of California Computer Products, Inc.)

13.4 USE OF THE SYSTEM

The following examples and explanations deal specifically with the computer system at Texas A&M University and may not be literally applicable to another hardware or software configuration (Fig. 13.9). However, the general principles covered throughout the remainder of this section are applicable to most systems.

The system referred to above consists of the following components.

HARDWARE:

- CPU (Central Processing Unit)—Northstar Advantage (Z-80 based) microcomputer with 64K RAM; Two 5.25″ floppy discs; 240 × 640 points raster-scan resolution 9″ × 7″ screen).
- Plotter—Houston Instrument DMP-3 HIPLOT single pen, 11″ × 8.5″ flatbed.
- Digitizer—Houston Instrument HIPAD 11″ × 8.5″ resolution to 0.005 inch.

SOFTWARE: Microsoft's MBASIC and CP/M operating system. Hardware drivers written in Z-80 machine code. (MBASIC is the trademark of Microsoft Inc., CP/M is the trademark of Digital Research Inc.)

Basic Functions

The basic functions of this system are discussed below.

TYPING IN THE PROGRAM: All typing, done on the keyboard while the computer is not executing a program, is displayed on the CRT screen by the computer, so that the user can see what is being received by the computer. One concept that should be remembered is that a line of program is accepted only if it begins with a number; otherwise it is considered to be a command and is executed immediately. (This concept will be discussed further in Section 13.5.) As illustrated in

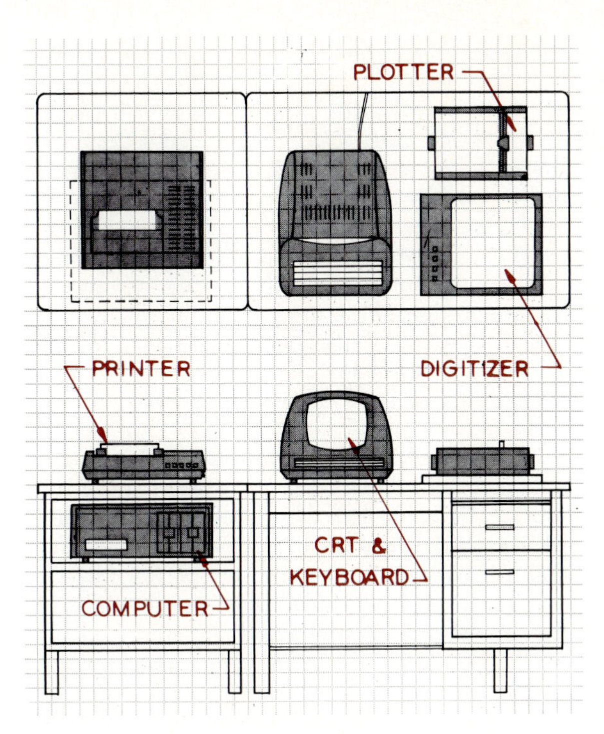

Fig. 13.9 A typical computer graphics work station used at Texas A&M University. This system consists of microcomputer, a cathode-ray tube, a plotter, and digitizer. (Courtesy of Alan Kent.)

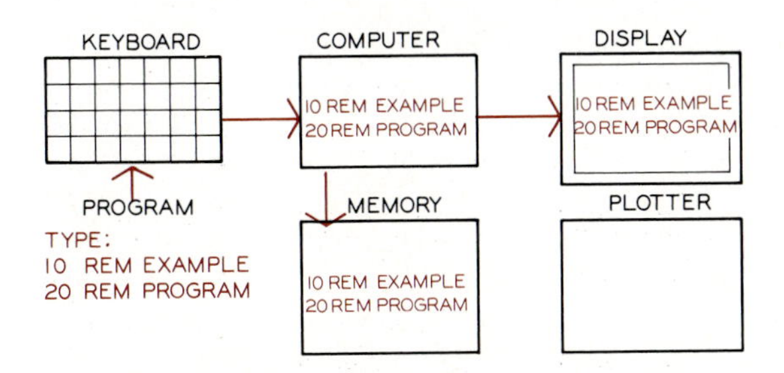

Fig. 13.10 Typing on the keyboard of a terminal feeds the program from the keyboard to the computer, and into its memory. At the same time, the input program is displayed on the screen as it is put in.

Fig. 13.10 a line entered with a line number is displayed on the screen and is also retained in the computer memory.

LISTING TO THE DISPLAY: Typing the command LIST on the keyboard (remember that commands such as this one do not have line numbers) will cause the computer to display the lines of program

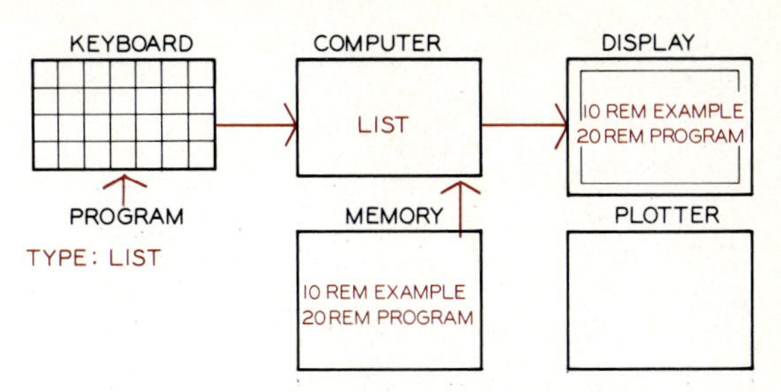

Fig. 13.11 A listing of a program on the display can be called from memory by typing in the command, LIST, on the keyboard. The program is called from memory by the computer and is then listed on the display.

it has in memory on the display screen. This concept is illustrated by Fig. 13.11.

LISTING TO THE PRINTER: Figure 13.12 shows the use of the command LLIST to cause the program listing to be made on the printer instead of on the display screen. As with the LIST command, only the lines of the program currently in memory are displayed. The extra L before the LIST command tells the computer that the output is to be sent to the printer.

GRAPHICS TO THE DISPLAY AND PLOTTER: Graphics can be drawn by the computer, either on the display or on the plotter. The plotting program determines which of these two is to receive the generated graphics, by the value of the variable P.OUT%. If the value of P.OUT% = 0, the graphics are displayed on the screen. If the value of P.OUT% = 1 the graphics are plotted on the plotter (Fig. 13.13 and Fig. 13.14).

After the RUN command is given and before the program is run by the computer, all variables

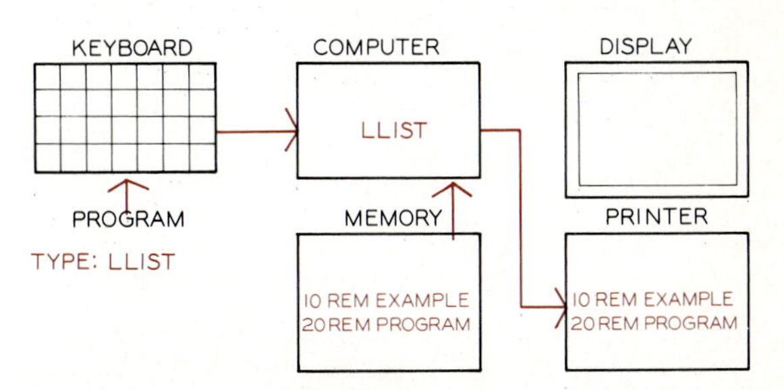

Fig. 13.12 A print of the program can be obtained by typing the command, LLIST, into the keyboard, which causes the computer to call the program from memory and then prints it out at the printer.

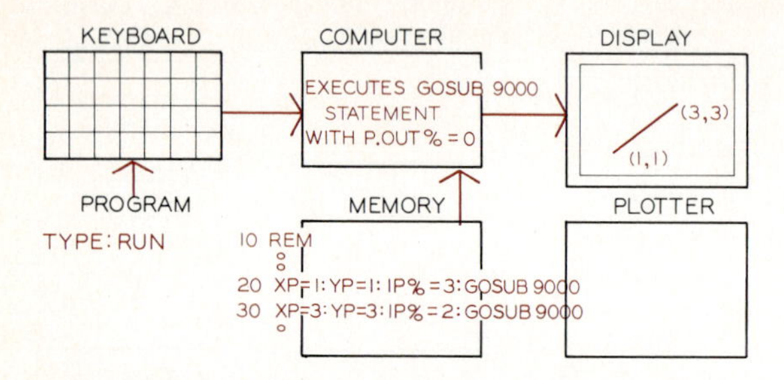

Fig. 13.13 A graphics display can be obtained at the display screen when a program has the statement P.OUT% = 0. The command RUN is typed into the keyboard, and the computer commands the display to plot the graphical output from the program.

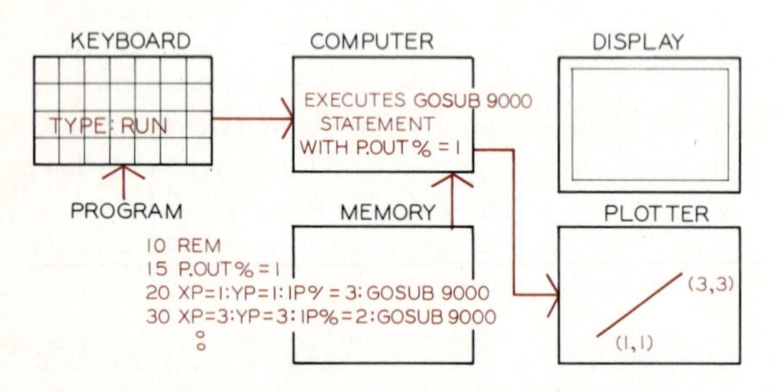

Fig. 13.14 A graphics plot can be obtained at the plotter when the program has the statement P.OUT% = 1. The command RUN is typed into the keyboard, and the computer directs the plotter to plot the graphical output.

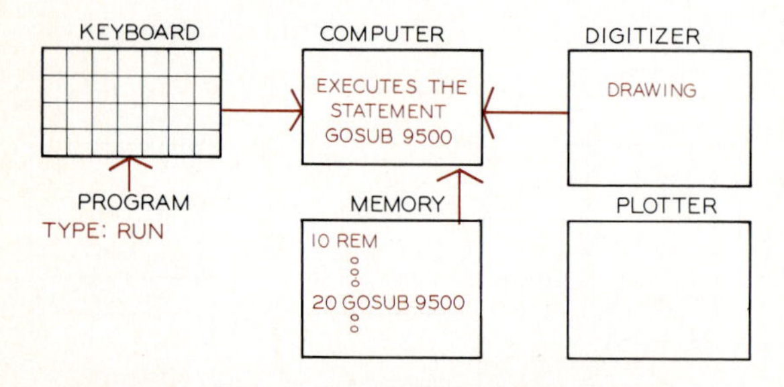

Fig. 13.15 Data from a drawing can be fed into the computer by means of a digitizer board by using a pen to digitize points on the drawing. The digitized points are transferred to the computer and this data is input into the existing program by the statement GOSUB 9500. The command RUN, input at the keyboard, runs the program.

are assigned a value of zero. This means that unless a program specifically assigns P.OUT% a value other than zero, all graphics are displayed on the screen. Once P.OUT% is set equal to one to send graphics to the plotter, it must be reassigned to the value of zero to display graphics on the screen.

The convention of using the variable, P.OUT%, to control the graphics output is dependent upon the subroutine used at the Engineering Design Graphics Department of Texas A&M University. This variable will not be compatible with all subroutines and systems.

GETTING DATA FROM THE DIGITIZER: Data is obtained from the digitizer by executing the subroutine, GOSUB 9500, which defines the variables DIGI.X, DIGI.Y, and DIGI.P%. The values of DIGI.X and DIGI.Y are the coordinates of the digitizer pen, and DIGI.P% has the value of either 2 or 3 for pen down and up commands. As shown in Fig. 13.15, the RUN command causes the computer to execute the program statements. Executing the subroutine, GOSUB 9500, causes the coordinates and pen value to be accepted from the digitizer.

13.5 BASIC PROGRAMMING RULES

The following example programs were written in a language called BASIC, which is different from but has many similarities to the FORTRAN language. Some elementary aspects of BASIC programming are covered in this article.

A *program* is a sequence of instructions and specifications that will be received by a computer, causing it to perform the operations desired by the programmer. Each statement must be given a number that is an integer between 0 and 65535. For example, a statement may be written as 30 X = X + 1, where 30 is the statement number and X = X + 1 is the statement.

Any statement that does not have a line number is interpreted as a BASIC command to be performed immediately rather than a BASIC statement to become part of a program and performed later, when the program is run.

The following rules apply to a statement that is entered after a line number:

1. If a line number of a newly entered statement is the same as a previously entered statement,

then the new line will replace the old line with the same line number.

2. If a line containing only a line number and no statement is entered, then any previously entered statement with the same line number will be deleted from the program.
3. If the line number of a newly entered statement is different from any other line in the program, the statement will be added to the program. When the statements are listed by the computer, they will be arranged in ascending order (i.e., 10, 20, 30, etc.).

Some BASIC *commands* should not be given line numbers and will be executed immediately by BASIC. These commands cause BASIC to operate on the current program or to perform some system function. Listed below are some of the commonly used commands and their functions:

```
LIST—CAUSES THE CURRENT PROGRAM
TO BE LISTED ON THE TERMINAL
DEVICE {USUALLY A CRT}.

NEW—CAUSES THE CURRENT PROGRAM
TO BE ERASED.

SAVE ''PROGRAM NAME HERE''—
SAVES THE PROGRAM UNDER THE GIVEN
NAME THAT IS ENCLOSED IN
QUOTATIONS. EXAMPLE: SAVE
''PROG1''

LOAD ''PROGRAM NAME HERE''—
TRANSFERS A PROGRAM BY NAME, IN
QUOTATIONS, FROM THE DISC INTO
MEMORY AND MAKES IT THE CURRENT
PROGRAM. EXAMPLE: LOAD
''PROG1''

KILL ''PROGRAM NAME HERE''—
REMOVES THE NAMED PROGRAM, IN
QUOTATIONS, FROM THE DISC.
EXAMPLE: KILL ''PROG1''

FILES—PRODUCES A LIST OF THE
PROGRAMS THAT ARE STORED ON A
PARTICULAR DISC.

RUN—BEGINS THE EXECUTION OF THE
CURRENT PROGRAM.
```

All BASIC programs are executed (set into operation) by the command RUN, which is typed using the keyboard of the terminal. The program will continue to run until one of the following events occurs:

1. The statement STOP is executed,
2. An invalid statement is executed, or
3. The program runs out of sequential line numbers to execute.

Variable Names Alphabetic or numeric characters up to 40 characters long, provided there are no spaces, can be used for variable names. Periods can be used effectively to separate words such as: PROG.ONE. Variable names cannot be identical to commands such as LIST, RUN, etc.

Arrays Subscripted variables can be assigned any valid variable name followed by a number in parentheses. The letter, or letter and number combination, is the name of the array and the number of parentheses locates the element within the array and is called a subscript. For example, the tenth element in array B would be B(10) and the sixth element in array F3 would be F3(6).

To allocate space in the computer's memory for an array, the DIM (dimensioning) statement is used at the beginning of a BASIC program. A statement such as 10 DIM (50), Y(50), X9(10), P(2,10), would allocate 50 elements for the array of X, 50 elements for the array Y, 10 elements for the variable X9, and 2 groups of 10 elements each for the variable P.

Remark statements are used to document a program within the program for future reference, but they have no effect on the function of the program when it is run. The REMARK statement can be shortened to REM when written in the program. You will note in example programs in the following sections that REM statements are essential to understanding the logic of the programmer as he or she writes the program.

Examples of REM statements are given below:

```
10 REM THIS IS AN EXAMPLE OF
20 REM BASIC REMARK STATEMENTS
30 REM THAT ARE IGNORED DURING
40 REM EXECUTION
50 REM OF THE BASIC PROGRAM
```

Mathematical Functions The version of BASIC that is presented in this chapter has eight mathematical functions that can be used in numerical expressions. These functions are listed below:

- ABS (numerical values here)—Returns the absolute value of the numerical expression. *Example:* ABS(3) = 3, ABS(−3) = 3, and ABS(0) = 0.
- SGN (numerical expression here)—Returns 1, 0, or −1, which indicates whether the numerical expression is positive, zero-valued, or negative, respectively. *Example:* SGN (10) = 1, SGN (0) = 0, and SGN (−3.2) = −1.
- INT (numerical expression here)—Returns the greatest integer value that is less than or equal to the value of the numerical expression. *Example:* INT (3) = 3, INT (3.9) = 3, and INT (−3.5) = −4.
- LOG (numerical expression here)—Returns an approximation of the natural logarithm of the value of the numerical expression. If LOG is called with an argument value less than or equal to zero, a program error will occur. *Example:* LOG (1) = 0, LOG (7) = 1.945901, and LOG (0.1) = −2.3025851.
- EXP (numerical expression here)—Returns an approximation to the value of the base *e* raised to the power of the numerical expression. *Example:* EXP (0) = 1, EXP (2) = 7.3890562, EXP (−2.3025851) = 0.1, and EXP (1) = 2.7182817.
- SQR (numerical expression here)—Returns an approximation of the positive square root of the numerical expression. A program error will occur if this function is called with a negative argument. *Example:* SQR (0) = 0, SQR (10) = 3.1622776, and SQR (0.3) = 0.54772256.
- SIN (numerical expression here)— Computes an approximation of the trigonometric sine of the value of the numerical expression. The answers must be given in radians rather than degrees. (Note that 2×Pi radians = 360 degrees). *Example:* SIN (0) = 0, and SIN (3.1415926/2) = 1.
- COS (numerical expression here)—COS computes an approximation of the trigonometric cosine of the value of the numerical expression, which must be specified in radians. *Example:* COS (0) = 1, and COS (3.1415926/2) = 0.
- ATN (numerical expression here)—Computes an approximation of the trigonometric arctangent function. The angular value that is returned is expressed in radians. *Example:* ATN (5) = 1.3734007, and ATN (1.7) = 1.0390722.

The BASIC language has some similarities to FORTRAN. Table 13.1 lists some of the similarities.

Table 13.1 SIMILARITIES OF BASIC TO FORTRAN

Statement type	Basic	Fortran
Unconditional Branch	GOTO line #	GO TO statement #
Conditional Branch	IF X>Y THEN statement ELSE statement	IF (X.GT.Y) statement
Subroutine call	GOSUB line #	CALL program name (arguments)
Subroutines	line # statements # statements line # RETURN	SUBROUTINE name (parameters) statements RETURN
Loops	line # FOR J = 0 TO 100 STEP 2 statements line # NEXT J	DO line # J = 0,100,2 statements line # CONTINUE

13.6 THE BASIC PLOT STATEMENT

To command the plotter to draw or move to a given point, a statement of the format shown in Fig. 13.16 is used. The X- and Y-coordinates of a point are given as XP and YP. The pen position, IP%, and the plotting subroutine, GOSUB 9000, complete the plotting commands. This statement communicates with a device, either a plotter or a CRT, in order for a point to be located and a line drawn to it.

The coordinates of XP and YP can be given at the right of the equal sign as numbers or as defined variables. For example, XP can be expressed as XP = 4.1 or as XP = A + 2. The coordinates XP = 0 and YP = 0 are customarily located at the lower left corner of the plotting area. The colons between the values of XP, YP, IP%, and GOSUB 9000 permit you to list these four statements on a single line. Colons can be used throughout any program to write more than one statement per line.

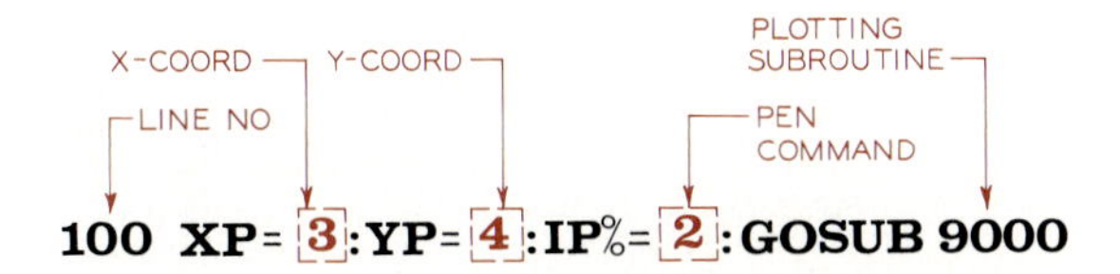

Fig. 13.16 This is the plot statement that is used in a program to command the plotter either on the screen or on the plotter. The *line number* is required on all BASIC program statements. *XP* and *YP* are the desired X- and Y-coordinates to which the pen is to move. IP% gives the desired pen position: 2 for down and 3 for up. GOSUB 9000 is the *plotting subroutine* that commands the movement of the pen.

Pen positions can be assigned IP% of 0, 2, or 3. When IP% = 2, the pen is lowered to the surface and a line is drawn. When IP% = 3, the pen is raised and a line is not drawn. When IP% = 0, the origin of XP = 0, YP = 0 is reestablished and the pen is moved to the lower left corner of the plotting surface and clears the screen.

Pen commands lift and lower the pen in much the same manner as you lift your pencil when making a drawing.

The plotting area of a 11″ × 8.5″ flatbed printer is 10″ × 7″, while the same proportional area on the CRT screen is approximately 7.14″ × 5.00″. A scale factor of 0.71 (71% of full size) is used to represent measurements on the screen that is only 71 percent of the area of the plotting surface. The scale factor of 0.71 is included within the plotting subroutine, GOSUB 9000, used in the examples that are illustrated in the following sections.

13.7 PLOTTING POINTS

An example of a point plotted on a screen is shown in Fig. 13.17. The rectangular outline represents the 7 × 5 inch screen of the CRT.

The first two statements, numbers 10 and 20, are REM (remark) statements that are used to document the program, but these statements are not executed when the program is run by the computer. The first two executable statements, lines 30 and 40, set the X- and Y-coordinates at the lower left corner of the screen with the pen in the up position.

Statement 50 moves the pen to X = 7 and Y = 5 with the pen up. Statement 60 lowers the pen by using IP% = 2, which marks the point.

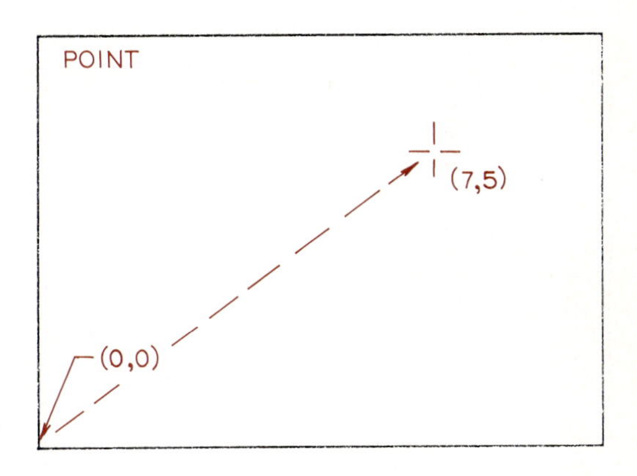


```
10 REM PROGRAM TO PLOT A POINT
20 REM POINT A ---> (X=7,Y=5)
30 IP%=0 : GOSUB 9000
40 XP=0 : YP=0 : IP%=3 : GOSUB 9000
50 XP=7 : YP=5 : IP%=3 : GOSUB 9000
60 XP=7 : YP=5 : IP%=2 : GOSUB 9000
70 XP=0 : YP=0 : IP%=0 : GOSUB 9000
80 STOP
90 END
```

Fig. 13.17 A program for plotting a point on the screen of the CRT. The information that appears in color will not appear on the screen, but it has been added for clarity.

Statement 70 moves the pen back to the origin with the pen up, and the last two statements stop and end the program.

13.8 PLOTTING LINES

A program has been written in Fig. 13.18 for drawing a line from A to B when the coordinates of each point are known. The coordinates are given in the remark statements, 100, 110, and 120.

The pen is initialized at 0,0 by statements 130 and 140. Statement 150 directs the pen to point A(6,5) with the pen up (IP% = 3). Statement 160 lowers the pen and moves it to point B at coordinates (2,3) and statement 170 moves the pen back to the point of origin.

A program has been written in Fig. 13.19 to illustrate how the line CD can be drawn on the screen, and how line AB can be drawn by the plotter. By inserting the statement in line 190, P.OUT% = 1, the program commands the computer to output the line on the plotter. When AB has been plotted, line 280 P.OUT% = 0 commands the computer to output the line CD on the display screen of the CRT. This technique will

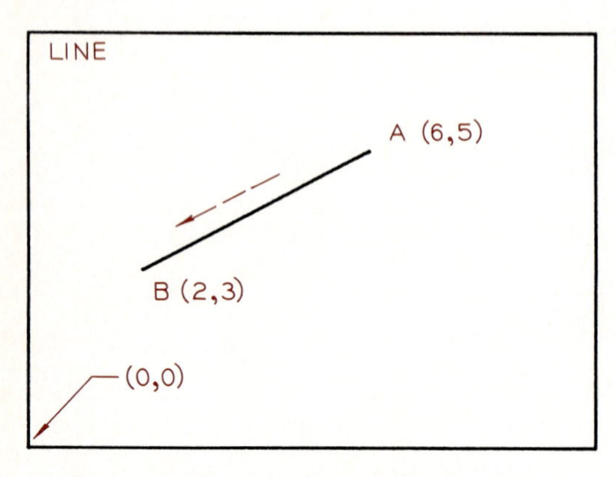


```
100 REM PORGRAM TO DRAW A LINE FROM A TO B
110 REM POINT A ---> (X=6,Y=5)
120 REM POINT B ---> (X=2,Y=3)
130 IP%=0 : GOSUB 9000
140 XP=0 : YP=0 : IP%=3 : GOSUB 9000
150 XP=6 : YP=5 : IP%=3 : GOSUB 9000
160 XP=2 : YP=3 : IP%=2 : GOSUB 9000
170 XP=0 : YP=0 : IP%=3 : GOSUB 9000
180 STOP
190 END
```

Fig. 13.18 A program and the resulting plot on the screen for drawing a line from point A to B.

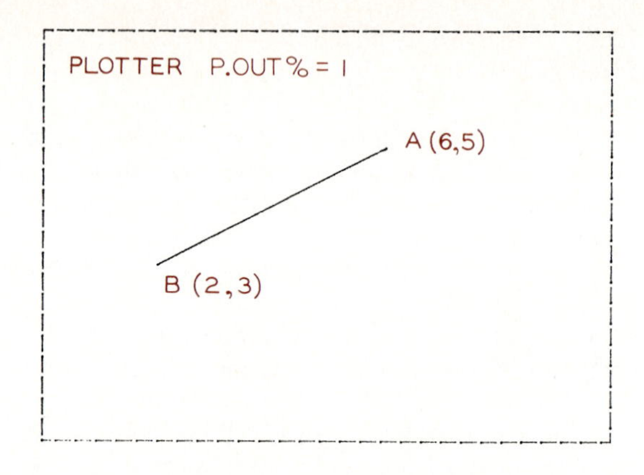

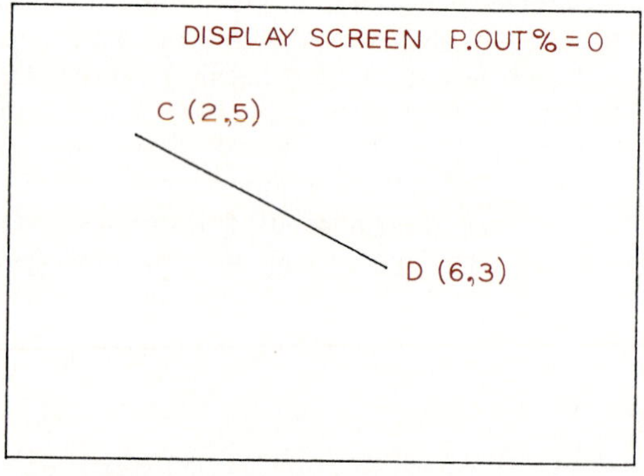


```
100 REM PORGRAM TO DRAW A LINE FROM A TO B
110 REM ON THE PLOTTER AND DRAW A LINE FROM
120 REM C TO D ON THE SCREEN.
130 REM POINT A ---> (X=6,Y=5)
140 REM POINT B ---> (X=2,Y=3)
150 REM POINT C ---> (X=2,Y=5)
160 REM POINT D ---> (X=6,Y=3)
170 REM ------------------------
180 REM SELECT PLOTTER BY DEFINING P.OUT%=1.
190 P.OUT%=1
200 REM NOW PLOT LINE
210 IP%=0 : GOSUB 9000
220 XP=0 : YP=0 : IP%=3 : GOSUB 9000
230 XP=6 : YP=5 : IP%=3 : GOSUB 9000
240 XP=2 : YP=3 : IP%=2 : GOSUB 9000
250 XP=0 : YP=0 : IP%=3 : GOSUB 9000
260 REM NOW SELECT DISPLAY SCREEN
270 REM BY REDEFINING P.OUT%=0
280 P.OUT%=0
290 IP%=0 : GOSUB 9000
300 XP=0 : YP=0 : IP%=3 : GOSUB 9000
310 XP=2 : YP=5 : IP%=3 : GOSUB 9000
320 XP=6 : YP=3 : IP%=2 : GOSUB 9000
330 XP=0 : YP=0 : IP%=3 : GOSUB 9000
340 STOP
350 END
```

Fig. 13.19 The program commands the computer to output the drawing in "hard-copy" form by using the command, 190 P.OUT% = 1. The program then directs the output back to the display screen by the command, 280 P.OUT% = 0.

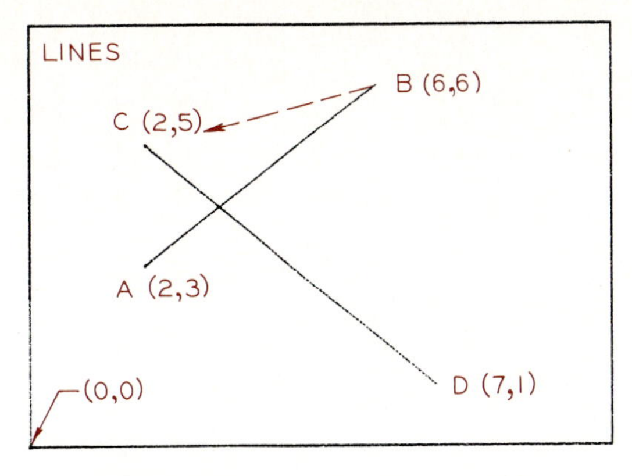


```
100 REM PROGRAM TO DRAW TWO LINES AB & CD
110 REM POINT A & B ---> XA,YA & XB,YB
120 REM POINT A & B ---> XC,YC & XD,YD
130 XA=2 : YA=3
140 XB=6 : YB=6
150 XC=2 : YC=5
160 XD=7 : YD=1
170 IP%=0 : GOSUB 9000
180 XP=0 : YP=0 : IP%=3 : GOSUB 9000
190 REM PLOT LINE AB
200 XP=XA : YP=YA : IP%=3 : GOSUB 9000
210 XP=XB : YP=YB : IP%=2 : GOSUB 9000
220 REM PLOT LINE CD
230 XP=XC : YP=YC : IP%=3 : GOSUB 9000
240 XP=XD : YP=YD : IP%=2 : GOSUB 9000
250 REM RETURN TO ORGIN
260 XP=0 : YP=0 : IP%=3 : GOSUB 9000
270 STOP
280 END
```

Fig. 13.20 A program for drawing lines AB and CD on the display screen.

work on any problem by using the two commands P.OUT% = 0 and P.OUT% = 1.

The same principles are used in the program shown in Fig. 13.20 where two lines are drawn: AB and CD. Note that instead of using numerical coordinates for the ends of the lines, variables (XA, YA, XB, etc.) are used with numerical values assigned to them as shown in statements 130 through 160. Variables XA and YB are used in the plot statement in line 200. This technique of using variable names enables you to define the variables at the beginning of the program without changing the variables in the plot statement when other values are substituted for different coordinates.

A line can be drawn using trigonometric functions when the following are given: length, angle of inclination, and the coordinates of one end.

In Fig. 13.21, end A is located at XA, YA, where XA = 2 and YA = 3 (statements 180 and 190). The 35-degree angle of inclination is converted to radians in statement 230.

The end at B is found by computing the XB-value as the cosine of the angle times the length of the line plus the X-component of point A(XA) (statement 250). The YB component is the sine of the angle times the length of the line plus the Y component of point A, (YA). These trigonometric values are substituted into the plot statements in lines 300 and 310.

13.9 PLOTTING RECTANGLES

The plotting of a rectangle is done by a program that is written in the same format as was used to plot a series of lines in Fig. 13.22. The example in Fig. 13.22 gives the coordinates of each of the four

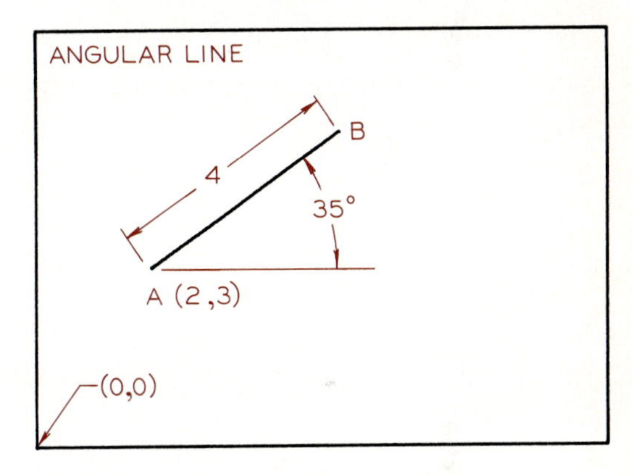


```
100 REM PROGRAM TO DRAW A LINE
110 REM GIVEN ONE END POINT, ITS LENGTH,
120 REM AND ITS ANGLE OF ROTATION
130 REM ==================================
140 REM POINT A -------------> XA,YA
150 REM POINT B -------------> (COMPUTED)
160 REM LENGTH --------------> LENGTH (INCHES)
170 REM ANGLE OF ROTATION ---> ANGLE (DEGREES)
180 XA = 2
190 YA = 3
200 LENGTH = 4
210 ANGLE = 35
220 REM CONVERT ANGLE TO RADIANS
230 ANGLE = ANGLE * 3.1416 / 180
240 REM COMPUTE THE COORDINATES OF POINT B
250 XB = XA + LENGTH * COS(ANGLE)
260 YB = YA + LENGTH * SIN(ANGLE)
270 REM PLOT THE LINE FROM A TO B
280 IP%=0 : GOSUB 9000
290 XP=0 : YP=0 : IP%=3 : GOSUB 9000
300 XP=XA : YP=YA : IP%=3 : GOSUB 9000
310 XP=XB : YP=YB : IP%=2 : GOSUB 9000
320 XP=0 : YP=0 : IP%=3 : GOSUB 9000
330 STOP
340 END
```

Fig. 13.21 A line can be drawn on the display screen by using its trigonometric functions when its length, angle of inclination, and one point on it are known.

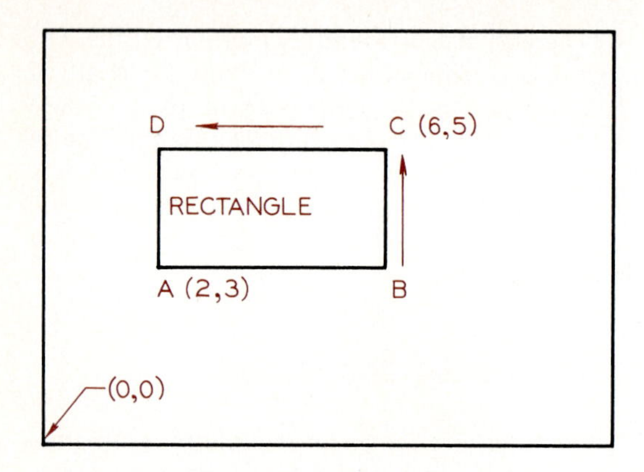


```
100 REM PROGRAM TO DRAW A RECTANGLE
110 REM GIVEN COORDINATES OF TWO CORNERS
120 REM POINT A AND POINT C
130 REM ================================
140 REM POINT A ---> XA,YA
150 REM POINT B ---> XC,YA
160 REM POINT C ---> XC,YC
170 REM POINT D ---> XA,YC
180 XA = 2 : YA = 3
190 XC = 6 : YC = 5
200 IP%=0 : GOSUB 9000
210 XP=0 : YP=0 : IP%=3 : GOSUB 9000
220 REM PLOT TO POINT A
230 XP=XA : YP=YA : IP%=3 : GOSUB 9000
240 REM PLOT TO POINT B
250 XP=XC : YP=YA : IP%=2 : GOSUB 9000
260 REM PLOT TO POINT C
270 XP=XC : YP=YC : IP%=2 : GOSUB 9000
280 REM PLOT TO POINT D
290 XP=XA : YP=YC : IP%=2 : GOSUB 9000
300 REM PLOT TO POINT A
310 XP=XA : YP=YA : IP%=2 : GOSUB 9000
320 REM RETURN TO ORIGIN
330 XP=0 : YP=0 : IP%=3 : GOSUB 9000
340 STOP
350 END
```

Fig. 13.22 A program and a display of its output for drawing a rectangle when the coordinates of each corner are known.

corners of a rectangle in terms of variables XA, YA, XC, and YC in lines 180 and 190.

After the pen is initialized at 0, 0, it is directed to point A (XA, YA) with the pen up in line 230. Statements 250, 270, 290, and 310 successively move the pen to points B, C, D, and back to A. You can see that a variety of different rectangles can be drawn by changing the numerical values of the variables in statements 180 and 190.

Inclined rectangles can be constructed when the following are known: The coordinates of one corner, the height and width, and the angle of inclination from 0 to 360 degrees. In the example in Fig. 13.23, these known values are given in statements 190 through 220.

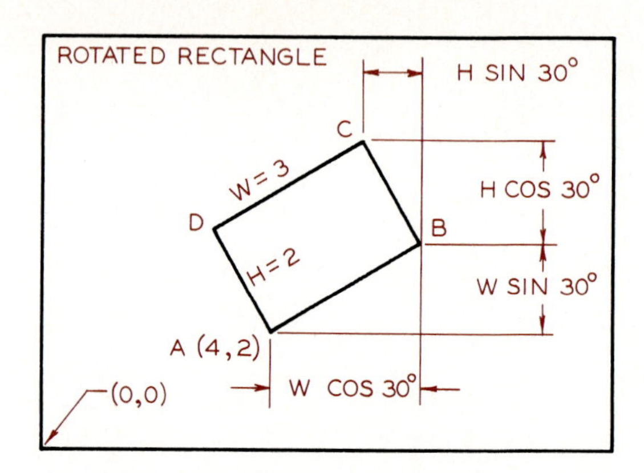


```
100 REM PROGRAM TO DRAW A RECTANGLE
110 REM GIVEN THE COORDINATE OF THE POINT A,
120 REM THE HEIGHT, THE WIDTH, AND THE ROTATION
130 REM ABOUT THE POINT A IN DEGREES
140 REM ========================================
150 REM POINT A ---------------------> XA,YA
160 REM HEIGHT ----------------------> R.HEIGHT
170 REM WIDTH -----------------------> R.WIDTH
180 REM ROTATION ABOUT POINT A -----> R.ANGLE
190 XA = 4 : YA = 2
200 R.HEIGHT = 2
210 R.WIDTH = 3
220 R.ANGLE = 30
230 REM CONVERT R.ANGLE TO RADIANS
240 R.ANGLE = R.ANGLE * 3.1416 / 180
250 REM COMPUTE POINT B
260 XB = XA + R.WIDTH * COS(R.ANGLE)
270 YB = YA + R.WIDTH * SIN(R.ANGLE)
280 REM COMPUTE POINT C
290 XC = XB - R.HEIGHT * SIN(R.ANGLE)
300 YC = YB + R.HEIGHT * COS(R.ANGLE)
310 REM COMPUTE POINT D
320 XD = XA - R.HEIGHT * SIN(R.ANGLE)
330 YD = YA + R.HEIGHT * COS(R.ANGLE)
340 IP%=0 : GOSUB 9000
350 XP=0 : YP=0 : IP%=3 : GOSUB 9000
360 REM PLOT TO POINT A
370 XP=XA : YP=YA : IP%=3 : GOSUB 9000
380 REM PLOT TO POINT B
390 XP=XB : YP=YB : IP%=2 : GOSUB 9000
400 REM PLOT TO POINT C
410 XP=XC : YP=YC : IP%=2 : GOSUB 9000
420 REM PLOT TO POINT D
430 XP=XD : YP=YD : IP%=2 : GOSUB 9000
440 REM PLOT TO POINT A
450 XP=XA : YP=YA : IP%=2 : GOSUB 9000
460 REM RETURN TO ORIGIN
470 XP=0 : YP=0 : IP%=3 : GOSUB 9000
480 STOP
490 END
```

Fig. 13.23 A program and a display of its output for drawing a rectangle by expressing its dimensions in variables that can be easily changed for other variations of size and rotation.

The angle is converted to radians in line 240. The coordinates at corner B are XB and YB, and these are computed in lines 260 and 270 as the cosine and sine of the 30-degree angle multiplied by the width of the rectangle. Similarly, the coordinates of corner C are found using the sine and cosine of the height of the rectangle in statements 290 and 300. The coordinates of corner D are defined in lines 320 and 330.

The variables for each corner are sequentially inserted into the plot statement in statements 370 through 450 to drive the pen to each of the four corners. The use of variables makes it possible to change any or all of the variables in statements 190 through 220 and thereby obtain a different rectangle with a minimum of programming modification.

13.10 PLOTTING POLYGONS

A regular polygon, having a number of sides each of the same length, can be constructed using the program shown in Fig. 13.24 when the following are given: The coordinates of the center, the distance from the center to each corner (RADIUS), and the number of sides (NSIDES). This information is given in lines 170 through 190.

The coordinates of corner 1 are found in statements 260 and 270 where distance RADIUS is multiplied by the sine and cosine of the angle between each corner point and each is added to the coordinates of the center of the polygon. An IF statement is used in line 290 that tells the plotter to lift the pen when the angle is at zero degrees (horizontal) and to move to the first calculated point. A FOR NEXT loop is used between statements 220 and 310 where the steps of angle ANGLE are specified to be 360/NSIDES, which is 30° in this example.

The values of each set of X- and Y-coordinates are computed and plotted for each successive 30° angle through 360° and back to the point of beginning where the pen is then lifted and the loop is ended. The pen is then directed by statement 320 to return to the origin, 0, 0.

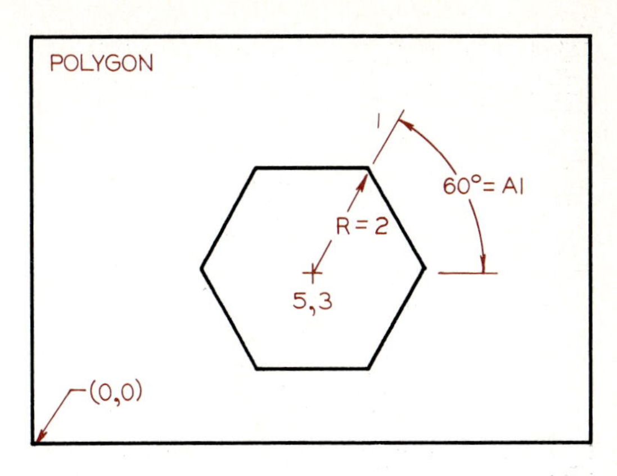


```
100 REM PROGRAM TO DRAW A POLYGON
110 REM GIVEN CENTER, RADIUS,
120 REM AND NUMBER OF SIDES.
130 REM =========================
140 REM CENTER ---> XCENTER , YCENTER
150 REM RADIUS ---> RADIUS
160 REM #SIDES ---> NSIDES
170 XCENTER=5 : YCENTER=3
180 RADIUS = 2
190 NSIDES = 6
200 IP%=0 : GOSUB 9000
210 XP=0 : YP=0 : IP%=3 : GOSUB 9000
220 FOR ANGLE = 0 TO 360 STEP 360/NSIDES
230    REM CONVERT ANGLE TO RADIANS
240    ANGLE.R = ANGLE * 3.1416 / 180
250    REM COMPUTE COORDINATES OF POINT X,Y
260    XP = XCENTER + RADIUS * COS(ANGLE.R)
270    YP = YCENTER + RADIUS * SIN(ANGLE.R)
280    REM MOVE PEN UP IF ANGLE = 0
290    IF ANGLE = 0 THEN IP%=3 ELSE IP%=2
300    GOSUB 9000
310 NEXT ANGLE
320 XP=0 : YP=0 : IP%=3 : GOSUB 9000
330 STOP
340 END
```

Fig. 13.24 A program and its resulting display for drawing regular polygons.

13.11 CIRCLES

The program for drawing a circle is identical to the one used to draw regular polygons. The only exception is the smallness of the size of the steps of the angle from corner to corner of the polygon. The circle is really no more than a polygon with many small sides.

Like the polygon, the following must be given: The coordinates of the center (XCENTER and YCENTER) and the radius R, as given in statements 170 and 180 of the program in Fig. 13.25. The angular steps are given in the FOR NEXT loop as 5° increments and the resulting polygon approximates a circle.

As the circle is drawn larger, the angular steps can be specified in smaller increments. Small circles can be drawn with slightly larger increments, and therefore plotting time can be reduced without affecting the appearance of the plotted circle.

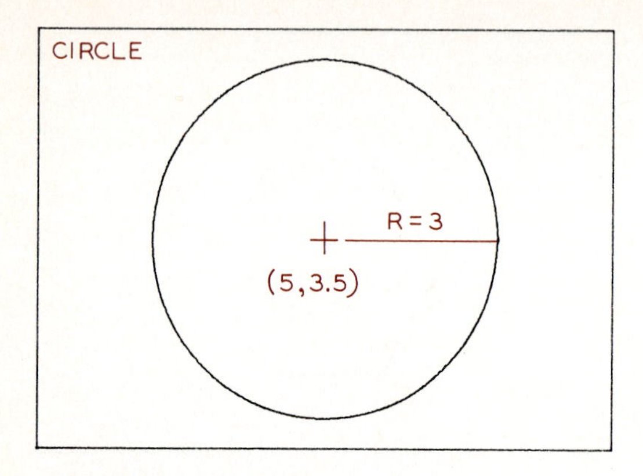


```
100 REM PROGRAM TO DRAW A CIRCLE
110 REM GIVEN THE COORDINATES OF THE
120 REM CENTER AND THE LENGTH OF THE
125 REM RADIUS
130 REM =========================
140 REM CENTER ---> XCENTER , YCENTER
150 REM RADIUS ---> RADIUS
170 XCENTER=5 : YCENTER=3.5
180 RADIUS = 3
190 IP%=0 : GOSUB 9000
200 FOR ANGLE = 0 TO 360 STEP 5
210   REM CONVERT ANGLE TO RADIANS
220   ANGLE.R = ANGLE * 3.1416 / 180
230   REM COMPUTE COORDINATES OF POINT X,Y
240   XP = XCENTER + RADIUS * COS(ANGLE.R)
250   YP = YCENTER + RADIUS * SIN(ANGLE.R)
260   REM MOVE PEN UP IF ANGLE = 0
270   IF ANGLE = 0 THEN IP%=3 ELSE IP%=2
280   GOSUB 9000
290 NEXT ANGLE
300 XP=0 : YP=0 : IP%=3 : GOSUB 9000
310 STOP
320 END
```

Fig. 13.25 A program for drawing a circle is the same one used to draw regular polygons with the only difference being the number of sides that are plotted.

13.12 PROGRAMS WITH DATA STATEMENTS

An alternative method of programming the coordinates of points is the uses of DATA statements as given in lines 160 through 200 of Fig. 13.26. The three numbers in each DATA statement represent X, Y, and pen values for each point of a triangle, and the coordinates are listed in this order for each point.

A READ statement is given in lines 260, 280, and 290. Once read, the values are transferred to the plot statement in line 310 where the values are executed. Statement 320 sends the program back to line 260 where the second set of data is read. This loop continues until the data in line 210 is encountered and the program is then ended.

13.13 SUBROUTINES

A main program can be written that will recall previously written programs, called **subroutines.** Subroutines are loaded from the disc and appended to the main program, using the MERGE, rather than the LOAD command.

An example main program and subroutine 600 is used to draw a rectangle in Fig. 13.27. The main program gives the following specifics: The initial point, the coordinates of a corner (XA and YA), the height and width (R.HEIGHT and R.WIDTH), and the angle of inclination (R.ANGLE). The subroutine is called by statement 220 GOSUB 600.

The variable values given in the main program are transferred to the subroutine where the

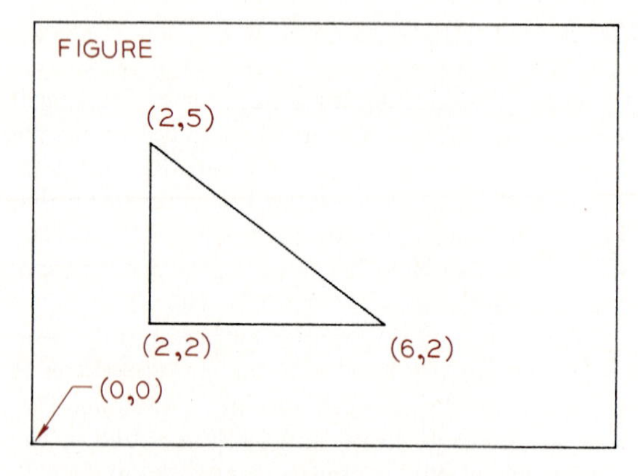


```
100 REM PROGRAM TO DRAW A FIGURE
110 REM GIVEN X,Y AND PEN COORDINATES
120 REM IN "DATA" STATEMENTS
130 REM ==============================
140 REM LAST DATA VALUE IS ALWAYS 999
150 REM DEFINE DATA
160 DATA 2,2,3
170 DATA 6,2,2
180 DATA 2,5,2
190 DATA 2,2,2
200 DATA 0,0,3
210 DATA 999
220 REM PLOT DATA
230 IP%=0 : GOSUB 9000
240 XP=0 : YP=0 : IP%=3 : GOSUB 9000
250 REM ----- READ AND PLOT LOOP -----
260 READ XP
270 IF XP=999 THEN STOP
280 READ YP
290 READ IP%
300 REM PLOT POINT
310 GOSUB 9000
320 GOTO 260
330 REM ----- END OF LOOP -----
340 END
```

Fig. 13.26 This program utilizes points of a triangle that are fed to the program by DATA statements as shown in lines 160 through 210.

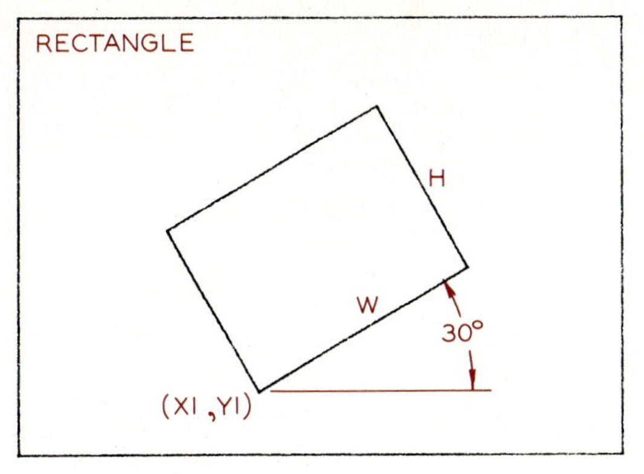


```
100 REM --- MAIN ROUTINE ---
110 REM DRAWS A RECTANGLE BY CALLING THE
120 REM RECTANGLE SUBROUTINE AT LINE 600
130 REM ==== STEP 1 ====
140 IP%=0 : GOSUB 9000
150 XP=0 : YP=0 : IP%=3 : GOSUB 9000
160 REM ==== STEP 2 ====
170 REM DEFINE VARIABLES NEEDED BY THE
180 REM RECTANGLE SUBROUTINE
190 XA = 4 : YA = 1 : R.HEIGHT = 3
200 R.WIDTH = 4 : R.ANGLE = 30
210 REM CALL THE RECTANGLE SUBROUTINE
220 GOSUB 600
230 STOP
240 END
```

```
600 REM PROGRAM TO DRAW A RECTANGLE
610 REM GIVEN THE COORDINATE OF THE POINT A,
620 REM THE HEIGHT, THE WIDTH, AND THE ROTATION
630 REM ABOUT THE POINT A IN DEGREES
640 REM ========================================
650 REM POINT A ---------------------> XA,YA
660 REM HEIGHT ----------------------> R.HEIGHT
670 REM WIDTH -----------------------> R.WIDTH
680 REM ROTATION ABOUT POINT A -----> R.ANGLE
690 REM CONVERT R.ANGLE TO RADIANS
700 R.ANGLE = R.ANGLE * 3.1416 / 180
710 REM COMPUTE POINT B
720 XB = XA + R.WIDTH * COS(R.ANGLE)
730 YB = YA + R.WIDTH * SIN(R.ANGLE)
740 REM COMPUTE POINT C
750 XC = XB - R.HEIGHT * SIN(R.ANGLE)
760 YC = YB + R.HEIGHT * COS(R.ANGLE)
770 REM COMPUTE POINT D
780 XD = XA - R.HEIGHT * SIN(R.ANGLE)
790 YD = YA + R.HEIGHT * COS(R.ANGLE)
800 REM PLOT TO POINT A
810 XP=XA : YP=YA : IP%=3 : GOSUB 9000
820 REM PLOT TO POINT B
830 XP=XB : YP=YB : IP%=2 : GOSUB 9000
840 REM PLOT TO POINT C
850 XP=XC : YP=YC : IP%=2 : GOSUB 9000
860 REM PLOT TO POINT D
870 XP=XD : YP=YD : IP%=2 : GOSUB 9000
880 REM PLOT TO POINT A
890 XP=XA : YP=YA : IP%=2 : GOSUB 9000
900 REM RETURN TO ORIGIN
910 RETURN
920 END
```

Fig. 13.27 A subroutine is used in conjunction with a main program for drawing this rectangle. The subroutine 600 is appended to the main program, and control is then returned to the main program once the subroutine has been utilized.

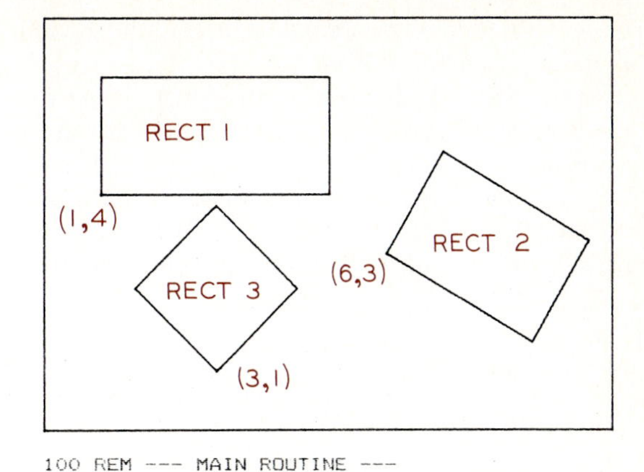


```
100 REM --- MAIN ROUTINE ---
110 REM DRAWS 3 RECTANGLES BY CALLING THE
120 REM RECTANGLE SUBROUTINE AT LINE 600
130 IP%=0 : GOSUB 9000
140 XP=0 : YP=0 : IP%=3 : GOSUB 9000
150 REM DRAW RECTANGLE 1
160 XA=1 : YA=4 : R.HEIGHT=2 : R.WIDTH=4
170 R.ANGLE=0 : GOSUB 600
180 REM DRAW RECTANGLE 2
190 XA=6 : YA=3 : R.HEIGHT=2 : R.WIDTH=3
200 R.ANGLE=-30 : GOSUB 600
210 REM DRAW RECTANGLE 3
220 XA=3 : YA=1 : R.HEIGHT=2 : R.WIDTH=2
230 R.ANGLE=45 : GOSUB 600
240 REM RETURN TO ORIGIN
250 XP=0 : YP=0 : IP%=3 : GOSUB 9000
260 END
```

```
600 REM PROGRAM TO DRAW A RECTANGLE
610 REM GIVEN THE COORDINATE OF THE POINT A,
620 REM THE HEIGHT, THE WIDTH, AND THE ROTATION
630 REM ABOUT THE POINT A IN DEGREES
640 REM ========================================
650 REM POINT A ---------------------> XA,YA
660 REM HEIGHT ----------------------> R.HEIGHT
670 REM WIDTH -----------------------> R.WIDTH
680 REM ROTATION ABOUT POINT A -----> R.ANGLE
690 REM CONVERT R.ANGLE TO RADIANS
700 R.ANGLE = R.ANGLE * 3.1416 / 180
710 REM COMPUTE POINT B
720 XB = XA + R.WIDTH * COS(R.ANGLE)
730 YB = YA + R.WIDTH * SIN(R.ANGLE)
740 REM COMPUTE POINT C
750 XC = XB - R.HEIGHT * SIN(R.ANGLE)
760 YC = YB + R.HEIGHT * COS(R.ANGLE)
770 REM COMPUTE POINT D
780 XD = XA - R.HEIGHT * SIN(R.ANGLE)
790 YD = YA + R.HEIGHT * COS(R.ANGLE)
800 REM PLOT TO POINT A
810 XP=XA : YP=YA : IP%=3 : GOSUB 9000
820 REM PLOT TO POINT B
830 XP=XB : YP=YB : IP%=2 : GOSUB 9000
840 REM PLOT TO POINT C
850 XP=XC : YP=YC : IP%=2 : GOSUB 9000
860 REM PLOT TO POINT D
870 XP=XD : YP=YD : IP%=2 : GOSUB 9000
880 REM PLOT TO POINT A
890 XP=XA : YP=YA : IP%=2 : GOSUB 9000
900 REM RETURN TO ORIGIN
910 RETURN
920 END
```

Fig. 13.28 In this plot, you can see that the main program has called subroutine 600 (GOSUB 600) three times for drawing three rectangles. The dimensions of the rectangles are given in the main program.

other values of the rectangle are calculated and plotted. When the subroutine has been executed, it returns to the main program (line 910) where the main program resumes its control over the next steps of the program.

A subroutine can be called several times and in combination with other subroutines by the main program. Fig. 13.28 is an example of subrou-

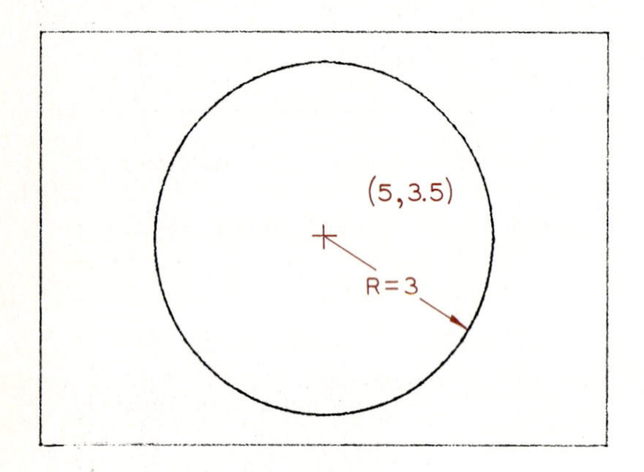


```
100 REM --- MAIN PROGRAM ---
110 REM PROGRAM TO DRAW A CIRCLE BY
120 REM CALLING THE CIRCLE SUBROUTINE
130 REM AT LINE 800.
140 REM ==== STEP 1 ====
150 REM INITIALIZE
160 IP%=0 : GOSUB 9000
170 REM ==== STEP 2 ====
180 REM DEFINE THE VARIABLES NEEDED BY
190 REM THE CIRCLE SUBROUTINE
200 XCENTER=5 : YCENTER=3.5 : RADIUS=3
210 REM ==== STEP 3 ====
220 REM CALL THE CIRCLE SUBROUTINE
230 GOSUB 800
240 STOP
250 END


800 REM SUBROUTINE TO DRAW A CIRCLE
810 REM GIVEN THE COORDINATES OF THE
820 REM CENTER AND THE LENGTH OF THE
830 REM RADIUS
840 REM =========================
850 REM CENTER ---> XCENTER , YCENTER
860 REM RADIUS ---> RADIUS
870 FOR ANGLE = 0 TO 360 STEP 5
880    REM CONVERT ANGLE TO RADIANS
890    ANGLE.R = ANGLE * 3.1416 / 180
900    REM COMPUTE COORDINATES OF POINT X,Y
910    XP = XCENTER + RADIUS * COS(ANGLE.R)
920    YP = YCENTER + RADIUS * SIN(ANGLE.R)
930    IF ANGLE = 0 THEN IP%=3 ELSE IP%=2
940    GOSUB 9000
950 NEXT ANGLE
960 XP=0 : YP=0 : IP%=3 : GOSUB 9000
970 RETURN
980 END
```

SUBROUTINE

Fig. 13.29 Subroutine 800 is called by the main program to plot a circle. Control is then returned to the main program.

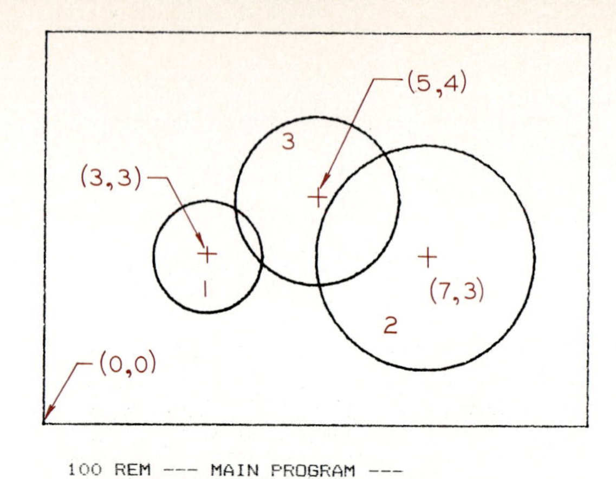


```
100 REM --- MAIN PROGRAM ---
110 REM PROGRAM TO DRAW 3 CIRCLES BY
120 REM CALLING THE CIRCLE SUBROUTINE
130 REM AT LINE 800.
140 IP%=0 : GOSUB 9000
150 REM DRAW CIRCLE 1
160 XCENTER=3 : YCENTER=3
170 RADIUS=1 : GOSUB 800
180 REM DRAW CIRCLE 2
190 XCENTER=7 : YCENTER=3
200 RADIUS=2 : GOSUB 800
210 REM DRAW CIRCLE 3
220 XCENTER=5 : YCENTER=4
230 RADIUS=1.5 : GOSUB 800
240 REM RETURN TO ORIGIN
250 XP=0 : YP=0 : IP%=3 : GOSUB 9000
260 END


800 REM SUBROUTINE TO DRAW A CIRCLE
810 REM GIVEN THE COORDINATES OF THE
820 REM CENTER AND THE LENGTH OF THE
830 REM RADIUS
840 REM =========================
850 REM CENTER ---> XCENTER , YCENTER
860 REM RADIUS ---> RADIUS
870 FOR ANGLE = 0 TO 360 STEP 5
880    REM CONVERT ANGLE TO RADIANS
890    ANGLE.R = ANGLE * 3.1416 / 180
900    REM COMPUTE COORDINATES OF POINT X,Y
910    XP = XCENTER + RADIUS * COS(ANGLE.R)
920    YP = YCENTER + RADIUS * SIN(ANGLE.R)
930    IF ANGLE = 0 THEN IP%=3 ELSE IP%=2
940    GOSUB 9000
950 NEXT ANGLE
960 XP=0 : YP=0 : IP%=3 : GOSUB 9000
970 RETURN
980 END
```

Fig. 13.30 Subroutine 800 (circle) is called three times by the main program to draw the different circles.

tine 600 that is called three times for drawing different rectangles, 1, 2, and 3.

Rectangle 1 is drawn by using subroutine 600 and the command statement in lines 160 and 170 of the main program. The subroutine returns to the main program where new data is given and the subroutine is called once more. Finally, rectangle 3 is drawn using the data given in lines 220 and 230 of the main program and subroutine 600.

Circle subroutines can be called in the same manner as rectangular subroutines. In Fig. 13.29 a

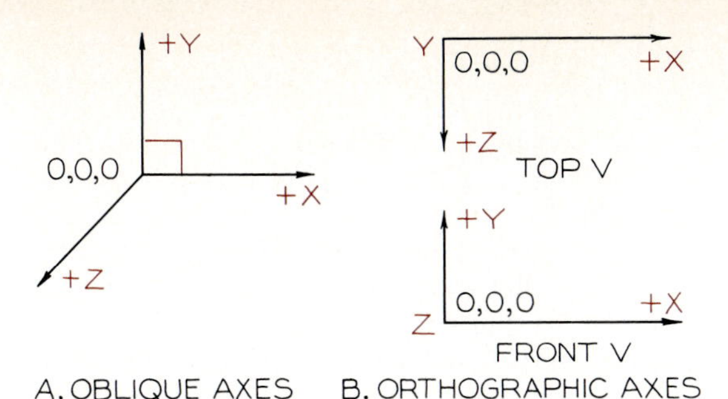

Fig. 13.31 The axes of an oblique drawing are shown in part A. The same axes are shown orthographically in part B where (0, 0, 0) is the origin.

main program is used to define the values of the coordinates of the center and the radius of the circle (line 200). Line 230 calls the subroutine by the statement GOSUB 800.

The circle subroutine is called three times in Fig. 13.30 to draw three different circles. The main program gives the data for each circle and calls the subroutine in lines 160 through 230. Circles 1, 2, and 3 are drawn each time the subroutine 800 is called. The subroutine returns its control to the main program after the third circle has been drawn.

Subroutines of various types for drawing different geometric shapes can be accessed. However, they must be listed in ascending order where the subroutine with the largest number is listed last.

13.14 OBLIQUES

The program for plotting an oblique pictorial can be written by using X-, Y-, and Z-axes as shown in Fig. 13.31, with the origin at the rear of the position of the oblique. Three points are shown plot-

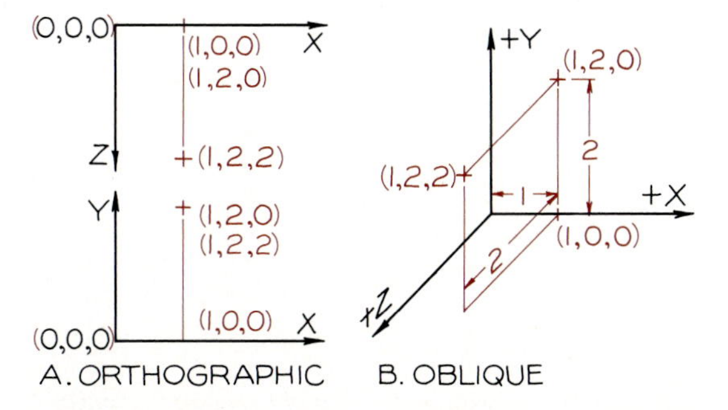

Fig. 13.32 The orthographic coordinates of three points are shown on the X-, X-, and Z-axes at A. The points shown at A are shown plotted in oblique at B.

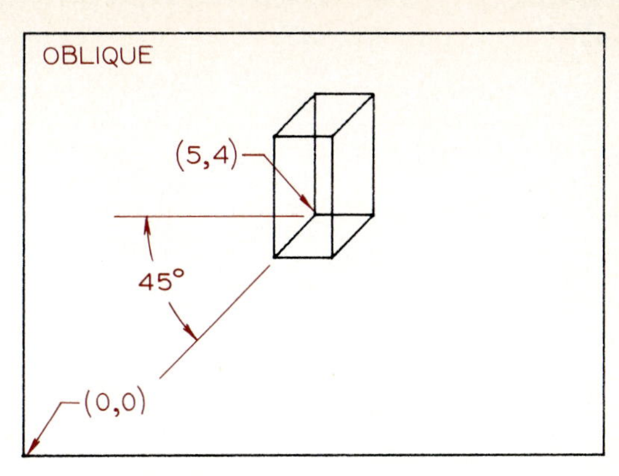


```
100 REM PROGRAM TO DRAW A FIGURE GIVEN THE
110 REM COORDINATES FOR ITS POINTS ON THE
120 REM X,Y, & Z AXIS.  THESE VALUES ARE
130 REM TRANSLATED ONTO AN OBLIQUE AXIS
140 REM BEFORE THEY ARE PLOTTED.  THE
150 REM COORDINATES ARE DEFINED WITH "DATA"
160 REM STATEMENTS CONTAINING X,Y,Z & PEN
170 REM VALUES.  X.ORG AND Y.ORG ARE THE
180 REM X & Y LOCATION OF THE ORIGIN OF THE
190 REM FIGURE.  Z.SCALE IS THE Z-AXIS
200 REM SCALE FACTOR.  SCALE IS THE OVERALL
210 REM SCALE FACTOR. ANGLE IS THE Z-AXIS
220 REM ANGLE. THE LAST DATA VALUE IS
230 REM ALWAYS 999.
240 REM DEFINE ORIGIN, SCALE AND ROTATION
250 X.ORG=5 : Y.ORG=4
260 Z.SCALE=.5 : SCALE=1 : ANGLE = 45
270 DATA 0,0,0,3,1,0,0,2,1,2,0,2,1,2,2,2
280 DATA 1,0,2,2,0,0,2,2,0,0,0,2,0,2,0,2
290 DATA 0,2,2,2,0,0,2,2,1,0,2,3,1,0,0,2
300 DATA 0,2,2,3,1,2,2,2,0,2,0,3,1,2,0,2
310 DATA 0,0,0,3,999
320 IP%=0 : GOSUB 9000
330 REM CONVERT ANGLE TO RADIANS
340 ANGLE = ANGLE * 3.1416 /180
350 REM ------ READ LOOP ------
360 READ XP
370 IF XP = 999 THEN STOP
380 READ YP
390 READ ZP
400 READ IP%
410 XP=X.ORG+((XP-(ZP*COS(ANGLE)*Z.SCALE)))*SCALE
420 YP=Y.ORG+((YP-(ZP*SIN(ANGLE)*Z.SCALE)))*SCALE
430 REM PLOT XP,YP & IP%
440 GOSUB 9000
450 GOTO 360
460 REM ------ END OF LOOP ------
470 END
```

Fig. 13.33 A program and its plot that converts given data points into an oblique drawing.

ted on these axes in Fig. 13.32 with their coordinates given at part A. If these views were drawn in a top and front view the points would appear orthographically as shown in Fig. 13.32B.

The program in Fig. 13.33 gives the coordinates of the origin of the oblique (X.ORG and Y.ORG), the scale factor (Z.SCALE) of the Z-axis, the overall scale factor (SCALE), and the angle of the receding axis (ANGLE). If Z.SCALE = 1, the oblique will be a cavalier oblique; if S = 0.5, the oblique will be a cabinet oblique. Scale factors

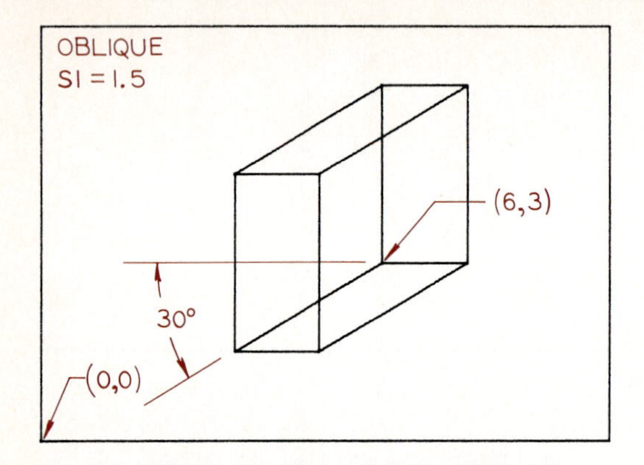


```
100 REM PROGRAM TO DRAW A FIGURE GIVEN THE
110 REM COORDINATES FOR ITS POINTS ON THE
120 REM X,Y, & Z AXIS.  THESE VALUES ARE
130 REM TRANSLATED ONTO AN OBLIQUE AXIS
140 REM BEFORE THEY ARE PLOTTED.  THE
150 REM COORDINATES ARE DEFINED WITH "DATA"
160 REM STATEMENTS CONTAINING X,Y,Z & PEN
170 REM VALUES.  X.ORG AND Y.ORG ARE THE
180 REM X & Y LOCATION OF THE ORIGIN OF THE
190 REM FIGURE.  Z.SCALE IS THE Z-AXIS
200 REM SCALE FACTOR.  SCALE IS THE OVERALL
210 REM SCALE FACTOR. ANGLE IS THE Z-AXIS
220 REM ANGLE. THE LAST DATA VALUE IS
230 REM ALWAYS 999.
240 REM DEFINE ORIGIN, SCALE AND ROTATION
250 X.ORG=6 : Y.ORG=3
260 Z.SCALE=1 : SCALE=1.5 : ANGLE = 30
270 DATA 0,0,0,3,1,0,0,2,1,2,0,2,1,2,2,2
280 DATA 1,0,2,2,0,0,2,2,0,0,0,2,0,2,0,2
290 DATA 0,2,2,2,0,0,2,2,1,0,2,3,1,0,0,2
300 DATA 0,2,2,3,1,2,2,2,0,2,0,3,1,2,0,2
310 DATA 0,0,0,3,999
320 IP%=0 : GOSUB 9000
330 REM CONVERT ANGLE TO RADIANS
340 ANGLE = ANGLE * 3.1416 /180
350 REM ------ READ LOOP ------
360 READ XP
370 IF XP = 999 THEN STOP
380 READ YP
390 READ ZP
400 READ IP%
410 XP=X.ORG+((XP-(ZP*COS(ANGLE)*Z.SCALE)))*SCALE
420 YP=Y.ORG+((YP-(ZP*SIN(ANGLE)*Z.SCALE)))*SCALE
430 REM PLOT XP,YP & IP%
440 GOSUB 9000
450 GOTO 360
460 REM ------ END OF LOOP ------
470 END
```

Fig. 13.34 A cavalier oblique has been plotted by the accompanying program. Note that the drawing has been enlarged since a scale factor (SCALE) of 1.5 was given.

(SCALE) are used to reduce the size of a figure so it will fit on the screen's area.

Data statements are used in lines 270 through 310 to give the X-, Y-, Z-, and P-values of each point of the oblique. The data statements are read in statements 360, 380, 390, and 400; calculations are made in statements 410 and 420; the points are plotted in statements 440; and then the program returns to statement 360 for the next values. This is repeated until all of the points have been read and plotted.

A cavalier oblique is drawn by the program in Fig. 13.34. The scale factor of the receding axis, Z.SCALE = 1, means it will be full size, which is the characteristic of the cavalier oblique. The same data points that were given in the previous example are drawn in this oblique and the angle of the receding axis is changed from 45° to 30°.

Note the method of converting Z-values into X- and Y-coordinates in statements 410 and 420. Since points on a sheet or on a screen are located using only X- and Y-coordinates, the Z-coordinates must be converted by these formulas. A loop is used between lines 360 and 450 for reading, calculating, and plotting the points.

13.15 AXONOMETRICS AND ISOMETRICS

An axonometric pictorial can be programmed and plotted by revolving the views of the part about the Y-axis and then the X-axis by the desired angles as shown orthographically in Fig. 13.35. The resulting front view is the axonometric pictorial of the object. Notice the assignments of positive and negative directions of rotation in each view.

A program and the plot of an axonometric that has been revolved −30° about the Y-axis in a negative direction, and revolved 30° in a positive di-

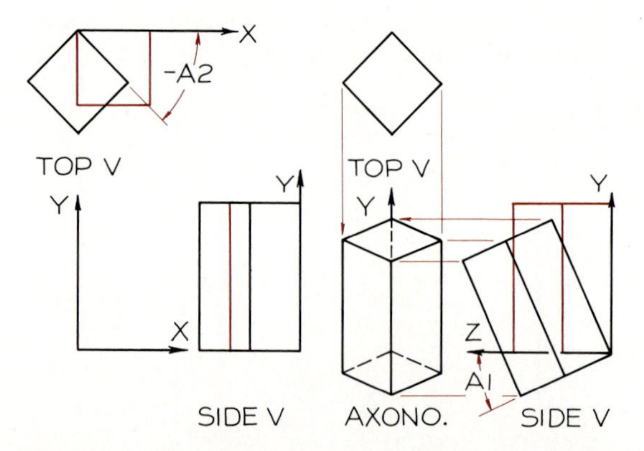

Fig. 13.35 An axonometric view of an object is found by revolving the top view about the Y-axis and the side view about the X-axis. The resulting front view is an axonometric projection.

rection about the X-axis are shown in Fig. 13.36. In line 270, the coordinates of the origin of the figure (X.ORG, Y.ORG), the two angles of rotation (X.ANGLE and Y.ANGLE), and the scale factor (SCALE) are given.

The data statements, lines 270 through 310, give the coordinates of the points of the pictorial that are to be drawn. The equations in lines 410, 420, and 430 determine the axonometric coordinates of the points that are to be plotted. Each of the data sets is looped through these equations until all have been calculated and plotted.

Since the three types of axonometrics are isometrics, dimetrics, and trimetrics, you can see that this program can be used for a variety of axonometric pictorials. An isometric projection is shown plotted in Fig. 13.37. The program that was

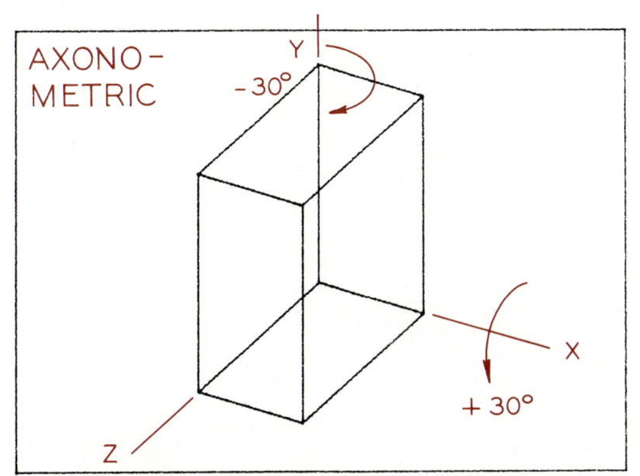


```
100 REM PROGRAM TO DRAW A FIGURE GIVEN THE
110 REM COORDINATES FOR ITS POINTS ON THE
120 REM X,Y, & Z AXIS.  THESE VALUES ARE
130 REM TRANSLATED ONTO AN AXONOMETRIC AXIS
140 REM BEFORE THEY ARE PLOTTED.   THE
150 REM COORDINATES ARE DEFINED WITH "DATA"
160 REM STATEMENTS CONTAINING X,Y,Z & PEN
170 REM VALUES.   X.ORG AND Y.ORG ARE THE
180 REM X & Y LOCATION OF THE ORIGIN OF THE
190 REM FIGURE.   SCALE IS THE OVERALL SCALE
200 REM FACTOR.   X.ANGLE IS THE ROTATION
210 REM ABOUT THE X AXIS.   Y.ANGLE IS THE
220 REM ROTATION ABOUT THE Y AXIS.
230 REM ===================================
240 REM DEFINE ORIGIN, SCALE AND ROTATION
250 X.ORG=5 : Y.ORG=3 : SCALE=2
260 X.ANGLE=30 : Y.ANGLE=-30
270 DATA 0,0,0,3,1,0,0,2,1,2,0,2,1,2,2,2
280 DATA 1,0,2,2,0,0,2,2,0,0,0,2,0,2,0,2
290 DATA 0,2,2,2,0,0,2,2,1,0,2,3,1,0,0,2
300 DATA 0,2,2,3,1,2,2,2,0,2,0,3,1,2,0,2
310 DATA 0,0,0,3,999
320 REM PLOT DATA
330 IP%=0 : GOSUB 9000
340 REM CONVERT ANGLE TO RADIANS
350 X.ANGLE = X.ANGLE * 3.1416 / 180
360 Y.ANGLE = Y.ANGLE * 3.1416 / 180
370 REM ------ READ LOOP ------
380 READ XP
390 IF XP = 999 THEN STOP
400 READ YP,ZP,IP%
410 YP=XP*SIN(Y.ANGLE)*SIN(X.ANGLE)+YP*COS(X.ANGLE)
420 YP=(YP-ZP*COS(Y.ANGLE)*SIN(X.ANGLE))*SCALE+Y.ORG
430 XP=(XP*COS(Y.ANGLE)+ZP*SIN(Y.ANGLE))*SCALE+X.ORG
440 REM PLOT XP,YP & IP%
450 GOSUB 9000
460 GOTO 380
470 REM ------ END OF LOOP ------
480 END
```

Fig. 13.36 This program rotates the top and side views by 30° to give this axonometric pictorial.

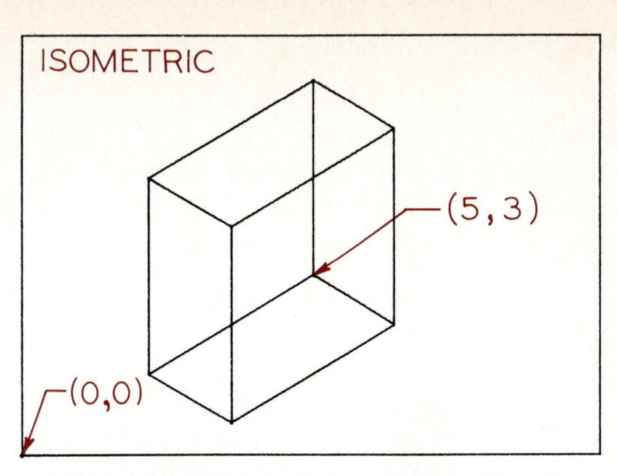


```
100 REM PROGRAM TO DRAW A FIGURE GIVEN THE
110 REM COORDINATES FOR ITS POINTS ON THE
120 REM X,Y, & Z AXIS.   THESE VALUES ARE
130 REM TRANSLATED ONTO AN ISOMETRIC AXIS
140 REM BEFORE THEY ARE PLOTTED.   THE
150 REM COORDINATES ARE DEFINED WITH "DATA"
160 REM STATEMENTS CONTAINING X,Y,Z & PEN
170 REM VALUES.   X.ORG AND Y.ORG ARE THE
180 REM X & Y LOCATION OF THE ORIGIN OF THE
190 REM FIGURE.   SCALE IS THE OVERALL SCALE
200 REM FACTOR.   X.ANGLE IS THE ROTATION
210 REM ABOUT THE X AXIS(-45).   Y.ANGLE IS
220 REM THE ROTATION ABOUT THE Y AXIS(35.3).
230 REM ===================================
240 REM DEFINE ORIGIN, SCALE AND ROTATION
250 X.ORG=5 : Y.ORG=3 : SCALE=2
260 X.ANGLE=35.3 : Y.ANGLE=-45
270 DATA 0,0,0,3,1,0,0,2,1,2,0,2,1,2,2,2
280 DATA 1,0,2,2,0,0,2,2,0,0,0,2,0,2,0,2
290 DATA 0,2,2,2,0,0,2,2,1,0,2,3,1,0,0,2
300 DATA 0,2,2,3,1,2,2,2,0,2,0,3,1,2,0,2
310 DATA 0,0,0,3,999
320 REM PLOT DATA
330 IP%=0 : GOSUB 9000
340 REM CONVERT ANGLE TO RADIANS
350 X.ANGLE = X.ANGLE * 3.1416 / 180
360 Y.ANGLE = Y.ANGLE * 3.1416 / 180
370 REM ------ READ LOOP ------
380 READ XP
390 IF XP = 999 THEN STOP
400 READ YP,ZP,IP%
410 YP=XP*SIN(Y.ANGLE)*SIN(X.ANGLE)+YP*COS(X.ANGLE)
420 YP=(YP-ZP*COS(Y.ANGLE)*SIN(X.ANGLE))*SCALE+Y.ORG
430 XP=(XP*COS(Y.ANGLE)+ZP*SIN(Y.ANGLE))*SCALE+X.ORG
440 REM PLOT XP,YP & IP%
450 GOSUB 9000
460 GOTO 380
470 REM ------ END OF LOOP ------
480 END
```

Fig. 13.37 The axonometric program can be used to draw an isometric pictorial by revolving the top view and the side view so the point view of a cube's diagonal will be found in the axonometric view.

used for this pictorial was the one used to draw the dimetric in Fig. 13.36. The only difference was the two angles of rotation.

13.16 THE BASIC CHARACTER-PLOTTING FUNCTION

Much like the basic plotting subroutine located at line number 9000 and called using GOSUB 9000, characters are plotted using a subroutine located at 9300 that is called using GOSUB 9300 (Fig. 13.38). However, before the character subroutine can be called, an initialization subroutine must be called using GOSUB 9400.

The initialization subroutine performs certain functions needed before the character subroutine can work properly. The initializing subroutine, called GOSUB 9400, should be executed only once per program.

Similar to the plotting subroutine, certain key variables must be assigned values before calling the character-plotting subroutine. These are CHAR$, which should be assigned the string of characters to be plotted; CHAR.X and CHAR.Y, which should be assigned the X- and Y-coordinate locations of the lower left corner of the first character string; CHAR.H, which should be assigned a value representing the height of the characters in inches; and CHAR.A, which should be assigned the degrees of rotation of the string about the lower left corner of the first character of the string (Fig. 13.39).

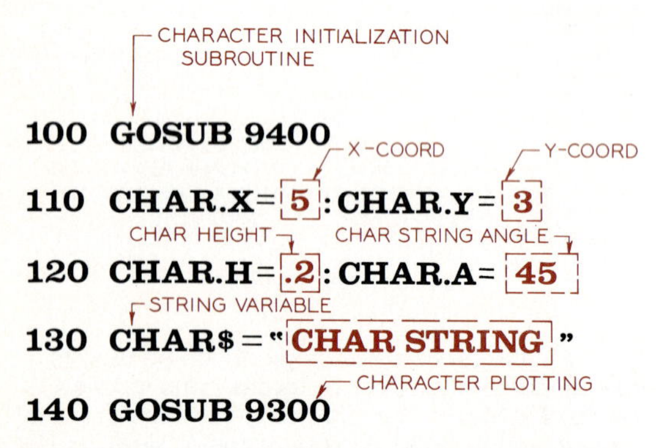

Fig. 13.38 This plot statement is used for writing text or characters. GOSUB 9400 is the *character initialization subroutine*, which loads the characters and is called only once per program. CHAR.X and CHAR.Y are the X- and Y-coordinates of the lower left corner of a string of characters that are to be plotted. CHAR.H is the desired height of the characters in inches. CHAR.A is the angle of rotation of a string of characters in a counterclockwise direction from the horizontal. CHAR$ is assigned the string of characters that are to be plotted. The character plotting subroutine is called with GOSUB 9300.

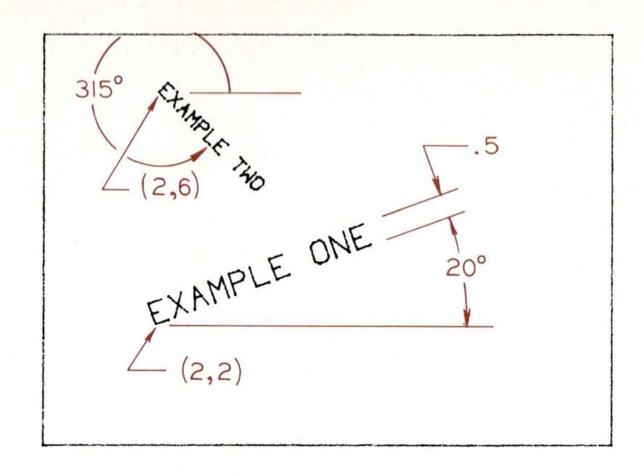


```
100 REM PROGRAM TO SHOW HOW CHARACTERS ARE
110 REM PLOTTED USING THE CHARACTER PLOTTING
120 REM SUBROUTINE.
130 REM ===================================
140 REM INITIALIZE
150 IP%=0 : GOSUB 9000 : GOSUB 9400
160 REM NOW DEFINE THE CHARACTER STRING TO BE
170 REM PLOTTED, IT'S ORIGIN, HEIGHT
180 REM AND ROTATION.
190 CHAR$="EXAMPLE ONE"
200 REM DEFINE ORIGIN
210 CHAR.X=2 : CHAR.Y=2
220 REM DEFINE DEFINE ROTATION AND HEIGHT.
230 CHAR.A=20 : CHAR.H=.5
240 REM NOW PLOT THE STRING
250 GOSUB 9300
260 REM NOW PLOT EXAMPLE STRING 2
270 CHAR.X=2 : CHAR.Y=6 : CHAR.H=.3
280 CHAR.A=315 : CHAR$="EXAMPLE TWO"
290 GOSUB 9300
300 REM RETURN TO ORIGIN
310 XP=0 : YP=0 : IP%=3 : GOSUB 9000
320 STOP
330 END
```

Fig. 13.39 An example program and its output on a display screen is shown where the words, "EXAMPLE ONE," are plotted.

13.17 ADVANCED APPLICATIONS

Using the concepts of computer graphics learned thus far, one might imagine how the following more advanced applications were created. All of these plotted examples are the output of programs using the ideas of circles, rectangles, or simple lines and points.

Bar Graph The bar graph in Fig. 13.40 was generated from information about the number of bars, the maximum bar value, X-axis and graph titles. Looking at the graph in parts rather than as a whole, it can be seen that the bars are just simple

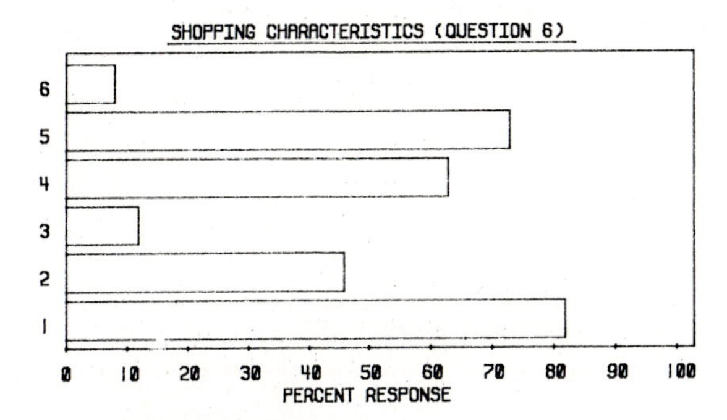

SHOPPING CHARACTERISTICS (QUESTION 6)
6
5
4
3
2
1
0 10 20 30 40 50 60 70 80 90 100
PERCENT RESPONSE
1. IS FOR DISCOUNT STORES.
2. IS FOR T.V. SPECIALITY STORES.
3. IS FOR WHOLESALE OUTLETS.
4. IS FOR DEPARTMENT STORES.
5. IS FOR RETAIL OUTLETS.
6. IS FOR MAIL ORDER CATALOGUE OR ADVERTISEMENTS.

Fig. 13.40 A bar graph is the result of using a number of plot statements and rectangle subroutines in addition to the text plotting statement.

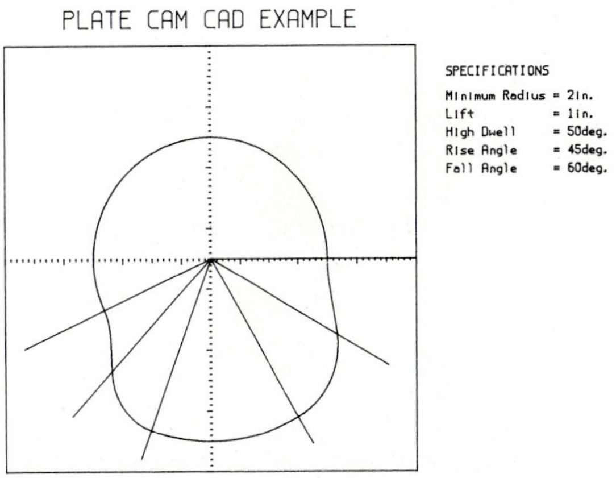

PLATE CAM CAD EXAMPLE
SPECIFICATIONS
Minimum Radius = 2in.
Lift = 1in.
High Dwell = 50deg.
Rise Angle = 45deg.
Fall Angle = 60deg.

Fig. 13.41 This cam profile is an example of an advanced computer graphics plotting problem.

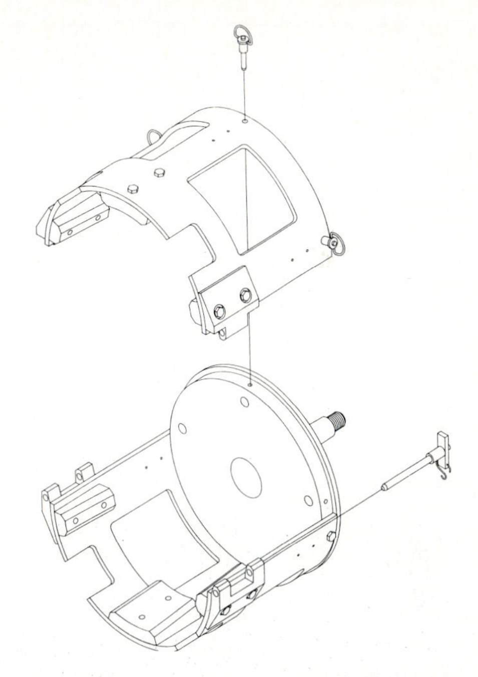

Fig. 13.42 A computer-drawn pictorial of an assembly of several parts. (Courtesy of Texas Instruments.)

rectangles. The tick marks along the X-axis are single lines and the text was plotted using the FNC character function.

Putting these parts together in the right places created the bar graph in Fig. 13.40.

Plate Cam The plate cam example in Fig. 13.41 is a very good example of computer-aided design of different shapes of cams. The one shown, designed for a knife-edge follower, was developed from the given specifications. This cam design was plotted by a computer program that used only the specifications shown.

The program is based upon a circle routine with the radius of the circle defined according to the step angle and a parabolic function of the lift and the rise and fall angles. This program used an axis-drawing subroutine to plot the boarder and tick-marks and the FNC character plotting functions to draw the text.

A more advanced example of computer graphics is the pictorial in Fig. 13.42 that illustrates an assembly of parts. This drawing was developed by Texas Instruments.

PROBLEMS

The problems below can be assigned to be written, programmed, and run on a computer graphics system. When a system is not available, the programs can be written and then plotted by hand on a 7.5" × 10" grid to simulate a display screen. When plotted by hand, it is suggested that the programs and plots be done on a Size AV sheet or sheets.

1. Write a program that will plot the lines given below. Obtain a printout of your program and a plot of the lines when your program has run.

Problem	Line	
A	A (2,2)	B (6,6)
B	B (1,6)	C (7,2)
C	E (3,5)	F (8,3)
D	G(8,5)	H (3,3)

2. Write a program that will plot the horizontally positioned rectangles from the given information. Obtain a printout of your program and a plot of the rectangles when your program has run.

Problem	Lower Left Corner	Height	Width
A	(2,2)	4"	6"
B	(1,1)	6"	5"
C	(3,1)	2"	3"
D	(2,1)	5"	4"

3. Write a program that will plot the angular lines below by using trigonometric functions of the angle of inclination, and the length of the line. Obtain a printout of your program and plot the lines when your program has run.

Problem	Lower Left Point	Length	Angle with Horizontal
A	(1,1)	5"	30°
B	(3,1)	4"	45°
C	(2,2)	3"	135°
D	(5,2)	4"	195°

4. Write a program that will plot the rectangles that incline with the horizontal at the angles given in the following table. Use variables similar to those illustrated in Fig. 13.23. Obtain a printout of your program and plot the rectangles when your program has run.

Problem	Lower Left Corner	Weight	Width	Angle
A	(2,2)	4"	5"	15°
B	(1,1)	5"	4"	35°
C	(3,1)	2"	3.4"	40°
D	(2,1)	4.5"	4"	28°

5. Write a program that will plot regular polygons using the specifications given in the table below for each. Refer to Fig. 13.24. Obtain a printout of your program and plot the solutions when your program has run.

Problem	No. of Sides	Center	Radius
A	4	(5,4)	2.5"
B	6	(5,4)	2.75"
C	8	(5,4)	2.00"
D	10	(5,4)	2.25"

6. Write a program that will plot circles using the specifications given in Problem 5 by increasing the number of sides of the polygons to approximate the circles. Obtain a printout of your program and plot the solutions when your program has run.

7. Write a program that will use data points to plot the front views of any of the assigned objects shown in Fig. 13.43. Place the lower left corner at (2,2). Obtain a printout of your program and plot the solutions when your program has run.

8. Write a main program and a subroutine for plotting the rectangles assigned in Problem 4. Refer to Fig. 13.27. Obtain a printout of your program and a plot of the solutions when the program has run.

9. Write a main program and a subroutine for plotting the circles assigned in Problem 6. Refer to Fig. 13.29. Obtain a printout of your program and a plot of the solutions when the program has run.

10. By referring to Fig. 13.33, write a similar program to plot one of the oblique pictorials below using the orthographic views of the objects in Fig. 13.43 and the supplementary data in Table 13.2.

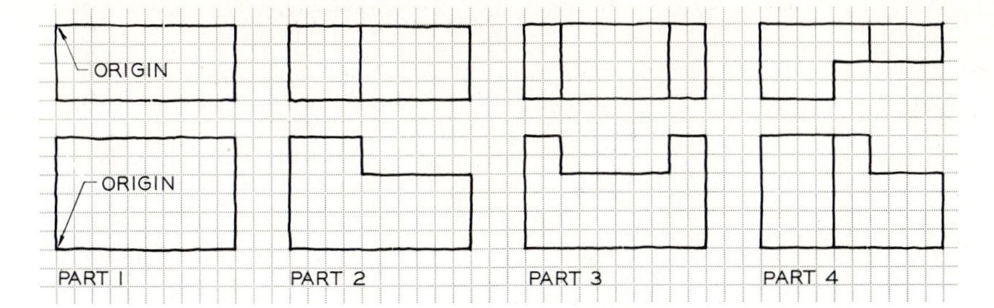

Fig. 13.43 Orthographic views of problems where each square is equal to 0.50 inches. The same corner point in the top and front views is the origin for each problem.

Obtain a printout of your program and a plot of the solutions when the program has run.

Table 13.2

Part No.	Origin	Overall Scale Factor	Z-axis Scale Factor	Angle
1	(5,3)	0.5	0.5	30°
2	(5,3)	0.65	1.00	45°
3	(5,3)	0.40	0.75	36°
4	(5,3)	0.70	0.60	25°

11. By referring to Fig. 13.36, write a similar program to plot one of the axonometric pictorials using the orthographic views of the objects in Fig. 13.43 and the supplementary data in Table 13.3.

Obtain a printout of your program and a plot of the solutions when the program has run.

Table 13.3

Part	Origin	Overall Scale Factor	Rotation about Y-axis	Rotation about X-axis
1	(5,3)	0.40	−20°	40°
2	(5,3)	0.50	−45°	35.5°
3	(5,3)	0.60	−50°	15°
4	(5,3)	0.37	−15°	20°

12. Write a program similar to the one given in Fig. 13.38 that will plot three lines of text (of your choice) that will be plotted at different angles and will not overlap.

APPENDIXES

APPENDIXES CONTENTS

APPENDIXES

Appendix 1 CONVERSION TABLES

Length conversions		
Angstrom units	$\times 1 \times 10^{-10}$	= meters
	$\times 1 \times 10^{-4}$	= microns
	$\times 1.650\,763\,73 \times 10^{-4}$	= wavelengths of orange-red line of krypton 86
Cables	× 120	= fathoms
	× 720	= feet
	× 219.456	= meters
Fathoms	× 6	= feet
	× 1.828 8	= meters
Feet	× 12	= inches
	× 0.3048	= meters
Furlongs	× 660	= feet
	× 201.168	= meters
	× 220	= yards
Inches	$\times 2.54 \times 10^{8}$	= Angstroms
	× 25.4	= millimeters
	$\times 8.333\,33 \times 10^{-2}$	= feet
Kilometers	$\times 3.280\,839 \times 10^{3}$	= feet
	× 0.62	= miles
	× 0.539 956	= nautical miles
	× 0.621 371	= statute miles
	$\times 1.093\,613 \times 10^{3}$	= yards
Light-years	$\times 9.460\,55 \times 10^{12}$	= kilometers
	$\times 5.878\,51 \times 10^{12}$	= statute miles
Meters	$\times 1 \times 10^{10}$	= Angstroms
	× 3.280 839 9	= feet
	× 39.370 079	= inches
	× 1.093 61	= yards
Microns	$\times 10^{4}$	= Angstroms
	$\times 10^{-4}$	= centimeters
	$\times 10^{-6}$	= meters
Nautical Miles (International)	× 8.439 049	= cables
	$\times 6.076\,115\,49 \times 10^{3}$	= feet
	$\times 1.852 \times 10^{3}$	= meters
	× 1.150 77	= statute miles

Appendix 1 CONVERSION TABLES (Cont.)

Length conversions

Statute Miles	× 5.280 × 10^{3}	= feet
	× 8	= furlongs
	× 6.336 0 × 10^{4}	= inches
	× 1.609 34	= kilometers
	× 8.689 7 × 10^{-1}	= nautical miles
Miles	× 10^{-3}	= inches
	× 2.54 × 10^{-2}	= millimeters
	× 25.4	= micrometers
	× 0.61	= kilometers
Yards	× 3	= feet
	× 9.144 × 10^{-1}	= meters
Feet/hour	× 3.048 × 10^{-4}	= kilometers/hour
	× 1.645 788 × 10^{-4}	= knots
Feet/minute	× 0.3048	= meters/minute
	× 5.08 × 10^{-3}	= meters/second
Feet/second	× 1.097 28	= kilometers/hour
	× 18.288	= meters/minute
Kilometers/hour	× 3.280 839 × 10^{3}	= feet/hour
	× 54.680 66	= feet/minute
	× 0.277 777	= meters/second
	× 0.621 371	= miles/hour
Kilometers/minute	× 3.280 839 × 10^{3}	= feet/minute
	× 37.282 27	= miles/hour
Knots	× 6.076 115 × 10^{3}	= feet/hour
	× 101.268 5	= feet/minute
	× 1.687 809	= feet/second
	× 1.852	= kilometers/hour
	× 30.866	= meters/minute
	× 0.514 4	= meters/second
	× 1.150 77	= statute miles/hour
Meters/hour	× 3.280 839	= feet/hour
	× 88	= feet/minute
	× 1.466	= feet/second
	× 1 × 10^{-3}	= kilometers/hour
	× 1.667 × 10^{-2}	= meters/minute
Feet/second^{2}	× 1.097 28	= kilometers/hour/second
	× 0.304 8	= meters/second^{2}

Area conversions

Acres	× 4.046 85 × 10^{-3}	= square kilometers
	× 4.046 856 × 10^{3}	= square meters
	× 4.356 0 × 10^{4}	= square feet
Ares	× 2.471 053 8 × 10^{-2}	= acres
	× 1	= square dekameters
	× 10^{2}	= square meters
Barns	× 1 × 10^{-28}	= square meters
Circular mils	× 1 × 10^{-6}	= circular inches
	× 5.067 074 8 × 10^{-4}	= square millimeters
	× 0.785 398 1	= square mils

Cont.

Appendix 1 CONVERSION TABLES (Cont.)

Area conversions		
Hectares	× 2.471 05	= acres
	× 10^2	= ares
	× 10^4	= square meters
Square feet	× 2.295 684 × 10^{-5}	= acres
	× 9.290 3 × 10^{-4}	= ares
	× 144	= square inches
	× 9.290 304 × 10^{-2}	= square meters
Square inches	× 1.273 239 5 × 10^6	= circular mils
	× 6.944 4 × 10^{-3}	= square feet
	× 6.451 6 × 10^{-4}	= square meters
Square kilometers	× 247.105 38	= acres
	× 1.076 391 0 × 10^7	= square feet
	× 1.000	= cubic meters
	× 1.307 950 6	= cubic yards
	× 219.969	= imperial gallons
Liters	× 10^3	= cubic centimeters
	× 1.000 × 10^6	= cubic millimeters
	× 1.000 × 10^{-3}	= cubic meters
	× 61.023 74	= cubic inches
	× 3.531 5 × 10^{-2}	= cubic feet
	× 1.307 95 × 10^{-3}	= cubic yards
	× 0.22	= gallons
	× 0.219 969	= imperial gallons
	× 0.879 877	= imperial quarts
Imperial pints	× 0.125	= imperial gallons
	× 0.568 261	= liters
	× 20	= imperial fluid ounces
	× 0.5	= imperial quarts
	× 568.260 9	= cubic centimeters
Imperial quarts	× 1.136 52 × 10^3	= cubic centimeters
	× 69.354 8	= cubic inches
	× 1.136 522 8	= liters

Power conversions		
British Thermal Units/hour	× 2.928 7 × 10^{-4}	= kilowatts
	× 0.292 875	= watts
BTU/minute	× 1.757 25 × 10^{-2}	= kilowatts
BTU/pound	× 2.324 4	= joules/gram
BTU/second	× 1.413 91	= horsepower
	× 107.514	= kilogrammeters/second
	× 1.054 35	= kilowatts
	× 1.054 35 × 10^3	= watts
Foot-pound-force/hour	× 5.050 × 10^{-7}	= horsepower
	× 3.766 16 × 10^{-7}	= kilowatts
Foot-pound-force/ minute	× 3.030 303 × 10^{-5}	= horsepower
	× 2.259 70 × 10^{-2}	= joules/second
	× 2.259 70 × 10^{-5}	= kilowatts
Horsepower	× 42.435 6	= BTU/minute
	× 550	= footpounds/second
	× 0.746	= kilowatts
	× 746	= joules/second

Appendix 1 CONVERSION TABLES (Cont.)

Power conversions

Kilogrammeters/second	$\times$ 9.806 65	= watts
Kilowatts	$\times$ 3.414 43 $\times 10^{3}$	= BTU/hour
	$\times$ 2.655 22 $\times 10^{6}$	= footpounds/hour
	$\times$ 4.425 37 $\times 10^{4}$	= footpounds/minute
	$\times$ 737.562	= footpounds/second
	$\times$ 1.019 726 $\times 10^{7}$	= gramcentimeters/second
	$\times$ 1.341 02	= horsepower
	$\times$ 3.6 $\times 10^{6}$	= joules/hour
	$\times 10^{3}$	= joules/second
	$\times$ 3.671 01 $\times 10^{5}$	= kilogrammeters/hour
	$\times$ 999.835	= international watt
Watts	$\times$ 44.253 7	= footpounds/minute
	$\times$ 1.341 02 $\times 10^{-3}$	= horsepower
	$\times$ 1	= joules/second

Time conversions

(No attempt has been made in this brief treatment to correlate solar, mean solar, sidereal, and mean sidereal days.)

Mean solar days	$\times$ 24	= mean solar hours
Mean solar hours	$\times$ 3.600 $\times 10^{3}$	= mean solar seconds
	$\times$ 60	= mean solar minutes

Angle conversions

Degrees	$\times$ 60	= minutes
	$\times$ 1.745 329 3 $\times 10^{-2}$	= radians
Degrees/foot	$\times$ 5.726 145 $\times 10^{-4}$	= radians/centimeter
Degrees/minute	$\times$ 2.908 8 $\times 10^{-4}$	= radians/second
	$\times$ 4.629 629 $\times 10^{-5}$	= revolutions/second
Degrees/second	$\times$ 1.745 329 3 $\times 10^{-2}$	= radians/second
	$\times$ 0.166	= revolutions/minute
	$\times$ 2.77 $\times 10^{-3}$	= revolutions/second
Minutes	$\times$ 1.667 $\times 10^{-2}$	= degrees
	$\times$ 2.908 8 $\times 10^{-4}$	= radians
	$\times$ 60	= seconds
Radians	$\times$ 0.159 154	= circumferences
	$\times$ 57.295 77	= degrees
	$\times$ 3.437 746 $\times 10^{3}$	= minutes
Seconds	$\times$ 2.777 $\times 10^{-4}$	= degrees
	$\times$ 1.667 $\times 10^{-2}$	= minutes
	$\times$ 4.848 136 8 $\times 10^{-6}$	= radians
Steradians	$\times$ 0.159 154 9	= hemispheres
	$\times$ 7.957 74 $\times 10^{-2}$	= spheres
	$\times$ 0.636 619 7	= spherical right angles

Mass conversions

Grains	$\times$ 6.479 8 $\times 10^{-2}$	= grams
	$\times$ 2.285 71 $\times 10^{-3}$	= avoirdupois ounces

Cont.

Appendix 1 CONVERSION TABLES (Cont.)

Mass conversions		
Grams	$\times$ 15.432 358	= grains
	$\times$ 3.527 396 $\times 10^{-2}$	= avoirdupois ounces
	$\times$ 2.204 62 $\times 10^{-3}$	= avoirdupois pounds
Kilograms	$\times$ 564.383 4	= avoirdupois drams
	$\times$ 2.204 622 6	= avoirdupois pounds
	$\times$ 2.2	= pounds
	$\times$ 9.842 065 $\times 10^{-4}$	= long tons
	$\times 10^{-3}$	= metric tons
	$\times$ 1.102 31 $\times 10^{-3}$	= short tons
Avoirdupois ounces	$\times$ 28.349 5	= grams
	$\times$ 6.25 $\times 10^{-2}$	= avoirdupois pounds
	$\times$ 0.911 458	= troy ounces
Avoirdupois pounds	$\times$ 256	= drams
	$\times$ 4.535 923 7 $\times 10^{2}$	= grams
	$\times$ 0.453 592 4	= kilograms
	$\times$ 16	= ounces
Long tons	$\times$ 2.24 $\times 10^{3}$	= avoirdupois pounds
	$\times$ 1.106 046 9	= metric tons
	$\times$ 1.12	= short tons
Metric tons	$\times 10^{3}$	= kilograms
	$\times$ 2.204 622 $\times 10^{3}$	= avoirdupois pounds
Short tons	$\times$ 2 $\times 10^{3}$	= avoirdupois pounds
	$\times$ 907.184 74	= kilograms

Force conversions		
Dynes	$\times 10^{-5}$	= newtons
Newtons	$\times 10^{5}$	= dynes
	$\times$ 0.224 808	= pounds-force
Pounds	$\times$ 4.448 22	= newtons

Energy conversions		
British Thermal Units (thermochemical)	$\times$ 1.054 35 $\times 10^{3}$	= joules
	$\times$ 2.928 27 $\times 10^{-4}$	= kilowatthours
	$\times$ 1.054 35 $\times 10^{3}$	= wattseconds
Foot-pound-force	$\times$ 1.355 818 0	= joules
	$\times$ 0.138 255	= kilogramforce-meters
	$\times$ 3.766 16 $\times 10^{-7}$	= kilowatthours
	$\times$ 1.355 818 0	= newtonmeters
Joules	$\times$ 9.484 5 $\times 10^{-4}$	= British Thermal Units
	$\times$ 0.737 562	= foot-pounds-force
	$\times$ 0.101 971 6	= kilogramforce-meters
	$\times$ 2.777 7 $\times 10^{-7}$	= kilowatthours
	$\times$ 1	= wattseconds
Kilogramforce-meters	$\times$ 9.287 7 $\times 10^{-3}$	= British Thermal Units
	$\times$ 7.233 01	= foot-pounds-force
	$\times$ 9.806 65	= joules
	$\times$ 9.806 65	= newtonmeters
	$\times$ 2.724 0 $\times 10^{-3}$	= watthours

Appendix 1 CONVERSION TABLES (Cont.)

Energy conversions		
Kilowatthours	$\times$ 3.409 52 $\times 10^3$	= British Thermal Units
	$\times$ 2.655 22 $\times 10^6$	= foot-pounds-force
	$\times$ 1.341 02	= horsepowerhours
	$\times$ 3.6 $\times 10^6$	= joules
	$\times$ 3.670 98 $\times 10^5$	= kilogramforce-meters
Newtonmeters	$\times$ 0.101 971	= kilogramforce-meters
	$\times$ 0.737 562	= poundforce-feet
Watthours	$\times$ 3.414 43	= British Thermal Units
	$\times$ 2.655 22 $\times 10^3$	= foot-pounds-force
	$\times$ 3.6 $\times 10^3$	= joules
	$\times$ 3.670 98 $\times 10^2$	= kilogramforce-meters

Pressure conversions		
Atmospheres	$\times$ 1.013 25	= bars
	$\times$ 1.033 23 $\times 10^3$	= grams/square centimeter
	$\times$ 1.033 23 $\times 10^7$	= grams/square meter
	$\times$ 14.696 0	= pounds/square inch
	$\times$ 760	= torrs
	$\times$ 101	= kilopascals
Bars	$\times$ 0.986 923	= atmospheres
	$\times 10^6$	= baryes
	$\times$ 1.019 716 $\times 10^7$	= grams/square meter
	$\times$ 1.019 716 $\times 10^4$	= kilogramsforce/square meter
	$\times$ 14.503 8	= poundsforce/square inch
Baryes	$\times 10^{-6}$	= bars
Inches of mercury	$\times$ 3.386 4 $\times 10^{-2}$	= bars
	$\times$ 345.316	= kilogramsforce/square meter
	$\times$ 70.726 2	= poundsforce/square foot
Pascal	$\times$ 1	= newton/square meter

Appendix 2 LOGARITHMS OF NUMBERS

N	0	1	2	3	4	5	6	7	8	9
1.0	.0000	.0043	.0086	.0128	.0170	.0212	.0253	.0294	.0334	.0374
1.1	.0414	.0453	.0492	.0531	.0569	.0607	.0645	.0682	.0719	.0755
1.2	.0792	.0828	.0864	.0899	.0934	.0969	.1004	.1038	.1072	.1106
1.3	.1139	.1173	.1206	.1239	.1271	.1303	.1335	.1367	.1399	.1430
1.4	.1461	.1492	.1523	.1553	.1584	.1614	.1644	.1673	.1703	.1732
1.5	.1761	.1790	.1818	.1847	.1875	.1903	.1931	.1959	.1987	.2014
1.6	.2041	.2068	.2095	.2122	.2148	.2175	.2201	.2227	.2253	.2279
1.7	.2304	.2330	.2355	.2380	.2405	.2430	.2455	.2480	.2504	.2529
1.8	.2553	.2577	.2601	.2625	.2648	.2672	.2695	.2718	.2742	.2765
1.9	.2788	.2810	.2833	.2856	.2878	.2900	.2923	.2945	.2967	.2989
2.0	.3010	.3032	.3054	.3075	.3096	.3118	.3139	.3160	.3181	.3201
2.1	.3222	.3243	.3263	.3284	.3304	.3324	.3345	.3365	.3385	.3404
2.2	.3424	.3444	.3464	.3483	.3502	.3522	.3541	.3560	.3579	.3598
2.3	.3617	.3636	.3655	.3674	.3692	.3711	.3729	.3747	.3766	.3784
2.4	.3802	.3820	.3838	.3856	.3874	.3892	.3909	.3927	.3945	.3962
2.5	.3979	.3997	.4014	.4031	.4048	.4065	.4082	.4099	.4116	.4133
2.6	.4150	.4166	.4183	.4200	.4216	.4232	.4249	.4265	.4281	.4298
2.7	.4314	.4330	.4346	.4362	.4378	.4393	.4409	.4425	.4440	.4456
2.8	.4472	.4487	.4502	.4518	.4533	.4548	.4564	.4579	.4594	.4609
2.9	.4624	.4639	.4654	.4669	.4683	.4698	.4713	.4728	.4742	.4757
3.0	.4771	.4786	.4800	.4814	.4829	.4843	.4857	.4871	.4886	.4900
3.1	.4914	.4928	.4942	.4955	.4969	.4983	.4997	.5011	.5024	.5038
3.2	.5051	.5065	.5079	.5092	.5105	.5119	.5132	.5145	.5159	.5172
3.3	.5185	.5198	.5211	.5224	.5237	.5250	.5263	.5276	.5289	.5302
3.4	.5315	.5328	.5340	.5353	.5366	.5378	.5391	.5403	.5416	.5428
3.5	.5441	.5453	.5465	.5478	.5490	.5502	.5514	.5527	.5539	.5551
3.6	.5563	.5575	.5587	.5599	.5611	.5623	.5635	.5647	.5658	.5670
3.7	.5682	.5694	.5705	.5717	.5729	.5740	.5752	.5763	.5775	.5786
3.8	.5798	.5809	.5821	.5832	.5843	.5855	.5866	.5877	.5888	.5899
3.9	.5911	.5922	.5933	.5944	.5955	.5966	.5977	.5988	.5999	.6010
4.0	.6021	.6031	.6042	.6053	.6064	.6075	.6085	.6096	.6107	.6117
4.1	.6128	.6138	.6149	.6160	.6170	.6180	.6191	.6201	.6212	.6222
4.2	.6232	.6243	.6253	.6263	.6274	.6284	.6294	.6304	.6314	.6325
4.3	.6335	.6345	.6355	.6365	.6375	.6385	.6395	.6405	.6415	.6425
4.4	.6435	.6444	.6454	.6464	.6474	.6484	.6493	.6503	.6513	.6522
4.5	.6532	.6542	.6551	.6561	.6571	.6580	.6590	.6599	.6609	.6618
4.6	.6628	.6637	.6646	.6656	.6665	.6675	.6684	.6693	.6702	.6712
4.7	.6721	.6730	.6739	.6749	.6758	.6767	.6776	.6785	.6794	.6803
4.8	.6812	.6821	.6830	.6839	.6848	.6857	.6866	.6875	.6884	.6893
4.9	.6902	.6911	.6920	.6928	.6937	.6946	.6955	.6964	.6972	.6981
5.0	.6990	.6998	.7007	.7016	.7024	.7033	.7042	.7050	.7059	.7067
5.1	.7076	.7084	.7093	.7101	.7110	.7118	.7126	.7135	.7143	.7152
5.2	.7160	.7168	.7177	.7185	.7193	.7202	.7210	.7218	.7226	.7235
5.3	.7243	.7251	.7259	.7267	.7275	.7284	.7292	.7300	.7308	7316
5.4	.7324	.7332	.7340	.7348	.7356	.7364	.7372	.7380	.7388	.7396
N	0	1	2	3	4	5	6	7	8	9

Appendix 2 LOGARITHMS OF NUMBERS (Cont.)

N	0	1	2	3	4	5	6	7	8	9
5.5	.7404	.7412	.7419	.7427	.7435	.7443	.7451	.7459	.7466	.7474
5.6	.7482	.7490	.7497	.7505	.7513	.7520	.7528	.7536	.7543	.7551
5.7	.7559	.7566	.7574	.7582	.7589	.7597	.7604	.7612	.7619	.7627
5.8	.7634	.7642	.7649	.7657	.7664	.7672	.7679	.7686	.7694	.7701
5.9	.7709	.7716	.7723	.7731	.7738	.7745	.7752	.7760	.7767	.7774
6.0	.7782	.7789	.7796	.7803	.7810	.7818	.7825	.7832	.7839	.7846
6.1	.7853	.7860	.7868	.7875	.7882	.7889	.7896	.7903	.7910	.7917
6.2	.7924	.7931	.7938	.7945	.7952	.7959	.7966	.7973	.7980	.7987
6.3	.7993	.8000	.8007	.8014	.8021	.8028	.8035	.8041	.8048	.8055
6.4	.8062	.8069	.8075	.8082	.8089	.8096	.8102	.8109	.8116	.8122
6.5	.8129	.8136	.8142	.8149	.8156	.8162	.8169	.8176	.8182	.8189
6.6	.8195	.8202	.8209	.8215	.8222	.8228	.8235	.8241	.8248	.8254
6.7	.8261	.8267	.8274	.8280	.8287	.8293	.8299	.8306	.8312	.8319
6.8	.8325	.8331	.8338	.8344	.8351	.8357	.8363	.8370	.8376	.8382
6.9	.8388	.8395	.8401	.8407	.8414	.8420	.8426	.8432	.8439	.8445
7.0	.8451	.8457	.8463	.8470	.8476	.8482	.8488	.8494	.8500	.8506
7.1	.8513	.8519	.8525	.8531	.8537	.8543	.8549	.8555	.8561	.8567
7.2	.8573	.8579	.8585	.8591	.8597	.8603	.8609	.8615	.8621	.8627
7.3	.8633	.8639	.8645	.8651	.8657	.8663	.8669	.8675	.8681	.8686
7.4	.8692	.8698	.8704	.8710	.8716	.8722	.8727	.8733	.8739	.8745
7.5	.8751	.8756	.8762	.8768	.8774	.8779	.8785	.8791	.8797	.8802
7.6	.8808	.8814	.8820	.8825	.8831	.8837	.8842	.8848	.8854	.8859
7.7	.8865	.8871	.8876	.8882	.8887	.8893	.8899	.8904	.8910	.8915
7.8	.8921	.8927	.8932	.8938	.8943	.8949	.8954	.8960	.8965	.8971
7.9	.8976	.8982	.8987	.8993	.8998	.9004	.9009	.9015	.9020	.9025
8.0	.9031	.9036	.9042	.9047	.9053	.9058	.9063	.9069	.9074	.9079
8.1	.9085	.9090	.9096	.9101	.9106	.9112	.9117	.9122	.9128	.9133
8.2	.9138	.9143	.9149	.9154	.9159	.9165	.9170	.9175	.9180	.9186
8.3	.9191	.9196	.9201	.9206	.9212	.9217	.9222	.9227	.9232	.9238
8.4	.9243	.9248	.9253	.9258	.9263	.9269	.9274	.9279	.9284	.9289
8.5	.9294	.9299	.9304	.9309	.9315	.9320	.9325	.9330	.9335	.9340
8.6	.9345	.9350	.9355	.9360	.9365	.9370	.9375	.9380	.9385	.9390
8.7	.9395	.9400	.9405	.9410	.9415	.9420	.9425	.9430	.9435	.9440
8.8	.9445	.9450	.9455	.9460	.9465	.9469	.9474	.9479	.9484	.9489
8.9	.9494	.9499	.9504	.9509	.9513	.9518	.9523	.9528	.9533	.9538
9.0	.9542	.9547	.9552	.9557	.9562	.9566	.9571	.9576	.9581	.9586
9.1	.9590	.9595	.9600	.9605	.9609	.9614	.9619	.9624	.9628	.9633
9.2	.9638	.9643	.9647	.9652	.9657	.9661	.9666	.9671	.9675	.9680
9.3	.9685	.9689	.9694	.9699	.9703	.9708	.9713	.9717	.9722	.9727
9.4	.9731	.9736	.9741	.9745	.9750	.9754	.9759	.9763	.9768	.9773
9.5	.9777	.9782	.9786	.9791	.9795	.9800	.9805	.9809	.9814	.9818
9.6	.9823	.9827	.9832	.9836	.9841	.9845	.9850	.9854	.9859	.9863
9.7	.9868	.9872	.9877	.9881	.9886	.9890	.9894	.9899	.9903	.9908
9.8	.9912	.9917	.9921	.9926	.9930	.9934	.9939	.9943	.9948	.9952
9.9	.9956	.9961	.9965	.9969	.9974	.9978	.9983	.9987	.9991	.9996
N	0	1	2	3	4	5	6	7	8	9

Appendix 3 VALUES OF TRIGONOMETRIC FUNCTIONS

Degrees	Radians	Sine	Tangent	Cotangent	Cosine		
0° 00′	.0000	.0000	.0000		1.0000	1.5708	90° 00′
10′	.0029	.0029	.0029	343.77	1.0000	1.5679	50′
20′	.0058	.0058	.0058	171.89	1.0000	1.5650	40′
30′	.0087	.0087	.0087	114.59	1.0000	1.5621	30′
40′	.0116	.0116	.0116	85.940	.9999	1.5592	20′
50′	.0145	.0145	.0145	68.750	.9999	1.5563	10′
1° 00′	.0175	.0175	.0175	57.290	.9998	1.5533	89° 00′
10′	.0204	.0204	.0204	49.104	.9998	1.5504	50′
20′	.0233	.0233	.0233	42.964	.9997	1.5475	40′
30′	.0262	.0262	.0262	38.188	.9997	1.5446	30′
40′	.0291	.0291	.0291	34.368	.9996	1.5417	20′
50′	.0320	.0320	.0320	31.242	.9995	1.5388	10′
2° 00′	.0349	.0349	.0349	28.636	.9994	1.5359	88° 00′
10′	.0378	.0378	.0378	26.432	.9993	1.5330	50′
20′	.0407	.0407	.0407	24.542	.9992	1.5301	40′
30′	.0436	.0436	.0437	22.904	.9990	1.5272	30′
40′	.0465	.0465	.0466	21.470	.9989	1.5243	20′
50′	.0495	.0494	.0495	20.206	.9988	1.5213	10′
3° 00′	.0524	.0523	.0524	19.081	.9986	1.5184	87° 00′
10′	.0553	.0552	.0553	18.075	.9985	1.5155	50′
20′	.0582	.0581	.0582	17.169	.9983	1.5126	40′
30′	.0611	.0610	.0612	16.350	.9981	1.5097	30′
40′	.0640	.0640	.0641	15.605	.9980	1.5068	20′
50′	.0669	.0669	.0670	14.924	.9978	1.5039	10′
4° 00′	.0698	.0698	.0699	14.301	.9976	1.5010	86° 00′
10′	.0727	.0727	.0729	13.727	.9974	1.4981	50′
20′	.0756	.0756	.0758	13.197	.9971	1.4952	40′
30′	.0785	.0785	.0787	12.706	.9969	1.4923	30′
40′	.0814	.0814	.0816	12.251	.9967	1.4893	20′
50′	.0844	.0843	.0846	11.826	.9964	1.4864	10′
5° 00′	.0873	.0872	.0875	11.430	.9962	1.4835	85° 00′
10′	.0902	.0901	.0904	11.059	.9959	1.4806	50′
20′	.0931	.0929	.0934	10.712	.9957	1.4777	40′
30′	.0960	.0958	.0963	10.385	.9954	1.4748	30′
40′	.0989	.0987	.0992	10.078	.9951	1.4719	20′
50′	.1018	.1016	.1022	9.7882	.9948	1.4690	10′
6° 00′	.1047	.1045	.1051	9.5144	.9945	1.4661	84° 00′
10′	.1076	.1074	.1080	9.2553	.9942	1.4632	50′
20′	.1105	.1103	.1110	9.0098	.9939	1.4603	40′
30′	.1134	.1132	.1139	8.7769	.9936	1.4573	30′
40′	.1164	.1161	.1169	8.5555	.9932	1.4544	20′
50′	.1193	.1190	.1198	8.3450	.9929	1.4515	10′
7° 00′	.1222	.1219	.1228	8.1443	.9925	1.4486	83° 00′
10′	.1251	.1248	.1257	7.9530	.9922	1.4457	50′
20′	.1280	.1276	.1287	7.7704	.9918	1.4428	40′
30′	.1309	.1305	.1317	7.5958	.9914	1.4399	30′
40′	.1338	.1334	.1346	7.4287	.9911	1.4370	20′
50′	.1367	.1363	.1376	7.2687	.9907	1.4341	10′
8° 00′	.1396	.1392	.1405	7.1154	.9903	1.4312	82° 00′
10′	.1425	.1421	.1435	6.9682	.9899	1.4283	50′
20′	.1454	.1449	.1465	6.8269	.9894	1.4254	40′
30′	.1484	.1478	.1495	6.6912	.9890	1.4224	30′
40′	.1513	.1507	.1524	6.5606	.9886	1.4195	20′
50′	.1542	.1536	.1554	6.4348	.9881	1.4166	10′
9° 00′	.1571	.1564	.1584	6.3138	.9877	1.4137	81° 00′
		Cosine	Cotangent	Tangent	Sine	Radians	Degrees

Appendix 3 VALUES OF TRIGONOMETRIC FUNCTIONS (Cont.)

Degrees	Radians	Sine	Tangent	Cotangent	Cosine		
9° 00′	.1571	.1564	.1584	6.3138	.9877	1.4137	81° 00′
10′	.1600	.1593	.1614	6.1970	.9872	1.4108	50′
20′	.1629	.1622	.1644	6.0844	.9868	1.4079	40′
30′	.1658	.1650	.1673	5.9758	.9863	1.4050	30′
40′	.1687	.1679	.1703	5.8708	.9858	1.4021	20′
50′	.1716	.1708	.1733	5.7694	.9853	1.3992	10′
10° 00′	.1745	.1736	.1763	5.6713	.9848	1.3963	80° 00′
10′	.1774	.1765	.1793	5.5764	.9843	1.3934	50′
20′	.1804	.1794	.1823	5.4845	.9838	1.3904	40′
30′	.1833	.1822	.1853	5.3955	.9833	1.3875	30′
40′	.1862	.1851	.1883	5.3093	.9827	1.3846	20′
50′	.1891	.1880	.1914	5.2257	.9822	1.3817	10′
11° 00′	.1920	.1908	.1944	5.1446	.9816	1.3788	79° 00′
10′	.1949	.1937	.1974	5.0658	.9811	1.3759	50′
20′	.1978	.1965	.2004	4.9894	.9805	1.3730	40′
30′	.2007	.1994	.2035	4.9152	.9799	1.3701	30′
40′	.2036	.2022	.2065	4.8430	.9793	1.3672	20′
50′	.2065	.2051	.2095	4.7729	.9787	1.3643	10′
12° 00′	.2094	.2079	.2126	4.7046	.9781	1.3614	78° 00′
10′	.2123	.2108	.2156	4.6382	.9775	1.3584	50′
20′	.2153	.2136	.2186	4.5736	.9769	1.3555	40′
30′	.2182	.2164	.2217	4.5107	.9763	1.3526	30′
40′	.2211	.2193	.2247	4.4494	.9757	1.3497	20′
50′	.2240	.2221	.2278	4.3897	.9750	1.3468	10′
13° 00′	.2269	.2250	.2309	4.3315	.9744	1.3439	77° 00′
10′	.2298	.2278	.2339	4.2747	.9737	1.3410	50′
20′	.2327	.2306	.2370	4.2193	.9730	1.3381	40′
30′	.2356	.2334	.2401	4.1653	.9724	1.3352	30′
40′	.2385	.2363	.2432	4.1126	.9717	1.3323	20′
50′	.2414	.2391	.2462	4.0611	.9710	1.3294	10′
14° 00′	.2443	.2419	.2493	4.0108	.9703	1.3265	76° 00′
10′	.2473	.2447	.2524	3.9617	.9696	1.3235	50′
20′	.2502	.2476	.2555	3.9136	.9689	1.3206	40′
30′	.2531	.2504	.2586	3.8667	.9681	1.3177	30′
40′	.2560	.2532	.2617	3.8208	.9674	1.3148	20′
50′	.2589	.2560	.2648	3.7760	.9667	1.3119	10′
15° 00′	.2618	.2588	.2679	3.7321	.9659	1.3090	75° 00′
10′	.2647	.2616	.2711	3.6891	.9652	1.3061	50′
20′	.2676	.2644	.2742	3.6470	.9644	1.3032	40′
30′	.2705	.2672	.2773	3.6059	.9636	1.3003	30′
40′	.2734	.2700	.2805	3.5656	.9628	1.2974	20′
50′	.2763	.2728	.2836	3.5261	.9621	1.2945	10′
16° 00′	.2793	.2756	.2867	3.4874	.9613	1.2915	74° 00′
10′	.2822	.2784	.2899	3.4495	.9605	1.2886	50′
20′	.2851	.2812	.2931	3.4124	.9596	1.2857	40′
30′	.2880	.2840	.2962	3.3759	.9588	1.2828	30′
40′	.2909	.2868	.2994	3.3402	.9580	1.2799	20′
50′	.2938	.2896	.3026	3.3052	.9572	1.2770	10′
17° 00′	.2967	.2924	.3057	3.2709	.9563	1.2741	73° 00′
10′	.2996	.2952	.3089	3.2371	.9555	1.2712	50′
20′	.3025	.2979	.3121	3.2041	.9546	1.2683	40′
30′	.3054	.3007	.3153	3.1716	.9537	1.2654	30′
40′	.3083	.3035	.3185	3.1397	.9528	1.2625	20′
50′	.3113	.3062	.3217	3.1084	.9520	1.2595	10′
18° 00′	.3142	.3090	.3249	3.0777	.9511	1.2566	72° 00′
		Cosine	Cotangent	Tangent	Sine	Radians	Degrees

Cont.

Appendix 3 VALUES OF TRIGONOMETRIC FUNCTIONS (Cont.)

Degrees	Radians	Sine	Tangent	Cotangent	Cosine		
18° 00′	.3142	.3090	.3249	3.0777	.9511	1.2566	72° 00′
10′	.3171	.3118	.3281	3.0475	.9502	1.2537	50′
20′	.3200	.3145	.3314	3.0178	.9492	1.2508	40′
30′	.3229	.3173	.3346	2.9887	.9483	1.2479	30′
40′	.3258	.3201	.3378	2.9600	.9474	1.2450	20′
50′	.3287	.3228	.3411	2.9319	.9465	1.2421	10′
19° 00′	.3316	.3256	.3443	2.9042	.9455	1.2392	71° 00′
10′	.3345	.3283	.3476	2.8770	.9446	1.2363	50′
20′	.3374	.3311	.3508	2.8502	.9436	1.2334	40′
30′	.3403	.3338	.3541	2.8239	.9426	1.2305	30′
40′	.3432	.3365	.3574	2.7980	.9417	1.2275	20′
50′	.3462	.3393	.3607	2.7725	.9407	1.2246	10′
20° 00′	.3491	.3420	.3640	2.7475	.9397	1.2217	70° 00′
10′	.3520	.3448	.3673	2.7228	.9387	1.2188	50′
20′	.3549	.3475	.3706	2.6985	.9377	1.2159	40′
30′	.3578	.3502	.3739	2.6746	.9367	1.2130	30′
40′	.3607	.3529	.3772	2.6511	.9356	1.2101	20′
50′	.3636	.3557	.3805	2.6279	.9346	1.2072	10′
21° 00′	.3665	.3584	.3839	2.6051	.9336	1.2043	69° 00′
10′	.3694	.3611	.3872	2.5826	.9325	1.2014	50′
20′	.3723	.3638	.3906	2.5605	.9315	1.1985	40′
30′	.3752	.3665	.3939	2.5386	.9304	1.1956	30′
40′	.3782	.3692	.3973	2.5172	.9293	1.1926	20′
50′	.3811	.3719	.4006	2.4960	.9283	1.1897	10′
22° 00′	.3840	.3746	.4040	2.4751	.9272	1.1868	68° 00′
10′	.3869	.3773	.4074	2.4545	.9261	1.1839	50′
20′	.3898	.3800	.4108	2.4342	.9250	1.1810	40′
30′	.3927	.3827	.4142	2.4142	.9239	1.1781	30′
40′	.3956	.3854	.4176	2.3945	.9228	1.1752	20′
50′	.3985	.3881	.4210	2.3750	.9216	1.1723	10′
23° 00′	.4014	.3907	.4245	2.3559	.9205	1.1694	67° 00′
10′	.4043	.3934	.4279	2.3369	.9194	1.1665	50′
20′	.4072	.3961	.4314	2.3183	.9182	1.1636	40′
30′	.4102	.3987	.4348	2.2998	.9171	1.1606	30′
40′	.4131	.4014	.4383	2.2817	.9159	1.1577	20′
50′	.4160	.4041	.4417	2.2637	.9147	1.1548	10′
24° 00′	.4189	.4067	.4452	2.2460	.9135	1.1519	66° 00′
10′	.4218	.4094	.4487	2.2286	.9124	1.1490	50′
20′	.4247	.4120	.4522	2.2113	.9112	1.1461	40′
30′	.4276	.4147	.4557	2.1943	.9100	1.1432	30′
40′	.4305	.4173	.4592	2.1775	.9088	1.1403	20′
50′	.4334	.4200	.4628	2.1609	.9075	1.1374	10′
25° 00′	.4363	.4226	.4663	2.1445	.9063	1.1345	65° 00′
10′	.4392	.4253	.4699	2.1283	.9051	1.1316	50′
20′	.4422	.4279	.4734	2.1123	.9038	1.1286	40′
30′	.4451	.4305	.4770	2.0965	.9026	1.1257	30′
40′	.4480	.4331	.4806	2.0809	.9013	1.1228	20′
50′	.4509	.4358	.4841	2.0655	.9001	1.1199	10′
26° 00′	.4538	.4384	.4877	2.0503	.8988	1.1170	64° 00′
10′	.4567	.4410	.4913	2.0353	.8975	1.1141	50′
20′	.4596	.4436	.4950	2.0204	.8962	1.1112	40′
30′	4625	.4462	.4986	2.0057	.8949	1.1083	30′
40′	.4654	.4488	.5022	1.9912	.8936	1.1054	20′
50′	.4683	.4514	.5059	1.9768	.8923	1.1025	10′
27° 00′	.4712	.4540	.5095	1.9626	.8910	1.0996	63° 00′
		Cosine	Cotangent	Tangent	Sine	Radians	Degrees

Appendix 3 VALUES OF TRIGONOMETRIC FUNCTIONS (Cont.)

Degrees	Radians	Sine	Tangent	Cotangent	Cosine		
27° 00′	.4712	.4540	.5095	1.9626	.8910	1.0996	**63° 00′**
10′	.4741	.4566	.5132	1.9486	.8897	1.0966	50′
20′	.4771	.4592	.5169	1.9347	.8884	1.0937	40′
30′	.4800	.4617	.5206	1.9210	.8870	1.0908	30′
40′	.4829	.4643	.5243	1.9074	.8857	1.0879	20′
50′	.4858	.4669	.5280	1.8940	.8843	1.0850	10′
28° 00′	.4887	.4695	.5317	1.8807	.8829	1.0821	**62° 00′**
10′	.4916	.4720	.5354	1.8676	.8816	1.0792	50′
20′	.4945	.4746	.5392	1.8546	.8802	1.0763	40′
30′	.4974	.4772	.5430	1.8418	.8788	1.0734	30′
40′	.5003	.4797	.5467	1.8291	.8774	1.0705	20′
50′	.5032	.4823	.5505	1.8165	.8760	1.0676	10′
29° 00′	.5061	.4848	.5543	1.8040	.8746	1.0647	**61° 00′**
10′	.5091	.4874	.5581	1.7917	.8732	1.0617	50′
20′	.5120	.4899	.5619	1.7796	.8718	1.0588	40′
30′	.5149	.4924	.5658	1.7675	.8704	1.0559	30′
40′	.5178	.4950	.5696	1.7556	.8689	1.0530	20′
50′	.5207	.4975	.5735	1.7437	.8675	1.0501	10′
30° 00′	.5236	.5000	.5774	1.7321	.8660	1.0472	**60° 00′**
10′	.5265	.5025	.5812	1.7205	.8646	1.0443	50′
20′	.5294	.5050	.5851	1.7090	.8631	1.0414	40′
30′	.5323	.5075	.5890	1.6977	.8616	1.0385	30′
40′	.5352	.5100	.5930	1.6864	.8601	1.0356	20′
50′	.5381	.5125	.5969	1.6753	.8587	1.0327	10′
31° 00′	.5411	.5150	.6009	1.6643	.8572	1.0297	**59° 00′**
10′	.5440	.5175	.6048	1.6534	.8557	1.0268	50′
20′	.5469	.5200	.6088	1.6426	.8542	1.0239	40′
30′	.5498	.5225	.6128	1.6319	.8526	1.0210	30′
40′	.5527	.5250	.6168	1.6212	.8511	1.0181	20′
50′	.5556	.5275	.6208	1.6107	.8496	1.0152	10′
32° 00′	.5585	.5299	.6249	1.6003	.8480	1.0123	**58° 00′**
10′	.5614	.5324	.6289	1.5900	.8465	1.0094	50′
20′	.5643	.5348	.6330	1.5798	.8450	1.0065	40′
30′	.5672	.5373	.6371	1.5697	.8434	1.0036	30′
40′	.5701	.5398	.6412	1.5597	.8418	1.0007	20′
50′	.5730	.5422	.6453	1.5497	.8403	.9977	10′
33° 00′	.5760	.5446	.6494	1.5399	.8387	.9948	**57° 00′**
10′	.5789	.5471	.6536	1.5301	.8371	.9919	50′
20′	.5818	.5495	.6577	1.5204	.8355	.9890	40′
30′	.5847	.5519	.6619	1.5108	.8339	.9861	30′
40′	.5876	.5544	.6661	1.5013	.8323	.9832	20′
50′	.5905	.5568	.6703	1.4919	.8307	.9803	10′
34° 00′	.5934	.5592	.6745	1.4826	.8290	.9774	**56° 00′**
10′	.5963	.5616	.6787	1.4733	.8274	.9745	50′
20′	.5992	.5640	.6830	1.4641	.8258	.9716	40′
30′	.6021	.5664	.6873	1.4550	.8241	.9687	30′
40′	.6050	.5688	.6916	1.4460	.8225	.9657	20′
50′	.6080	.5712	.6959	1.4370	.8208	.9628	10′
35° 00′	.6109	.5736	.7002	1.4281	.8192	.9599	**55° 00′**
10′	.6138	.5760	.7046	1.4193	.8175	.9570	50′
20′	.6167	.5783	.7089	1.4106	.8158	.9541	40′
30′	.6196	.5807	.7133	1.4019	.8141	.9512	30′
40′	.6225	.5831	.7177	1.3934	.8124	.9483	20′
50′	.6254	.5854	.7221	1.3848	.8107	.9454	10′
36° 00′	.6283	.5878	.7265	1.3764	.8090	.9425	**54° 00′**
		Cosine	Cotangent	Tangent	Sine	Radians	Degrees

Cont.

Appendix 3 VALUES OF TRIGONOMETRIC FUNCTIONS (Cont.)

Degrees	Radians	Sine	Tangent	Cotangent	Cosine		
36° 00′	.6283	.5878	.7265	1.3764	.8090	.9425	54° 00′
10′	.6312	.5901	.7310	1.3680	.8073	.9396	50′
20′	.6341	.5925	.7355	1.3597	.8056	.9367	40′
30′	.6370	.5948	.7400	1.3514	.8039	.9338	30′
40′	.6400	.5972	.7445	1.3432	.8021	.9308	20′
50′	.6429	.5995	.7490	1.3351	.8004	.9279	10′
37° 00′	.6458	.6018	.7536	1.3270	.7986	.9250	53° 00′
10′	.6487	.6041	.7581	1.3190	.7969	.9221	50′
20′	.6516	.6065	.7627	1.3111	.7951	.9192	40′
30′	.6545	.6088	.7673	1.3032	.7934	.9163	30′
40′	.6574	.6111	.7720	1.2954	.7916	.9134	20′
50′	.6603	.6134	.7766	1.2876	.7898	.9105	10′
38° 00′	.6632	.6157	.7813	1.2799	.7880	.9076	52° 00′
10′	.6661	.6180	.7860	1.2723	.7862	.9047	50′
20′	.6690	.6202	.7907	1.2647	.7844	.9018	40′
30′	.6720	.6225	.7954	1.2572	.7826	.8988	30′
40′	.6749	.6248	.8002	1.2497	.7808	.8959	20′
50′	.6778	.6271	.8050	1.2423	.7790	.8930	10′
39° 00′	.6807	.6293	.8098	1.2349	.7771	.8901	51° 00′
10′	.6836	.6316	.8146	1.2276	.7753	.8872	50′
20′	.6865	.6338	.8195	1.2203	.7735	.8843	40′
30′	.6894	.6361	.8243	1.2131	.7716	.8814	30′
40′	.6923	.6383	.8292	1.2059	.7698	.8785	20′
50′	.6952	.6406	.8342	1.1988	.7679	.8756	10′
40° 00′	.6981	.6428	.8391	1.1918	.7660	.8727	50° 00′
10′	.7010	.6450	.8441	1.1847	.7642	.8698	50′
20′	.7039	.6472	.8491	1.1778	.7623	.8668	40′
30′	.7069	.6494	.8541	1.1708	.7604	.8639	30′
40′	.7098	.6517	.8591	1.1640	.7585	.8610	20′
50′	.7127	.6539	.8642	1.1571	.7566	.8581	10′
41° 00′	.7156	.6561	.8693	1.1504	.7547	.8552	49° 00′
10′	.7185	.6583	.8744	1.1436	.7528	.8523	50′
20′	.7214	.6604	.8796	1.1369	.7509	.8494	40′
30′	.7243	.6626	.8847	1.1303	.7490	.8465	30′
40′	.7272	.6648	.8899	1.1237	.7470	.8436	20′
50′	.7301	.6670	.8952	1.1171	.7451	.8407	10′
42° 00′	.7330	.6691	.9004	1.1106	.7431	.8378	48° 00′
10′	.7359	.6713	.9057	1.1041	.7412	.8348	50′
20′	.7389	.6734	.9110	1.0977	.7392	.8319	40′
30′	.7418	.6756	.9163	1.0913	.7373	.8290	30′
40′	.7447	.6777	.9217	1.0850	.7353	.8261	20′
50′	.7476	.6799	.9271	1.0786	.7333	.8232	10′
43° 00′	.7505	.6820	.9325	1.0724	.7314	.8203	47° 00′
10′	.7534	.6841	.9380	1.0661	.7294	.8174	50′
20′	.7563	.6862	.9435	1.0599	.7274	.8145	40′
30′	.7592	.6884	.9490	1.0538	.7254	.8116	30′
40′	.7621	.6905	.9545	1.0477	.7234	.8087	20′
50′	.7650	.6926	.9601	1.0416	.7214	.8058	10′
44° 00′	.7679	.6947	.9657	1.0355	.7193	.8029	46° 00′
10′	.7709	.6967	.9713	1.0295	.7173	.7999	50′
20′	.7738	.6988	.9770	1.0235	.7153	.7970	40′
30′	.7767	.7009	.9827	1.0176	.7133	.7941	30′
40′	.7796	.7030	.9884	1.0117	.7112	.7912	20′
50′	.7825	.7050	.9942	1.0058	.7092	.7883	10′
45° 00′	.7854	.7071	1.0000	1.0000	.7071	.7854	45° 00′
		Cosine	Cotangent	Tangent	Sine	Radians	Degrees

Appendix 4 WEIGHTS AND MEASURES

UNITED STATES SYSTEM

LINEAR MEASURE

Inches	Feet	Yards	Rods	Furlongs	Miles
1.0 =	.08333 =	.02778 =	.0050505 =	.00012626 =	.00001578
12.0 =	1.0 =	.33333 =	.0606061 =	.00151515 =	.00018939
36.0 =	3.0 =	1.0 =	.1818182 =	.00454545 =	.00056818
198.0 =	16.5 =	5.5 =	1.0 =	.025 =	.003125
7920.0 =	660.0 =	220.0 =	40.0 =	1.0 =	.125
63360.0 =	5280.0 =	1760.0 =	320.0 =	8.0 =	1.0

SQUARE AND LAND MEASURE

Sq. Inches	Square Feet	Square Yards	Sq. Rods	Acres	Sq. Miles
1.0 =	.006944 =	.000772			
144.0 =	1.0 =	.111111			
1296.0 =	9.0 =	1.0 =	.03306 =	.000207	
39204.0 =	272.25 =	30.25 =	1.0 =	.00625 =	.0000098
	43560.0 =	4840.0 =	160.0 =	1.0 =	.0015625
		3097600.0 =	102400.0 =	640.0 =	1.0

AVOIRDUPOIS WEIGHTS

Grains	Drams	Ounces	Pounds	Tons
1.0 =	.03657 =	.002286 =	.000143 =	.0000000714
27.34375 =	1.0 =	.0625 =	.003906 =	.00000195
437.5 =	16.0 =	1.0 =	.0625 =	.00003125
7000.0 =	256.0 =	16.0 =	1.0 =	.0005
14000000.0 =	512000.0 =	32000.0 =	2000.0 =	1.0

DRY MEASURE

Pints	Quarts	Pecks	Cubic Feet	Bushels
1.0 =	.5 =	.0625 =	.01945 =	.01563
2.0 =	1.0 =	.125 =	.03891 =	.03125
16.0 =	8.0 =	1.0 =	.31112 =	.25
51.42627 =	25.71314 =	3.21414 =	1.0 =	.80354
64.0 =	32.0 =	4.0 =	1.2445 =	1.0

LIQUID MEASURE

Gills	Pints	Quarts	U. S. Gallons	Cubic Feet
1.0 =	.25 =	.125 =	.03125 =	.00418
4.0 =	1.0 =	.5 =	.125 =	.01671
8.0 =	2.0 =	1.0 =	.250 =	.03342
32.0 =	8.0 =	4.0 =	1.0 =	.1337
			7.48052 =	1.0

METRIC SYSTEM

UNITS

Length—Meter : Mass—Gram : Capacity—Liter

for pure water at 4°C. (39.2°F.)

1 cubic decimeter or 1 liter = 1 kilogram

1000 Milli{*meters* (mm), *grams* (mg), *liters* (ml)} = 100 Centi{*meters* (cm), *grams* (cg), *liters* (cl)} = 10 Deci{*meters* (dm), *grams* (dg), *liters* (dl)} = 1{meter, gram, liter}

1000{meters, grams, liters} = 100 Deka{*meters* (dkm), *grams* (dkg), *liters* (dkl)} = 10 Hecto{*meters* (hm), *grams* (hg), *liters* (hl)} = 1 Kilo{*meter* (km), *gram* (kg), *liter* (kl)}

1 Metric Ton	=	1000 Kilograms
100 Square Meters	=	1 Are
100 Ares	=	1 Hectare
100 Hectares	=	1 Square Kilometer

DECIMAL EQUIVALENTS — INCH-MILLIMETER CONVERSION TABLE

1/2	1/4	1/8	1/16	1/32	1/64	Decimals	Millimeters	1/2	1/4	1/8	1/16	1/32	1/64	Decimals	Millimeters
					1	.015625	.396875						33	.515625	13.096875
				1		.031250	.793750					17		.531250	13.493750
					3	.046875	1.190625						35	.546875	13.890625
			1			.062500	1.587500				9			.562500	14.287500
					5	.078125	1.984375						37	.578125	14.684375
				3		.093750	2.381250					19		.593750	15.081250
					7	.109375	2.778125						39	.609375	15.478125
		1				.125000	3.175000			5				.625000	15.875000
					9	.140625	3.571875						41	.640625	16.271875
				5		.156250	3.968750					21		.656250	16.668750
					11	.171875	4.365625						43	.671875	17.065625
			3			.187500	4.762500				11			.687500	17.462500
					13	.203125	5.159375						45	.703125	17.859375
				7		.218750	5.556250					23		.718750	18.256250
					15	.234375	5.953125						47	.734375	18.653125
	1					.250000	6.350000		3					.750000	19.050000
					17	.265625	6.746875						49	.765625	19.446875
				9		.281250	7.143750					25		.781250	19.843750
					19	.296875	7.540625						51	.796875	20.240625
			5			.312500	7.937500				13			.812500	20.637500
					21	.328125	8.334375						53	.828125	21.034375
				11		.343750	8.731250					27		.843750	21.431250
					23	.359375	9.128125						55	.859375	21.828125
		3				.375000	9.525000			7				.875000	22.225000
					25	.390625	9.921875						57	.890625	22.621875
				13		.406250	10.318750					29		.906250	23.018750
					27	.421875	10.715625						59	.921875	23.415625
			7			.437500	11.112500				15			.937500	23.812500
					29	.453125	11.509375						61	.953125	24.209375
				15		.468750	11.906250					31		.968750	24.606250
					31	.484375	12.303125						63	.984375	25.003125
1						.500000	12.700000	2	4	8	16	32	64	1.000000	25.400000

Appendix 5 DECIMAL EQUIVALENTS AND TEMPERATURE CONVERSION (Cont.)

TEMPERATURE CONVERSION

-210 to 0			1 to 25			26 to 50			51 to 75			76 to 100			101 to 340			341 to 490			491 to 750		
C.	C. or F.	F.	C.	C. or F.	F.	C.	C. or F.	F.	C.	C. or F.	F.	C.	C. or F.	F.	C.	C. or F.	F.	C.	C. or F.	F.	C.	C. or F.	F.
-134	**-210**	-346	-17.2	**1**	33.8	-3.33	**26**	78.8	10.6	**51**	123.8	24.4	**76**	168.8	43	**110**	230	177	**350**	662	260	**500**	932
-129	**-200**	-328	-16.7	**2**	35.6	-2.78	**27**	80.6	11.1	**52**	125.6	25.0	**77**	170.6	49	**120**	248	182	**360**	680	266	**510**	950
-123	**-190**	-310	-16.1	**3**	37.4	-2.22	**28**	82.4	11.7	**53**	127.4	25.6	**78**	172.4	54	**130**	266	188	**370**	698	271	**520**	968
-118	**-180**	-292	-15.6	**4**	39.2	-1.67	**29**	84.2	12.2	**54**	129.2	26.1	**79**	174.2	60	**140**	284	193	**380**	716	277	**530**	986
-112	**-170**	-274	-15.0	**5**	41.0	-1.11	**30**	86.0	12.8	**55**	131.0	26.7	**80**	176.0	66	**150**	302	199	**390**	734	282	**540**	1004
-107	**-160**	-256	-14.4	**6**	42.8	-0.56	**31**	87.8	13.3	**56**	132.8	27.2	**81**	177.8	71	**160**	320	204	**400**	752	288	**550**	1022
-101	**-150**	-238	-13.9	**7**	44.6	0	**32**	89.6	13.9	**57**	134.6	27.8	**82**	179.6	77	**170**	338	210	**410**	770	293	**560**	1040
-95.6	**-140**	-220	-13.3	**8**	46.4	0.56	**33**	91.4	14.4	**58**	136.4	28.3	**83**	181.4	82	**180**	356	216	**420**	788	299	**570**	1058
-90.0	**-130**	-202	-12.8	**9**	48.2	1.11	**34**	93.2	15.0	**59**	138.2	28.9	**84**	183.2	88	**190**	374	221	**430**	806	304	**580**	1076
-84.4	**-120**	-184	-12.2	**10**	50.0	1.67	**35**	95.0	15.6	**60**	140.0	29.4	**85**	185.0	93	**200**	392	227	**440**	824	310	**590**	1094
-78.9	**-110**	-166	-11.7	**11**	51.8	2.22	**36**	96.8	16.1	**61**	141.8	30.0	**86**	186.8	99	**210**	410	232	**450**	842	316	**600**	1112
-73.3	**-100**	-148	-11.1	**12**	53.6	2.78	**37**	98.6	16.7	**62**	143.6	30.6	**87**	188.6	100	**212**	413	238	**460**	860	321	**610**	1130
-67.8	**-90**	-130	-10.6	**13**	55.4	3.33	**38**	100.4	17.2	**63**	145.4	31.1	**88**	190.4	104	**220**	428	243	**470**	878	327	**620**	1148
-62.2	**-80**	-112	-10.0	**14**	57.2	3.89	**39**	102.2	17.8	**64**	147.2	31.7	**89**	192.2	110	**230**	446	249	**480**	896	332	**630**	1166
-56.7	**-70**	-94	-9.44	**15**	59.0	4.44	**40**	104.0	18.3	**65**	149.0	32.2	**90**	194.0	116	**240**	464	254	**490**	914	338	**640**	1184
-51.1	**-60**	-76	-8.89	**16**	60.8	5.00	**41**	105.8	18.9	**66**	150.8	32.8	**91**	195.8	121	**250**	482				343	**650**	1202
-45.6	**-50**	-58	-8.33	**17**	62.6	5.56	**42**	107.6	19.4	**67**	152.6	33.3	**92**	197.6	127	**260**	500				349	**660**	1220
-40.0	**-40**	-40	-7.78	**18**	64.4	6.11	**43**	109.4	20.0	**68**	154.4	33.9	**93**	199.4	132	**270**	518				354	**670**	1238
-34.4	**-30**	-22	-7.22	**19**	66.2	6.67	**44**	111.2	20.6	**69**	156.2	34.4	**94**	201.2	138	**280**	536				360	**680**	1256
-28.9	**-20**	-4	-6.67	**20**	68.0	7.22	**45**	113.0	21.1	**70**	158.0	35.0	**95**	203.0	143	**290**	554				366	**690**	1274
-23.3	**-10**	14	-6.11	**21**	69.8	7.78	**46**	114.8	21.7	**71**	159.8	35.6	**96**	204.8	149	**300**	572				371	**700**	1292
-17.8	**0**	32	-5.56	**22**	71.6	8.33	**47**	116.6	22.2	**72**	161.6	36.1	**97**	206.6	154	**310**	590				377	**710**	1310
			-5.00	**23**	73.4	8.89	**48**	118.4	22.8	**73**	163.4	36.7	**98**	208.4	160	**320**	608				382	**720**	1328
			-4.44	**24**	75.2	9.44	**49**	120.2	23.3	**74**	165.2	37.2	**99**	210.2	166	**330**	626				388	**730**	1346
			-3.89	**25**	77.0	10.0	**50**	122.0	23.9	**75**	167.0	37.8	**100**	212.0	171	**340**	644				393	**740**	1364
																					399	**750**	1382

NOTE:—The numbers in bold face type refer to the temperature either in degrees Centigrade or Fahrenheit which it is desired to convert into the other scale. If converting from Fahrenheit degrees to Centigrade degrees the equivalent temperature will be found in the left column, while if converting from degrees Centigrade to degrees Fahrenheit, the answer will be found in the column on the right.

$$°F = \frac{9}{5}(°C) + 32$$

$$°C = \frac{5}{9}(°F - 32)$$

INTERPOLATION FACTORS

C.		F.	C.		F.
0.56	**1**	1.8	3.33	**6**	10.8
1.11	**2**	3.6	3.89	**7**	12.6
1.67	**3**	5.4	4.44	**8**	14.4
2.22	**4**	7.2	5.00	**9**	16.2
2.78	**5**	9.0	5.56	**10**	18.0

Appendix 6 WEIGHTS AND SPECIFIC GRAVITIES

Substance	Weight Lb. per Cu. Ft.	Specific Gravity
METALS, ALLOYS, ORES		
Aluminum, cast, hammered	165	2.55-2.75
Brass, cast, rolled	534	8.4-8.7
Bronze, 7.9 to 14% Sn	509	7.4-8.9
Bronze, aluminum	481	7.7
Copper, cast, rolled	556	8.8-9.0
Copper ore, pyrites	262	4.1-4.3
Gold, cast, hammered	1205	19.25-19.3
Iron, cast, pig	450	7.2
Iron, wrought	485	7.6-7.9
Iron, spiegel-eisen	468	7.5
Iron, ferro-silicon	437	6.7-7.3
Iron ore, hematite	325	5.2
Iron ore, hematite in bank	160-180	....
Iron ore, hematite loose	130-160	....
Iron ore, limonite	237	3.6-4.0
Iron ore, magnetite	315	4.9-5.2
Iron slag	172	2.5-3.0
Lead	710	11.37
Lead ore, galena	465	7.3-7.6
Magnesium, alloys	112	1.74-1.83
Manganese	475	7.2-8.0
Manganese ore, pyrolusite	259	3.7-4.6
Mercury	849	13.6
Monel Metal	556	8.8-9.0
Nickel	565	8.9-9.2
Platinum, cast, hammered	1330	21.1-21.5
Silver, cast, hammered	656	10.4-10.6
Steel, rolled	490	7.85
Tin, cast, hammered	459	7.2-7.5
Tin ore, cassiterite	418	6.4-7.0
Zinc, cast, rolled	440	6.9-7.2
Zinc ore, blende	253	3.9-4.2
VARIOUS SOLIDS		
Cereals, oats ... bulk	32	....
Cereals, barley ... bulk	39	....
Cereals, corn, rye ... bulk	48	....
Cereals, wheat ... bulk	48	....
Hay and Straw ... bales	20	....
Cotton, Flax, Hemp	93	1.47-1.50
Fats	58	0.90-0.97
Flour, loose	28	0.40-0.50
Flour, pressed	47	0.70-0.80
Glass, common	156	2.40-2.60
Glass, plate or crown	161	2.45-2.72
Glass, crystal	184	2.90-3.00
Leather	59	0.86-1.02
Paper	58	0.70-1.15
Potatoes, piled	42	....
Rubber, caoutchouc	59	0.92-0.96
Rubber goods	94	1.0-2.0
Salt, granulated, piled	48	....
Saltpeter	67	....
Starch	96	1.53
Sulphur	125	1.93-2.07
Wool	82	1.32

Substance	Weight Lb. per Cu. Ft.	Specific Gravity
TIMBER, U. S. SEASONED		
Moisture Content by Weight:		
Seasoned timber 15 to 20%		
Green timber up to 50%		
Ash, white, red	40	0.62-0.65
Cedar, white, red	22	0.32-0.38
Chestnut	41	0.66
Cypress	30	0.48
Fir, Douglas spruce	32	0.51
Fir, eastern	25	0.40
Elm, white	45	0.72
Hemlock	29	0.42-0.52
Hickory	49	0.74-0.84
Locust	46	0.73
Maple, hard	43	0.68
Maple, white	33	0.53
Oak, chestnut	54	0.86
Oak, live	59	0.95
Oak, red, black	41	0.65
Oak, white	46	0.74
Pine, Oregon	32	0.51
Pine, red	30	0.48
Pine, white	26	0.41
Pine, yellow, long-leaf	44	0.70
Pine, yellow, short-leaf	38	0.61
Poplar	30	0.48
Redwood, California	26	0.42
Spruce, white, black	27	0.40-0.46
Walnut, black	38	0.61
VARIOUS LIQUIDS		
Alcohol, 100%	49	0.79
Acids, muriatic 40%	75	1.20
Acids, nitric 91%	94	1.50
Acids, sulphuric 87%	112	1.80
Lye, soda 66%	106	1.70
Oils, vegetable	58	0.91-0.94
Oils, mineral, lubricants	57	0.90-0.93
Water, 4°C. max. density	62.428	1.0
Water, 100°C.	59.830	0.9584
Water, ice	56	0.88-0.92
Water, snow, fresh fallen	8	.125
Water, sea water	64	1.02-1.03
GASES		
Air, 0°C. 760 mm.	.08071	1.0
Ammonia	.0478	0.5920
Carbon dioxide	.1234	1.5291
Carbon monoxide	.0781	0.9673
Gas, illuminating	.028-.036	0.35-0.45
Gas, natural	.038-.039	0.47-0.48
Hydrogen	.00559	0.0693
Nitrogen	.0784	0.9714
Oxygen	.0892	1.1056

The specific gravities of solids and liquids refer to water at 4°C., those of gases to air at 0°C. and 760 mm. pressure. The weights per cubic foot are derived from average specific gravities, except where stated that weights are for bulk, heaped or loose material, etc.

(Courtesy of the American Institute of Steel Construction.)

Appendix 6 WEIGHTS AND SPECIFIC GRAVITIES (Cont.)

Substance	Weight Lb. per Cu. Ft.	Specific Gravity
ASHLAR MASONRY		
Granite, syenite, gneiss	165	2.3-3.0
Limestone, marble	160	2.3-2.8
Sandstone, bluestone	140	2.1-2.4
MORTAR RUBBLE MASONRY		
Granite, syenite, gneiss	155	2.2-2.8
Limestone, marble	150	2.2-2.6
Sandstone, bluestone	130	2.0-2.2
DRY RUBBLE MASONRY		
Granite, syenite, gneiss	130	1.9-2.3
Limestone, marble	125	1.9-2.1
Sandstone, bluestone	110	1.8-1.9
BRICK MASONRY		
Pressed brick	140	2.2-2.3
Common brick	120	1.8-2.0
Soft brick	100	1.5-1.7
CONCRETE MASONRY		
Cement, stone, sand	144	2.2-2.4
Cement, slag, etc.	130	1.9-2.3
Cement, cinder, etc.	100	1.5-1.7
VARIOUS BUILDING MATERIALS		
Ashes, cinders	40-45	....
Cement, portland, loose	90	....
Cement, portland, set	183	2.7-3.2
Lime, gypsum, loose	53-64	....
Mortar, set	103	1.4-1.9
Slags, bank slag	67-72	....
Slags, bank screenings	98-117	....
Slags, machine slag	96	....
Slags, slag sand	49-55	....
EARTH, ETC., EXCAVATED		
Clay, dry	63	....
Clay, damp, plastic	110	....
Clay and gravel, dry	100	....
Earth, dry, loose	76	....
Earth, dry, packed	95	....
Earth, moist, loose	78	....
Earth, moist, packed	96	....
Earth, mud, flowing	108	....
Earth, mud, packed	115	....
Riprap, limestone	80-85	....
Riprap, sandstone	90	....
Riprap, shale	105	....
Sand, gravel, dry, loose	90-105	....
Sand, gravel, dry, packed	100-120	....
Sand, gravel, dry, wet	118-120	....
EXCAVATIONS IN WATER		
Sand or gravel	60	....
Sand or gravel and clay	65	....
Clay	80	....
River mud	90	....
Soil	70	....
Stone riprap	65	....
MINERALS		
Asbestos	153	2.1-2.8
Barytes	281	4.50
Basalt	184	2.7-3.2
Bauxite	159	2.55
Borax	109	1.7-1.8
Chalk	137	1.8-2.6
Clay, marl	137	1.8-2.6
Dolomite	181	2.9
Feldspar, orthoclase	159	2.5-2.6
Gneiss, serpentine	159	2.4-2.7
Granite, syenite	175	2.5-3.1
Greenstone, trap	187	2.8-3.2
Gypsum, alabaster	159	2.3-2.8
Hornblende	187	3.0
Limestone, marble	165	2.5-2.8
Magnesite	187	3.0
Phosphate rock, apatite	200	3.2
Porphyry	172	2.6-2.9
Pumice, natural	40	0.37-0.90
Quartz, flint	165	2.5-2.8
Sandstone, bluestone	147	2.2-2.5
Shale, slate	175	2.7-2.9
Soapstone, talc	169	2.6-2.8
STONE, QUARRIED, PILED		
Basalt, granite, gneiss	96	....
Limestone, marble, quartz	95	....
Sandstone	82	....
Shale	92	....
Greenstone, hornblende	107	....
BITUMINOUS SUBSTANCES		
Asphaltum	81	1.1-1.5
Coal, anthracite	97	1.4-1.7
Coal, bituminous	84	1.2-1.5
Coal, lignite	78	1.1-1.4
Coal, peat, turf, dry	47	0.65-0.85
Coal, charcoal, pine	23	0.28-0.44
Coal, charcoal, oak	33	0.47-0.57
Coal, coke	75	1.0-1.4
Graphite	131	1.9-2.3
Paraffine	56	0.87-0.91
Petroleum	54	0.87
Petroleum, refined	50	0.79-0.82
Petroleum, benzine	46	0.73-0.75
Petroleum, gasoline	42	0.66-0.69
Pitch	69	1.07-1.15
Tar, bituminous	75	1.20
COAL AND COKE, PILED		
Coal, anthracite	47-58	....
Coal, bituminous, lignite	40-54	....
Coal, peat, turf	20-26	....
Coal, charcoal	10-14	....
Coal, coke	23-32	....

The specific gravities of solids and liquids refer to water at 4°C., those of gases to air at 0°C. and 760 mm. pressure. The weights per cubic foot are derived from average specific gravities, except where stated that weights are for bulk, heaped or loose material, etc.

Appendix 7 WIRE AND SHEET METAL GAGES

WIRE AND SHEET METAL GAGES IN DECIMALS OF AN INCH

Name of Gage	United States Standard Gage*		The United States Steel Wire Gage	American or Brown & Sharpe Wire Gage	New Birmingham Standard Sheet & Hoop Gage	British Imperial or English Legal Standard Wire Gage	Birmingham or Stubs Iron Wire Gage	Name of Gage
Principal Use	Uncoated Steel Sheets and Light Plates		Steel Wire except Music Wire	Non-Ferrous Sheets and Wire	Iron and Steel Sheets and Hoops	Wire	Strips, Bands, Hoops and Wire	Principal Use
Gage No.	Weight Oz. per Sq. Ft.	Approx. Thickness Inches	Thickness, Inches					Gage No.
7/0's			.4900		.6666	.500		7/0's
6/0's			.4615	.5800	.625	.464		6/0's
5/0's			.4305	.5165	.5883	.432	.500	5/0's
4/0's			.3938	.4600	.5416	.400	.454	4/0's
3/0's			.3625	.4096	.500	.372	.425	3/0's
2/0's			.3310	.3648	.4452	.348	.380	2/0's
0			.3065	.3249	.3964	.324	.340	0
1			.2830	.2893	.3532	.300	.300	1
2			.2625	.2576	.3147	.276	.284	2
3	160	.2391	.2437	.2294	.2804	.252	.259	3
4	150	.2242	.2253	.2043	.250	.232	.238	4
5	140	.2092	.2070	.1819	.2225	.212	.220	5
6	130	.1943	.1920	.1620	.1981	.192	.203	6
7	120	.1793	.1770	.1443	.1764	.176	.180	7
8	110	.1644	.1620	.1285	.1570	.160	.165	8
9	100	.1495	.1483	.1144	.1398	.144	.148	9
10	90	.1345	.1350	.1019	.1250	.128	.134	10
11	80	.1196	.1205	.0907	.1113	.116	.120	11
12	70	.1046	.1055	.0808	.0991	.104	.109	12
13	60	.0897	.0915	.0720	.0882	.092	.095	13
14	50	0747	.0800	.0641	.0785	.080	.083	14
15	45	.0673	.0720	.0571	.0699	.072	.072	15
16	40	.0598	.0625	.0508	.0625	.064	.065	16
17	36	.0538	.0540	.0453	.0556	.056	.058	17
18	32	.0478	.0475	.0403	.0495	.048	.049	18
19	28	.0418	.0410	.0359	.0440	.040	.042	19
20	24	.0359	.0348	.0320	.0392	.036	.035	20
21	22	.0329	.0318	.0285	.0349	.032	.032	21
22	20	.0299	.0286	.0253	.0313	.028	.028	22
23	18	.0269	.0258	.0226	.0278	.024	.025	23
24	16	.0239	.0230	.0201	.0248	.022	.022	24
25	14	.0209	.0204	.0179	.0220	.020	.020	25
26	12	.0179	.0181	.0159	.0196	.018	.018	26
27	11	.0164	.0173	.0142	.0175	.0164	.016	27
28	10	.0149	.0162	.0126	.0156	.0148	.014	28
29	9	.0135	.0150	.0113	.0139	.0136	.013	29
30	8	.0120	.0140	.0100	.0123	.0124	.012	30
31	7	.0105	.0132	.0089	.0110	.0116	.010	31
32	6.5	.0097	.0128	.0080	.0098	.0108	.009	32
33	6	.0090	.0118	.0071	.0087	.0100	.008	33
34	5.5	.0082	.0104	.0063	.0077	.0092	.007	34
35	5	.0075	.0095	.0056	.0069	.0084	.005	35
36	4.5	.0067	.0090	.0050	.0061	.0076	.004	36
37	4.25	.0064	.0085	.0045	.0054	.0068		37
38	4	.0060	.0080	.0040	.0048	.0060		38
39			.0075	.0035	.0043	.0052		39
40			.0070	.0031	.0039	.0048		40

* U. S. Standard Gage is officially a weight gage, in oz. per sq. ft. as tabulated. The Approx. Thickness shown is the "Manufacturers' Standard" of the American Iron and Steel Institute, based on steel as weighing 501.81 lbs. per cu. ft. (489.6 true weight plus 2.5 percent for average over-run in area and thickness). The A.I.S.I. standard nomenclature for flat rolled carbon steel is as follows:

Widths, Inches	Thicknesses, Inch							
	0.2500 and thicker	0.2499 to 0.2031	0.2030 to 0.1875	0.1874 to 0.0568	0.0567 to 0.0344	0.0343 to 0.0255	0.0254 to 0.0142	0.0141 and thinner
To 3½ incl.	Bar	Bar	Strip	Strip	Strip	Strip	Sheet	Sheet
Over 3½ to 6 incl.	Bar	Bar	Strip	Strip	Strip	Sheet	Sheet	Sheet
" 6 to 12 "	Plate	Strip	Strip	Strip	Sheet	Sheet	Sheet	Sheet
" 12 to 32 "	Plate	Sheet	Sheet	Sheet	Sheet	Sheet	Sheet	Black Plate
" 32 to 48 "	Plate	Sheet	Sheet	Sheet	Sheet	Sheet	Sheet	Sheet
" 48	Plate	Plate	Plate	Sheet	Sheet	Sheet	Sheet	—

Appendix 8 PIPING SYMBOLS

TYPE OF FITTING		DOUBLE LINE CONVENTION					SINGLE LINE CONVENTION					FLOW DIAGRAM
		FLANGED	SCREWED	B & S	WELDED	SOLDERED	FLANGED	SCREWED	B & S	WELDED	SOLDERED	
1	Joint											
2	Joint - Expansion											
3	Union											
4	Sleeve											
5	Reducer											
6	Reducer - Eccentric											
7	Reducing Flange											
8	Bushing											
9	Elbow - 45°											
10	Elbow - 90°											
11	Elbow - Long radius	L.R	L.R		L.R	L.R	L.R	L.R		L.R	L.R	
12	Elbow - (turned up)											
13	Elbow - (turned down)											
14	Elbow - Side outlet (outlet up)											
15	Elbow - Side outlet (outlet down)											
16	Elbow - Base											
17	Elbow - Double branch											
18	Elbow - Reducing											
19	Lateral											
20	Tee											
21	Tee - Single sweep											

Cont.

Appendix 8 PIPING SYMBOLS (Cont.)

TYPE OF FITTING		DOUBLE LINE CONVENTION					SINGLE LINE CONVENTION					FLOW DIAGRAM
		FLANGED	SCREWED	B & S	WELDED	SOLDERED	FLANGED	SCREWED	B & S	WELDED	SOLDERED	
22	Tee - Double sweep											
23	Tee - (outlet up)											
24	Tee - (outlet down)											
25	Tee - Side outlet (outlet up)											
26	Tee - Side outlet (outlet down)											
27	Cross											
28	Valve - Globe											
29	Valve - Angle											
30	Valve - Motor operated globe											Motor operated
31	Valve - Gate											
32	Valve - Angle gate											
33	Valve - Motor operated gate											Motor operated
34	Valve - Check											
35	Valve - Angle check											
36	Valve - Safety											
37	Valve - Angle safety											
38	Valve - Quick opening											
39	Valve - Float operating											
40	Stop Cock											

Appendix 9 AMERICAN STANDARD TAPER PIPE THREADS, NPT[1]

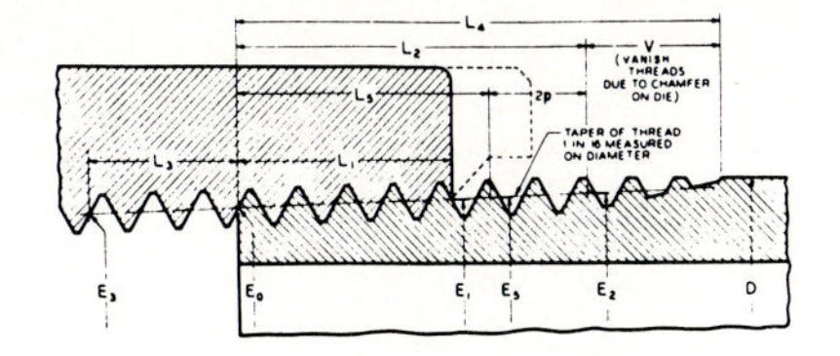

1	2	3	4	5	6	7	8	9	10	11
					Hand-Tight Engagement			Effective Thread, External		
					Length[2] L_1			Length L_2		Dia E_2
Nominal Pipe Size	Outside Diameter of Pipe D	Threads per Inch n	Pitch of Thread p	Pitch Diameter at Beginning of External Thread E_0	In.	Thds	Dia E_1	In.	Thds	In.
$\frac{1}{16}$	0.3125	27	0.03704	0.27118	0.160	4.32	0.28118	0.2611	7.05	0.28750
$\frac{1}{8}$	0.405	27	0.03704	0.36351	0.180	4.86	0.37476	0.2639	7.12	0.38000
$\frac{1}{4}$	0.540	18	0.05556	0.47739	0.200	3.60	0.48989	0.4018	7.23	0.50250
$\frac{3}{8}$	0.675	18	0.05556	0.61201	0.240	4.32	0.62701	0.4078	7.34	0.63750
$\frac{1}{2}$	0.840	14	0.07143	0.75843	0.320	4.48	0.77843	0.5337	7.47	0.79179
$\frac{3}{4}$	1.050	14	0.07143	0.96768	0.339	4.75	0.98887	0.5457	7.64	1.00179
1	1.315	$11\frac{1}{2}$	0.08696	1.21363	0.400	4.60	1.23863	0.6828	7.85	1.25630
$1\frac{1}{4}$	1.660	$11\frac{1}{2}$	0.08696	1.55713	0.420	4.83	1.58338	0.7068	8.13	1.60130
$1\frac{1}{2}$	1.900	$11\frac{1}{2}$	0.08696	1.79609	0.420	4.83	1.82234	0.7235	8.32	1.84130
2	2.375	$11\frac{1}{2}$	0.08696	2.26902	0.436	5.01	2.29627	0.7565	8.70	2.31630
$2\frac{1}{2}$	2.875	8	0.12500	2.71953	0.682	5.46	2.76216	1.1375	9.10	2.79062
3	3.500	8	0.12500	3.34062	0.766	6.13	3.38850	1.2000	9.60	3.41562
$3\frac{1}{2}$	4.000	8	0.12500	3.83750	0.821	6.57	3.88881	1.2500	10.00	3.91562
4	4.500	8	0.12500	4.33438	0.844	6.75	4.38712	1.3000	10.40	4.41562
5	5.563	8	0.12500	5.39073	0.937	7.50	5.44929	1.4063	11.25	5.47862
6	6.625	8	0.12500	6.44609	0.958	7.66	6.50597	1.5125	12.10	6.54062
8	8.625	8	0.12500	8.43359	1.063	8.50	8.50003	1.7125	13.70	8.54062
10	10.750	8	0.12500	10.54531	1.210	9.68	10.62094	1.9250	15.40	10.66562
12	12.750	8	0.12500	12.53281	1.360	10.88	12.61781	2.1250	17.00	12.66562
14 OD	14.000	8	0.12500	13.77500	1.562	12.50	13.87262	2.2500	18.90	13.91562
16 OD	16.000	8	0.12500	15.76250	1.812	14.50	15.87575	2.4500	19.60	15.91562
18 OD	18.000	8	0.12500	17.75000	2.000	16.00	17.87500	2.6500	21.20	17.91562
20 OD	20.000	8	0.12500	19.73750	2.125	17.00	19.87031	2.8500	22.80	19.91562
24 OD	24.000	8	0.12500	23.71250	2.375	19.00	23.86094	3.2500	26.00	23.91562

All dimensions are given in inches.

[1] The basic dimensions of the American Standard Taper Pipe Thread are given in inches to four or five decimal places. While this implies a greater degree of precision than is ordinarily attained, these dimensions are the basis of gage dimensions and are so expressed for the purpose of eliminating errors in computations.

[2] Also length of thin ring gage and length from gaging notch to small end of plug gage.

(Courtesy of ANSI; B2.1–1960.)

Appendix 10 AMERICAN STANDARD 250-LB CAST IRON FLANGED FITTINGS

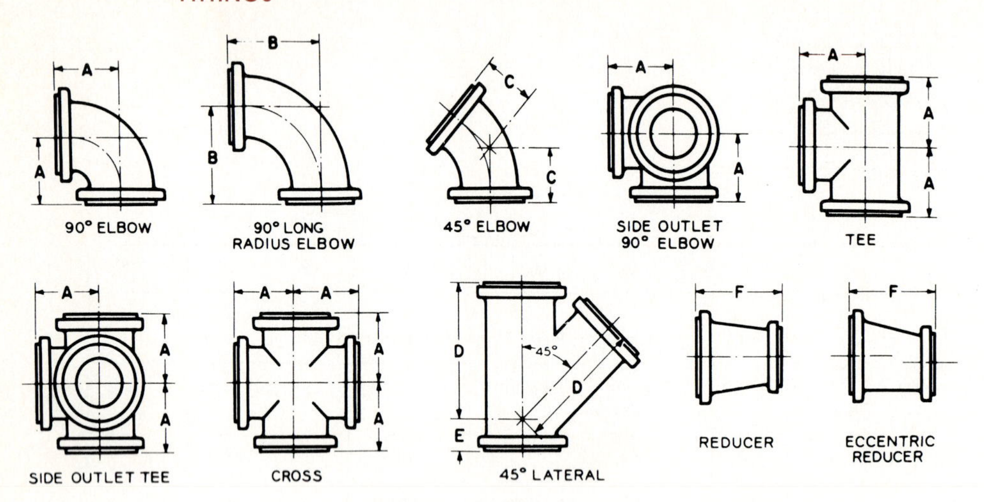

Dimensions of 250-lb Cast Iron Flanged Fittings

Nominal Pipe Size	Flanges			Fittings		Straight					
	Dia of Flange	Thick-ness of Flange (Min)	Dia of Raised Face	Inside Dia of Fittings (Min)	Wall Thick-ness	Center to Face 90 Deg Elbow Tees, Crosses and True "Y" A	Center to Face 90 Deg Long Radius Elbow B	Center to Face 45 Deg Elbow C	Center to Face Lateral D	Short Center to Face True "Y" and Lateral E	Face to Face Reducer F
1	4 7/8	11/16	2 11/16	1	7/16	4	5	2	6 1/2	2	....
1 1/4	5 1/4	3/4	3 1/16	1 1/4	7/16	4 1/4	5 1/2	2 1/2	7 1/4	2 1/4	....
1 1/2	6 1/8	13/16	3 9/16	1 1/2	7/16	4 1/2	6	2 3/4	8 1/2	2 1/2	....
2	6 1/2	7/8	4 3/16	2	7/16	5	6 1/2	3	9	2 1/2	5
2 1/2	7 1/2	1	4 15/16	2 1/2	1/2	5 1/2	7	3 1/2	10 1/2	2 1/2	5 1/2
3	8 1/4	1 1/8	5 11/16	3	9/16	6	7 3/4	3 1/2	11	3	6
3 1/2	9	1 3/16	6 5/16	3 1/2	9/16	6 1/2	8 1/2	4	12 1/2	3	6 1/2
4	10	1 1/4	6 15/16	4	5/8	7	9	4 1/2	13 1/2	3	7
5	11	1 3/8	8 5/16	5	11/16	8	10 1/4	5	15	3 1/2	8
6	12 1/2	1 7/16	9 11/16	6	3/4	8 1/2	11 1/2	5 1/2	17 1/2	4	9
8	15	1 5/8	11 15/16	8	13/16	10	14	6	20 1/2	5	11
10	17 1/2	1 7/8	14 1/16	10	15/16	11 1/2	16 1/2	7	24	5 1/2	12
12	20 1/2	2	16 7/16	12	1	13	19	8	27 1/2	6	14
14	23	2 1/8	18 15/16	13 1/4	1 1/8	15	21 1/2	8 1/2	31	6 1/2	16
16	25 1/2	2 1/4	21 1/16	15 1/4	1 1/4	16 1/2	24	9 1/2	34 1/2	7 1/2	18
18	28	2 3/8	23 5/16	17	1 3/8	18	26 1/2	10	37 1/2	8	19
20	30 1/2	2 1/2	25 9/16	19	1 1/2	19 1/2	29	10 1/2	40 1/2	8 1/2	20
24	36	2 3/4	30 5/16	23	1 5/8	22 1/2	34	12	47 1/2	10	24
30	43	3	37 3/16	29	2	27 1/2	41 1/2	15	....	....	30

All dimensions are given in inches.
(Courtesy of ANSI; B16.1–1967.)

Appendix 11 AMERICAN STANDARD 125-LB CAST IRON FLANGED FITTINGS

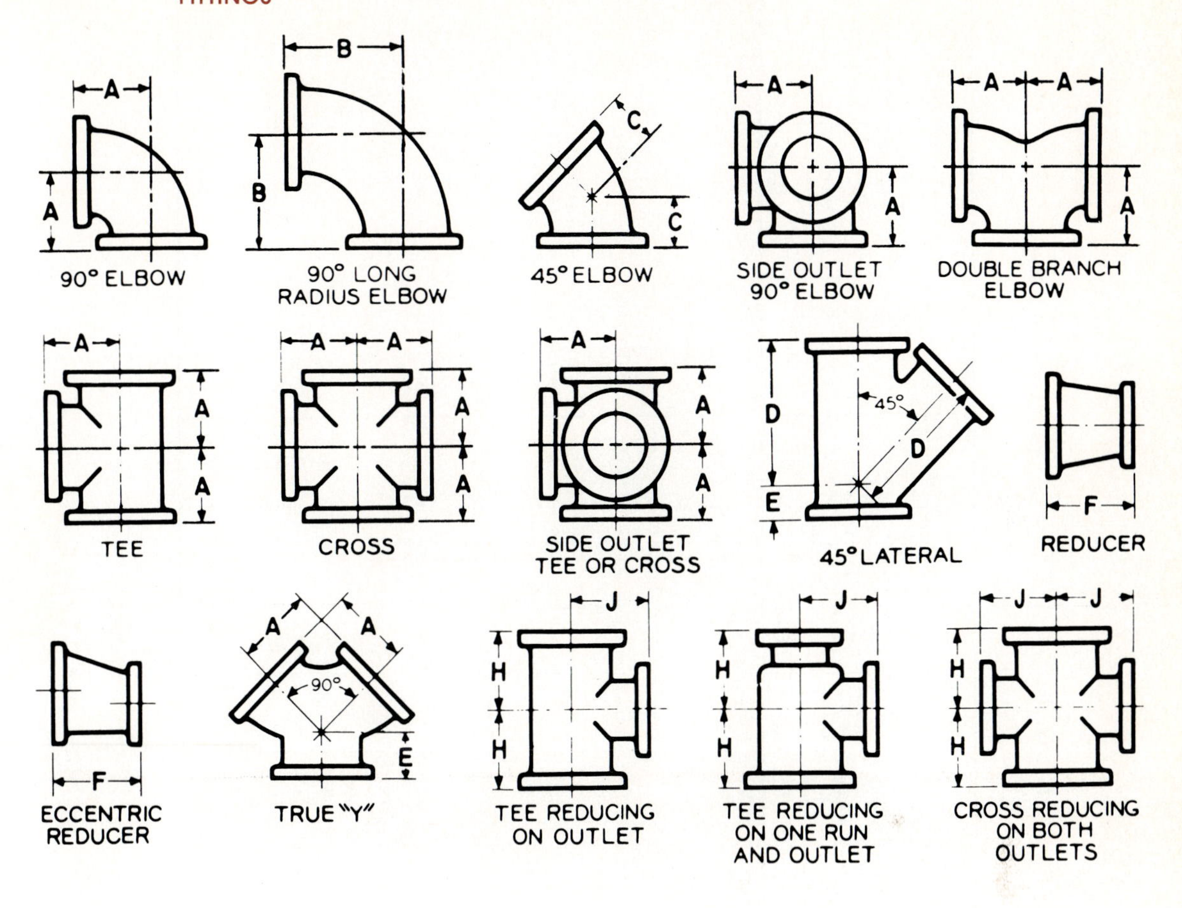

Appendix 11 AMERICAN STANDARD 125-LB CAST IRON FLANGED FITTINGS (Cont.)

Nominal Pipe Size	Flanges		General		Straight Fittings						Reducing Fittings (Short Body Patterns)		
											Tees and Crosses		
	Dia of Flange	Thickness of Flange (Min)	Inside Dia of Flange Fittings	Wall Thickness	Center to Face 90 deg Elbow Tees, Crosses True "Y" and Double Branch Elbow A	Center to Face 90 deg Long Radius El-bow B	Center to Face 45 deg Elbow C	Center to Face Lateral D	Short Center to Face True "Y" and Lateral E	Face to Face Reducer F	Size of Outlet and Smaller	Center to Face Run H	Center to Face Outlet or Side Outlet J
1	4¼	7/16	1	5/16	3½	5	1¾	5¾	1¾	. . .			
1¼	4 5/8	½	1¼	5/16	3¾	5½	2	6¼	1¾	. . .			
1½	5	9/16	1½	5/16	4	6	2¼	7	2				
2	6	5/8	2	5/16	4½	6½	2½	8	2½	5			
2½	7	11/16	2½	5/16	5	7	3	9½	2½	5½			
3	7½	¾	3	3/8	5½	7¾	3	10	3	6			
3½	8½	13/16	3½	7/16	6	8½	3½	11½	3	6½			
4	9	15/16	4	½	6½	9	4	12	3	7			
5	10	15/16	5	½	7½	10¼	4½	13½	3½	8			
6	11	1	6	9/16	8	11½	5	14½	3½	9			
8	13½	1 1/8	8	5/8	9	14	5½	17½	4½	11			
10	16	1 3/16	10	¾	11	16½	6½	20½	5	12			
12	19	1¼	12	13/16	12	19	7½	24½	5½	14			
14	21	1 3/8	14	7/8	14	21½	7½	27	6	16			
16	23½	1 7/16	16	1	15	24	8	30	6½	18			
18	25	1 9/16	18	1 1/16	16½	26½	8½	32	7	19	12	13	15½
20	27½	1 11/16	20	1 1/8	18	29	9½	35	8	20	14	14	17
24	32	1 7/8	24	1¼	22	34	11	40½	9	24	16	15	19
30	38¾	2 1/8	30	1 7/16	25	41½	15	49	10	30	20	18	23
36	46	2 3/8	36	1 5/8	28*	49	18	. . .	. . .	36	24	20	26
42	53	2 5/8	42	1 13/16	31*	56½	21	. . .	. . .	42	24	23	30
48	59½	2¾	48	2	34*	64	24	. . .	. . .	48	30	26	34

All reducing tees and crosses, sizes 16 in. and smaller, shall have same center to face dimensions as straight size fittings, corresponding to the size of the largest opening.

All dimensions are given in inches.
(Courtesy of ANSI; B16.1–1967.)

Appendix 12 AMERICAN STANDARD 125-LB CAST IRON FLANGES*

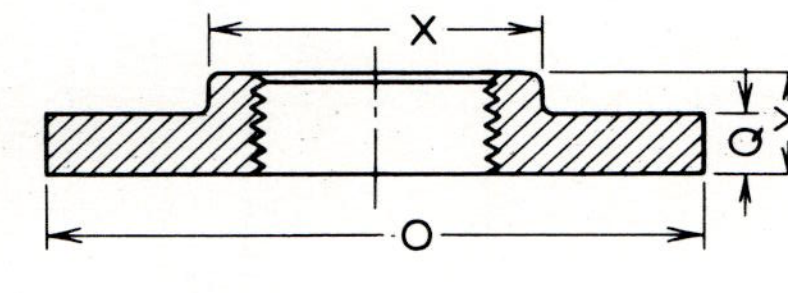

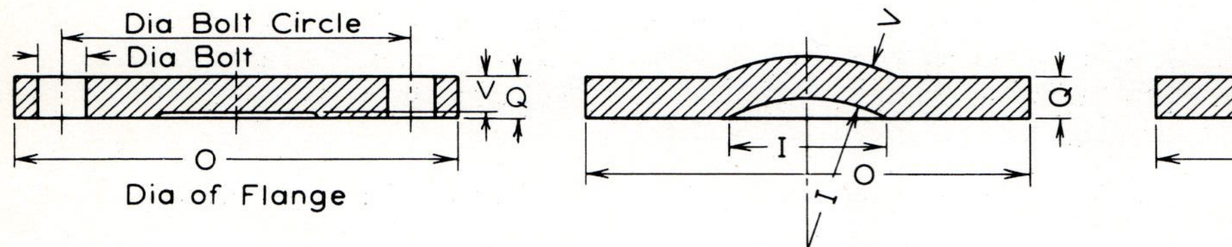

Size *I*	*O*	*Q*	*V*	*X*	*Y*	Dia. Bolt Circle	No. of Bolts	Dia. Bolts	Dia. Bolt Holes	Length of Bolts
1	4 1/4	7/16	—	1 15/16	0.68	3 1/8	4	1/2	5/8	1 3/4
1 1/4	4 5/8	1/2	—	2 5/16	0.76	3 1/2	4	1/2	5/8	2
1 1/2	5	9/16	—	2 9/16	0.87	3 7/8	4	1/2	5/8	2
2	6	5/8	—	3 1/16	1.00	4 3/4	4	5/8	3/4	2 1/4
2 1/2	7	11/16	—	3 9/16	1.14	5 1/2	4	5/8	3/4	2 1/2
3	7 1/2	3/4	—	4 1/4	1.20	6	4	5/8	3/4	2 1/2
3 1/2	8 1/2	13/16	—	4 13/16	1.25	7	8	5/8	3/4	2 3/4
4	9	15/16	—	5 5/16	1.30	7 1/2	8	5/8	3/4	3
5	10	15/16	—	6 7/16	1.41	8 1/2	8	3/4	7/8	3 1/4
6	11	1	—	7 9/16	1.51	9 1/2	8	3/4	7/8	3 1/4
8	13 1/2	1 1/8	—	9 11/16	1.71	11 3/4	8	3/4	7/8	3 1/2
10	16	1 3/16	—	11 15/16	1.93	14 1/4	12	7/8	1	3 3/4
12	19	1 1/4	12/16	14 1/16	2.13	17	12	7/8	1	3 3/4
14 O.D.	21	1 3/8	7/8	15 3/8	2.25	18 3/4	12	1	1 1/8	4 1/4
16 O.D.	23 1/2	1 7/16	1	17 1/2	2.45	21 1/4	16	1 1/8	1 1/8	4 1/2
18 O.D.	25	1 9/16	1 1/16	19 5/8	2.65	22 3/4	16	1 1/8	1 1/4	4 3/4

All dimensions in inches.

* Extracted from American Standards, "Cast-Iron Pipe Flanges and Flanged Fittings" (ANSI B16.1), with the permission of the publisher, The American Society of Mechanical Engineers.

Appendix 13 AMERICAN NATIONAL STANDARD 125-LB CAST IRON SCREWED FITTINGS*

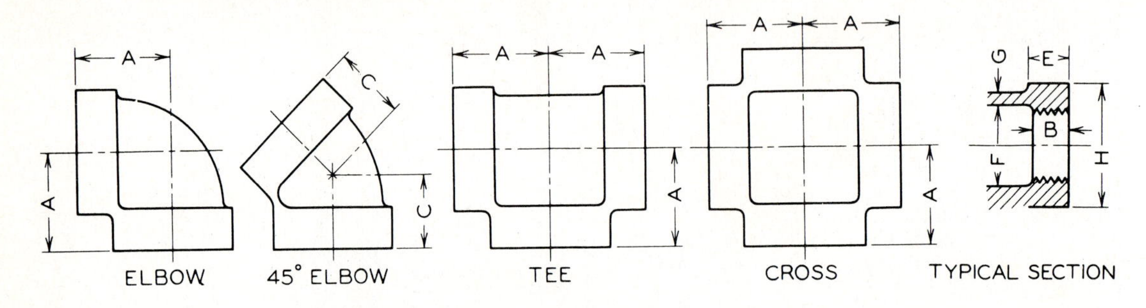

Nominal Pipe Size	A	C	B	E	F		G	H
			Min	Min	Min	Max	Min	Min
¼	0.81	0.73	0.32	0.38	0.540	0.584	0.110	0.93
⅜	0.95	0.80	0.36	0.44	0.675	0.719	0.120	1.12
½	1.12	0.88	0.43	0.50	0.840	0.897	0.130	1.34
¾	1.31	0.98	0.50	0.56	1.050	1.107	0.155	1.63
1	1.50	1.12	0.58	0.62	1.315	1.385	0.170	1.95
1¼	1.75	1.29	0.67	0.69	1.660	1.730	0.185	2.39
1½	1.94	1.43	0.70	0.75	1.900	1.970	0.200	2.68
2	2.25	1.68	0.75	0.84	2.375	2.445	0.220	3.28
2½	2.70	1.95	0.92	0.94	2.875	2.975	0.240	3.86
3	3.08	2.17	0.98	1.00	3.500	3.600	0.260	4.62
3½	3.42	2.39	1.03	1.06	4.000	4.100	0.280	5.20
4	3.79	2.61	1.08	1.12	4.500	4.600	0.310	5.79
5	4.50	3.05	1.18	1.18	5.563	5.663	0.380	7.05
6	5.13	3.46	1.28	1.28	6.625	6.725	0.430	8.28
8	6.56	4.28	1.47	1.47	8.625	8.725	0.550	10.63
10	8.08	5.16	1.68	1.68	10.750	10.850	0.690	13.12
12	9.50	5.97	1.88	1.88	12.750	12.850	0.800	15.47
14 O.D.	10.40	—	2.00	2.00	14.000	14.100	0.880	16.94
16 O.D.	11.82	—	2.20	2.20	16.000	16.100	1.000	19.30

All dimensions in inches.

* Extracted from American National Standards, "Cast-Iron Screwed Fittings, 125- and 250-lb" (ANSI B16.4), with the permission of the publisher, The American Society of Mechanical Engineers.

Appendix 14 AMERICAN NATIONAL STANDARD UNIFIED INCH SCREW THREADS (UN AND UNR THREAD FORM)*

Sizes			Threads per Inch											
			Series with Graded Pitches			Series with Constant Pitches								
Primary	Secondary	Basic Major Diameter	Coarse UNC	Fine UNF	Extra Fine UNEF	4UN	6UN	8UN	12UN	16UN	20UN	28UN	32UN	Sizes
0		0.0600	—	80	—	—	—	—	—	—	—	—	—	0
	1	0.0730	64	72	—	—	—	—	—	—	—	—	—	1
2		0.0860	56	64	—	—	—	—	—	—	—	—	—	2
	3	0.0990	48	56	—	—	—	—	—	—	—	—	—	3
4		0.1120	40	48	—	—	—	—	—	—	—	—	—	4
5		0.1250	40	44	—	—	—	—	—	—	—	—	—	5
6		0.1380	32	40	—	—	—	—	—	—	—	—	UNC	6
8		0.1640	32	36	—	—	—	—	—	—	—	—	UNC	8
10		0.1900	24	32	—	—	—	—	—	—	—	—	UNF	10
	12	0.2160	24	28	32	—	—	—	—	—	—	UNF	UNEF	12
$\frac{1}{4}$		0.2500	20	28	32	—	—	—	—	—	UNC	UNF	UNEF	$\frac{1}{4}$
$\frac{5}{16}$		0.3125	18	24	32	—	—	—	—	—	20	28	UNEF	$\frac{5}{16}$
$\frac{3}{8}$		0.3750	16	24	32	—	—	—	—	UNC	20	28	UNEF	$\frac{3}{8}$
$\frac{7}{16}$		0.4375	14	20	28	—	—	—	—	16	UNF	UNEF	32	$\frac{7}{16}$
$\frac{1}{2}$		0.5000	13	20	28	—	—	—	—	16	UNF	UNEF	32	$\frac{1}{2}$
$\frac{9}{16}$		0.5625	12	18	24	—	—	—	UNC	16	20	28	32	$\frac{9}{16}$
$\frac{5}{8}$		0.6250	11	18	24	—	—	—	12	16	20	28	32	$\frac{5}{8}$
	$\frac{11}{16}$	0.6875	—	—	24	—	—	—	12	16	20	28	32	$\frac{11}{16}$
$\frac{3}{4}$		0.7500	10	16	20	—	—	—	12	UNF	UNEF	28	32	$\frac{3}{4}$
	$\frac{13}{16}$	0.8125	—	—	20	—	—	—	12	16	UNEF	28	32	$\frac{13}{16}$
$\frac{7}{8}$		0.8750	9	14	20	—	—	—	12	16	UNEF	28	32	$\frac{7}{8}$
	$\frac{15}{16}$	0.9375	—	—	20	—	—	—	12	16	UNEF	28	32	$\frac{15}{16}$
1		1.0000	8	12	20	—	—	UNC	UNF	16	UNEF	28	32	1
	$1\frac{1}{16}$	1.0625	—	—	18	—	—	8	12	16	20	28	—	$1\frac{1}{16}$
$1\frac{1}{8}$		1.1250	7	12	18	—	—	8	UNF	16	20	28	—	$1\frac{1}{8}$
	$1\frac{3}{16}$	1.1875	—	—	18	—	—	8	12	16	20	28	—	$1\frac{3}{16}$
$1\frac{1}{4}$		1.2500	7	12	18	—	—	8	UNF	16	20	28	—	$1\frac{1}{4}$
	$1\frac{5}{16}$	1.3125	—	—	18	—	—	8	12	16	20	28	—	$1\frac{5}{16}$
$1\frac{3}{8}$		1.3750	6	12	18	—	UNC	8	UNF	16	20	28	—	$1\frac{3}{8}$
	$1\frac{7}{16}$	1.4375	—	—	18	—	6	8	12	16	20	28	—	$1\frac{7}{16}$
$1\frac{1}{2}$		1.5000	6	12	18	—	UNC	8	UNF	16	20	28	—	$1\frac{1}{2}$
	$1\frac{9}{16}$	1.5625	—	—	18	—	6	8	12	16	20	—	—	$1\frac{9}{16}$
$1\frac{5}{8}$		1.6250	—	—	18	—	6	8	12	16	20	—	—	$1\frac{5}{8}$
	$1\frac{11}{16}$	1.6875	—	—	18	—	6	8	12	16	20	—	—	$1\frac{11}{16}$
$1\frac{3}{4}$		1.7500	5	—	—	—	6	8	12	16	20	—	—	$1\frac{3}{4}$
	$1\frac{13}{16}$	1.8125	—	—	—	—	6	8	12	16	20	—	—	$1\frac{13}{16}$
$1\frac{7}{8}$		1.8750	—	—	—	—	6	8	12	16	20	—	—	$1\frac{7}{8}$
	$1\frac{15}{16}$	1.9375	—	—	—	—	6	8	12	16	20	—	—	$1\frac{15}{16}$
2		2.0000	$4\frac{1}{2}$	—	—	—	6	8	12	16	20	—	—	2
	$2\frac{1}{8}$	2.1250	—	—	—	—	6	8	12	16	20	—	—	$2\frac{1}{8}$
$2\frac{1}{4}$		2.2500	$4\frac{1}{2}$	—	—	—	6	8	12	16	20	—	—	$2\frac{1}{4}$
	$2\frac{3}{8}$	2.3750	—	—	—	—	6	8	12	16	20	—	—	$2\frac{3}{8}$
$2\frac{1}{2}$		2.5000	4	—	—	UNC	6	8	12	16	20	—	—	$2\frac{1}{2}$
	$2\frac{5}{8}$	2.6250	—	—	—	4	6	8	12	16	20	—	—	$2\frac{5}{8}$
$2\frac{3}{4}$		2.7500	4	—	—	UNC	6	8	12	16	20	—	—	$2\frac{3}{4}$
	$2\frac{7}{8}$	2.8750	—	—	—	4	6	8	12	16	20	—	—	$2\frac{7}{8}$

* Series designation shown indicates the UN thread form; however, the UNR thread form may be specified by substituting UNR in place of UN in all designations for external use only.

Cont.

Appendix 14 AMERICAN NATIONAL STANDARD UNIFIED INCH SCREW THREADS (UN AND UNR THREAD FORM)* (Cont.)

Sizes			Threads per Inch											
			Series with Graded Pitches			Series with Constant Pitches								
Primary	Secondary	Basic Major Diameter	Coarse UNC	Fine UNF	Extra Fine UNEF	4UN	6UN	8UN	12UN	16UN	20UN	28UN	32UN	Sizes
3		3.0000	4	—	—	UNC	6	8	12	16	20	—	—	3
	$3\frac{1}{8}$	3.1250	—	—	—	4	6	8	12	16	—	—	—	$3\frac{1}{8}$
$3\frac{1}{4}$		3.2500	4	—	—	UNC	6	8	12	16	—	—	—	$3\frac{1}{4}$
	$3\frac{3}{8}$	3.3750	—	—	—	4	6	8	12	16	—	—	—	$3\frac{3}{8}$
$3\frac{1}{2}$		3.5000	4	—	—	UNC	6	8	12	16	—	—	—	$3\frac{1}{2}$
	$3\frac{5}{8}$	3.6250	—	—	—	4	6	8	12	16	—	—	—	$3\frac{5}{8}$
$3\frac{3}{4}$		3.7500	4	—	—	UNC	6	8	12	16	—	—	—	$3\frac{3}{4}$
	$3\frac{7}{8}$	3.8750	—	—	—	4	6	8	12	16	—	—	—	$3\frac{7}{8}$
4		4.0000	4	—	—	UNC	6	8	12	16	—	—	—	4
	$4\frac{1}{8}$	4.1250	—	—	—	4	6	8	12	16	—	—	—	$4\frac{1}{8}$
$4\frac{1}{4}$		4.2500	—	—	—	4	6	8	12	16	—	—	—	$4\frac{1}{4}$
	$4\frac{3}{8}$	4.3750	—	—	—	4	6	8	12	16	—	—	—	$4\frac{3}{8}$
$4\frac{1}{2}$		4.5000	—	—	—	4	6	8	12	16	—	—	—	$4\frac{1}{2}$
	$4\frac{5}{8}$	4.6250	—	—	—	4	6	8	12	16	—	—	—	$4\frac{5}{8}$
$4\frac{3}{4}$		4.7500	—	—	—	4	6	8	12	16	—	—	—	$4\frac{3}{4}$
	$4\frac{7}{8}$	4.8750	—	—	—	4	6	8	12	16	—	—	—	$4\frac{7}{8}$
5		5.0000	—	—	—	4	6	8	12	16	—	—	—	5
	$5\frac{1}{8}$	5.1250	—	—	—	4	6	8	12	16	—	—	—	$5\frac{1}{8}$
$5\frac{1}{4}$		5.2500	—	—	—	4	6	8	12	16	—	—	—	$5\frac{1}{4}$
	$5\frac{3}{8}$	5.3750	—	—	—	4	6	8	12	16	—	—	—	$5\frac{3}{8}$
$5\frac{1}{2}$		5.5000	—	—	—	4	6	8	12	16	—	—	—	$5\frac{1}{2}$
	$5\frac{5}{8}$	5.6250	—	—	—	4	6	8	12	16	—	—	—	$5\frac{5}{8}$
$5\frac{3}{4}$		5.7500	—	—	—	4	6	8	12	16	—	—	—	$5\frac{3}{4}$
	$5\frac{7}{8}$	5.8750	—	—	—	4	6	8	12	16	—	—	—	$5\frac{7}{8}$
6		6.0000	—	—	—	4	6	8	12	16	—	—	—	6

(Courtesy of ANSI; B1.1–1974.)

Appendix 15 TAP DRILL SIZES FOR AMERICAN NATIONAL AND UNIFIED COARSE AND FINE THREADS

$$p = \text{pitch} = \frac{1}{\text{No. thrd. per in.}}$$

$$d = \text{depth} = p \times .649519$$

$$f = \text{flat} = \frac{p}{8}$$

$$\text{pitch diameter} = d - \frac{.6495}{N}$$

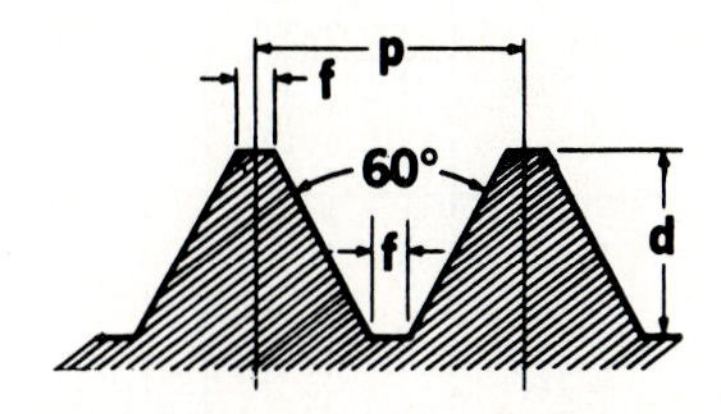

For Nos. 575 and 585 Screw Thread Micrometers

Size	Threads per inch NC UNC	Threads per inch NF UNF	Outside Diameter Inches	Pitch Diameter Inches	Root Diameter Inches	Tap Drill Approx. 75% Full Thread	Decimal Equiv. of Tap Drill
0	..	80	.0600	.0519	.0438	3/64	.0469
1	64	..	.0730	.0629	.0527	53	.0595
1	..	72	.0730	.0640	.0550	53	.0595
2	56	..	.0860	.0744	.0628	50	.0700
2	..	64	.0860	.0759	.0657	50	.0700
3	48	..	.0990	.0855	.0719	47	.0785
3	..	56	.0990	.0874	.0758	46	.0810
4	40	..	.1120	.0958	.0795	43	.0890
4	..	48	.1120	.0985	.0849	42	.0935
5	40	..	.1250	.1088	.0925	38	.1015
5	..	44	.1250	.1102	.0955	37	.1040
6	32	..	.1380	.1177	.0974	36	.1065
6	..	40	.1380	.1218	.1055	33	.1130
8	32	..	.1640	.1437	.1234	29	.1360
8	..	36	.1640	.1460	.1279	29	.1360
10	24	..	.1900	.1629	.1359	26	.1470
10	..	32	.1900	.1697	.1494	21	.1590
12	24	..	.2160	.1889	.1619	16	.1770
12	..	28	.2160	.1928	.1696	15	.1800
1/4	20	..	.2500	.2175	.1850	7	.2010
1/4	..	28	.2500	.2268	.2036	3	.2130
5/16	18	..	.3125	.2764	.2403	F	.2570
5/16	..	24	.3125	.2854	.2584	I	.2720
3/8	16	..	.3750	.3344	.2938	5/16	.3125
3/8	..	24	.3750	.3479	.3209	Q	.3320
7/16	14	..	.4375	.3911	.3447	U	.3680
7/16	..	20	.4375	.4050	.3726	25/64	.3906
1/2	13	..	.5000	.4500	.4001	27/64	.4219
1/2	..	20	.5000	.4675	.4351	29/64	.4531
9/16	12	..	.5625	.5084	.4542	31/64	.4844
9/16	..	18	.5625	.5264	.4903	33/64	.5156
5/8	11	..	.6250	.5660	.5069	17/32	.5312
5/8	..	18	.6250	.5889	.5528	37/64	.5781
3/4	10	..	.7500	.6850	.6201	21/32	.6562
3/4	..	16	.7500	.7094	.6688	11/16	.6875
7/8	9	..	.8750	.8028	.7307	49/64	.7656
7/8	..	14	.8750	.8286	.7822	13/16	.8125

Cont.

Appendix 15 TAP DRILL SIZES FOR AMERICAN NATIONAL AND UNIFIED COARSE AND FINE THREADS (Cont.)

Size	Threads per inch NC UNC	Threads per inch NF UNF	Outside Diameter Inches	Pitch Diameter Inches	Root Diameter Inches	Tap Drill Approx. 75% Full Thread	Decimal Equiv. of Tap Drill
1	8	..	1.0000	.9188	.8376	7/8	.8750
1	..	12	1.0000	.9459	.8917	59/64	.9219
1 1/8	7	..	1.1250	1.0322	.9394	63/64	.9844
1 1/8	..	12	1.1250	1.0709	1.0168	1 3/64	1.0469
1 1/4	7	..	1.2500	1.1572	1.0644	1 7/64	1.1094
1 1/4	..	12	1.2500	1.1959	1.1418	1 11/64	1.1719
1 3/8	6	..	1.3750	1.2667	1.1585	1 7/32	1.2187
1 3/8	..	12	1.3750	1.3209	1.2668	1 19/64	1.2969
1 1/2	6	..	1.5000	1.3917	1.2835	1 11/32	1.3437
1 1/2	..	12	1.5000	1.4459	1.3918	1 27/64	1.4219
1 3/4	5	..	1.7500	1.6201	1.4902	1 9/16	1.5625
2	4 1/2	..	2.0000	1.8557	1.7113	1 25/32	1.7812
2 1/4	4 1/2	..	2.2500	2.1057	1.9613	2 1/32	2.0313
2 1/2	4	..	2.5000	2.3376	2.1752	2 1/4	2.2500
2 3/4	4	..	2.7500	2.5876	2.4252	2 1/2	2.5000
3	4	..	3.0000	3.8376	2.6752	2 3/4	2.7500
3 1/4	4	..	3.2500	3.0876	2.9252	3	3.0000
3 1/2	4	..	3.5000	3.3376	3.1752	3 1/4	3.2500
3 3/4	4	..	3.7500	3.5876	3.4252	3 1/2	3.5000
4	4	..	4.0000	3.3786	3.6752	3 3/4	3.7500

(Courtesy of the L. S. Starrett Company.)

Appendix 16 LENGTH OF THREAD ENGAGEMENT GROUPS

Nominal Size Diam.		Pitch P	Length of Thread Engagement			
			Group S	Group N		Group L
Over	To and Incl		To and Incl	Over	To and Incl	Over
1.5	2.8	0.2	0.5	0.5	1.5	1.5
		0.25	0.6	0.6	1.9	1.9
		0.35	0.8	0.8	2.6	2.6
		0.4	1	1	3	3
		0.45	1.3	1.3	3.8	3.8
2.8	5.6	0.35	1	1	3	3
		0.5	1.5	1.5	4.5	4.5
		0.6	1.7	1.7	5	5
		0.7	2	2	6	6
		0.75	2.2	2.2	6.7	6.7
		0.8	2.5	2.5	7.5	7.5
5.6	11.2	0.75	2.4	2.4	7.1	7.1
		1	3	3	9	9
		1.25	4	4	12	12
		1.5	5	5	15	15
11.2	22.4	1	3.8	3.8	11	11
		1.25	4.5	4.5	13	13
		1.5	5.6	5.6	16	16
		1.75	6	6	18	18
		2	8	8	24	24
		2.5	10	10	30	30

Nominal Size Diam.		Pitch P	Length of Thread Engagement			
			Group S	Group N		Group L
Over	To and Incl		To and Incl	Over	To and Incl	Over
22.4	45	1	4	4	12	12
		1.5	6.3	6.3	19	19
		2	8.5	8.5	25	25
		3	12	12	36	36
		3.5	15	15	45	45
		4	18	18	53	53
		4.5	21	21	63	63
45	90	1.5	7.5	7.5	22	22
		2	9.5	9.5	28	28
		3	15	15	45	45
		4	19	19	56	56
		5	24	24	71	71
		5.5	28	28	85	85
		6	32	32	95	95
90	180	2	12	12	36	36
		3	18	18	53	53
		4	24	24	71	71
		6	36	36	106	106
180	355	3	20	20	60	60
		4	26	26	80	80
		6	40	40	118	118

All dimensions are given in millimeters. (Courtesy of ISO Standards.)

Appendix 17 ISO METRIC SCREW THREAD STANDARD SERIES

Nominal Size Dia. (mm)			Pitches (mm)														Nominal Size Dia. (mm)
Column[a]			Series with Graded Pitches		Series with Constant Pitches												
1	2	3	Coarse	Fine	6	4	3	2	1.5	1.25	1	0.75	0.5	0.35	0.25	0.2	
0.25			0.075	—	—	—	—	—	—	—	—	—	—	—	—	—	0.25
0.3			0.08	—	—	—	—	—	—	—	—	—	—	—	—	—	0.3
	0.35		0.09	—	—	—	—	—	—	—	—	—	—	—	—	—	0.35
0.4			0.1	—	—	—	—	—	—	—	—	—	—	—	—	—	0.4
	0.45		0.1	—	—	—	—	—	—	—	—	—	—	—	—	—	0.45
0.5			0.125	—	—	—	—	—	—	—	—	—	—	—	—	—	0.5
	0.55		0.125	—	—	—	—	—	—	—	—	—	—	—	—	—	0.55
0.6			0.15	—	—	—	—	—	—	—	—	—	—	—	—	—	0.6
	0.7		0.175	—	—	—	—	—	—	—	—	—	—	—	—	—	0.7
0.8			0.2	—	—	—	—	—	—	—	—	—	—	—	—	—	0.8
	0.9		0.225	—	—	—	—	—	—	—	—	—	—	—	—	—	0.9
			0.25	—	—	—	—	—	—	—	—	—	—	—	—	0.2	1
	1.1		0.25	—	—	—	—	—	—	—	—	—	—	—	—	0.2	1.1
1.2			0.25	—	—	—	—	—	—	—	—	—	—	—	—	0.2	1.2
	1.4		0.3	—	—	—	—	—	—	—	—	—	—	—	—	0.2	1.4
1.6			0.35	—	—	—	—	—	—	—	—	—	—	—	—	0.2	1.6
	1.8		0.35	—	—	—	—	—	—	—	—	—	—	—	—	0.2	1.8
2			0.4	—	—	—	—	—	—	—	—	—	—	—	0.25	—	2
	2.2		0.45	—	—	—	—	—	—	—	—	—	—	—	0.25	—	2.2
2.5			0.45	—	—	—	—	—	—	—	—	—	—	0.35	—	—	2.5
3			0.5	—	—	—	—	—	—	—	—	—	—	0.35	—	—	3
	3.5		0.6	—	—	—	—	—	—	—	—	—	—	0.35	—	—	3.5
4			0.7	—	—	—	—	—	—	—	—	—	0.5	—	—	—	4
	4.5		0.75	—	—	—	—	—	—	—	—	—	0.5	—	—	—	4.5
5			0.8	—	—	—	—	—	—	—	—	—	0.5	—	—	—	5
		5.5	—	—	—	—	—	—	—	—	—	—	0.5	—	—	—	5.5
6			1	—	—	—	—	—	—	—	—	0.75	—	—	—	—	6
		7	1	—	—	—	—	—	—	—	—	0.75	—	—	—	—	7
8			1.25	1	—	—	—	—	—	—	1	0.75	—	—	—	—	8
		9	1.25	—	—	—	—	—	—	—	1	0.75	—	—	—	—	9
10			1.5	1.25	—	—	—	—	—	1.25	1	0.75	—	—	—	—	10
		11	1.5	—	—	—	—	—	—	—	1	0.75	—	—	—	—	11
12			1.75	1.25	—	—	—	—	1.5	1.25	1	—	—	—	—	—	12
	14		2	1.5	—	—	—	—	1.5	1.25[b]	1	—	—	—	—	—	14
		15	—	—	—	—	—	—	1.5	—	1	—	—	—	—	—	15
16			2	1.5	—	—	—	—	1.5	—	1	—	—	—	—	—	16
		17	—	—	—	—	—	—	1.5	—	1	—	—	—	—	—	17
	18		2.5	1.5	—	—	—	2	1.5	—	1	—	—	—	—	—	18
20			2.5	1.5	—	—	—	2	1.5	—	1	—	—	—	—	—	20
	22		2.5	1.5	—	—	—	2	1.5	—	1	—	—	—	—	—	22

[a] Thread diameter should be selected from columns 1, 2 or 3, with preference being in that order.
[b] Pitch 1.25 mm in combination with diameter 14 mm has been included for sparkplug applications.
[c] Diameter 35 mm has been included for bearing locknut applications.

The use of pitches shown in parentheses should be avoided wherever possible.

The pitches enclosed in the bold frame, together with the corresponding nominal diameters in columns 1 and 2, are those combinations which have been established by ISO Recommendations as a selected "coarse" and "fine" series for commercial fasteners.

Appendix 17 ISO METRIC SCREW THREAD STANDARD SERIES (Cont.)

Nominal Size Dia. (mm)			Pitches (mm)															Nominal Size Dia. (mm)
Column[a]			Series with Graded Pitches		Series with Constant Pitches												
1	2	3	Coarse	Fine	6	4	3	2	1.5	1.25	1	0.75	0.5	0.35	0.25	0.2	
24			3	2	—	—	—	2	1.5	—	1	—	—	—	—	—	24
		25	—	—	—	—	—	2	1.5	—	1	—	—	—	—	—	25
		26	—	—	—	—	—	—	1.5	—	1	—	—	—	—	—	26
	27		3	2	—	—	—	2	1.5	—	1	—	—	—	—	—	27
		28	—	—	—	—	—	2	1.5	—	1	—	—	—	—	—	28
30			3.5	2	—	—	(3)	2	1.5	—	1	—	—	—	—	—	30
		32	—	—	—	—	—	2	1.5	—	—	—	—	—	—	—	32
	33		3.5	2	—	—	(3)	2	1.5	—	—	—	—	—	—	—	33
		35[c]	—	—	—	—	—	—	1.5	—	—	—	—	—	—	—	35[c]
36			4	3	—	—	—	2	1.5	—	—	—	—	—	—	—	36
		38	—	—	—	—	—	—	1.5	—	—	—	—	—	—	—	38
	39		4	3	—	—	—	2	1.5	—	—	—	—	—	—	—	39
		40	—	—	—	—	3	2	1.5	—	—	—	—	—	—	—	40
42			4.5	3	—	4	3	2	1.5	—	—	—	—	—	—	—	42
	45		4.5	3	—	4	3	2	1.5	—	—	—	—	—	—	—	45
48			5	3	—	4	3	2	1.5	—	—	—	—	—	—	—	48
		50	—	—	—	—	3	2	1.5	—	—	—	—	—	—	—	50
	52		5	3	—	4	3	2	1.5	—	—	—	—	—	—	—	52
		55	—	—	—	4	3	2	1.5	—	—	—	—	—	—	—	55
56			5.5	4	—	4	3	2	1.5	—	—	—	—	—	—	—	56
		58	—	—	—	4	3	2	1.5	—	—	—	—	—	—	—	58
	60		5.5	4	—	4	3	2	1.5	—	—	—	—	—	—	—	60
		62	—	—	—	4	3	2	1.5	—	—	—	—	—	—	—	62
64			6	4	—	4	3	2	1.5	—	—	—	—	—	—	—	64
		65	—	—	—	4	3	2	1.5	—	—	—	—	—	—	—	65
	68		6	4	—	4	3	2	1.5	—	—	—	—	—	—	—	68
		70	—	—	6	4	3	2	1.5	—	—	—	—	—	—	—	70
72			—	—	6	4	3	2	1.5	—	—	—	—	—	—	—	72
		75	—	—	—	4	3	2	1.5	—	—	—	—	—	—	—	75
	76		—	—	6	4	3	2	1.5	—	—	—	—	—	—	—	76
		78	—	—	—	—	—	2	—	—	—	—	—	—	—	—	78
80			—	—	6	4	3	2	1.5	—	—	—	—	—	—	—	80
		82	—	—	—	—	—	2	—	—	—	—	—	—	—	—	82
	85		—	—	6	4	3	2	—	—	—	—	—	—	—	—	85
90			—	—	6	4	3	2	—	—	—	—	—	—	—	—	90

Cont.

Appendix 17 ISO METRIC SCREW THREAD STANDARD SERIES (Cont.)

Nominal Size Dia. (mm)			Pitches (mm)																Nom-inal Size Dia. (mm)
Column[a]			Series with Graded Pitches		Series with Constant Pitches												
1	2	3	Coarse	Fine	6	4	3	2	1.5	1.25	1	0.75	0.5	0.35	0.25	0.2	
	95		—	—	6	4	3	2	—	—	—	—	—	—	—	—	95
100			—	—	6	4	3	2	—	—	—	—	—	—	—	—	100
	105		—	—	6	4	3	2	—	—	—	—	—	—	—	—	105
110			—	—	6	4	3	2	—	—	—	—	—	—	—	—	110
	115		—	—	6	4	3	2	—	—	—	—	—	—	—	—	115
	120		—	—	6	4	3	2	—	—	—	—	—	—	—	—	120
125			—	—	6	4	3	2	—	—	—	—	—	—	—	—	125
	130		—	—	6	4	3	2	—	—	—	—	—	—	—	—	130
		135	—	—	6	4	3	2	—	—	—	—	—	—	—	—	135
140			—	—	6	4	3	2	—	—	—	—	—	—	—	—	140
		145	—	—	6	4	3	2	—	—	—	—	—	—	—	—	145
	150		—	—	6	4	3	2	—	—	—	—	—	—	—	—	150
		155	—	—	6	4	3	—	—	—	—	—	—	—	—	—	155
160			—	—	6	4	3	—	—	—	—	—	—	—	—	—	160
		165	—	—	6	4	3	—	—	—	—	—	—	—	—	—	165
	170		—	—	6	4	3	—	—	—	—	—	—	—	—	—	170
		175	—	—	6	4	3	—	—	—	—	—	—	—	—	—	175
180			—	—	6	4	3	—	—	—	—	—	—	—	—	—	180
		185	—	—	6	4	3	—	—	—	—	—	—	—	—	—	185
	190		—	—	6	4	3	—	—	—	—	—	—	—	—	—	190
		195	—	—	6	4	3	—	—	—	—	—	—	—	—	—	195
200			—	—	6	4	3	—	—	—	—	—	—	—	—	—	200
		205	—	—	6	4	3	—	—	—	—	—	—	—	—	—	205
	210		—	—	6	4	3	—	—	—	—	—	—	—	—	—	210
220			—	—	6	4	3	—	—	—	—	—	—	—	—	—	220
		225	—	—	6	4	3	—	—	—	—	—	—	—	—	—	225
		230	—	—	6	4	3	—	—	—	—	—	—	—	—	—	230
		235	—	—	6	4	3	—	—	—	—	—	—	—	—	—	235
	240		—	—	6	4	3	—	—	—	—	—	—	—	—	—	240
		245	—	—	6	4	3	—	—	—	—	—	—	—	—	—	245
250			—	—	6	4	3	—	—	—	—	—	—	—	—	—	250
		255	—	—	6	4	—	—	—	—	—	—	—	—	—	—	255
	260		—	—	6	4	—	—	—	—	—	—	—	—	—	—	260
		265	—	—	6	4	—	—	—	—	—	—	—	—	—	—	265
		270	—	—	6	4	—	—	—	—	—	—	—	—	—	—	270
		275	—	—	6	4	—	—	—	—	—	—	—	—	—	—	275
280			—	—	6	4	—	—	—	—	—	—	—	—	—	—	280
		285	—	—	6	4	—	—	—	—	—	—	—	—	—	—	285
		290	—	—	6	4	—	—	—	—	—	—	—	—	—	—	290
		295	—	—	6	4	—	—	—	—	—	—	—	—	—	—	295
	300		—	—	6	4	—	—	—	—	—	—	—	—	—	—	300

[1] Thread diameter should be selected from columns 1, 2, or 3; with preference being in that order.

Appendix 18 SQUARE AND ACME THREADS

Size	Threads per Inch	Size	Threads per Inch
$\frac{3}{8}$	12	2	$2\frac{1}{2}$
$\frac{7}{16}$	10	$2\frac{1}{4}$	2
$\frac{1}{2}$	10	$2\frac{1}{2}$	2
$\frac{9}{16}$	8	$2\frac{3}{4}$	2
$\frac{5}{8}$	8	3	$1\frac{1}{2}$
$\frac{3}{4}$	6	$3\frac{1}{4}$	$1\frac{1}{2}$
$\frac{7}{8}$	5	$3\frac{1}{2}$	$1\frac{1}{3}$
1	5	$3\frac{3}{4}$	$1\frac{1}{3}$
$1\frac{1}{8}$	4	4	$1\frac{1}{3}$
$1\frac{1}{4}$	4	$4\frac{1}{4}$	$1\frac{1}{3}$
$1\frac{1}{2}$	3	$4\frac{1}{2}$	1
$1\frac{3}{4}$	$2\frac{1}{2}$	over $4\frac{1}{2}$	1

Appendix 19 AMERICAN STANDARD SQUARE BOLTS AND NUTS

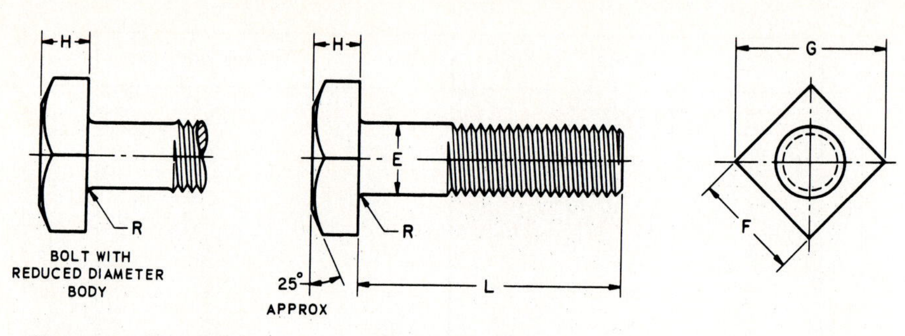

Dimensions of Square Bolts

Nominal Size or Basic Product Dia		Body Dia E	Width Across Flats F			Width Across Corners G		Height H			Radius of Fillet R
		Max	Basic	Max	Min	Max	Min	Basic	Max	Min	Max
1/4	0.2500	0.260	3/8	0.3750	0.362	0.530	0.498	11/64	0.188	0.156	0.031
5/16	0.3125	0.324	1/2	0.5000	0.484	0.707	0.665	13/64	0.220	0.186	0.031
3/8	0.3750	0.388	9/16	0.5625	0.544	0.795	0.747	1/4	0.268	0.232	0.031
7/16	0.4375	0.452	5/8	0.6250	0.603	0.884	0.828	19/64	0.316	0.278	0.031
1/2	0.5000	0.515	3/4	0.7500	0.725	1.061	0.995	21/64	0.348	0.308	0.031
5/8	0.6250	0.642	15/16	0.9375	0.906	1.326	1.244	27/64	0.444	0.400	0.062
3/4	0.7500	0.768	1 1/8	1.1250	1.088	1.591	1.494	1/2	0.524	0.476	0.062
7/8	0.8750	0.895	1 5/16	1.3125	1.269	1.856	1.742	19/32	0.620	0.568	0.062
1	1.0000	1.022	1 1/2	1.5000	1.450	2.121	1.991	21/32	0.684	0.628	0.093
1 1/8	1.1250	1.149	1 11/16	1.6875	1.631	2.386	2.239	3/4	0.780	0.720	0.093
1 1/4	1.2500	1.277	1 7/8	1.8750	1.812	2.652	2.489	27/32	0.876	0.812	0.093
1 3/8	1.3750	1.404	2 1/16	2.0625	1.994	2.917	2.738	29/32	0.940	0.872	0.093
1 1/2	1.5000	1.531	2 1/4	2.2500	2.175	3.182	2.986	1	1.036	0.964	0.093

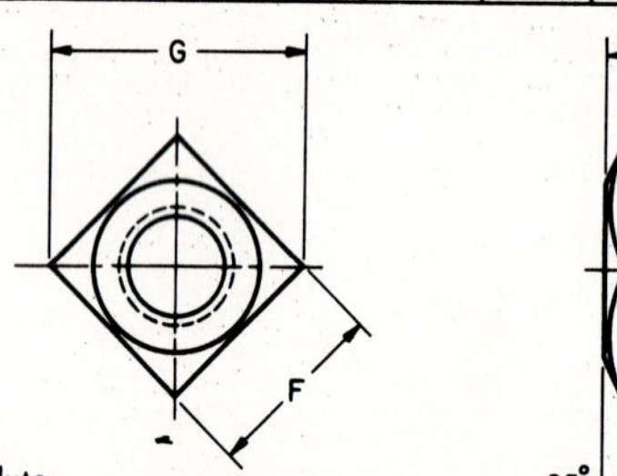

Dimensions of Square Nuts

Nominal Size or Basic Major Dia of Thread		Width Across Flats F			Width Across Corners G		Thickness H		
		Basic	Max	Min	Max	Min	Basic	Max	Min
1/4	0.2500	7/16	0.4375	0.425	0.619	0.584	7/32	0.235	0.203
5/16	0.3125	9/16	0.5625	0.547	0.795	0.751	17/64	0.283	0.249
3/8	0.3750	5/8	0.6250	0.606	0.884	0.832	21/64	0.346	0.310
7/16	0.4375	3/4	0.7500	0.728	1.061	1.000	3/8	0.394	0.356
1/2	0.5000	13/16	0.8125	0.788	1.149	1.082	7/16	0.458	0.418
5/8	0.6250	1	1.0000	0.969	1.414	1.330	35/64	0.569	0.525
3/4	0.7500	1 1/8	1.1250	1.088	1.591	1.494	21/32	0.680	0.632
7/8	0.8750	1 5/16	1.3125	1.269	1.856	1.742	49/64	0.792	0.740
1	1.0000	1 1/2	1.5000	1.450	2.121	1.991	7/8	0.903	0.847
1 1/8	1.1250	1 11/16	1.6875	1.631	2.386	2.239	1	1.030	0.970
1 1/4	1.2500	1 7/8	1.8750	1.812	2.652	2.489	1 3/32	1.126	1.062
1 3/8	1.3750	2 1/16	2.0625	1.994	2.917	2.738	1 13/64	1.237	1.169
1 1/2	1.5000	2 1/4	2.2500	2.175	3.182	2.986	1 5/16	1.348	1.276

(Courtesy of ANSI; B18.2.1–1965 and ANSI; B18.2.2–1965.)

Appendix 20 AMERICAN STANDARD HEXAGON HEAD BOLTS AND NUTS

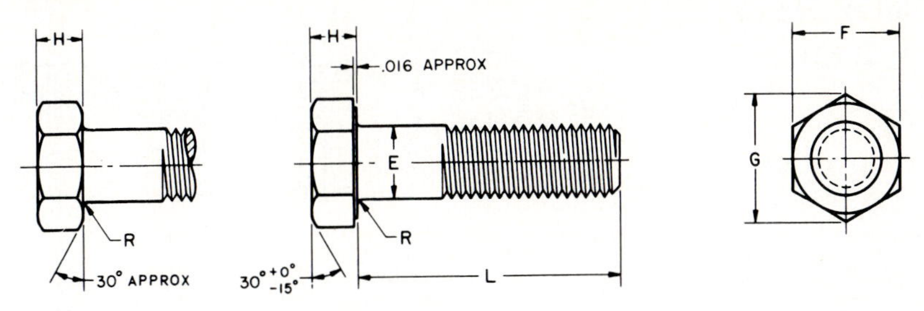

Dimensions of Hex Cap Screws (Finished Hex Bolts)

Nominal Size or Basic Product Dia	Body Dia E		Width Across Flats F			Width Across Corners G		Height H			Radius of Fillet R	
	Max	Min	Basic	Max	Min	Max	Min	Basic	Max	Min	Max	Min
1/4 0.2500	0.2500	0.2450	7/16	0.4375	0.428	0.505	0.488	5/32	0.163	0.150	0.025	0.015
5/16 0.3125	0.3125	0.3065	1/2	0.5000	0.489	0.577	0.557	13/64	0.211	0.195	0.025	0.015
3/8 0.3750	0.3750	0.3690	9/16	0.5625	0.551	0.650	0.628	15/64	0.243	0.226	0.025	0.015
7/16 0.4375	0.4375	0.4305	5/8	0.6250	0.612	0.722	0.698	9/32	0.291	0.272	0.025	0.015
1/2 0.5000	0.5000	0.4930	3/4	0.7500	0.736	0.866	0.840	5/16	0.323	0.302	0.025	0.015
9/16 0.5625	0.5625	0.5545	13/16	0.8125	0.798	0.938	0.910	23/64	0.371	0.348	0.045	0.020
5/8 0.6250	0.6250	0.6170	15/16	0.9375	0.922	1.083	1.051	25/64	0.403	0.378	0.045	0.020
3/4 0.7500	0.7500	0.7410	1 1/8	1.1250	1.100	1.299	1.254	15/32	0.483	0.455	0.045	0.020
7/8 0.8750	0.8750	0.8660	1 5/16	1.3125	1.285	1.516	1.465	35/64	0.563	0.531	0.065	0.040
1 1.0000	1.0000	0.9900	1 1/2	1.5000	1.469	1.732	1.675	39/64	0.627	0.591	0.095	0.060
1 1/8 1.1250	1.1250	1.1140	1 11/16	1.6875	1.631	1.949	1.859	11/16	0.718	0.658	0.095	0.060
1 1/4 1.2500	1.2500	1.2390	1 7/8	1.8750	1.812	2.165	2.066	25/32	0.813	0.749	0.095	0.060
1 3/8 1.3750	1.3750	1.3630	2 1/16	2.0625	1.994	2.382	2.273	27/32	0.878	0.810	0.095	0.060
1 1/2 1.5000	1.5000	1.4880	2 1/4	2.2500	2.175	2.598	2.480	15/16	0.974	0.902	0.095	0.060
1 3/4 1.7500	1.7500	1.7380	2 5/8	2.6250	2.538	3.031	2.893	1 3/32	1.134	1.054	0.095	0.060
2 2.0000	2.0000	1.9880	3	3.0000	2.900	3.464	3.306	1 7/32	1.263	1.175	0.095	0.060
2 1/4 2.2500	2.2500	2.2380	3 3/8	3.3750	3.262	3.897	3.719	1 3/8	1.423	1.327	0.095	0.060
2 1/2 2.5000	2.5000	2.4880	3 3/4	3.7500	3.625	4.330	4.133	1 17/32	1.583	1.479	0.095	0.060
2 3/4 2.7500	2.7500	2.7380	4 1/8	4.1250	3.988	4.763	4.546	1 11/16	1.744	1.632	0.095	0.060
3 3.0000	3.0000	2.9880	4 1/2	4.5000	4.350	5.196	4.959	1 7/8	1.935	1.815	0.095	0.060

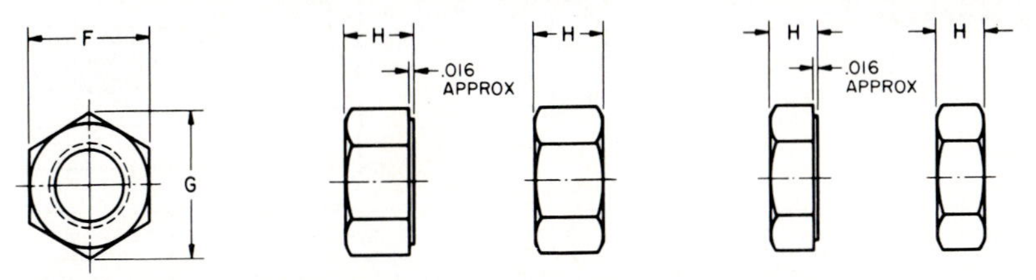

Dimensions of Hex Nuts and Hex Jam Nuts

Nominal Size or Basic Major Dia of Thread		Width Across Flats F			Width Across Corners G		Thickness Hex Nuts H			Thickness Hex Jam Nuts H		
		Basic	Max	Min	Max	Min	Basic	Max	Min	Basic	Max	Min
1/4	0.2500	7/16	0.4375	0.428	0.505	0.488	7/32	0.226	0.212	5/32	0.163	0.150
5/16	0.3125	1/2	0.5000	0.489	0.577	0.557	17/64	0.273	0.258	3/16	0.195	0.180
3/8	0.3750	9/16	0.5625	0.551	0.650	0.628	21/64	0.337	0.320	7/32	0.227	0.210
7/16	0.4375	11/16	0.6875	0.675	0.794	0.768	3/8	0.385	0.365	1/4	0.260	0.240
1/2	0.5000	3/4	0.7500	0.736	0.866	0.840	7/16	0.448	0.427	5/16	0.323	0.302
9/16	0.5625	7/8	0.8750	0.861	1.010	0.982	31/64	0.496	0.473	5/16	0.324	0.301
5/8	0.6250	15/16	0.9375	0.922	1.083	1.051	35/64	0.559	0.535	3/8	0.387	0.363
3/4	0.7500	1 1/8	1.1250	1.088	1.299	1.240	41/64	0.665	0.617	27/64	0.446	0.398
7/8	0.8750	1 5/16	1.3125	1.269	1.516	1.447	3/4	0.776	0.724	31/64	0.510	0.458
1	1.0000	1 1/2	1.5000	1.450	1.732	1.653	55/64	0.887	0.831	35/64	0.575	0.519
1 1/8	1.1250	1 11/16	1.6875	1.631	1.949	1.859	31/32	0.999	0.939	39/64	0.639	0.579
1 1/4	1.2500	1 7/8	1.8750	1.812	2.165	2.066	1 1/16	1.094	1.030	23/32	0.751	0.687
1 3/8	1.3750	2 1/16	2.0625	1.994	2.382	2.273	1 11/64	1.206	1.138	25/32	0.815	0.747
1 1/2	1.5000	2 1/4	2.2500	2.175	2.598	2.480	1 9/32	1.317	1.245	27/32	0.880	0.808

(Courtesy of ANSI; B18.2.1–1965 and ANSI; B18.2.2–1965.)

Appendix 21 FILLISTER HEAD AND ROUND HEAD CAP SCREWS

Fillister Head Cap Screws

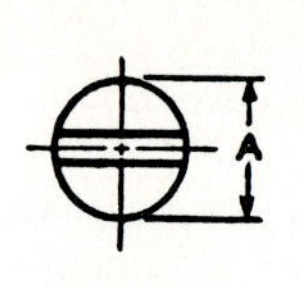

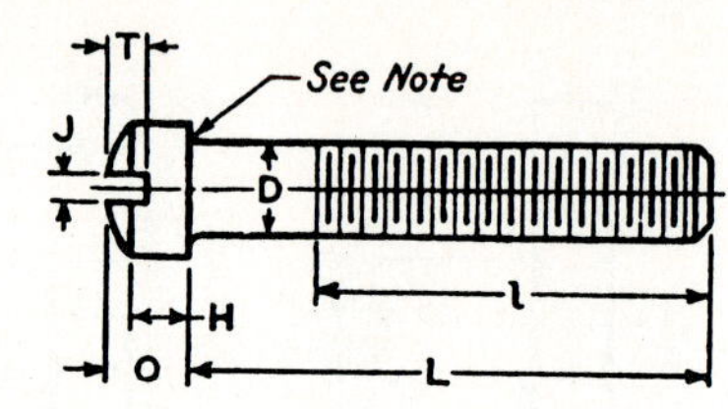

Nominal Size	D		A		H		O		J		T	
	Body Diameter		Head Diameter		Height of Head		Total Height of Head		Width of Slot		Depth of Slot	
	Max	Min	Max	Min	Max	Min	Max	Min	Max	Min	Max	Min
1/4	0.250	0.245	0.375	0.363	0.172	0.157	0.216	0.194	0.075	0.064	0.097	0.077
5/16	0.3125	0.307	0.437	0.424	0.203	0.186	0.253	0.230	0.084	0.072	0.115	0.090
3/8	0.375	0.369	0.562	0.547	0.250	0.229	0.314	0.284	0.094	0.081	0.142	0.112
7/16	0.4375	0.431	0.625	0.608	0.297	0.274	0.368	0.336	0.094	0.081	0.168	0.133
1/2	0.500	0.493	0.750	0.731	0.328	0.301	0.413	0.376	0.106	0.091	0.193	0.153
9/16	0.5625	0.555	0.812	0.792	0.375	0.346	0.467	0.427	0.118	0.102	0.213	0.168
5/8	0.625	0.617	0.875	0.853	0.422	0.391	0.521	0.478	0.133	0.116	0.239	0.189
3/4	0.750	0.742	1.000	0.976	0.500	0.466	0.612	0.566	0.149	0.131	0.283	0.223
7/8	0.875	0.866	1.125	1.098	0.594	0.556	0.720	0.668	0.167	0.147	0.334	0.264
1	1.000	0.990	1.312	1.282	0.656	0.612	0.803	0.743	0.188	0.166	0.371	0.291

All dimensions are given in inches.

The radius of the fillet at the base of the head:
For sizes 1/4 to 3/8 in. incl. is 0.016 min and 0.031 max,
7/16 to 9/16 in. incl. is 0.016 min and 0.047 max,
5/8 to 1 in. incl. is 0.031 min and 0.062 max.

Round Head Cap Screws

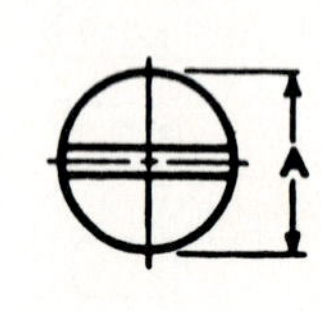

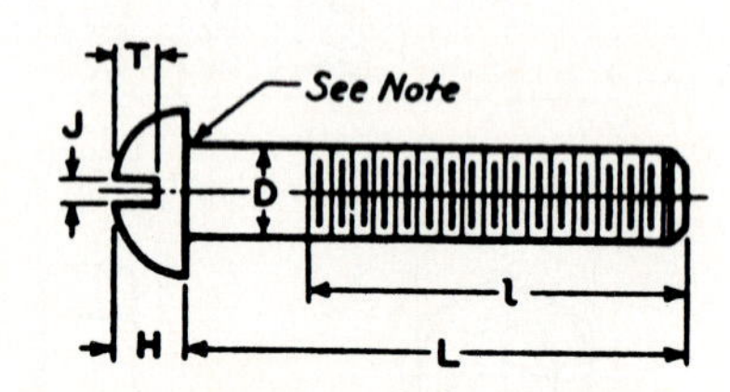

Nominal Size	D		A		H		J		T	
	Body Diameter		Head Diameter		Height of Head		Width of Slot		Depth of Slot	
	Max	Min	Max	Min	Max	Min	Max	Min	Max	Min
1/4	0.250	0.245	0.437	0.418	0.191	0.175	0.075	0.064	0.117	0.097
5/16	0.3125	0.307	0.562	0.540	0.245	0.226	0.084	0.072	0.151	0.126
3/8	0.375	0.369	0.625	0.603	0.273	0.252	0.094	0.081	0.168	0.138
7/16	0.4375	0.431	0.750	0.725	0.328	0.302	0.094	0.081	0.202	0.167
1/2	0.500	0.493	0.812	0.786	0.354	0.327	0.106	0.091	0.218	0.178
9/16	0.5625	0.555	0.937	0.909	0.409	0.378	0.118	0.102	0.252	0.207
5/8	0.625	0.617	1.000	0.970	0.437	0.405	0.133	0.116	0.270	0.220
3/4	0.750	0.742	1.250	1.215	0.546	0.507	0.149	0.131	0.338	0.278

All dimensions are given in inches.

Radius of the fillet at the base of the head:
For sizes 1/4 to 3/8 in. incl. is 0.016 min and 0.031 max,
7/16 to 9/16 in. incl..is 0.016 min and 0.047 max,
5/8 to 1 in..incl. is 0.031 min and 0.062 max.

(Courtesy of ANSI; B18.6.2–1956.)

Appendix 22 FLAT HEAD CAP SCREWS

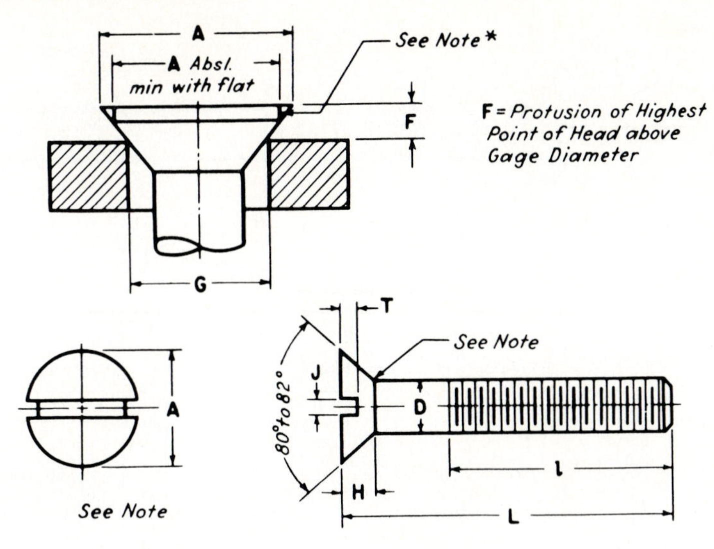

Nominal Size	D		A			G	H	J		T		F	
	Body Diameter		Head Diameter			Gaging Diameter	Height of Head	Width of Slot		Depth of Slot		Protrusion Above Gaging Diameter	
	Max	Min	Max	Min	Absolute Min with Flat		Average	Max	Min	Max	Min	Max	Min
1/4	0.250	0.245	0.500	0.477	0.452	0.4245	0.140	0.075	0.064	0.068	0.045	0.0452	0.0307
5/16	0.3125	0.307	0.625	0.598	0.567	0.5376	0.177	0.084	0.072	0.086	0.057	0.0523	0.0354
3/8	0.375	0.369	0.750	0.720	0.682	0.6507	0.210	0.094	0.081	0.103	0.068	0.0594	0.0401
7/16	0.4375	0.431	0.8125	0.780	0.736	0.7229	0.210	0.094	0.081	0.103	0.068	0.0649	0.0448
1/2	0.500	0.493	0.875	0.841	0.791	0.7560	0.210	0.106	0.091	0.103	0.068	0.0705	0.0495
9/16	0.5625	0.555	1.000	0.962	0.906	0.8691	0.244	0.118	0.102	0.120	0.080	0.0775	0.0542
5/8	0.625	0.617	1.125	1.083	1.020	0.9822	0.281	0.133	0.116	0.137	0.091	0.0846	0.0588
3/4	0.750	0.742	1.375	1.326	1.251	1.2085	0.352	0.149	0.131	0.171	0.115	0.0987	0.0682
7/8	0.875	0.866	1.625	1.568	1.480	1.4347	0.423	0.167	0.147	0.206	0.138	0.1128	0.0776
1	1.000	0.990	1.875	1.811	1.711	1.6610	0.494	0.188	0.166	0.240	0.162	0.1270	0.0870
1 1/8	1.125	1.114	2.062	1.992	1.880	1.8262	0.529	0.196	0.178	0.257	0.173	0.1401	0.0964
1 1/4	1.250	1.239	2.312	2.235	2.110	2.0525	0.600	0.211	0.193	0.291	0.197	0.1542	0.1056
1 3/8	1.375	1.363	2.562	2.477	2.340	2.2787	0.665	0.226	0.208	0.326	0.220	0.1684	0.1151
1 1/2	1.500	1.488	2.812	2.720	2.570	2.5050	0.742	0.258	0.240	0.360	0.244	0.1825	0.1245

All dimensions are given in inches.

The maximum and minimum head diameters, A, are extended to the theoretical sharp corners.

The radius of the fillet at the base of the head shall not exceed 0.4 Max. D.

*Edge of head may be flat as shown or slightly rounded.

(Courtesy of ANSI; B18.6.2–1956.)

Appendix 23 MACHINE SCREWS

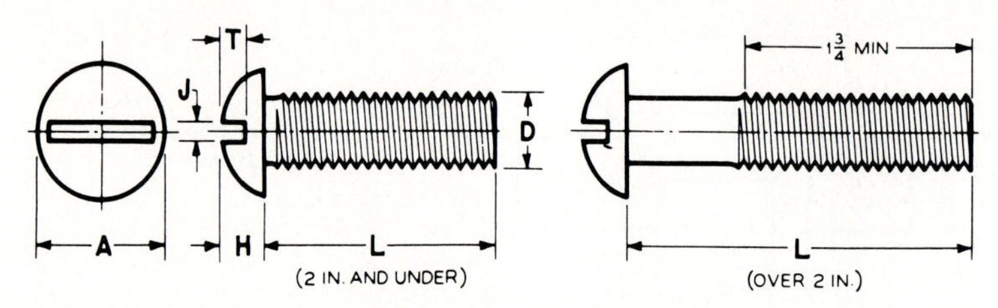

Dimensions of Slotted Round Head Machine Screws

Nominal Size	D	A		H		J		T	
	Diameter of Screw	Head Diameter		Head Height		Width of Slot		Depth of Slot	
	Basic	Max	Min	Max	Min	Max	Min	Max	Min
0	0.0600	0.113	0.099	0.053	0.043	0.023	0.016	0.039	0.029
1	0.0730	0.138	0.122	0.061	0.051	0.026	0.019	0.044	0.033
2	0.0860	0.162	0.146	0.069	0.059	0.031	0.023	0.048	0.037
3	0.0990	0.187	0.169	0.078	0.067	0.035	0.027	0.053	0.040
4	0.1120	0.211	0.193	0.086	0.075	0.039	0.031	0.058	0.044
5	0.1250	0.236	0.217	0.095	0.083	0.043	0.035	0.063	0.047
6	0.1380	0.260	0.240	0.103	0.091	0.048	0.039	0.068	0.051
8	0.1640	0.309	0.287	0.120	0.107	0.054	0.045	0.077	0.058
10	0.1900	0.359	0.334	0.137	0.123	0.060	0.050	0.087	0.065
12	0.2160	0.408	0.382	0.153	0.139	0.067	0.056	0.096	0.073
1/4	0.2500	0.472	0.443	0.175	0.160	0.075	0.064	0.109	0.082
5/16	0.3125	0.590	0.557	0.216	0.198	0.084	0.072	0.132	0.099
3/8	0.3750	0.708	0.670	0.256	0.237	0.094	0.081	0.155	0.117
7/16	0.4375	0.750	0.707	0.328	0.307	0.094	0.081	0.196	0.148
1/2	0.5000	0.813	0.766	0.355	0.332	0.106	0.091	0.211	0.159
9/16	0.5625	0.938	0.887	0.410	0.385	0.118	0.102	0.242	0.183
5/8	0.6250	1.000	0.944	0.438	0.411	0.133	0.116	0.258	0.195
3/4	0.7500	1.250	1.185	0.547	0.516	0.149	0.131	0.320	0.242

All dimensions are given in inches.

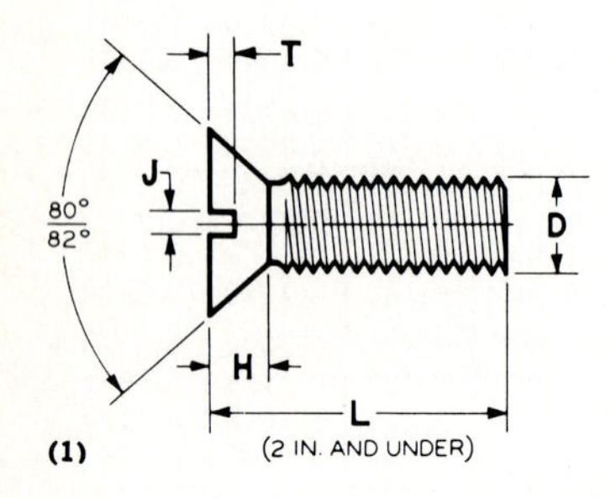

(1)

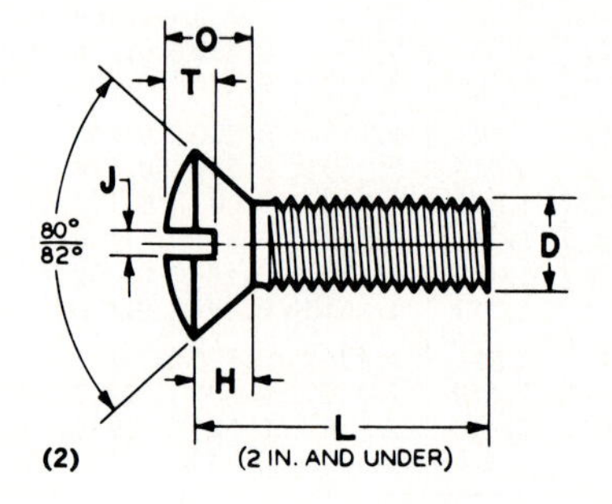

(2)

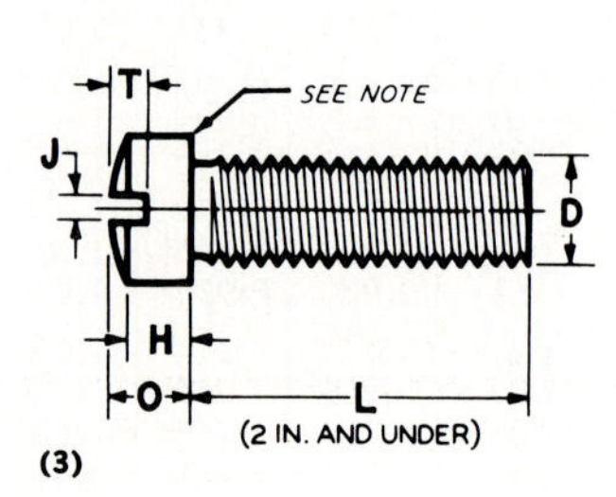

(3)

Three other common forms of machine screws are shown above: (1) flat head, (2) oval head, and (3) fillister head. Although dimension tables are not given for these three types of machine screws in this text, their general dimensions are closely related to those shown in the table above. Additional information on these screws can be obtained from ANSI; B18.6.3–1962.

(Courtesy of ANSI; B18.6.3–1962.)

Appendix 24 AMERICAN STANDARD MACHINE SCREWS

(The proportions of the screws can be found by multiplying the major diameter, D, by the factors given below.)

Flat Head

	Maximum	Minimum
A	2.04D + .003	1.84D
H	.619D − .002	.552D − .007
J	.182D + .020	.176 D + .010
T	.288D − .002	.192D − .002
θ	82°	80°

Round Head

	Maximum	Minimum
A	1.887D	1.813D − .010
H	.636D + .015	.624D + .005
J	.182D + .020	.176 D + .010
T	.362D + .017	.268D + .013
Profile of head is semi-elliptical		

Oval Head

	Maximum	Minimum
A	2.04D + .003	1.84D
H	.619D − .002	.552D − .007
J	.182D + .020	.176 D + .010
O	.923D + .001	.820D − .008
T	.556D − .003	.460D − .003
θ	82°	80°

Fillister Head

	Maximum	Minimum
A	1.670D − .004	1.610D − .014
H	.620D + .010	.582D + .005
J	.182D + .020	.176 D + .010
O	.940D + .002	.820D − .008
T	.440D − .001	.374D − .011

Appendix 25 AMERICAN STANDARD MACHINE TAPERS*

No. of Taper	Taper per Foot (Basic)	Origin of Series	No. of Taper	Taper per Foot (Basic)	Origin of Series	No. of Taper	Taper per Foot (Basic)	Origin of Series	No. of Taper	Taper per Foot (Basic)	Origin of Series
0.239	0.50200	Brown & Sharpe	*	0.62326	Morse	250	0.750	¾ in. per ft.	600	0.750	¾ in. per ft.
.299	.50200	Brown & Sharpe	4½	.62400	Morse	300	.750	¾ in. per ft.	800	0.750	¾ in. per ft.
.375	.50200	Brown & Sharpe	5	.63151	Morse	350	.750	¾ in. per ft.	1000	0.750	¾ in. per ft.
1	.59858	Morse	6	.62565	Morse	400	.750	¾ in. per ft.	1200	0.750	¾ in. per ft.
2	.59941	Morse	7	.62400	Morse	450	.750	¾ in. per ft.			
3	.60235	Morse	200	.750	¾ in. per ft.	500	.750	¾ in. per ft.			

All dimensions in inches.

* Extracted from American Standards, "Machine Tapers, Self-Holding and Steep Taper Series" (ASA B5,10-1960), with the permission of the publisher, The American Society of Mechanical Engineers.

Appendix 26 AMERICAN NATIONAL STANDARD SQUARE HEAD SET SCREWS (ANSI B18.6.2)

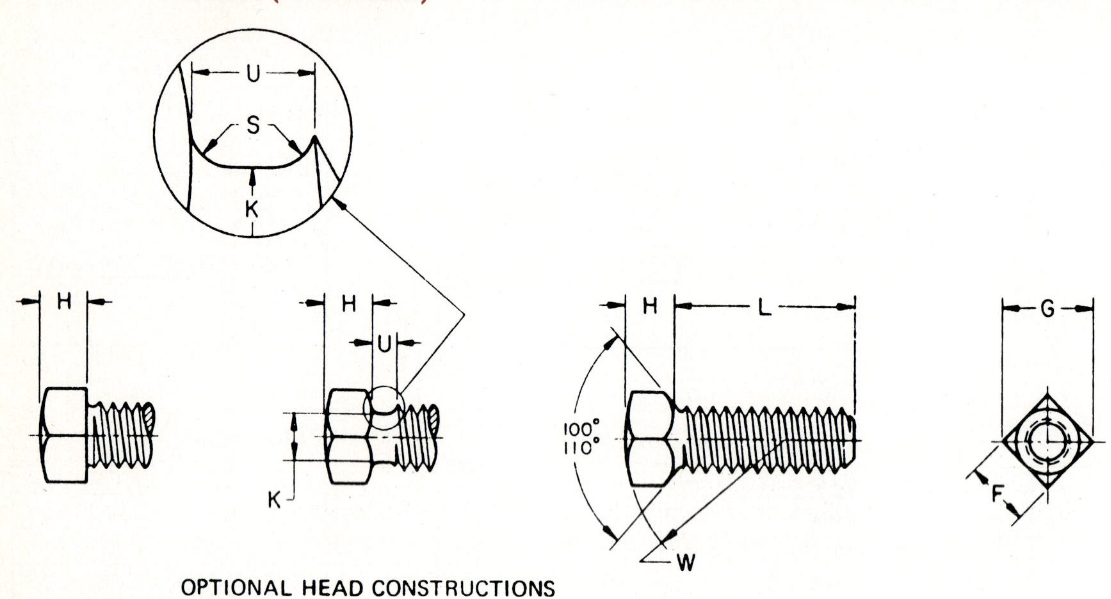

OPTIONAL HEAD CONSTRUCTIONS

Nominal Size[1] or Basic Screw Diameter		F		G		H		K		S	U	W
		Width Across Flats		Width Across Corners		Head Height		Neck Relief Diameter		Neck Relief Fillet Radius	Neck Relief Width	Head Radius
		Max	Min	Max	Min	Max	Min	Max	Min	Max	Min	Min
10	0.1900	0.188	0.180	0.265	0.247	0.148	0.134	0.145	0.140	0.027	0.083	0.48
1/4	0.2500	0.250	0.241	0.354	0.331	0.196	0.178	0.185	0.170	0.032	0.100	0.62
5/16	0.3125	0.312	0.302	0.442	0.415	0.245	0.224	0.240	0.225	0.036	0.111	0.78
3/8	0.3750	0.375	0.362	0.530	0.497	0.293	0.270	0.294	0.279	0.041	0.125	0.94
7/16	0.4375	0.438	0.423	0.619	0.581	0.341	0.315	0.345	0.330	0.046	0.143	1.09
1/2	0.5000	0.500	0.484	0.707	0.665	0.389	0.361	0.400	0.385	0.050	0.154	1.25
9/16	0.5625	0.562	0.545	0.795	0.748	0.437	0.407	0.454	0.439	0.054	0.167	1.41
5/8	0.6250	0.625	0.606	0.884	0.833	0.485	0.452	0.507	0.492	0.059	0.182	1.56
3/4	0.7500	0.750	0.729	1.060	1.001	0.582	0.544	0.620	0.605	0.065	0.200	1.88
7/8	0.8750	0.875	0.852	1.237	1.170	0.678	0.635	0.731	0.716	0.072	0.222	2.19
1	1.0000	1.000	0.974	1.414	1.337	0.774	0.726	0.838	0.823	0.081	0.250	2.50
1 1/8	1.1250	1.125	1.096	1.591	1.505	0.870	0.817	0.939	0.914	0.092	0.283	2.81
1 1/4	1.2500	1.250	1.219	1.768	1.674	0.966	0.908	1.064	1.039	0.092	0.283	3.12
1 3/8	1.3750	1.375	1.342	1.945	1.843	1.063	1.000	1.159	1.134	0.109	0.333	3.44
1 1/2	1.5000	1.500	1.464	2.121	2.010	1.159	1.091	1.284	1.259	0.109	0.333	3.75

[1] Where specifying nominal size in decimals, zeros preceding decimal and in the fourth decimal place shall be omitted.

Appendix 27 AMERICAN NATIONAL STANDARD POINTS FOR SQUARE HEAD SET SCREWS (ANSI B18.6.2)

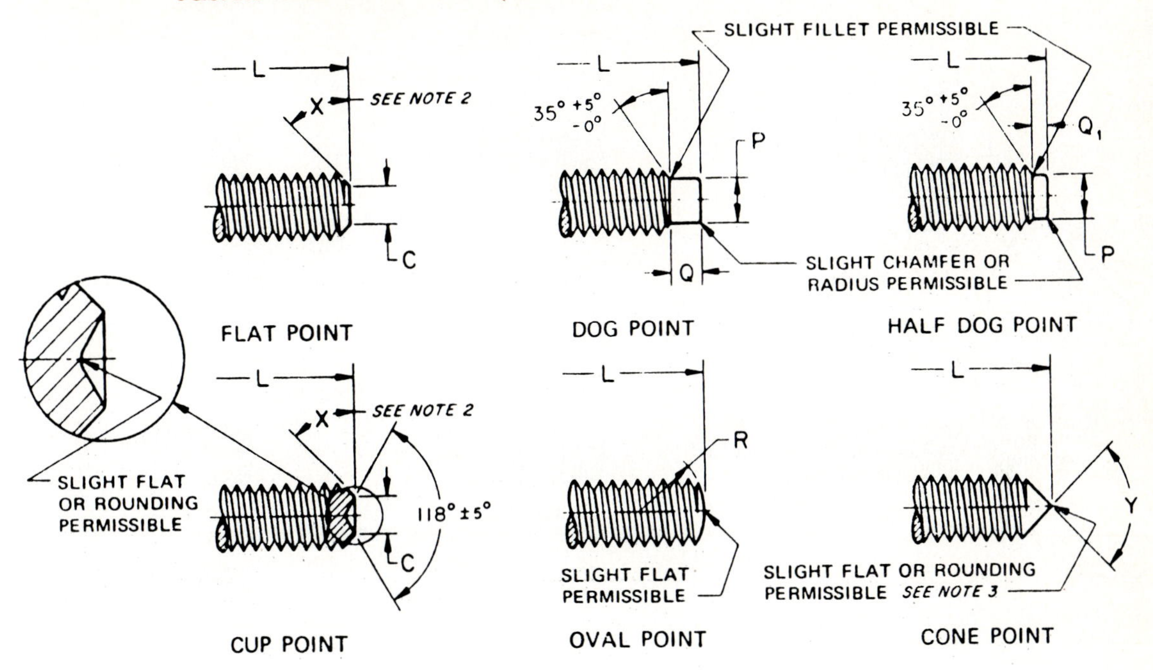

Nominal Size[1] or Basic Screw Diameter	C		P		Q		Q_1		R	Y
	Cup and Flat Point Diameters		Dog and Half Dog Point Diameters		Point Length				Oval Point Radius	Cone Point Angle 90° ±2° For These Nominal Lengths or Longer; 118° ±2° For Shorter Screws
					Dog		Half Dog		+0.031 -0.000	
	Max	Min	Max	Min	Max	Min	Max	Min		
10 0.1900	0.102	0.088	0.127	0.120	0.095	0.085	0.050	0.040	0.142	1/4
1/4 0.2500	0.132	0.118	0.156	0.149	0.130	0.120	0.068	0.058	0.188	5/16
5/16 0.3125	0.172	0.156	0.203	0.195	0.161	0.151	0.083	0.073	0.234	3/8
3/8 0.3750	0.212	0.194	0.250	0.241	0.193	0.183	0.099	0.089	0.281	7/16
7/16 0.4375	0.252	0.232	0.297	0.287	0.224	0.214	0.114	0.104	0.328	1/2
1/2 0.5000	0.291	0.270	0.344	0.334	0.255	0.245	0.130	0.120	0.375	9/16
9/16 0.5625	0.332	0.309	0.391	0.379	0.287	0.275	0.146	0.134	0.422	5/8
5/8 0.6250	0.371	0.347	0.469	0.456	0.321	0.305	0.164	0.148	0.469	3/4
3/4 0.7500	0.450	0.425	0.562	0.549	0.383	0.367	0.196	0.180	0.562	7/8
7/8 0.8750	0.530	0.502	0.656	0.642	0.446	0.430	0.227	0.211	0.656	1
1 1.0000	0.609	0.579	0.750	0.734	0.510	0.490	0.260	0.240	0.750	1 1/8
1 1/8 1.1250	0.689	0.655	0.844	0.826	0.572	0.552	0.291	0.271	0.844	1 1/4
1 1/4 1.2500	0.767	0.733	0.938	0.920	0.635	0.615	0.323	0.303	0.938	1 1/2
1 3/8 1.3750	0.848	0.808	1.031	1.011	0.698	0.678	0.354	0.334	1.031	1 5/8
1 1/2 1.5000	0.926	0.886	1.125	1.105	0.760	0.740	0.385	0.365	1.125	1 3/4

[1] Where specifying nominal size in decimals, zeros preceding decimal and in the fourth decimal place shall be omitted.

[2] Point angle X shall be 45° plus 5°, minus 0°, for screws of nominal lengths equal to or longer than those listed in Column Y, and 30° minimum for screws of shorter nominal lengths.

[3] The extent of rounding or flat at apex of cone point shall not exceed an amount equivalent to 10 per cent of the basic screw diameter.

Appendix 28 AMERICAN NATIONAL STANDARD SLOTTED HEADLESS SET SCREWS (ANSI B18.6.2)

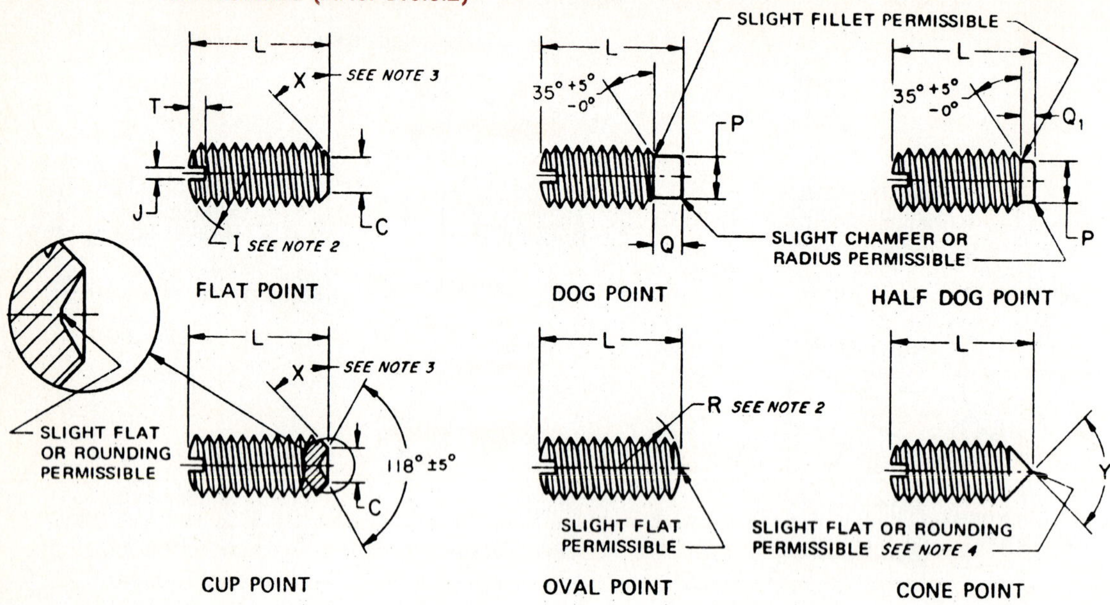

Nominal Size[1] or Basic Screw Diameter	I[2] Crown Radius	J Slot Width		T Slot Depth		C Cup and Flat Point Diameters		P Dog Point Diameters		Q Point Length Dog		Q1 Point Length Half Dog		R[2] Oval Point Radius	Y Cone Point Angle 90° ±2° For These Nominal Lengths or Longer; 118° ±2° For Shorter Screws
	Basic	Max	Min	Max	Min	Max	Min	Max	Min	Max	Min	Max	Min	Basic	
0 0.0600	0.060	0.014	0.010	0.020	0.016	0.033	0.027	0.040	0.037	0.032	0.028	0.017	0.013	0.045	5/64
1 0.0730	0.073	0.016	0.012	0.020	0.016	0.040	0.033	0.049	0.045	0.040	0.036	0.021	0.017	0.055	3/32
2 0.0860	0.086	0.018	0.014	0.025	0.019	0.047	0.039	0.057	0.053	0.046	0.042	0.024	0.020	0.064	7/64
3 0.0990	0.099	0.020	0.016	0.028	0.022	0.054	0.045	0.066	0.062	0.052	0.048	0.027	0.023	0.074	1/8
4 0.1120	0.112	0.024	0.018	0.031	0.025	0.061	0.051	0.075	0.070	0.058	0.054	0.030	0.026	0.084	5/32
5 0.1250	0.125	0.026	0.020	0.036	0.026	0.067	0.057	0.083	0.078	0.063	0.057	0.033	0.027	0.094	3/16
6 0.1380	0.138	0.028	0.022	0.040	0.030	0.074	0.064	0.092	0.087	0.073	0.067	0.038	0.032	0.104	3/16
8 0.1640	0.164	0.032	0.026	0.046	0.036	0.087	0.076	0.109	0.103	0.083	0.077	0.043	0.037	0.123	1/4
10 0.1900	0.190	0.035	0.029	0.053	0.043	0.102	0.088	0.127	0.120	0.095	0.085	0.050	0.040	0.142	1/4
12 0.2160	0.216	0.042	0.035	0.061	0.051	0.115	0.101	0.144	0.137	0.115	0.105	0.060	0.050	0.162	5/16
1/4 0.2500	0.250	0.049	0.041	0.068	0.058	0.132	0.118	0.156	0.149	0.130	0.120	0.068	0.058	0.188	5/16
5/16 0.3125	0.312	0.055	0.047	0.083	0.073	0.172	0.156	0.203	0.195	0.161	0.151	0.083	0.073	0.234	3/8
3/8 0.3750	0.375	0.068	0.060	0.099	0.089	0.212	0.194	0.250	0.241	0.193	0.183	0.099	0.089	0.281	7/16
7/16 0.4375	0.438	0.076	0.068	0.114	0.104	0.252	0.232	0.297	0.287	0.224	0.214	0.114	0.104	0.328	1/2
1/2 0.5000	0.500	0.086	0.076	0.130	0.120	0.291	0.270	0.344	0.334	0.255	0.245	0.130	0.120	0.375	9/16
9/16 0.5625	0.562	0.096	0.086	0.146	0.136	0.332	0.309	0.391	0.379	0.287	0.275	0.146	0.134	0.422	5/8
5/8 0.6250	0.625	0.107	0.097	0.161	0.151	0.371	0.347	0.469	0.456	0.321	0.305	0.164	0.148	0.469	3/4
3/4 0.7500	0.750	0.134	0.124	0.193	0.183	0.450	0.425	0.562	0.549	0.383	0.367	0.196	0.180	0.562	7/8

[1] Where specifying nominal size in decimals, zeros preceding decimal and in the fourth decimal place shall be omitted.

[2] Tolerance on radius for nominal sizes up to and including 5 (0.125 in.) shall be plus 0.015 in. and minus 0.000, and for larger sizes, plus 0.031 in. and minus 0.000. Slotted ends on screws may be flat at option of manufacturer.

[3] Point angle X shall be 45° plus 5°, minus 0°, for screws of nominal lengths equal to or longer than those listed in Column Y, and 30° minimum for screws of shorter nominal lengths.

[4] The extent of rounding or flat at apex of cone point shall not exceed an amount equivalent to 10 per cent of the basic screw diameter.

Appendix 29 TWIST DRILL SIZES

Number Size Drills

Size	Drill Diameter		Size	Drill Diameter		Size	Drill Diameter		Size	Drill Diameter	
	Inches	mm		Inches	mm		Inches	mm		Inches	mm
1	0.2280	5.7912	21	0.1590	4.0386	41	0.0960	2.4384	61	0.0390	0.9906
2	0.2210	5.6134	22	0.1570	3.9878	42	0.0935	2.3622	62	0.0380	0.9652
3	0.2130	5.4102	23	0.1540	3.9116	43	0.0890	2.2606	63	0.0370	0.9398
4	0.2090	5.3086	24	0.1520	3.8608	44	0.0860	2.1844	64	0.0360	0.9144
5	0.2055	5.2197	25	0.1495	3.7973	45	0.0820	2.0828	65	0.0350	0.8890
6	0.2040	5.1816	26	0.1470	3.7338	46	0.0810	2.0574	66	0.0330	0.8382
7	0.2010	5.1054	27	0.1440	3.6576	47	0.0785	1.9812	67	0.0320	0.8128
8	0.1990	5.0800	28	0.1405	3.5560	48	0.0760	1.9304	68	0.0310	0.7874
9	0.1960	4.9784	29	0.1360	3.4544	49	0.0730	1.8542	69	0.0292	0.7417
10	0.1935	4.9149	30	0.1285	3.2639	50	0.0700	1.7780	70	0.0280	0.7112
11	0.1910	4.8514	31	0.1200	3.0480	51	0.0670	1.7018	71	0.0260	0.6604
12	0.1890	4.8006	32	0.1160	2.9464	52	0.0635	1.6129	72	0.0250	0.6350
13	0.1850	4.6990	33	0.1130	2.8702	53	0.0595	1.5113	73	0.0240	0.6096
14	0.1820	4.6228	34	0.1110	2.8194	54	0.0550	1.3970	74	0.0225	0.5715
15	0.1800	4.5720	35	0.1100	2.7940	55	0.0520	1.3208	75	0.0210	0.5334
16	0.1770	4.4958	36	0.1065	2.7051	56	0.0465	1.1684	76	0.0200	0.5080
17	0.1730	4.3942	37	0.1040	2.6416	57	0.0430	1.0922	77	0.0180	0.4572
18	0.1695	4.3053	38	0.1015	2.5781	58	0.0420	1.0668	78	0.0160	0.4064
19	0.1660	4.2164	39	0.0995	2.5273	59	0.0410	1.0414	79	0.0145	0.3638
20	0.1610	4.0894	40	0.0980	2.4892	60	0.0400	1.0160	80	0.0135	0.3429

Metric Drill Sizes

Preferred sizes are in color type. Decimal-inch equivalents are for reference only.

Drill Diameter		Drill Diameter		Drill Diameter		Drill Diameter		Drill Diameter		Drill Diameter		Drill Diameter	
mm	in.	mm	in.	mm	in.	mm	in.	mm	in.	mm	in.	mm	in.
.40	.0157	1.03	.0406	2.20	.0866	5.00	.1969	10.00	.3937	21.50	.8465	48.00	1.8898
.42	.0165	1.05	.0413	2.30	.0906	5.20	.2047	10.30	.4055	22.00	.8661	50.00	1.9685
.45	.0177	1.08	.0425	2.40	.0945	5.30	.2087	10.50	.4134	23.00	.9055	51.50	2.0276
.48	.0189	1.10	.0433	2.50	.0984	5.40	.2126	10.80	.4252	24.00	.9449	53.00	2.0866
.50	.0197	1.15	.0453	2.60	.1024	5.60	.2205	11.00	.4331	25.00	.9843	54.00	2.1260
.52	.0205	1.20	.0472	2.70	.1063	5.80	.2283	11.50	.4528	26.00	1.0236	56.00	2.2047
.55	.0217	1.25	.0492	2.80	.1102	6.00	.2362	12.00	.4724	27.00	1.0630	58.00	2.2835
.58	.0228	1.30	.0512	2.90	.1142	6.20	.2441	12.50	.4921	28.00	1.1024	60.00	2.3622
.60	.0236	1.35	.0531	3.00	.1181	6.30	.2480	13.00	.5118	29.00	1.1417		
.62	.0244	1.40	.0551	3.10	.1220	6.50	.2559	13.50	.5315	30.00	1.1811		
.65	.0256	1.45	.0571	3.20	.1260	6.70	.2638	14.00	.5512	31.00	1.2205		
.68	.0268	1.50	.0591	3.30	.1299	6.80	.2677	14.50	.5709	32.00	1.2598		
.70	.0276	1.55	.0610	3.40	.1339	6.90	.2717	15.00	.5906	33.00	1.2992		
.72	.0283	1.60	.0630	3.50	.1378	7.10	.2795	15.50	.6102	34.00	1.3386		
.75	.0295	1.65	.0650	3.60	.1417	7.30	.2874	16.00	.6299	35.00	1.3780		
.78	.0307	1.70	.0669	3.70	.1457	7.50	.2953	16.50	.6496	36.00	1.4173		
.80	.0315	1.75	.0689	3.80	.1496	7.80	.3071	17.00	.6693	37.00	1.4567		
.82	.0323	1.80	.0709	3.90	.1535	8.00	.3150	17.50	.6890	38.00	1.4961		
.85	.0335	1.85	.0728	4.00	.1575	8.20	.3228	18.00	.7087	39.00	1.5354		
.88	.0346	1.90	.0748	4.10	.1614	8.50	.3346	18.50	.7283	40.00	1.5748		
.90	.0354	1.95	.0768	4.20	.1654	8.80	.3465	19.00	.7480	41.00	1.6142		
.92	.0362	2.00	.0787	4.40	.1732	9.00	.3543	19.50	.7677	42.00	1.6535		
.95	.0374	2.05	.0807	4.50	.1772	9.20	.3622	20.00	.7874	43.50	1.7126		
.98	.0386	2.10	.0827	4.60	.1811	9.50	.3740	20.50	.8071	45.00	1.7717		
1.00	.0394	2.15	.0846	4.80	.1890	9.80	.3858	21.00	.8268	46.50	1.8307		

(Courtesy of General Motors Corporation.)

Appendix 29 TWIST DRILL SIZES (Cont'd)

Letter Size Drills

Size	Drill Diameter		Size	Drill Diameter		Size	Drill Diameter		Size	Drill Diameter	
	Inches	mm		Inches	mm		Inches	mm		Inches	mm
A	0.234	5.944	H	0.266	6.756	O	0.316	8.026	V	0.377	9.576
B	0.238	6.045	I	0.272	6.909	P	0.323	8.204	W	0.386	9.804
C	0.242	6.147	J	0.277	7.036	Q	0.332	8.433	X	0.397	10.084
D	0.246	6.248	K	0.281	7.137	R	0.339	8.611	Y	0.404	10.262
E	0.250	6.350	L	0.290	7.366	S	0.348	8.839	Z	0.413	10.490
F	0.257	6.528	M	0.295	7.493	T	0.358	9.093			
G	0.261	6.629	N	0.302	7.601	U	0.368	9.347			

Appendix 30 STRAIGHT PINS

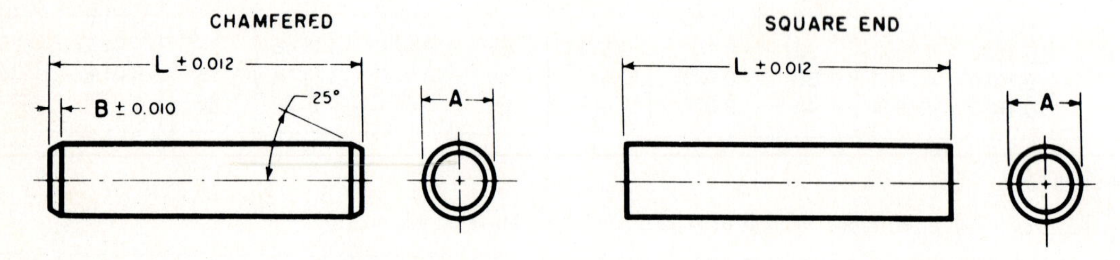

Nominal Diameter	Diameter A		Chamfer B
	Max	Min	
0.062	0.0625	0.0605	0.015
0.094	0.0937	0.0917	0.015
0.109	0.1094	0.1074	0.015
0.125	0.1250	0.1230	0.015
0.156	0.1562	0.1542	0.015
0.188	0.1875	0.1855	0.015
0.219	0.2187	0.2167	0.015
0.250	0.2500	0.2480	0.015
0.312	0.3125	0.3095	0.030
0.375	0.3750	0.3720	0.030
0.438	0.4375	0.4345	0.030
0.500	0.500	0.4970	0.030

All dimensions are given in inches.

These pins must be straight and free from burrs or any other defects that will affect their serviceability.

(Courtesy of ANSI; B5.20–1958.)

Appendix 31 WOODRUFF KEYS

USA STANDARD

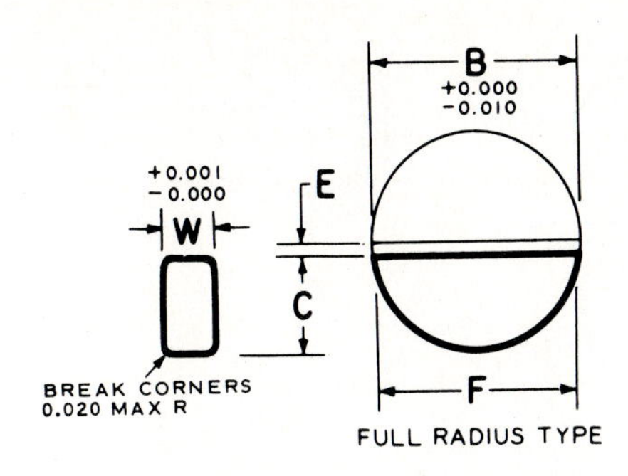

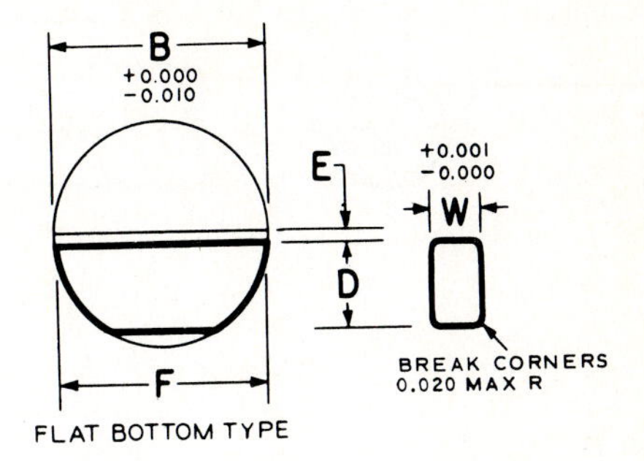

Woodruff Keys

Key No.	Nominal Key Size W × B	Actual Length F +0.000-0.010	Height of Key				Distance Below Center E
			C		D		
			Max	Min	Max	Min	
202	$\frac{1}{16} \times \frac{1}{4}$	0.248	0.109	0.104	0.109	0.104	$\frac{1}{64}$
202.5	$\frac{1}{16} \times \frac{5}{16}$	0.311	0.140	0.135	0.140	0.135	$\frac{1}{64}$
302.5	$\frac{3}{32} \times \frac{5}{16}$	0.311	0.140	0.135	0.140	0.135	$\frac{1}{64}$
203	$\frac{1}{16} \times \frac{3}{8}$	0.374	0.172	0.167	0.172	0.167	$\frac{1}{64}$
303	$\frac{3}{32} \times \frac{3}{8}$	0.374	0.172	0.167	0.172	0.167	$\frac{1}{64}$
403	$\frac{1}{8} \times \frac{3}{8}$	0.374	0.172	0.167	0.172	0.167	$\frac{1}{64}$
204	$\frac{1}{16} \times \frac{1}{2}$	0.491	0.203	0.198	0.194	0.188	$\frac{3}{64}$
304	$\frac{3}{32} \times \frac{1}{2}$	0.491	0.203	0.198	0.194	0.188	$\frac{3}{64}$
404	$\frac{1}{8} \times \frac{1}{2}$	0.491	0.203	0.198	0.194	0.188	$\frac{3}{64}$
305	$\frac{3}{32} \times \frac{5}{8}$	0.612	0.250	0.245	0.240	0.234	$\frac{1}{16}$
405	$\frac{1}{8} \times \frac{5}{8}$	0.612	0.250	0.245	0.240	0.234	$\frac{1}{16}$
505	$\frac{5}{32} \times \frac{5}{8}$	0.612	0.250	0.245	0.240	0.234	$\frac{1}{16}$
605	$\frac{3}{16} \times \frac{5}{8}$	0.612	0.250	0.245	0.240	0.234	$\frac{1}{16}$
406	$\frac{1}{8} \times \frac{3}{4}$	0.740	0.313	0.308	0.303	0.297	$\frac{1}{16}$
506	$\frac{5}{32} \times \frac{3}{4}$	0.740	0.313	0.308	0.303	0.297	$\frac{1}{16}$
606	$\frac{3}{16} \times \frac{3}{4}$	0.740	0.313	0.308	0.303	0.297	$\frac{1}{16}$
806	$\frac{1}{4} \times \frac{3}{4}$	0.740	0.313	0.308	0.303	0.297	$\frac{1}{16}$
507	$\frac{5}{32} \times \frac{7}{8}$	0.866	0.375	0.370	0.365	0.359	$\frac{1}{16}$
607	$\frac{3}{16} \times \frac{7}{8}$	0.866	0.375	0.370	0.365	0.359	$\frac{1}{16}$
707	$\frac{7}{32} \times \frac{7}{8}$	0.866	0.375	0.370	0.365	0.359	$\frac{1}{16}$
807	$\frac{1}{4} \times \frac{7}{8}$	0.866	0.375	0.370	0.365	0.359	$\frac{1}{16}$
608	$\frac{3}{16} \times 1$	0.992	0.438	0.433	0.428	0.422	$\frac{1}{16}$
708	$\frac{7}{32} \times 1$	0.992	0.438	0.433	0.428	0.422	$\frac{1}{16}$
808	$\frac{1}{4} \times 1$	0.992	0.438	0.433	0.428	0.422	$\frac{1}{16}$
1008	$\frac{5}{16} \times 1$	0.992	0.438	0.433	0.428	0.422	$\frac{1}{16}$
1208	$\frac{3}{8} \times 1$	0.992	0.438	0.433	0.428	0.422	$\frac{1}{16}$
609	$\frac{3}{16} \times 1\frac{1}{8}$	1.114	0.484	0.479	0.475	0.469	$\frac{5}{64}$
709	$\frac{7}{32} \times 1\frac{1}{8}$	1.114	0.484	0.479	0.475	0.469	$\frac{5}{64}$
809	$\frac{1}{4} \times 1\frac{1}{8}$	1.114	0.484	0.479	0.475	0.469	$\frac{5}{64}$
1009	$\frac{5}{16} \times 1\frac{1}{8}$	1.114	0.484	0.479	0.475	0.469	$\frac{5}{64}$

(Courtesy of ANSI; B17.2-1967.)

Appendix 32 WOODRUFF KEYS AND KEYSEATS

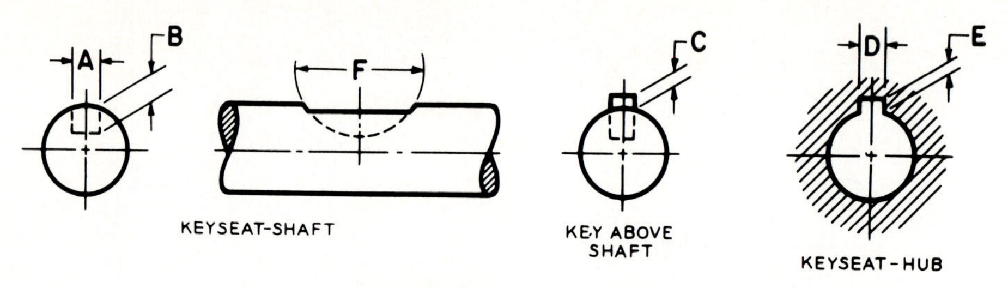

Keyseat Dimensions

Key Number	Nominal Size Key	Keyseat – Shaft					Key Above Shaft	Keyseat – Hub	
		Width A*		Depth B	Diameter F		Height C	Width D	Depth E
		Min	Max	+0.005 -0.000	Min	Max	+0.005 -0.005	+0.002 -0.000	+0.005 -0.000
202	1/16 × 1/4	0.0615	0.0630	0.0728	0.250	0.268	0.0312	0.0635	0.0372
202.5	1/16 × 5/16	0.0615	0.0630	0.1038	0.312	0.330	0.0312	0.0635	0.0372
302.5	3/32 × 5/16	0.0928	0.0943	0.0882	0.312	0.330	0.0469	0.0948	0.0529
203	1/16 × 3/8	0.0615	0.0630	0.1358	0.375	0.393	0.0312	0.0635	0.0372
303	3/32 × 3/8	0.0928	0.0943	0.1202	0.375	0.393	0.0469	0.0948	0.0529
403	1/8 × 3/8	0.1240	0.1255	0.1045	0.375	0.393	0.0625	0.1260	0.0685
204	1/16 × 1/2	0.0615	0.0630	0.1668	0.500	0.518	0.0312	0.0635	0.0372
304	3/32 × 1/2	0.0928	0.0943	0.1511	0.500	0.518	0.0469	0.0948	0.0529
404	1/8 × 1/2	0.1240	0.1255	0.1355	0.500	0.518	0.0625	0.1260	0.0685
305	3/32 × 5/8	0.0928	0.0943	0.1981	0.625	0.643	0.0469	0.0948	0.0529
405	1/8 × 5/8	0.1240	0.1255	0.1825	0.625	0.643	0.0625	0.1260	0.0685
505	5/32 × 5/8	0.1553	0.1568	0.1669	0.625	0.643	0.0781	0.1573	0.0841
605	3/16 × 5/8	0.1863	0.1880	0.1513	0.625	0.643	0.0937	0.1885	0.0997
406	1/8 × 3/4	0.1240	0.1255	0.2455	0.750	0.768	0.0625	0.1260	0.0685
506	5/32 × 3/4	0.1553	0.1568	0.2299	0.750	0.768	0.0781	0.1573	0.0841
606	3/16 × 3/4	0.1863	0.1880	0.2143	0.750	0.768	0.0937	0.1885	0.0997
806	1/4 × 3/4	0.2487	0.2505	0.1830	0.750	0.768	0.1250	0.2510	0.1310
507	5/32 × 7/8	0.1553	0.1568	0.2919	0.875	0.895	0.0781	0.1573	0.0841
607	3/16 × 7/8	0.1863	0.1880	0.2763	0.875	0.895	0.0937	0.1885	0.0997
707	7/32 × 7/8	0.2175	0.2193	0.2607	0.875	0.895	0.1093	0.2198	0.1153
807	1/4 × 7/8	0.2487	0.2505	0.2450	0.875	0.895	0.1250	0.2510	0.1310
608	3/16 × 1	0.1863	0.1880	0.3393	1.000	1.020	0.0937	0.1885	0.0997
708	7/32 × 1	0.2175	0.2193	0.3237	1.000	1.020	0.1093	0.2198	0.1153
808	1/4 × 1	0.2487	0.2505	0.3080	1.000	1.020	0.1250	0.2510	0.1310
1008	5/16 × 1	0.3111	0.3130	0.2768	1.000	1.020	0.1562	0.3135	0.1622
1208	3/8 × 1	0.3735	0.3755	0.2455	1.000	1.020	0.1875	0.3760	0.1935
609	3/16 × 1 1/8	0.1863	0.1880	0.3853	1.125	1.145	0.0937	0.1885	0.0997
709	7/32 × 1 1/8	0.2175	0.2193	0.3697	1.125	1.145	0.1093	0.2198	0.1153
809	1/4 × 1 1/8	0.2487	0.2505	0.3540	1.125	1.145	0.1250	0.2510	0.1310
1009	5/16 × 1 1/8	0.3111	0.3130	0.3228	1.125	1.145	0.1562	0.3135	0.1622

(Courtesy of ANSI; B17.2–1967.)

Appendix 33 STANDARD KEYS AND KEYWAYS

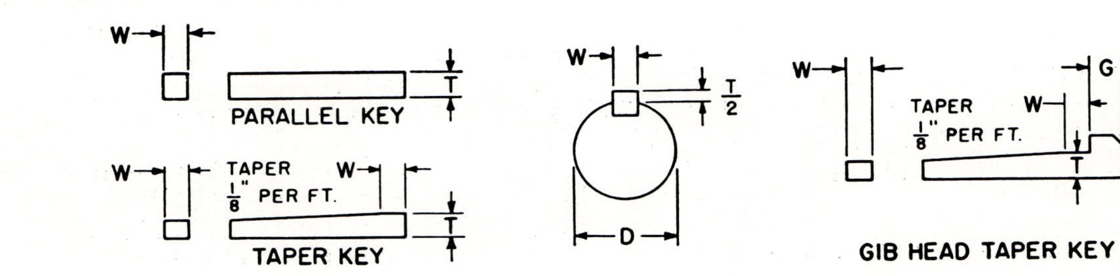

Sprocket Bore (= Shaft Diam.)	Keyway Dimensions — Inches				Key Dimensions — Inches					Gib Head Dimensions — Inches				Key Tolerances Taper and Gib Head	
	For Square Key		For Flat Key		Square		Flat		Tolerance on W and T	Square Key		Flat Key			
Inches D	Width W	Depth T/2	Width W	Depth T/2	Width W	Height T	Width W	Height T	(−)	H	G	H	G	W (−)	T (+)
1/2 — 9/16	1/8 x	1/16	1/8 x	3/64	1/8 x	1/8	1/8 x	3/32	0.002	1/4	7/32	3/16	1/8	0.002	0.002
5/8 — 7/8	3/16 x	3/32	3/16 x	1/16	3/16 x	3/16	3/16 x	1/8	0.002	5/16	9/32	1/4	3/16	0.002	0.002
15/16 — 1 1/4	1/4 x	1/8	1/4 x	3/32	1/4 x	1/4	1/4 x	3/16	0.002	7/16	11/32	5/16	1/4	0.002	0.002
1 5/16 — 1 3/8	5/16 x	5/32	5/16 x	1/8	5/16 x	5/16	5/16 x	1/4	0.002	9/16	13/32	3/8	5/16	0.002	0.002
1 7/16 — 1 3/4	3/8 x	3/16	3/8 x	1/8	3/8 x	3/8	3/8 x	1/4	0.002	11/16	15/32	7/16	3/8	0.002	0.002
1 13/16 — 2 1/4	1/2 x	1/4	1/2 x	3/16	1/2 x	1/2	1/2 x	3/8	0.0025	7/8	19/32	5/8	1/2	0.0025	0.0025
2 5/16 — 2 3/4	5/8 x	5/16	5/8 x	7/32	5/8 x	5/8	5/8 x	7/16	0.0025	1 1/16	23/32	3/4	5/8	0.0025	0.0025
2 7/8 — 3 1/4	3/4 x	3/8	3/4 x	1/4	3/4 x	3/4	3/4 x	1/2	0.0025	1 1/4	7/8	7/8	3/4	0.0025	0.0025
3 3/8 — 3 3/4	7/8 x	7/16	7/8 x	5/16	7/8 x	7/8	7/8 x	5/8	0.003	1 1/2	1	1 1/16	7/8	0.003	0.003
3 7/8 — 4 1/2	1 x	1/2	1 x	3/8	1 x	1	1 x	3/4	0.003	1 3/4	1 3/16	1 1/4	1	0.003	0.003
4 3/4 — 5 1/2	1 1/4 x	5/8	1 1/4 x	7/16	1 1/4 x	1 1/4	1 1/4 x	7/8	0.003	2	1 7/16	1 1/2	1 1/4	0.003	0.003
5 3/4 — 7 3/8	1 1/2 x	3/4	1 1/2 x	1/2	1 1/2 x	1 1/2	1 1/2 x	1	0.003	2 1/2	1 3/4	1 3/4	1 1/2	0.003	0.003
7 1/2 — 9 7/8	1 3/4 x	7/8	. .	. .	1 3/4 x	1 3/4	. .	. .	0.004	3	2	. .	. .	0.004	0.004
10 — 12 1/2	2 x	1	. .	. .	2 x	2	. .	. .	0.004	3 1/2	2 3/8	. .	. .	0.004	0.004

Standard Keyway Tolerances: Straight Keyway — Width (W) $^{+.005}_{-.000}$ Depth (T/2) $^{+.010}_{-.000}$

Taper Keyway — Width (W) $^{+.005}_{-.000}$ Depth (T/2) $^{+.000}_{-.010}$

Appendix 34 TAPER PINS

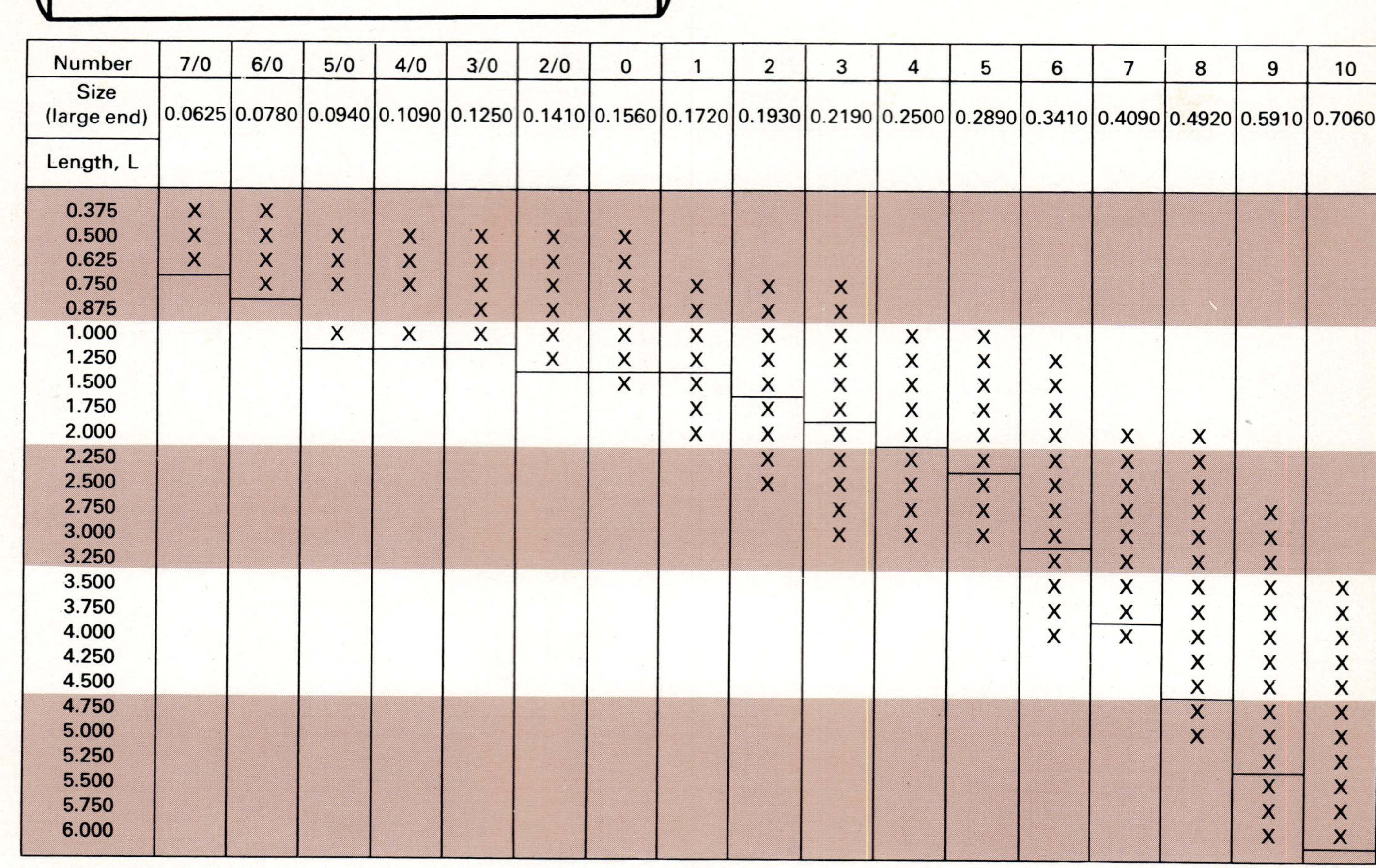

Number	7/0	6/0	5/0	4/0	3/0	2/0	0	1	2	3	4	5	6	7	8	9	10
Size (large end)	0.0625	0.0780	0.0940	0.1090	0.1250	0.1410	0.1560	0.1720	0.1930	0.2190	0.2500	0.2890	0.3410	0.4090	0.4920	0.5910	0.7060
Length, L																	
0.375	X	X															
0.500	X	X	X	X	X	X	X										
0.625	X	X	X	X	X	X	X										
0.750		X	X	X	X	X	X	X	X	X							
0.875					X	X	X	X	X	X							
1.000			X	X	X	X	X	X	X	X	X	X					
1.250						X	X	X	X	X	X	X	X				
1.500							X	X	X	X	X	X	X				
1.750								X	X	X	X	X	X				
2.000								X	X	X	X	X	X	X	X		
2.250									X	X	X	X	X	X	X		
2.500									X	X	X	X	X	X	X		
2.750										X	X	X	X	X	X	X	
3.000										X	X	X	X	X	X	X	
3.250													X	X	X	X	
3.500													X	X	X	X	X
3.750													X	X	X	X	X
4.000													X	X	X	X	X
4.250															X	X	X
4.500															X	X	X
4.750															X	X	X
5.000															X	X	X
5.250																X	X
5.500																X	X
5.750																X	X
6.000																X	X

All dimensions are given in inches.

Standard reamers are available for pins given above the line.

Pins Nos. 11 (size 0.8600), 12 (size 1.032), 13 (size 1.241), and 14 (1.523) are special sizes—hence their lengths are special.

To find small diameter of pin, multiply the length by 0.2083 and subtract the result from the large diameter.

(Courtesy of ANSI; B5.20–1958.)

Appendix 35 PLAIN WASHERS

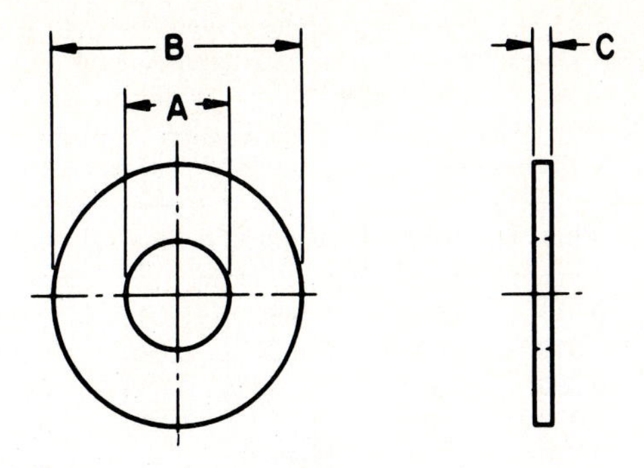

Dimensions of Preferred Sizes of Type A Plain Washers[a]

When specifying washers on drawings or in notes, give the inside diameter, outside diameter, and the thickness. *Example:* 0.938 × 1.750 × 0.134 TYPE A PLAIN WASHER.

Nominal Washer Size[b]			Inside Diameter A			Outside Diameter B			Thickness C		
			Basic	Tolerance Plus	Tolerance Minus	Basic	Tolerance Plus	Tolerance Minus	Basic	Max	Min
—	—		0.078	0.000	0.005	0.188	0.000	0.005	0.020	0.025	0.016
—	—		0.094	0.000	0.005	0.250	0.000	0.005	0.020	0.025	0.016
—	—		0.125	0.008	0.005	0.312	0.008	0.005	0.032	0.040	0.025
No. 6	0.138		0.156	0.008	0.005	0.375	0.015	0.005	0.049	0.065	0.036
No. 8	0.164		0.188	0.008	0.005	0.438	0.015	0.005	0.049	0.065	0.036
No. 10	0.190		0.219	0.008	0.005	0.500	0.015	0.005	0.049	0.065	0.036
$\frac{3}{16}$	0.188		0.250	0.015	0.005	0.562	0.015	0.005	0.049	0.065	0.036
No. 12	0.216		0.250	0.015	0.005	0.562	0.015	0.005	0.065	0.080	0.051
$\frac{1}{4}$	0.250	N	0.281	0.015	0.005	0.625	0.015	0.005	0.065	0.080	0.051
$\frac{1}{4}$	0.250	W	0.312	0.015	0.005	0.734[c]	0.015	0.007	0.065	0.080	0.051
$\frac{5}{16}$	0.312	N	0.344	0.015	0.005	0.688	0.015	0.007	0.065	0.080	0.051
$\frac{5}{16}$	0.312	W	0.375	0.015	0.005	0.875	0.030	0.007	0.083	0.104	0.064
$\frac{3}{8}$	0.375	N	0.406	0.015	0.005	0.812	0.015	0.007	0.065	0.080	0.051
$\frac{3}{8}$	0.375	W	0.438	0.015	0.005	1.000	0.030	0.007	0.083	0.104	0.064
$\frac{7}{16}$	0.438	N	0.469	0.015	0.005	0.922	0.015	0.007	0.065	0.080	0.051
$\frac{7}{16}$	0.438	W	0.500	0.015	0.005	1.250	0.030	0.007	0.083	0.104	0.064
$\frac{1}{2}$	0.500	N	0.531	0.015	0.005	1.062	0.030	0.007	0.095	0.121	0.074
$\frac{1}{2}$	0.500	W	0.562	0.015	0.005	1.375	0.030	0.007	0.109	0.132	0.086

[a] Preferred sizes are for the most part from series previously designated "Standard Plate" and "SAE." Where common sizes existed in the two series, the SAE size is designated "N" (narrow) and the Standard Plate "W" (wide). These sizes as well as all other sizes of Type A Plain Washers are to be ordered by ID, OD, and thickness dimensions.

[b] Nominal washer sizes are intended for use with comparable nominal screw or bolt sizes.

[c] The 0.734 in., 1.156 in., and 1.469 in. outside diameters avoid washers which could be used in coin-operated devices.

Cont.

Appendix 35 PLAIN WASHERS (Cont.)

Nominal Washer Size[b]			Inside Diameter A			Outside Diameter B			Thickness C		
				Tolerance			Tolerance				
			Basic	Plus	Minus	Basic	Plus	Minus	Basic	Max	Min
$\frac{9}{16}$	0.562	N	0.594	0.015	0.005	1.156[c]	0.030	0.007	0.095	0.121	0.074
$\frac{9}{16}$	0.562	W	0.625	0.015	0.005	1.469[c]	0.030	0.007	0.109	0.132	0.086
$\frac{5}{8}$	0.625	N	0.656	0.030	0.007	1.312	0.030	0.007	0.095	0.121	0.074
$\frac{5}{8}$	0.625	W	0.688	0.030	0.007	1.750	0.030	0.007	0.134	0.160	0.108
$\frac{3}{4}$	0.750	N	0.812	0.030	0.007	1.469	0.030	0.007	0.134	0.160	0.108
$\frac{3}{4}$	0.750	W	0.812	0.030	0.007	2.000	0.030	0.007	0.148	0.177	0.122
$\frac{7}{8}$	0.875	N	0.938	0.030	0.007	1.750	0.030	0.007	0.134	0.160	0.108
$\frac{7}{8}$	0.875	W	0.938	0.030	0.007	2.250	0.030	0.007	0.165	0.192	0.136
1	1.000	N	1.062	0.030	0.007	2.000	0.030	0.007	0.134	0.160	0.108
1	1.000	W	1.062	0.030	0.007	2.500	0.030	0.007	0.165	0.192	0.136
$1\frac{1}{8}$	1.125	N	1.250	0.030	0.007	2.250	0.030	0.007	0.134	0.160	0.108
$1\frac{1}{8}$	1.125	W	1.250	0.030	0.007	2.750	0.030	0.007	0.165	0.192	0.136
$1\frac{1}{4}$	1.250	N	1.375	0.030	0.007	2.500	0.030	0.007	0.165	0.192	0.136
$1\frac{1}{4}$	1.250	W	1.375	0.030	0.007	3.000	0.030	0.007	0.165	0.192	0.136
$1\frac{3}{8}$	1.375	N	1.500	0.030	0.007	2.750	0.030	0.007	0.165	0.192	0.136
$1\frac{3}{8}$	1.375	W	1.500	0.045	0.010	3.250	0.045	0.010	0.180	0.213	0.153
$1\frac{1}{2}$	1.500	N	1.625	0.030	0.007	3.000	0.030	0.007	0.165	0.192	0.136
$1\frac{1}{2}$	1.500	W	1.625	0.045	0.010	3.500	0.045	0.010	0.180	0.213	0.153
$1\frac{5}{8}$	1.625		1.750	0.045	0.010	3.750	0.045	0.010	0.180	0.213	0.153
$1\frac{3}{4}$	1.750		1.875	0.045	0.010	4.000	0.045	0.010	0.180	0.213	0.153
$1\frac{7}{8}$	1.875		2.000	0.045	0.010	4.250	0.045	0.010	0.180	0.213	0.153
2	2.000		2.125	0.045	0.010	4.500	0.045	0.010	0.180	0.213	0.153
$2\frac{1}{4}$	2.250		2.375	0.045	0.010	4.750	0.045	0.010	0.220	0.248	0.193
$2\frac{1}{2}$	2.500		2.625	0.045	0.010	5.000	0.045	0.010	0.238	0.280	0.210
$2\frac{3}{4}$	2.750		2.875	0.065	0.010	5.250	0.065	0.010	0.259	0.310	0.228
3	3.000		3.125	0.065	0.010	5.500	0.065	0.010	0.284	0.327	0.249

(Courtesy of ANSI; B27.2–1965.)

Appendix 36 LOCK WASHERS (ANSI B27.1)

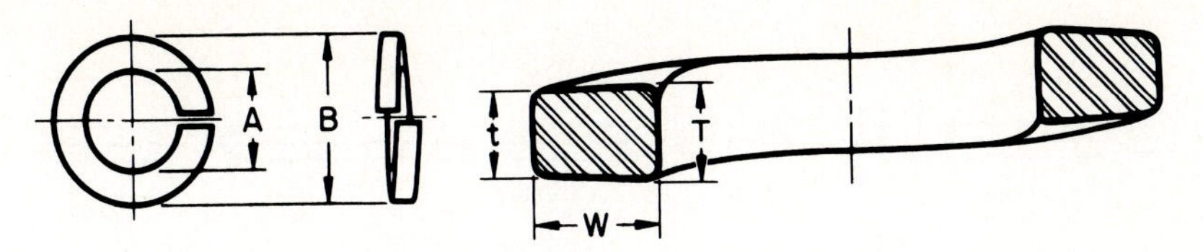

Dimensions of Regular* Helical Spring Lock Washers

Nominal Washer Size		Inside Diameter *A*		Outside Diameter *B*	Washer Section Width *W*	Washer Section Thickness $\frac{T+t}{2}$
		Min	Max	Max**	Min	Min
No. 2	0.086	0.088	0.094	0.172	0.035	0.020
No. 3	0.099	0.101	0.107	0.195	0.040	0.025
No. 4	0.112	0.115	0.121	0.209	0.040	0.025
No. 5	0.125	0.128	0.134	0.236	0.047	0.031
No. 6	0.138	0.141	0.148	0.250	0.047	0.031
No. 8	0.164	0.168	0.175	0.293	0.055	0.040
No. 10	0.190	0.194	0.202	0.334	0.062	0.047
No. 12	0.216	0.221	0.229	0.377	0.070	0.056
1/4	0.250	0.255	0.263	0.489	0.109	0.062
5/16	0.312	0.318	0.328	0.586	0.125	0.078
3/8	0.375	0.382	0.393	0.683	0.141	0.094
7/16	0.438	0.446	0.459	0.779	0.156	0.109
1/2	0.500	0.509	0.523	0.873	0.171	0.125
9/16	0.562	0.572	0.587	0.971	0.188	0.141
5/8	0.625	0.636	0.653	1.079	0.203	0.156
11/16	0.688	0.700	0.718	1.176	0.219	0.172
3/4	0.750	0.763	0.783	1.271	0.234	0.188
13/16	0.812	0.826	0.847	1.367	0.250	0.203
7/8	0.875	0.890	0.912	1.464	0.266	0.219
15/16	0.938	0.954	0.978	1.560	0.281	0.234
1	1.000	1.017	1.042	1.661	0.297	0.250
1 1/16	1.062	1.080	1.107	1.756	0.312	0.266
1 1/8	1.125	1.144	1.172	1.853	0.328	0.281
1 3/16	1.188	1.208	1.237	1.950	0.344	0.297
1 1/4	1.250	1.271	1.302	2.045	0.359	0.312
1 5/16	1.312	1.334	1.366	2.141	0.375	0.328
1 3/8	1.375	1.398	1.432	2.239	0.391	0.344
1 7/16	1.438	1.462	1.497	2.334	0.406	0.359
1 1/2	1.500	1.525	1.561	2.430	0.422	0.375

*Formerly designated Medium Helical Spring Lock Washers.

**The maximum outside diameters specified allow for the commercial tolerances on cold drawn wire.

Appendix 37 COTTER PINS

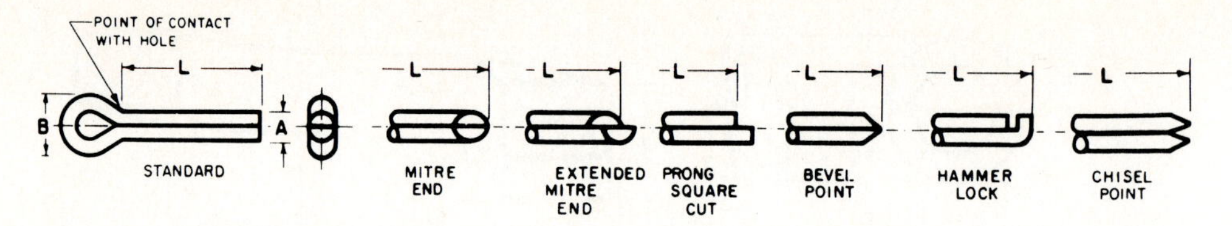

Nominal Diameter	Diameter A		Outside Eye Diameter B Min	Hole Sizes Recom-mended
	Max	Min		
0.031	0.032	0.028	$\frac{1}{16}$	$\frac{3}{64}$
0.047	0.048	0.044	$\frac{3}{32}$	$\frac{1}{16}$
0.062	0.060	0.056	$\frac{1}{8}$	$\frac{5}{64}$
0.078	0.076	0.072	$\frac{5}{32}$	$\frac{3}{32}$
0.094	0.090	0.086	$\frac{3}{16}$	$\frac{7}{64}$
0.109	0.104	0.100	$\frac{7}{32}$	$\frac{1}{8}$
0.125	0.120	0.116	$\frac{1}{4}$	$\frac{9}{64}$
0.141	0.134	0.130	$\frac{9}{32}$	$\frac{5}{32}$
0.156	0.150	0.146	$\frac{5}{16}$	$\frac{11}{64}$
0.188	0.176	0.172	$\frac{3}{8}$	$\frac{13}{64}$
0.219	0.207	0.202	$\frac{7}{16}$	$\frac{15}{64}$
0.250	0.225	0.220	$\frac{1}{2}$	$\frac{17}{64}$
0.312	0.280	0.275	$\frac{5}{8}$	$\frac{5}{16}$
0.375	0.335	0.329	$\frac{3}{4}$	$\frac{3}{8}$
0.438	0.406	0.400	$\frac{7}{8}$	$\frac{7}{16}$
0.500	0.473	0.467	1	$\frac{1}{2}$
0.625	0.598	0.590	$1\frac{1}{4}$	$\frac{5}{8}$
0.750	0.723	0.715	$1\frac{1}{2}$	$\frac{3}{4}$

All dimensions are given in inches.

A certain amount of leeway is permitted in the design of the head; however, the outside diameters given should be adhered to.

Prongs are to be parallel; ends shall not be open.

Points may be blunt, bevel, extended prong, mitre, etc., and purchaser may specify type required.

Lengths shall be measured as shown on the above illustration (L-dimension).

Cotter pins shall be free from burrs or any defects that will affect their serviceability.

(Courtesy of ANSI; B5.20–1958.)

INDEX

508.7
(42 FT. 4.7 IN.)
18.0
(1 FT. 6.0 IN.)
38.9
(3 FT. 2.9 IN.)
13.0
FT. 1.0 IN.)
52.0
(4 FT. 4.0 IN.)
77.9
(6 FT. 5.9 IN.)
444.0
(37 FT. 0.0 IN.)
44.0
(3 FT. 8.0 IN.)
118.9
(9 FT. 10.9 IN.)
282.0
(23 FT. 6.2 IN.)
MAX. FLAP HT.
(11 FT. 8.3 IN.)
120.4
(10 FT. 0.4 IN.)
5°
8 30' FLAPPING
8 30' FLAPPING
23.3 CLEARANC
(1 FT. 11.3 IN.)
123.8
(10 FT. 3.8 IN.)
74.0
(6 FT. 2.0 IN.)
52.0
(4 FT. 4.0 IN.)
15.0
GROUND LINE:
DESIGN GROSS WT – 4000 POUNDS
89.0
(7 FT. 5.0 IN.)
92.1 AT GROSS WT.
OF 4000 LBS.
(7 FT. 8.1 IN.)
72.0
(6 FT. 0.0 IN.)
154.2
(12 FT. 10.2 IN.)
STA
1.00
STATIC POSITION AT
DESIGN GROSS WEIGHT
30° UP FWD
2° UP
18'
Transport 30 Passengers
300 NM in 1 Hour and 5 Minute
Return with 30 Passengers Without Refueling
89'